Handbook of Techniques in
High-Pressure Research
and Engineering

Handbook of Techniques in High-Pressure Research and Engineering

by Daniil S. Tsiklis

Translation Edited and with a Preface by
Alfred Bobrowsky
Pressure Technology Corporation of America
Boalsburg, Pennsylvania

and

Materials Research Laboratory
The Pennsylvania State University
University Park, Pennsylvania

and

Consultant, Tem-Pres Research, Inc.
Advanced Projects Division
The Carborundum Company
State College, Pennsylvania

Translated from Russian

℗ PLENUM PRESS • NEW YORK • 1968

This translation is based on the third revised and expanded
edition, published by Khimiya Press in Moscow in 1965.

Translated from Russian by
Albert Peabody
Techtron, Glen Burnie, Maryland

Даниил Семенович Циклис
ТЕХНИКА ФИЗИКО-ХИМИЧЕСКИХ ИССЛЕДОВАНИЙ
ПРИ ВЫСОКИХ И СВЕРХВЫСОКИХ ДАВЛЕНИЯХ

TEKHNIKA FIZIKO-KHIMICHESKIKH ISSLEDOVANII
PRI VYSOKIKH I SVERKHVYSOKIKH DAVLENIYAKH

ISBN 978-1-4684-8362-8   ISBN 978-1-4684-8360-4 (eBook)
DOI 10.1007/978-1-4684-8360-4

Library of Congress Catalog Card Number 68-14854

© 1968 Plenum Press
Softcover reprint of the hardcover 1st edition 1968
A Division of Plenum Publishing Corporation
227 West 17 Street, New York, N. Y. 10011

Preface

The extent of experimentation with high pressures has become so great that it appears timely to publish a book in this field. The author, D. S. Tsiklis, is already known to persons working with high pressures as a sound reviewer and compiler, as from Bridgman's mention of him in "Physics of High Pressures," Bell & Co., 1949. The present book offers a wide scope of comparison of equipment and procedures used with high pressures. The original application of topics was to physics and chemistry, but it can be seen that the text material is equally useful in earth sciences and engineering. Some of the fields to which the subject matter is being applied are:

Synthesis of new phases under high pressures
Chemical reactions under high pressures
Measurements of physical properties of materials under high pressures
Rock mechanics
Metalworking under high pressures
Mechanical design associated with high pressures

It is believed that this book will serve as a sound general basis for experimentation with high pressure for many years.

The references in the book are up to date (1965) and large in number. The illustrations can serve as assembly drawings from which detail drawings can be made; for this reason, the figures in the English edition are reproduced to larger scale than in the original Russian.

The editor has placed occasional comments in brackets throughout the text wherever in his opinion there were questions as to the intent of the Russian language or wherever text material could be construed in different ways.

Footnotes have been added to the manuscript by the editor with the intention of comparing Russian and American practice, and sometimes in order to update the text in accordance with later experience.

Finally, the subject index has been expanded in order to increase the utility of reference of this volume.

A. Bobrowsky

Foreword

The second edition of this book was published in 1958. In the years following, new designs of high-pressure apparatus have been developed that allow pressures on the order of hundreds of thousands of atmospheres to be generated at very high temperatures, and in volumes far from microscopic. The invention of these apparatuses and the performance of investigations using them have led to the synthesis of diamonds and borazon also. Industrial production of these materials has begun. During this same time, the area of application of high pressures has expanded considerably.

All this could not help but influence the contents of this book. It should be noted that a huge number of works on the influence of pressure on various physical and chemical properties of materials, methods of investigation at high pressures, etc., were published during this period. The author has attempted to reflect the most valuable work which has become known to him from the literature and from his personal experience.

In this edition more attention is paid (than in the second edition) to problems of design of high and ultrahigh-pressure apparatus, and to measurement of pressure; problems of temperature measurement are also covered.

The chapters on methods of investigation of phase equilibria and compressibility have been rewritten. A new chapter has been written that deals with methods of optical, X-ray-diffraction, and electrical measurement. The material presented in the other chapters of this book also has been reworked.

The author thanks I. R. Krichevskii, the editor of the first edition of this book, whose efforts influenced the third edition as well. The author also thanks L. F. Vereshchagin, who helped greatly in the job of familiarization with works published by the Institute of High Pressure Physics of the Academy of Sciences of the USSR, as well as V. P. Butuzov, M. G. Gonikberg, V. A. Galaktionov, B. P. Demyashkevich, M. K. Zhokhovskii, A. I. Likhter, Yu. N. Ryabinin, and others who kindly presented him with reprints of their works and thereby facilitated the difficult task of the author.

The author thanks his readers in advance for all their good wishes and critical notes.

D. S. Tsiklis

CONTENTS

CHAPTER I. Construction Materials Used for Operations
   at High and Ultrahigh Pressures   1
Steels and Alloys   1
Heat Treatment   5
Testing of Materials   7
Machined Finishes.   9
Behavior of Steels, Metals, and Alloys Under Experimental
   Conditions   9
Packing Materials (Gaskets)   20
Insulation Materials   23
Pressure-Transmitting Media   27
Bibliography.   34

CHAPTER II. Design and Construction of High- and
   Ultrahigh-Pressure Apparatus   39
Design of Apparatus.   39
 Behavior of Material under Load   39
 Stress Present in Thick-Walled Vessels
   under Pressure.   44
 Equivalent Stress   46
 Determination of Thermal Stresses   51
 Simultaneous Action of Pressure and Temperature . .   54
Construction of Apparatus.   55
 Autofrettaged Vessels   55
 Multilayered Vessels   57
 Vessels with Variable External Mechanical Support. .   60
 Vessels with Hydraulic Support   65
 Wedge Vessels   66
 Tetrahedral Press   69

      Conical Dies (Bridgman Anvils).   70
      Belt Apparatus   73
      Piston Piezometer (Supported Piston and Cylinder). .   75
  Bibliography   76

CHAPTER III. Methods of Generating Pressure at Normal
      and Elevated Temperatures   79
  Compression of Gases   79
      Laboratory Compressors   79
      Additional Compression Stages   83
      Compressing Devices   85
      Other Methods of Compressing Gases.   87
  Compression of Liquids and Solids.   88
      Liquid Pumps and Compressors   88
      Sidorov Mechanical Press   100
      Pressure Intensifier.   101
      Anvil Apparatus   112
      Wedge Apparatus   117
      Belt Apparatus   120
      Small Hydraulic Presses   122
  Generation of High Pressures with Simultaneous
      Application of Shear Force   123
  Other Methods of Producing High Pressures   124
  Heating at High Pressures   127
  Bibliography   133

CHAPTER IV. Methods of Measuring High Pressure   137
  Pressure Units.   137
  Absolute Manometers   138
      Liquid Manometers   138
      Piston Manometers   140
  Relative Manometers.   164
      Manometers Using Elastic Properties
      of Materials.   164
      Electrical Manometers   169
      Ryabinin Crusher [Upsetting] Manometer.   184
  Differential Manometers   186
  Bibliography.   193

CHAPTER V. Measurement of the Flow and Temperature
      of a Compressed Medium   197
  Measurement of the Flow of a Compressed Medium
      High-Pressure Rheometers   197
  Measurement of Temperature at High Pressure   204
  Bibliography.   207

CHAPTER VI. Parts for High-Pressure Installations   209
    Valves   211
        Control Valves   211
        Shut-off Valves   213
    Seals for Laboratory Apparatus   222
        Seals with Compensated Area   222
        Unsupported-Area Closures   225
        Closures for Large-Diameter Apparatus   233
        Hydraulic Closures   234
    Glands   235
        Glands with Natural Sealing   235
        Packed Glands   239
    Intensifier-Piston Seals   243
        Small-Piston Seals   243
        Large-Piston Seals   247
    Check Valves   249
    Joints   250
        Packing   251
        Cones   251
        Lenses   252
        Lips   253
        Metal–Glass Joints   253
        Metal–Glass Solder Joints   256
    Electric-Insulation Techniques   257
        Low-Pressure Electrical Leads   257
        Unsupported-Area Electrical Leads   257
    Viewing Windows   263
    Bibliography   268

CHAPTER VII. Mixing and Circulation Under Pressure   271
    Mixing (Agitation)   271
        Mechanical Agitators   271
        Rotating and Rocking Apparatus   273
        Electromagnetic Stirrers   274
    Circulation   284
    Bibliography   289

CHAPTER VIII. General Laboratory High-Pressure Equipment .   291
    Buildings   291
    Equipment   292
        Gas Holders   292
        Compressors   293
        Gas Purification   293
        Installation of Apparatus   293
    Rules of Usage   298
    Bibliography   298

CHAPTER IX. Methods of Investigating Phase Equilibria
          at High Pressures   299
     Liquid-Gas Systems   301
     Solid-Liquid Systems   328
     Gas-Gas Systems   337
     Solid-Gas Systems   341
     Methods of Sampling for Analysis   344
     Methods of Analysis   345
          Volumetric Measurements   345
          Gravimetric Measurements   348
     Bibliography   351

CHAPTER X. Determination of Compressibility of Gases
          and Liquids   355
     Compressibility of Gases   356
          Measurement with Constant-Volume Piezometers . . .   356
          Measurement with Variable-Volume Piezometers . . .   365
          Measurement with Variable-Volume Piezometers
               with Variable Quantity of Material   393
     Compressibility of Liquids   397
          Measurement with Constant-Volume Piezometers . . .   397
          Measurement with Variable-Volume Piezometers . . .   398
          Measurement of Compressibility Using Pycnometers.   402
          Determination of Compressibility by Hydrostatic
               Weighing   403
     Bibliography   411

CHAPTER XI. Methods of Measuring Surface Tension
          and Wettability   415
     Measurement of Surface Tension at Liquid-Gas Boundary. .   415
          Capillary-Rise Methods   415
          Method of Maximum Pressure in Bubble   419
          Hanging-Drop Method   422
          Method of Drop Weight (Volume)   423
     Measurement of Contact Angles (of Wettability) at High
          Pressures   424
     Measurement of Surface Tension at Gas-Gas Boundary . . .   427
     Bibliography   429

CHAPTER XII. Optical, X-Ray and Electrical Measurements . . .   431
     Optical Measurements   431
     X-Ray Measurements   436
     Electrical Measurements   439
     Bibliography   443

APPENDICES 444
AUTHOR INDEX 455
SUBJECT INDEX 463

Chapter I

Construction Materials Used for Operations at High and Ultrahigh Pressures

The first stage in experimental investigation is the design, manufacture, and assembly of apparatus, instruments, and installations that will permit the necessary experiments to be performed. The design of apparatus depends to a great extent on the materials available to the investigator for manufacture. When investigations are to be performed at pressures reaching a half million atmospheres and at temperatures reaching several thousands of degrees, the experimenter needs materials with special qualities. In this chapter there are analyzed the properties of a number of materials (steels; packing, lubricating, and insulating materials; and pressure-transmitting materials) employed at high and ultrahigh pressures.

STEELS AND ALLOYS

The stronger the steel, the easier is the task of the designer and the smaller can the dimensions of apparatus of a given capacity be. The quality criterion for steel is the tensile strength; at the present time tensile strength is as high as 25,000 kg/cm^2 for certain steels.* Also, steels

---

* Editor's comment: Later in this book, the phenomenon of "pinching-off" will be discussed. It will be shown there that the male member of an unsupported-area seal is in a stress state equivalent to a uniaxial tensile stress of at least 115% of the pressure being sealed, submerged in an average compressive stress of about 75% of the pressure. If we accept as a viewpoint that mean normal stress does not greatly affect tensile strengths of steels (not the best approximation for highly hardened steels, though), then a male member successfully sealing a frequently-quoted pressure of

possess a characteristic ductility. Insufficiently ductile steel (low unit elongation) is unsuitable for construction of high-pressure apparatus.*

Appendices IV and V present types [1-5], composition, and heat treatment of steels which we have used for a long time.† All of these steels have strengths under 20,000 kg/cm^2. At temperatures of 200-300°C they can be used only up to pressures of 15,000-18,000 atm. It will be shown below that certain special measures can be taken to increase the operating pressure significantly.

Type St. 5 steel is used at pressures of not over 800-1000 atm. This steel usually is not heat treated. It is easily worked mechanically, and it is generally used in cases where a stronger steel is not needed. This steel is used for the manufacture of nipples, valve bodies, and other parts working at even higher pressure (up to 1500-2000 atm).

The group of steels including types 18KhNVA, 18KhNMA, 30KhMA, OKhNZM, 30KhGS, 40Kh, and 45KhNMFA are used at pressures above 1000 atm. They have different compositions but differ relatively little in their mechanical properties. When these steels are used at pressures up to 10,000 atm there is no need for special design solutions, which are necessary however if these steels are used at higher pressures, especially at pressures exceeding their yield strengths.

30,000 atmospheres is being subjected to a tensile stress of at least 35,500 kg/cm^2. Actually the editor and others have relatively routinely sealed pressures of 550,000 psi and occasionally pressures of 600,000 psi with unsupported-area hardened steel seals which consequently have manifested tensile strengths of at least 115% of these values (or, expressed in kg/cm^2, tensile strengths about 44,000 and 48,000 kg/cm^2 respectively).

* Editor's comment: This is an important point. When average normal stress is tensile, as is usually the case in portions of a simple pressure chamber under pressure, ductility of the chamber steel is lower than that of a tensile specimen in a standard uniaxial tensile test. Thus a material with nominally small ductility may lose that ductility completely when under biaxial tension. Further, size effects and anisotropy of the actual chamber material may also tend to reduce ductility to zero. A steel in a stress state with zero ductility is disadvantageous for two reasons: (1) brittle cracks propagate readily, and (2) tensile strength frequently drops in a marginally ductile material, more as the average normal stress becomes more tensile.

† Information can be found in literature [6] concerning mechanical properties and composition of various types of steels used in the European countries and in the USA.

TABLE 1. Properties of Cemented Carbides*

Group of cemented carbides	Type number	Alloy composition, %			Density, g/cm	Rockwell hardness (scale A)	Bending strength, kg/cm²	Compressive strength, kg/cm²
		Tungsten carbide	Cobalt	Titanium carbide				
Tungsten	VK2	98	2	–	15.0-15.4	90.0	10,000	–
	VK3	97	3	–	14.9-15.3	89.0	11,000	51,500
	VK6	94	6	–	14.6-15.0	88.0	12,000	43,000
	VK8	92	8	–	14.4-14.8	87.5	13,000	–
	VK15	85	15	–	13.9-14.1	86.0	16,000	–
Titanium-Tungsten	T5K10	85	9	6	12.3-19.2	88.5	11,500	–
	T15K6	79	6	15	11.0-11.7	90.0	11,000	42,500
	T30K4	66	4	30	9.5-9.8	92.0	9,000	–

*Temperature at which loss of strength begins is 1000°C.

TABLE 2. Mechanical Properties of Titanium Alloys*

Properties	VT-1 (pure titanium)	VT-2	OTCh-1	OTCh
Tensile strength, kg/cm²	4500-6000	5500-7000	6500-7500	7500-9000
Yield strength, kg/cm²	3800-5000	4600-6000	4700-6500	5500-6500
Unit elongation, %	25	20	20	15
Unit reduction in area, %	50	45	–	25
Young's modulus ($\times 10^{-6}$), kg/cm²	1.05-1.10	–	1.05	1.10-1.20
Shear modulus ($\times 10^{-5}$), kg/cm²	–	4-5	–	4
Thermal coefficient of linear expansion ($\times 10^{-6}$)	8.3	–	–	8.0

*The alloy containing titanium (1.5% Fe, 1.5% Cr, 1.2% Mo, and 5.5% Al [8]) has high strength $\sigma_B = 12{,}600$ kg/cm².

Steels types 1Kh18N9 and 1Kh18N9T are used at pressures up to 1000 atm and temperatures up to 500-600°C; steel type 2Kh13 is used at pressures up to 5000 atm (at the same temperatures) with small sizes of apparatus. Although steels type 1Kh18N9 and 1Kh18N9T are less ductile at high temperatures than steel type 2Kh13, the latter is considerably stronger.

Stainless steel type ÉI437 B (alloy Kh20N77T 2YuR) is even stronger and is suitable for the manufacture of apparatus working at pressures up to 10,000 atm and temperatures up to 500°C in corrosive media. Steels type 1Kh18N9, 1Kh18N9T, and ÉI437B are used to make nonmagnetic parts for designs with magnetic stirrers.

Steels types ShKh13 and ShKh15 are used to make parts which support large compressive loads. These are very hard but brittle steels; they are used to make plungers, pressure rings, and other such parts. Steel type KhVG can be recommended for this same purpose [5]. This tungsten steel has a very valuable quality — it is not distorted and does not change its volume upon heating; it is suitable for manufacture of parts to final dimension; these parts can be used immediately after heat treatment and light polishing.

It can be considered that a pressure of 15,000-17,000 atm is the limit for these steels. In order to make apparatus for operation at higher pressures, we must use steels with higher tensile strength or resort to special kinds of designs. Extremely hard metal carbides are used to manufacture parts of apparatus operating at pressures measured in tens or hundreds of kiloatmospheres. Thus, apparatus made of tungsten carbide cemented with cobalt (this alloy is known in the foreign literature as "Carboloy") can withstand pressures up to a half million atm. Our industry manufactures the cemented carbides [7] whose properties are presented in Table 1 for this purpose.

Table 1 indicates that the higher the cobalt content of the cemented carbide, the higher the bending strength and the lower the compressive strength. Pistons for ultrahigh-pressure apparatus are manufactured from carbides similar in composition to VK3.

Certain titanium alloys (Table 2) have good mechanical properties and high corrosion resistance.

It follows from the above that the strongest materials currently available are the cemented carbides. However, even their strength is considerably less than the theoretical maximum strength of metals found by calculation [9]. It is considered that

$$\sigma_{th} = \sqrt{E\gamma/a}$$

where σ_{th} is the theoretical maximum tensile strength; E is Young's modulus; γ is the formation energy of the new surface; α is the distance between atomic planes in the crystal lattice.

Putting values in the expression for σ_{th} (E = 10^{11} dyn/cm², $\gamma = 10^3$ erg/cm², $\alpha = 3 \cdot 10^{-8}$ cm), we see that $\sigma_{th} = 10^{11}$ dyn/cm² or 10^5 kg/cm², i.e., the theoretical strengths of metals are at least ten times greater than their actual strengths. This great difference is explained by the presence of defects (disruptions of order, dislocations) in the crystal lattice of the metals. Actually, single crystals of certain metals have been grown under special conditions to produce so-called "whiskers" — thin threads of monocrystals with a strength near the theoretical.

Apparently, defect-free crystals can be produced under high-pressure conditions [10]. Crystallization of steels under pressure permits lattice defects to be eliminated, heat treatment under pressure (which is called thermobaric treatment) prevents grain growth, and annealing under pressure allows the ductility of the metal to be retained. Also, thermomechanical treatment consisting of mechanical deformation (compression) at low temperatures of austenite with subsequent martensite conversion has succeeded in producing material [11, 12] with a tensile strength of 30,000-35,000 kg/cm² and a yield strength of 25,000-28,000 kg/cm² [13].

Melting under pressure allows new alloys to be produced, such as an alloy of titanium with magnesium (because the boiling point of magnesium falls lower than the melting point of titanium). Finally, new materials can be produced on the basis of sintering of diamond powder with a metal or alloy at pressures under which the process of graphitization of diamond is stopped or retarded.

HEAT TREATMENT

The strength of steels (with certain exceptions) is increased by low temperature tempering. The brittleness is increased, however. The problem is how to attain the greatest strength while keeping unit elongation no less than 8-10%. The higher the pressure for which the apparatus is designed, the stricter are the requirements on heat treatment.* Selection

---

*Editor's comment: The 8-10% nominal tensile ductility is a convenient rule-of-thumb that is frequently used. What is actually desired, however, is a steel that will remain ductile under whatever (tensile) conditions of average normal stress may exist in the specific component made of that steel when maximum pressure is applied to equipment containing the

of a heat treatment [14] to give a steel the qualities required to satisfy the demands of the designer depends on the type of steel, the dimensions and configuration of the parts, and, finally, on the goals set before the designer.

In order to be able to select a type of steel, we must know its hardenability, i.e., the depth of occurrence of the martensite or troostite-martensite structure, since this determines the maximum diameter of workpiece suitable for heat treatment. The quality of heat treatment is determined to a great extent by the complexity of the configuration of a part. The surfaces of small-diameter through-holes and blind openings (which do not permit free flow of cooling liquids) will be soft because of their slower cooling. It is therefore desirable to have through-holes only. Also, since the center of thick blocks of forged steel often contains fissures, it is expedient to avoid small diameter holes in the center of these blocks altogether.

The final finishing of parts, i.e., grinding to the proper dimensions, threading, etc., should be performed after heat treatment. Although this requires the use of high-quality cutting tools as well as of polishing, it eliminates the possibility of having the dimensions altered during subsequent heat treatment. If a high hardness of steel is used and if parts are manufactured to close tolerances, heat treatment is performed in a protective atmosphere in order to prevent formation of scale, and then the parts are polished.

In all cases attempts must be made to avoid sharp edges, angles, holes, and so forth on parts to be heat-treated, since these lead to the formation of internal fissures.

---

component. (For example, biaxial tensile stress generated in the wall of a pressurized chamber must not give rise to so greatly tensile an average normal stress that the steel of the wall becomes brittle). The only way to know the ductility of a steel under an average normal stress that is more tensile than in a standard tensile test is to conduct tensile tests over a suitable range of tensile average normal stresses. One way of doing this is to run tensile tests on notched specimens with notch stresses of varying severity. (It is well-known that isolated notches increase the tensile character of average normal stress in a tensile specimen.) This method of assessing ductility is given by S. I. Gubkin in his papers on plasticity and formability of metals (during the 1950's and earlier), and by the editor in "Formability Criteria for Metals," Collected Papers Southeastern Engineering Conference and Tool Exposition, ASTME, 1965. It sometimes suffices to conduct tensile tests at several pressures, and then to extrapolate to negative values of pressure, but this method is risky if extrapolation is carried too far.

It should be remembered that improper heat treatment can completely ruin a part.*

TESTING OF MATERIALS

When working with new types of steels, it is always desirable to perform testing of coupons prepared from the identical heat-treated piece which will be used for manufacture of the parts. With repeated usage of steel of the same type, simple hardness testing of parts will be found to be sufficient for quality control.

Hardness testing, where special equipment is not available, can be performed using a Pol'di tester.† This instrument is easy to use. However, for determination of the hardness of very hard steels, the data from calculation of the diameter of the impression must be extrapolated or converted from the table supplied with the instrument to yield a Brinell number. These tables are not accurate [15]. They result in errors of up to 20%. Therefore, the use of several standard samples of known hardnesses is recommended, checking them so long as the diameters of impressions on the standard sample and on the part being investigated do not correspond.

In Appendix II we present data for conversion of Brinell numbers to values of tensile strength.

---

* Editor's comment: This matter cannot be overemphasized. Components that must operate reliably at the highest pressures can be quality-checked for hardenability as follows, at least before the first time a new design or steel is used. The component, or a coupon cut from the same bar of steel, should be subjected to a "certified" heat treatment, i.e., to one where conditions of heat treatment are well known and reproducible. (Where a coupon is used instead of a component, the minimum linear through-dimension must be equal to the corresponding dimension in the actual component.) Indentation hardness of a suitable sort should be determined on a number of known traverses. The component should next be sectioned without being heated, then if necessary finished on the sectioned surface again without heating, and hardness traverses run on the sectioned surface. Spot checks of hardness can be made on the untouched outer surface to verify lack of heating during sectioning. If the hardness variation thus determined is much below the hardenability specification for the steel, the heat-treatment parameters or the physical conduct of the heat treatment should be examined to locate the difficulty. If the spatial variation of hardness is found to be satisfactory, components finished from the same steel may be certified-heat-treated without worry. Of course, any decarburized surface layers should be removed after heat treatment by a final grinding, prior to determining hardness, unless a heat treatment was used that produces no decarburization. This is especially necessary for "superficial" hardness testing.

† Editor's comment: Probably the Pol'di portable tester, hit with a hammer. Impression is measured optically.

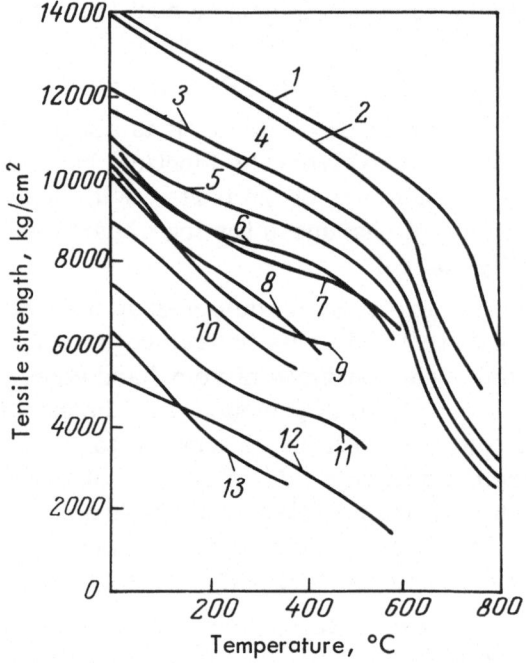

Fig. 1. Change in tensile strength of titanium alloys (alloys of Ti−Al−Cr−Fe−Si−B) with temperature: 1) AT-12; 2) AT-10; 3) AT-9; 4) AT-8; 5) AT-6; 6) AT-4; 7) VT-8; 8) VT-6; 9) VT-4; 10) VT-5; 11) OT-4; 12) TG-0; 13) VT-1.

Hardness testing can also be performed as follows [16]. A hardened steel cone with a peak angle of 90° is pressed into the surface of the material being tested, under a defined pressure. A rim (bead) is formed around the imprint; the boundary of this bead depends on the yield strength σ_S. This value is calculated from the formula

$$\sigma_S = \frac{G}{\pi a^2}$$

where G is the load on the cone; α is the bead radius.

In order to measure radius α, the extent of beading is determined by polishing the sample with a small piece of emery paper in a direction perpendicular to the coarsest preliminary polishing. Then, knowing the diameter of the impression, we can determine the yield strength by the following formula [17]:

$$\sigma_S = 91.5 \sqrt[3]{\frac{p - H\pi Dh}{D^2}}$$

TABLE 3. Young's Modulus of Carbon and Alloy Steels
at Various Temperatures

Temperature, °C	Young's modulus ($\times 10^{-6}$), kg/cm^2	Temperature, °C	Young's modulus ($\times 10^{-6}$), kg/cm^2	Temperature, °C	Young's modulus ($\times 10^{-6}$), kg/cm^2
Carbon steel		Carbon steel		Alloy steel	
20	2.10	350	1.84	400	1.75
100	2.05	400	1.77	425	1.73
150	2.00	425	1.71	450	1.70
200	1.96	450	1.67	475	1.61
250	1.93	475	1.55	500	1.58
300	1.88	500	1.48		

where p is the load on the sphere, H is the Brinell hardness number, D is the diameter of the sphere, and h is the depth of the impression.

These methods cannot be used to correctly measure the Rockwell hardness, since by the nature of the tests used, the hardness is determined by the work hardening during plastic deformation.

MACHINED FINISHES

Parts of high-pressure apparatus must be finished very carefully. The finishing of areas directly in contact with the medium creating the pressure, as well as finishing of threads, packing surfaces, etc., must be especially accurate. Burrs and grooves cannot be tolerated. Sharp angles should be smoothed off.

Special requirements are placed on threads. Threads in high-pressure apparatus are usually made short, i.e, with smoothed sharp angles, which prevents "seizing." Special attention must be turned to the centering of threads in holes that are sealed by rods, pins, or shafts. Methods of mechanical working of parts are described in the corresponding sections.

BEHAVIOR OF STEELS, METALS, AND ALLOYS
UNDER EXPERIMENTAL CONDITIONS

Influence of Temperature. At sufficiently high temperatures, all metals exhibit viscous behavior (creep). Also, high temperature leads to a change in the structure of steel.

Steels containing nickel, chromium, and a small quantity of tungsten have increased creep resistance. These steels are stable at 600°C and do not oxidize up to 900°C. Steels types 1Kh13, 2Kh13, ÉI-437B, 1Kh18N9, 1Kh18N9T, and Zh69 have high creep resistance. The last three types are stable up to 800°C. We present below the composition of a special heat-

TABLE 4. Mechanical Properties of Certain Carbon- and Low-Alloyed Steels at Various Temperatures

Steel	Tensile strength, kg/cm²			Elastic limit, kg/cm²			Unit elongation, %			Unit reduction in area, %			Impact strength, kg-m/cm²		
	+20°C	−80°C	−183°C	+20°C	−80°C	−183°C	+20°C	−80°C	−183°C	+20°C	−80°C	−183°C	+20°C	−80°C	−183°C
Carbon type A	3780	4530	8220	2810	3690	–	30.0	33.3	2.8	70.6	75.4	0.6	15.5	0.58	0.43
Carbon type B	4890	5480	–	3140	3950	–	27.2	28.4	–	65.0	55.3	–	7.39	0.58	0.39
Carbon	5900	–	9300	3500†	–	7100†	28.0	–	26.0	–	–	–	12.5	–	1.1
Chrome-moly	7600	8010*	11120	6130†	6380*	10380	19.1	22.5*	18.0	69.0	69.0*	41.0	16.7	–	0.7

* At −70°C.
† Yield strength.

TABLE 5. Mechanical Properties of Certain Austenitic Nickel Steels at +20 and −183°C

Chemical composition, wt. %				Tensile strength, kg/cm²		Elastic limit, kg/cm²		Unit elongation, %		Unit reduction in area,%		Impact strength, kg-m/cm²		Brinell hardness	
C	Mn	Ni	Si	+20°C	−183°C	+20°C	−183°C	+20°C	−183°C	+20°C	−183°C	+20°C	−183°C	+20°C	−183°C
0.57	6.88	13.00	0.31	7220	10990	3550	5990	81.8	46.0	76.0	48.0	39.2	15.0	184	227
0.04	0.29	85.18	0.02	5220	9050	4250	6400	27.6	31.6	72.0	61.0	21.5	14.0	144	220
0.12	0.94	42.06	0.24	5740	9250	3550	5160	35.5	40.9	70.0	62.0	22.7	25.0	153	205
0.2	18.8	8.1	0.43	7400	18500*	4000†	8600*	56.0	25.0*	53.5	30.5*	26-21	17-20	175	–

* At −255°C.
† Yield strength.

TABLE 6. Behavior of Various Materials at +20, −196, and −253°C

Material	Tensile strength, kg/cm²			Yield strength, kg/cm²			Unit reduction in area, %			Compression strength, kg/cm²		
	+20°C	−196°C	−253°C	+20°C	−196°C	−253°C	+20°C	−196°C	−253°C	+20°C	−196°C	−253°C*
St. 45, normalized	6200	10,700	14,100	3850	9700	—	53.0	38.0	0.0	12,450	15,700	14,100
18KhNVA (É-16)	12,600	17,000	18,000	10,500	13,100	15,100	63.0	54.0	3.1	22,400	27,000	18,600
St. 40Kh (tempered at 200°C)	22,400	24,800	27,200	18,500	22,000	—	37.0	2.0	1.6	19,700	25,300	27,700
1Kh18N9T (austenitic)	6600	15,500	17,900	2800	6400	7700	76.0	61.0	48.0	19,200	30,200	37,700
Bronze	4200	5900	6900	2700	3600	4200	48.0	50.0	45.0	7400	10,800	11,500
Duraluminum	4200	5000	7100	3000	4100	4900	38.0	25.0	21.0	6100	7400	8900

*S_{xT} is the resistance of ductile metals to brittle fracture ($\psi = 0$) [reduction in area is nil].

TABLE 7. Mechanical Properties of Nonferrous Metals and Alloys at +15, −80, and −180°C

Metal	Type†	Tensile strength, kg/cm²			Yield strength, kg/cm²			Unit elongation, %			Unit reduction in area, %			Impact strength, kg-m/cm²		
		+15°C	−80°C	−180°C	+15°C	−80°C	−180°C	+15°C	−80°C	−180°C	+15°C	−80°C	−180°C	+15°C	−80°C	−180°C
Copper	M1	2790	3670	4130	880	1000	1270	13.3	22.9	30.7	71.5	65.3	67.9	7.87	8.69	9.10
Annealed brass	L62	4000	—	5700	1160*	—	1700*	49.0	—	41.0	66.4	—	—	14.0	—	—
Bronze	BrA7	5000	3780	6750	2500*	1900*	2050*	60.0	31.0	29.0	70.0	30.0	30.0	—	—	—
Duralumi-num	D1	4110	4330	5330	1810	1790	2510	19.7	24.5	13.1	41.8	37.7	18.7	2.35	2.57	2.11

*Yield strength

† According to All-Union State Standards.

resistant steel [18] which is stable at pressures up to 2500 atm and temperatures up to 1000°C:

	Fe	C	Ni	Cr	Co	Mo	W	Nb
wt. %	23	1	30	20	20	3	2	1

Figure 1 shows the change in tensile strength of titanium alloys containing aluminum, chromium, iron, silicon, and boron, with temperature
[19, 20].

The strength of steels changes considerably with increasing temperature. Thus, for example, the tensile strength* of 18-8 chrome−nickel steel
decreases from 7000 to 4000 kg/cm^2 at 700°C and to 2000 kg/cm^2 at 800°C.
The modulus of elasticity (Table 3) also decreases [22, 23].

Steels behave differently at low temperatures [24]; their tensile
strengths increase somewhat, but their impact strengths decrease sharply.
Steels can be divided into two groups according to their behavior at low
temperatures: 1) carbon and low alloy steels; 2) alloy austenitic steels.

Steels of the first group show an increased tensile strength, yield
strength, elastic limit, and hardness and a reduced unit elongation, unit
reduction in area, and impact strength with reduction in temperature
(Table 4).

Steels containing nickel, copper, vanadium, molybdenum, manganese,
and chromium have increased impact strength at temperatures below 0°C;
those containing carbon have reduced impact strength below 0°C. Heat
treatment also affects the behavior of steels at low temperatures.

The second group includes the steels containing over 13% nickel,
chrome−nickel, and chrome−manganese steels, stainless steels (1Kh18N9),
etc. At low temperatures, these steels retain almost the same high ductility which they have at normal temperature. With reduced temperature, the
tensile strength limit, elastic yield strength, and hardness increase, while
(in most cases) the unit elongation decreases. The corresponding data for
these steels are presented in Table 5.

Table 6 shows the properties of materials of both groups at even
lower temperatures [24].

The property of steels and alloys of increasing their strength at low
temperatures has been used [25] in the design of apparatus for production

---

*Steel with the following composition also has high mechanical strength
 [21]: 0.05-0.09% C; 0.4% Si; 0.4% Mn; 16% Cr; 8-8.5% Ni; ~1.0% Al.

TABLE 8. Dependence of Shear Resistance on Pressure*

Material	$\dfrac{\tau_{50}}{\tau_{25}}$	$\dfrac{\tau_{100}}{\tau_{25}}$	$\dfrac{\tau_{125}}{\tau_{25}}$	$\dfrac{\tau_{150}}{\tau_{25}}$
Bismuth	1.7	3.0	3.5	—
Potassium	3.2	10.0	13.2	—
Lithium	2.0	6.5	11.0	—
Magnesium	2.3	5.4	7.5	—
Copper	1.9	4.5	6.5	—
Tin	2.2	7.8	10.4	12.3
Silver	1.9	3.7	6.3	8.8
ÉI437 steel	2.8	6.4	8.2	10.0
Tellurium	2.2	2.6	9.5	12.8
Zinc	1.9	4.2	5.3	—
Zirconium	2.0	3.4	4.3	—

* τ_{25}, τ_{50}, etc. are shear resistances at 25, 50 thousand atm., etc.

of pressures of 100,000 atm at the temperature of liquid air. At this temperature, the compressive strength of carboloy (the cemented carbide from which the apparatus was made) increases to 100,000 kg/cm^2.

The mechanical properties of nonferrous metals also increase at low temperatures. The strength, yield points, etc., increase.

The change in properties must be considered in selecting materials for packing and valves.

The properties of nonferrous metals at low temperatures are presented in Table 7.

Influence of Pressure. Under pressure, the plasticity (ductility) of steel increases considerably. Detailed investigations [26, 27] have shown that plastic deformation under hydrostatic pressure may reach immense magnitudes, and may increase almost indefinitely without rupture. Ordinary steel will withstand 300-fold reduction without rupture at a hydrostatic pressure of 25,000 atm. Rupture of a specimen under these conditions will result only from shear.

The nature of other types of deformation also changes under hydrostatic pressure. For example, shear deformation becomes qualitatively similar to plastic deformation in extension.

Tables 8 and 9 show data [28] indicating that the shear resistance increases by 8-12 times with an increase in pressure from 25,000 to 150,000 atm, whereas with an increase from 25,000 to 500,000 atm, shear resistance increases by a factor of 50.

TABLE 9. Shear Resistance of Various Materials at Various Pressures

Material	Shear resistance, kg/cm^2				
	at 25,000 atm	at 50,000 atm	at 75,000 atm	at 100,000 atm	at 500,000 atm
Silver chloride	450	900	1300	1800	14,000
Magnesium	750	1250	1800	2500	—
Armco iron	1850	3500	5150	6750	—
Katlenite	1620	3800	3650	7600	—
Pyrophyllite	1730	4500	7950	11,600	—
Graphite	2700	4800	6600	8200	135,000

During shear, a disruption which has begun may "heal itself." When a cylinder is twisted around a longitudinal axis, along which a compressive load has been applied, fissures "heal themselves" earlier than disruption can arise, and ductility so increases that the sample will not be ruptured at higher angles of rotation than can be tolerated under ordinary circumstances.

A punch can be pressed through a sheet of low carbon steel under hydrostatic pressure without formation of a burr,* a clean hole being formed.

Investigations [29] have shown that the hydrostatic pressure changes not only the terminal properties of the material (compressive strength and ductility), but also in most cases changes the entire course of the deformation [stress-strain] diagram [30].

The properties of metal carbides change considerably under pressure; Carboloy, which can withstand up to 75,000 kg/cm^2 in compression, is very brittle in bending or extension.† With all-around pressure of 25,000 atm, Carboloy is ductile and withstands great tensile stress, becoming an ideal material for manufacture of high-pressure apparatus.‡

---

*The burr is made by the punched slug [that fractures through the remaining thickness of the sheet after preliminary coining of the punch into the sheet]. In this case, the ductility of the steel is so great that the punch shears the steel aside without burring the material at all.

†For certain materials (such as the alloys AMK-2, AZh-6, BrS-30) the ratio of the compressive strength and tensile strength is greater than one, and for S415-32 cast iron it reaches [4] as high as 4. Hard steels types ShKh12 or KhVG are also stronger under compression than under tension.

‡The reasons for the increased ductility under high hydrostatic pressure were investigated in

The Action of Gases and Liquids. *Nitrogen* reacts chemically with iron at high temperatures (over 400°C) and high pressures [33], forming iron nitride. The crystal lattice of the iron is altered and its brittleness increases.

Hydrogen is one of the most dangerous gases for steel. A case is known in which hydrogen penetrated through the wall of a cylinder with an internal diameter of 0.6 cm and a wall thickness of 4.4 cm at 9000 atm with explosive velocity, although the metal had no visible cracks.

The action of hydrogen on steel results in an increase in the brittleness of the steel and in hydrogen embrittlement (corrosion). The increase in brittleness [34], i.e., the reduction of the strength without formation of microscopic fissures and alterations of the microstructure and composition, is caused by the adsorption of hydrogen on the surface of the metal, accumulation of hydrogen atoms at defects in the crystal lattice, and development of internal pressure.

Hydrogen embrittlement includes a reaction between hydrogen and carbon: the metal is decarburized and methane is formed. Its accumulation leads to the appearance of fissures and bulges.

Hydrogen corrosion of steel is decreased by a reduction in the carbon content. Chromium increases the resistance of steel to hydrogen corrosion. Still, the tensile strength of steel containing 0.06% carbon and 20% chromium under the continuous action of hydrogen at 475°C is reduced in one month from 7000 to 3600 kg/cm^2. The steel becomes considerably more brittle and its elongation is reduced from 53 to 6% [35]. Titanium, tungsten, and molybdenum in small quantities act like chromium. Thus, N. V. Ashmarin [36] has presented a composition of tungsten steel (0.53% carbon, 0.55% chromium, and 1.6% tungsten) which is immune to hydrogen embrittlement. Chrome−nickel steel containing 2.5% tungsten can withstand hydrogen pressure up to 1000 atm (at 550°C).

In Table 10 we present the properties of steels used especially for manufacture of apparatus operating at high temperatures under hydrogen pressure [37] (see also [38]).

Carbon Monoxide forms carbonyls with iron, nickel, and other metals. Pressure encourages the reaction. The carbonyls are also formed at room temperature. Copper lining is usually used for protection from carbon monoxide.

Helium also has a disruptive influence on steel. At a temperature of about 250°C formation of fissures is noted in steel parts which are in contact with helium under a pressure of several hundred atmospheres.

---

greater detail by B. I. Bersenev, L. F. Vereshchagin, Yu. N. Ryabinin, and L. D. Livshits [31] (see also [32]).

TABLE 10. Composition and Properties

Steel	Composition, %					Testing of samples at 20°C		
	C	Cr	Mo	W	V	Yield strength $\times 10^{-2}$, kg/cm^2	Tensile strength $\times 10^{-2}$, kg/cm^2	Elong-ation, %
a	0.11	3.2	0.30	—	—	40-46	54-56	20
3	0.10	3.2	—	—	0.11	~39	53	29
b	0.15-0.18	3.1	0.54-0.64	0.45-0.69	0.08-0.12	>45	60-70	22
e	0.18	2.6	0.52	—	0.38	82	90	18
6	0.24	2.9	0.26	—	0.61	67-73	79-84	18
W1	0.21	3.0	—	0.25	0.62	71-77	84-87	17
7	0.19	2.7	0.51	0.48	0.74	71-74	82-88	20
W2	0.22	3.0	—	0.22	0.86	71-76	84-87	18
V1	0.24	2.9	0.26	—	0.89	70-78	87-90	17
V2	0.23	2.9	0.27	0.24	0.92	68-70	83	17

* After 1000 h.
† 950°/A 2 h; 750°/A means 950°C, air, 2 hours; 750°C, air.

Mercury is the most dangerous material for metals. At a pressure of 6000 atm, mercury passes through the walls of the vessel [39]. The action of the mercury is that it penetrates into the pores (microscopic fissures) in the metal and then forms an amalgam on the clean surface of the metal, which leads to breakdown of the metal. In our experience, at a pressure of 10,000 atm certain parts of the apparatus which touched mercury were subject to amalgamation and fissuring, and the metal became so fragile that it could be broken by hand. In another case, a connecting pipe with an internal diameter of 3 mm and a wall thickness of 27 mm made of 18KhNVA steel burst explosively five minutes after the mercury pressure within it reached 9000 atm.

of Hydrogen- Resistant Steels [37]

Impact strength, kg-m/cm^2				Creep limit, kg/cm^2			Resistance to hydrogen pressure		Heat treatment
Thermally improved	after 500 h at			500℃	550℃	600℃	100 h 600℃ 700 atm	100 h 650℃ 700 atm	
	500℃	550℃	600℃						
27	—	27*	27*	11.5-13.5	─	—	Stable	Unstable	950°/A† 2 h 750°/A
30	—	—	30*	11.3	6.5	—	"	"	1050°/A 6 h 660°/A
30	—	—	30*	13-23	9-13	3-4.7	"	"	1050°/A 2 h 710°/A
15-17	—	—	—	36	24	—	"	"	1050°/A 3 h 700°/A
10-14	9	10-12	14	24.5	19.8	11.6	"	"	1050°/A 2 h 690°/A
9-15	7-8	8	11-12	27.4	19.0	10.0	"	"	1060°/A 2 h 700°/A
22	—	—	18-20*	24-28	20-22	13-15	Stable (1000 h)	"	1050°/A 2 h 700°/A
14-17	2-7	6-11	10-15	24.3	19.0	10.5	"	"	1050°/A 2 h 700°/A
7-12	7-8	8-11	13	28.1	20.0	12.6	"	"	1050°/A 2 h 700°/A
8-13	7-10	9-10	9-12	25	19.9	13.5	"	"	1050°/A 2 h 700°/A

N. T. Gudtsov and M. N. Govze [40] investigated the action of mercury on steel at high temperatures. According to the data of these authors, at 800°C mercury has practically no effect on steel after more than 1000 hours of contact and does not amalgamate even with stressed steel.

According to the data of T. A. Rebinder [41], liquid metal media cause a considerable reduction in strength when they are adsorbed onto the surface of the metal. In the presence of oxygen, the scale is removed from the surface of carbon steel under the action of mercury and the surface is freed for the formation of a new scale.

Mercury is rather frequently used in high-pressure apparatus as the medium transmitting the pressure and as a sealing or electrically-conduct-

TABLE 11. Mechanical Properties of Certain Nonferrous Metals and Alloys

Metal or alloy	Tensile strength (× 10⁻²), kg/cm²	Yield strength (× 10⁻²), kg/cm²	Unit elongation, %	Unit reduction in area, %	Modulus of elasticity, extension (× 10⁻²), kg/cm²	Melting point, °C	Density, g/cm³	Coefficient of linear expansion at 20°C (× 10⁶)	Thermal conductivity, cal/(g·cm·sec·deg)	Used up to temperature, °C
M4 copper										
Soft	23.0	4.9	50	—	11200	1080	8.89	17.3	0.43‡	800
Hard	48.5	34.3	10	—	11200					
Aluminum										
Soft	8–10	5–8	32–40	70–90	6000–7000	653	2.7	23.8	0.52	500
Hard	15–25	12–24	4–8	50–60	—					
Cast	9–12	—	—	—	—					
Lead	1.8	0.5	50	100	1700	327	11.34	29.5	0.089	250
Babbit (B90)	11.4*	4.3*	9.0	—	5700	342†	7.30	—	—	—
L80 Brass										
Soft	30–34	12	52	69.5	11600	1003	8.65	18.8	0.34	—
Hard	55–65	36	10	40.0	15350					
Beryllium bronze BrB2 [Be–copper alloy]										
Soft	40–60	22–35	30–35	—	11700	1084	8.6	18	0.25	—
Hard	75–100	85–90	0.1	—	13250					
Refined	130–135	128	1–2	—	—					
Zinc										
Cast	2–7	—	20	60–70	13000	419	7.14	35.4	0.268	350
Rolled	10–22	—	40–50	70–80	—					

*Upon compression.
† Beginning of hardening.
‡ M3.

TABLE 12. Properties of Certain Polymer Materials [44, 45, 48, 49]

Properties	Polyethylene (molecular wt. $(18-25) \cdot 10^3$]	Poly-propylene	Poly-isobutylene	Polyvinyl chloride	Polymethyl methacrylate	Nylon, (Poly-caproamide)	Teflon (fluoroplast-4)
Density, g/cm^3	0.922	0.9	1.32	1.39	1.19	1.14	2.2
Strength, kg/cm^2							
Tensile	120-150	300-500	45-65	500-600	560-700	420-780	140-320
Bending	120	—	—	1000-1200	—	—	—
Unit extension at rupture, %	400-600	400-800	475-550	85-100	4	50	50-100
Modulus of elasticity bending, kg/cm	1500-2000	—	—	31,500	31,500	$20-30 \cdot 10^3$	4200
Brinell hardness (3 kg, or sphere 2mm in diameter)	43-46*	—	—	15	15-18	—	1.0-2.0
Thermal conductivity, $(cal/sec \cdot cm \cdot deg)$ $\times 10^{-4}$	7	—	—	3.5	3.5	1.7	2.1
Softening temperature	108-115	—	—	90-95	110-112	180	—
Pour point, °C	—	—	—	180	—	264	350
Coefficient of linear expansion	$2.2 \cdot 10^{-4}$	—	—	$0.8 \cdot 10^{-4}$	$0.9 \cdot 10^{-4}$	10^{-4}	$0.8 \cdot 10^{-4}$

* According to Jones.

ing medium for contacts. Therefore the internal surface of the apparatus must be isolated from the action of the mercury by causing the mercury to contact only parts under hydrostatic pressure that are supported by the walls of the apparatus themselves (see Chapter III).

PACKING MATERIALS (GASKETS)

Packing materials for high-pressure apparatus must operate under very difficult conditions. Under strong compression, the packing material is in a state of plastic flow, filling all gaps in the sealing surfaces.

Various materials are used to make the packing, depending on the pressure level, sealing or packing method used, and the pressure-transmitting medium.

Copper. Copper packing seals well but withstands temperature fluctuations poorly, due to the high coefficient of thermal expansion of copper. The packing changes its configuration, and the seal must be repacked after every cycle of increasing and decreasing temperature. Copper is best used in seals of the unsupported-area type.

Long retention of copper as a sealing material under pressure causes it to harden. Of course, copper cannot be allowed to come into contact with mercury or corrosive materials.

Aluminum. Aluminum is somewhat softer than copper, but is more subject to seizing. Aluminum cannot be allowed to contact mercury.

Lead. Lead can be used only at temperatures up to 200°C. It is usually used in combination with copper or aluminum to make an initial seal, especially in unsupported-area seals.

Babbitt. Babbitt is an alloy of antimony, copper, tin, and lead with small additions of cadmium, nickel, tellurium, and arsenic, and is used for lining of bearings. Thanks to its ductility and lubricating properties, it can be conveniently used for packing of compressor piston glands, rods, valves, etc.*

Table 11 shows the properties of these and some other materials [43].

Fibra.† Fibra has limited usefulness as a packing material due to its low ductility. Its advantages are that it is not amalgamated by mercury and is not electrically conductive, due to which it can serve as insulation for electrical conductors in high-pressure vessels.

---

* Metallic sodium [42] is also used as packing for work at high pressures and temperatures on the order of 50–100°K.

†Tr. Note: a leatheroid material. Probably impregnated paper.

TABLE 13. Strength of Plastics at Low Temperatures [50]

Materials	Tensile strength, kg/cm^2				Young's modulus ($\times 10^{-4}$), kg/cm^2			
	at −196°C	at −120°C	at −75°C	at +20°C	at −196°C	at −120°C	at −75°C	at +20°C
Teflon	960	590	340	100	52	38	18	4.2
Polycaproamide	1960	1680	1400	660	77	56	42	32
Polyvinyl-chloride, rigid	1380	−	1200	540	77	−	38	36
Fluoroplast-3	1120	−	1100	440	58	−	43	18

TABLE 14. Mechanical Properties of Certain Packing
and Insulating Materials*

Material	Tensile strength ($\times 10^{-2}$), kg/cm^2	Compression strength ($\times 10^{-2}$), kg/cm^2
Leather		
vegetable tanned	2	−
chrome tanned	3	−
rawhide	3-12	−
Paronite (asbestos-filled elastomer)		
along fibers	3.2-4	−
across fibers	1.2-1.8	−
Oil stable rubber†	0.5-0.4	−
Fibra		
along fibers	5.5-4.7	} 18-32
across fibers	3.5-2.7	
Textolite (like Formica)	6.5-10	20-25
Porcelain	4-9	30-60

*Caoutchouc undergoes 70% compression when loaded to 60,000 kg/cm^2.
† The unit elongation at rupture of oil-stable rubber is 200-600%.

Leather.   Processed leather and rawhide are used to make packings (for glands) and collars used at very high pressures. Tanned leather is used primarily for packing glands of valves. If the leather is to come into contact with organic solvents, it is recommended that it be preliminarily boiled in glycerin. Rawhide is more elastic. We have used it for packing gas-pressure intensifier rods at pressures up to 10,000 atm. At higher pressures, the fatty substances contained in the leather are solidified, and the leather loses its plasticity and therefore its capacity to seal (it becomes harder than brass).

Polymer Materials.   It is very convenient to make packing out of resins of certain types of synthetic rubber which do not swell in gasoline and, especially, of various plasticized resins. Such polymer materials as polyvinyl chloride, polyethylene, and especially teflon (polytetrafluorethylene) are stable to corrosion and sufficiently plastic. They can be used up to comparatively high temperatures.

Teflon [44] (fluoroplast-4 [45]) has remarkable properties. It is stable to the action of all substances except metallic potassium and sodium. Teflon does not lose its ductility even at very low temperatures. For example, at 4°K, its plastic deformation reaches several percent [46]. When cooled samples are heated, their initial properties return. The strength of teflon increases with decreasing temperature. Thus, although the yield strength at 160°K is $0.5 \cdot 10^3$ kg/cm^2, at 4°K it increases to $2 \cdot 10^3$ kg/cm^2. The modulus of elasticity changes from $5 \cdot 10^4$ to $7 \cdot 10^4$ kg/cm^2. Thanks to these properties, teflon has been used in the manufacture of oxygen pump valve stems. The temperature coefficient of compressibility of teflon becomes negative at over 1000 atm pressure. This means that teflon packing expands with decreasing temperature (at high pressures) to improve sealing. Teflon, impregnated with molybdenum disulfide [46], can be used as a material for manufacture of piston glands for pressures up to 1000 atm.

Polytrifluorochlorethylene (fluoroplast-3), or gostaflon [47], has no less valuable properties; at high temperatures, it is considerably stronger than teflon. Whereas teflon samples will burst at a specific pressure of 100 kg/cm^2 at 60°C, and at 50 kg/cm^2 at 100°C, gostaflon samples of the same size are only plastically deformed at 140°C and 100 kg/cm^2 pressure.

The properties of certain types of plastics are given in Tables 12 and 13.

Polyethylene, which is now produced by low-pressure polymerization [51], has a tensile strength of 190-200 kg/cm^2.

The strength of polymethacrylate [52] is too low to be used for manufacture of transparent tubing for high-pressure work. Also, it is insufficiently stable to organic solvents.

Also, the properties of polymer materials change with pressure.* Thus, for example, the softening temperature of teflon increases [54] from 324°C at 69 atmospheres to 356°C at 207 atmospheres and 420°C at 615 atmospheres, which allows it to be used at higher temperatures. At the present time, the thermal properties of teflon have been well studied [55].

Asbestos. Asbestos is a wonderful material for manufacture of glands working at pressures up to 3000 atmospheres. Asbestos glands (made of pure asbestos) can withstand high temperatures. Asbestos is rather strong [56] (tensile strength 250-350 kg/cm^2).

---

* According to some data [53] the strength of polymer materials under compression in hydrostatic pressures up to 2000 atmospheres increases by $1\frac{1}{2}$ to 3 times.

Pure asbestos is, however, rarely used. Usually, corded asbestos, coated with powdered graphite mixed with oil to the consistency of a thick paste, is used. A gland of this type can be easily pressed. The presence of the graphite reduces friction in the gland.* If the gland is working under mercury pressure, the asbestos fiber is lubricated with a mastic composed of soft rubber boiled in lubricating oil or machine oil (50% solvent).

The mechanical properties of certain packing and insulating materials are presented in Table 14.

INSULATION MATERIALS

During investigations under high pressure, it is often necessary to introduce an insulated conductor into the apparatus. There are various types of electrical inlets, but in all such designs the insulating material must withstand high compressive and shear stresses. Therefore, it is very important to know how a given material acts under pressure. The primary requirements for an insulating material are high electrical insulating properties and sufficient strength.

Organic Insulators. For pressures on the order of 3000 atmospheres, ivory can serve as a good insulator. However, naturally, it, like other organic materials, cannot be used at temperatures over 100°C. Fibra is used for the same purposes in the same temperature range. We have obtained satisfactory results by using methyl methacrylate as an insulating material. It has good insulating properties and withstands high pressures (up to 20,000 atmospheres), but it cannot be used at temperatures over 80°C.

Glass. Glass is a good insulator. It is used for manufacture of electrical conductors operating at high temperatures and pressures [57]. A glass cone is melted-in hot. This structure will withstand pressures up to 3000 atmospheres and temperatures to 200°C.

Glass is also used as a material for manufacture of high-pressure apparatus. Thus, O. I. Leipunskii [58] performed investigations in glass capillary tubes at pressures up to 12,000 atm. The literature [59] contains a description of a special glass installation working at 100 atm pressure. Finally, glass windows are used for visual observation in high-pressure apparatus [60].

Depending on the composition of the glass involved, tensile strength limits vary from 500 to 850 kg/cm^2, compressive strength limits from 6100 to 12,300 kg/cm^2.

---

* Boron nitride or molybdenum sulfide can be used in place of the graphite.

TABLE 15. Compressibility of Glass at Various Temperatures
$$\Delta V/V_0 = - (aP + bP^2)*$$

Glass	$a \cdot 10^7$		$b \cdot 10^{12}$	
	30°C	70°C	30°C	70°C
Soda-lime with boron oxide	23.29	23.92	− 6.1	−10.1
Soda-potash	24.69	25.44	−22	−21.2
Pyrex	30.12	29.72	+ 6.1	+ 6.7
Lead-potash with high lead content	30.54	30.60	−24.3	−20.6
Lead-soda borosilicate with fluorides	27.78	27.71	+ 2.5	+ 3.8
Zinc-soda borosilicate	25.97	26.30	+ 4.0	+ 0.2

*P is the pressure in atmospheres.

TABLE 16. Strength of Glass Tubing

Tubing	Radius, mm		Glass thickness, mm	Rupture pressure, atm
	external	internal		
Capillary	5-7	0.24-1.0	4.76-6	430-1250
Thick wall	9-18	3-6	6-9	235-390
Thin wall	3.8-7.8	3.4-7.3	0.4-0.5	55-385

TABLE 17. Mechanical Properties of Quartz Glass and Porcelain

Material	Strength, kg/cm^2			
	compression	tensile	bending	bending shock
Quartz glass				
Nontransparent	3122	226	455	0.85
Transparent	6556	734	1131	1.08
Porcelain	2000-4000	300-350	600-700	1.2-1.5
Ordinary glass	6000-10000	350-870	400-800	−

TABLE 18. Resistance of Quartz Glass Tubing to Rupture from Hydrostatic Pressure

Transparent glass			Nontransparent glass		
internal diameter, mm	wall thickness, mm	rupture pressure, atm	internal diameter, mm	wall thickness, mm	rupture pressure, atm
4	0.5	83	15	2.5	15.0
5	1.0	150	57	9.0	10.0
7	2.0	220	80	11.0	13.0
8	1.0	100	200	13.0	7.0
9	2.0	190	370	14.0	3.0
10	1.0	70	−	−	−

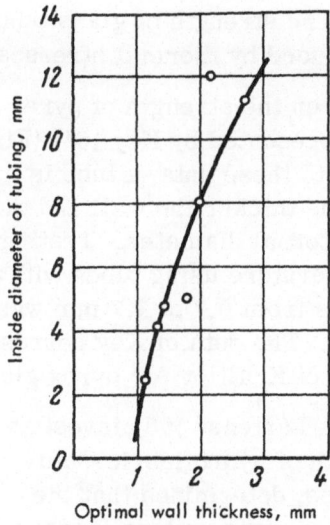

Fig. 2. Optimal wall thickness for glass
tubing of various diameters.

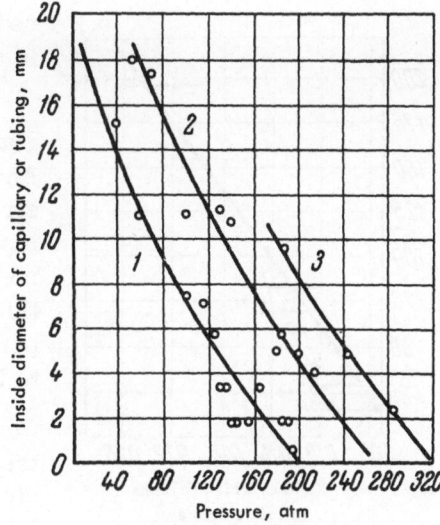

Fig. 3. Rupture pressure of glass tubing: 1), 2)
molybenum glass; 3) pyrex glass.

The mechanical properties of glass are:

Shear modulus G, kg/cm^2	311,000
Young's modulus E, kg/cm^2	715,000
Poisson's ratio, μ	0.143

The dependence of these quantities on the composition of the glass can be calculated by well-known methods [61]. The practice of working with glass piezometers often provides us with the necessary data concerning the compressibility of glass (Table 15).

Table 16 presents data on the strength of glass tubes used for manufacture of mercury manometers, high-pressure ampules, etc.

According to the data of D. I. Mendeleev [62], tubes with an external diameter of 11.5-12.5 mm and a wall thickness of 3-4 mm rupture at a pressure of 100-140 atm, while tubes with walls 2-3 mm thick rupture at a pressure of 140-200 atm.

The strength of glass tubes was investigated by Kh. M. Khalilov [63]. According to his data, there is an optimal wall thickness for every tubing diameter (Fig. 2), at which the tube will withstand the highest pressure (Fig. 3).*

If there are scratches on the surface of the glass, even though they may not be visible to the naked eye, the strength of tubing is considerably

---

* The data presented have been checked in the temperature range from 20 to 140°C.

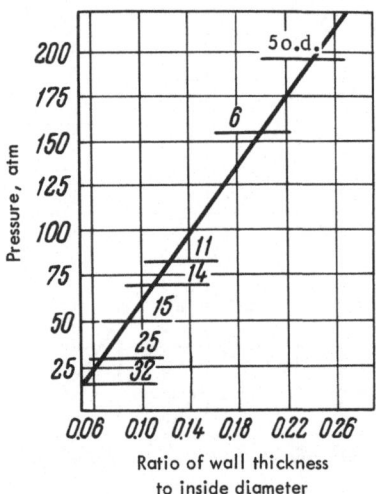

Fig. 4. Rupture pressure of pyrex glass tubing with wall thicknesses from 0.7 to 1.7 mm (o.d. = outside diameter).

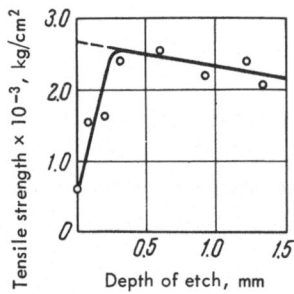

Fig. 5. Strength of glass as a function of etching depth.

reduced. The strength of glass tubing is also reduced by thermal stresses.

Data on the strength of pyrex glass are presented by Key [64] (Fig. 4). According to these data, a tube is stronger the thicker the wall and the less the internal diameter. Testing at room temperature using tubes with wall thicknesses from 0.7 to 1.7 mm was performed. The data of Key correspond to the data of Khalilov for pyrex glass.

G. M. Bartenev [65], investigating the strength of hydrofluoric-acid-etched glass, determined that the strength increases with an increase in the thickness of the etched layer up to 0.3 mm, then decreases (Fig. 5).

The considerable strengthening of the glass which takes place at first is explained by removal of surface defects or microfissures which extend to a certain depth only. With further etching, due to the ever greater distortion of the surface of the glass, its strength decreases somewhat. The strength of chemically homogeneous glass increases with etching, although it does not reach a value close to the strength of the internal layers (in the experiments of Bartenev, 2600 kg/cm^2). It is also known that glass fibers covered with a layer of softened glass [66, 67] (at 700°C) withstand stresses up to 9000 kg/cm^2. These data indicate great possibilities for the use of glass as a material for manufacture of high-pressure apparatus.

V. P. Pryanishnikov [68] presents the properties of quartz glass (Tables 17, 18). Transparent quartz is stronger than nontransparent quartz, and its strength is as great as that of ordinary glass, exceeding that of ordinary glass in bending.

Tubing of transparent quartz glass is as strong as tubing of pyrex glass (compare Fig. 3).

Porcelain. Porcelain insulation is rarely used, in spite of its excellent electrical insulating qualities and heat resistance. This is due to the brittleness of porcelain and the difficulty of working it.

Mica. The most suitable material for insulation of electrical conductors is mica. It has high compressive strength, possesses excellent electrical insulating properties (better than porcelain) and withstands high temperatures. The deficiency of mica is its brittleness, which causes it to be difficult to work. Good quality mica 0.03-mm thick can be rolled into 3-mm-diameter tubing without cracking [69]. This sort of mica cannot always be produced, however, in which case the methods described below can be used for production of reliable insulation.

When investigations are performed at pressures reaching hundreds of thousands of atmospheres, it is important to know the coefficients of friction (f) of the materials used as packing and insulating materials, and as pressure-transmitting media. This information is presented below for certain materials (mainly in powder form) at a pressure of 25,000 atm [70]:

	f		f
Iron oxides	0.71	Aluminum hydrosilicate	0.39
Zinc oxide	0.58	Kieselguhr [microcell]	0.39
Pumice	0.52	Calcium hydroxide	0.27
Chromium oxide	0.50	Boric acid	0.14
Pyrophyllite	0.25	Potassium chloride	0.12
Pyrophyllite		Sodium chloride	0.12
(massive)	0.47	Sheet mica	0.07
Lead oxide	0.46	Boron nitride	0.07
Manganese dioxide	0.46	Graphite	0.04
Titanium dioxide	0.45	Molybdenum disulfide	
Molydenum trioxide	0.42	[Molykote type Z]	0.04
Tin oxide	0.41	Silver chloride	0.03
Boron carbide	0.40	Sheet indium	0.01

PRESSURE-TRANSMITTING MEDIA

The selection of a pressure-transmitting medium is determined by the object of the investigations and the amount of pressure to be transmitted. For example, in investigations of the properties of an ordinary gas at pressures up to 1000-2000 atm, it is simplest to generate the pressure by pumping the gas being investigated into the pressure chamber. If the gas in question is a rare gas, it is more convenient to reach the required pressure by compressing the gas using some other medium such as mercury

TABLE 19. Viscosity of Some Oils at Various Temperatures and Pressures[*]

Oil	Temperature, °C	Viscosity in poises at pressures of:			
		1 atm	35 atm	700 atm	1000 atm
Aviation	22	14.5	–	56.0	–
	32	6.5	11.9	21.3	40.3
	60	1.75	2.92	–	–
Lubricating oil	25	10.4	17.0	–	40
	40	1.5	2.22	3.9	5.1
Machine oil	20	2.74	–	–	–
	33	1.1	2.27	–	8.1
	40	0.75	1.70	2.78	–
Upper cylinder	16	26.5	–	–	131.0

[*]According to the data of M. P. Volarovich [74].

TABLE 20. Viscosity of Transformer Oil at 20°C
According to Data of E. V. Zolotykh [75]

Pressure, atm	Viscosity, poise	Pressure, atm	Viscosity, poise
0	0.275	1500	7.568
200	0.428	2000	22.85
500	0.830	2500	68.98
1000	2.506	3000	208.3

TABLE 21. Coefficients for Calculation of Viscosity
According to the Formula of E. V. Zolotykh

Oil	η_0	k	η_0	k	η_0	k
	at 14°C		at 20°C		at 30°C	
Transformer	0.376	0.00241	0.275	0.00221	0.171	0.00201
Castor	17.6	0.00149	10.43	0.00142	4.79	0.00131
Vasoline	–	–	1.38	0.00269	0.733	0.00235

or another liquid, especially since it is much easier to seal an apparatus for work with liquids than for work with gases.*

At very high pressures, the pressure-transmitting medium may be solid sheet material with low coefficient of friction. During compression, quasi-hydrostatic stresses arise in such materials, which are transmitted to the sample contained between them. We must recall that at normal

* Editor's comment: Self-energizing seals (of which unsupported-area seals and O-rings are examples) seal both gases and liquids without any leakage.

TABLE 22. Relative Viscosity of Certain Materials and Logarithms of Relative Viscosity at Various Pressures

Material*	Pressure, atm	log η/η_0		$\dfrac{\eta_{30}}{\eta_{75}}$
		at 30°C	at 75°C	
Glycerin	1	0.000	$\bar{2}$.810	15.49
($\eta_0 = 3.8$)	1000	0.260	$\bar{1}$.023	17.26
	4000	0.936	$\bar{1}$.529	25.53
	8000	1.741	0.094	44.36
	12,000	–	0.628	–
Isopentane	1	0.000	$\bar{1}$.821	1.510
($\eta_0 = 0.00198$)	1000	0.344	0.193	1.416
	4000	0.894	0.715	1.510
	8000	1.431	1.179	1.786
	12,000	1.947	1.586	2.296
Kerosene	1	0.00	–	–
	1000	0.46	–	–
	4000	1.71	0.91	4.3
	8000	–	1.88	–
	12,000	–	2.80	–
n-Pentane	1	0.000	$\bar{1}$.811	1.545
($\eta_0 = 0.00220$)	1000	0.315	0.163	1.419
	4000	0.847	0.676	1.483
	8000	1.360	1.119	1.742
	12,000	1.846	1.493	2.254
Petroleum ether	1	0.00	–	–
	1000	0.30	–	–
	4000	0.93	0.56	2.34
	8000	1.59	1.06	3.39
	12,000	2.18	1.49	4.80
Ethyl alcohol	1	0.000	$\bar{1}$.657	2.203
($\eta_0 = 0.01003$)	1000	0.200	$\bar{1}$.873	2.123
	4000	0.617	0.289	2.128
	8000	1.023	0.634	2.449
	12,000	1.390	0.919	2.958

* η_0 is the viscosity in poise at 30°C and pressure 1 atm.

TABLE 23. Certain Properties of Mercury at High Pressures

Pressure, atm	Melting point, °C*	Change in volume upon melting, cm^3/g	Viscosity, poise	
			at 30°C	at 75°C
1	−38.2	0.002534	0.01516	0.01340
4000	−18.4	0.002512	0.01663	0.01463
8000	+1.8	0.002445	0.01822	0.01598
12,000	+21.9	0.002290	0.02007	0.01759

*M. K. Zhokhovskii investigated the course of the melting curve of mercury. Data presented in Chapter IV.

TABLE 24. Compressibility and Density of Mercury at 25°C and Pressures up to 2500 Bar [$k_P^{25} = (v_0^{25} - v_P^{25})/v_0^{25}$; ρ_P^{25} Is the Density at 25°C and Pressure P]

P bar	k_P^{25}	ρ_P^{25} g/cm^3	P bar	k_P^{25}	ρ_P^{25} g/cm^3
0	0.000000	13.53365	1500	0.005889	13.61382
500	0.002000	13.56077	2000	0.007781	13.63977
1000	0.003963	13.58749	2500	0.009639	13.6654

TABLE 25. Compressibility (k) of Methylpolysiloxanes (Silicone Oils) [72] [$k = \dfrac{1}{v_1} \cdot \dfrac{v_2 - v_1}{P_2 - P_1}$; v_1 and v_2 Are the Volumes of Liquids at Pressures P_1 and P_2]

P atm	$k \cdot 10^7$ for SS-I at			$k \cdot 10^7$ for SS-II at			$k \cdot 10^7$ for SS-III at			$k \cdot 10^7$ for SS-IV at
	20°C	30°C	40°C	20°C	30°C	40°C	20°C	30°C	40°C	20°C
1-187	1057	1118	1198	1030	1094	1153	1010	1061	1130	990
1-284	1006	1060	1134	979	1034	1085	963	1003	1062	930
1-380	962	1010	1074	930	980	1030	908	951	1009	881
1-477	915	961	1026	883	932	979	861	905	955	836
1-761	804	847	896	780	818	859	759	793	837	736
1-1014	732	771	810	713	748	781	692	726	758	674
1-1269	680	715	745	663	692	725	650	675	705	632
1-1525	642	666	695	625	649	676	616	638	661	605

TABLE 26. Properties of Methylpolysiloxanes

Liquid	Average molecular weights	Density (20°C), g/cm^3	Viscosity (20°C), centipoise
SS-I	1130	0.940	10.4
SS-II	6800	0.961	100
SS-III	13,900	0.971	272
SS-IV	43,000	0.974	4530

temperature and at pressures of 50,000-100,000 atm, both liquids and gases solidify.*

Liquid Pressure-Transmitting Media. A liquid used as a transmitting medium must be chemically inert and have a low freezing point. The viscosity of the liquid should change little over a wide temperature and pressure interval.

---

* Cf. [71] which presents melting curves for certain pressure-transmitting liquids and gases.

The silicone oils, or polysiloxanes [72], have these properties; for this reason, they are ever more widely used as pressure-transmitting media.

In order to generate pressures up to 5000 atm at temperatures from 0 to 100°C, mixtures of oil and kerosene can also be used. The role of the oil in this mixture is protection of leather packing from the action of kerosene. At temperatures below 0°C and pressures above 5000 atm, however, the kerosene and its mixtures with oil become so viscous that ball valves cease to operate and, in some cases (for example, when the pressure is increased to 9000-10,000 atm), the indications of a manganin manometer will be distorted due to the appearance of nonhydrostatic stresses in the coil.

The kerosene used as a compressing medium must be pure and free of moisture which otherwise may cause non-hydrostatic compression of the coils of the manganin manometer. The oil should have a low freezing point (transformer or refrigerant oil). The higher the pressure, the lower should be the fraction of oil in the mixture (1 :4 for 10,000 atm).

At higher pressures, light gasolines with very small admixtures of oil must be used; at 15,000-20,000 atm, pure gasoline can be used. However, at temperatures below 40-50°C, the viscosity of gasoline increases considerably and it is more expedient to use petroleum ether or isopentane. The properties of oils and gasolines have been described in the literature [73, 74]. The viscosities of oils at various temperatures and pressures are presented in Tables 19 and 20.

E. V. Zolotykh [75] presents an empirical formula which can be used to calculate the viscosity of transformer, castor, and vasoline oil depending on pressure:

$$\eta = \eta_0 e^{kP}$$

where η is the viscosity at the given temperature and pressure; η_0 is the viscosity at a pressure of 1 atm and given temperature; k is the proportionality coefficient; P is the pressure in atm.

The values of η_0 and k for various oils are presented in Table 21.

For glycerin at pressures up to 3500-4000 atm, E. V. Zolotykh developed the following relations:

$$\text{at } 20°C, \quad \eta = 5.12e^{0.000552P}$$
$$\text{at } 30°C, \quad \eta = 2.39e^{0.000524P}$$

where η is the viscosity, poise; P is the pressure, atm.

This same investigator studied [76] the viscosity of a number of liquids and their mixtures at pressures up to 10,000 atm.

TABLE 27. Relative Volumes of Certain Liquids at Various Pressures and Temperatures

Liquid	Relative volume of liquid at pressures					
	1 atm	1000 atm	2000 atm	4000 atm	8000 atm	10,000 atm
Gasoline [79] (B-70)						
25°C	1.0000	–	0.8820	0.8230	–	–
Water						
0°C	1.0000	0.9567	0.9248	0.8795	–	–
50°C	1.0119	0.9741	0.9439	0.8997	0.8407	0.8192
95°C	1.0395	0.9984	0.9661	0.9194	0.8577	0.8352
Glycerin	1.000	–	0.958	0.932	0.8930	0.866 (at 12,000 atm)
Kerosene [79](T-2)						
25°C	1.000	–	0.9022	0.8525	–	–
Isopentane						
0°C	1.0000	0.9028	0.8571	0.8025	–	–
50°C	1.0843	0.9415	0.8845	0.8222	0.7592	–
95°C	1.1741	0.9809	0.9119	0.8403	0.7706	–
n-Pentane						
0°C	1.0000	0.9021	0.8546	0.7997	0.7381	–
50°C	1.0837	0.9395	0.8820	0.8193	0.7520	–
95°C	1.1869	0.9768	0.9078	0.8371	0.7641	–
Mercury						
0°C	1.0000	0.99626	0.99261	0.98561	–	–
20°C	1.00362	0.99972	0.99593	0.98871	0.97608	0.86567 (at 12,000 atm)
Ethyl alcohol						
20°C	1.0212	0.9506	0.9081	0.8545	0.7917	0.7521 (at 12,000 atm)
80°C	1.0934	0.9944	0.9407	0.8787	0.8094	0.7682 (at 12,000 atm)

Table 22 presents the viscosity of certain material at pressures up to 12,000 atm.* This table includes materials which might be used as pressure-transmitting media and as solvents (see also [78]).†

Table 23 shows certain properties of mercury which is very often used as a filler for hydraulic valves separating the material being investigated from the pressure-transmitting media, and also is used as a liquid electrical contact.

In designing appratus, we must know how the volume of the pressure-transmitting medium changes with pressure. Tables 24, 25, 26, and 27 present data on the compressibility and density of certain pressure-transmitting media.

Table 24 presents data on the compressibility and density of mercury at 25°C and pressures up to 2500 bar [80].‡ Thermal expansion of mercury with constant pressure can be calculated according to the equation

$$\frac{v-v_0}{v_0}\cdot 10^8 = at\cdot 10^8 = 18144.01t + 70.16\cdot 10^{-2}t^2 + 28.625\cdot 10^{-4}t^3 + 2.617\cdot 10^{-6}t^4$$

where α is the coefficient of thermal expansion; t is the temperature. The equation is correct for temperatures from 0 to 350°C.

For calculation of the volume of mercury at temperatures from -30 to $+150$°C and pressures up to 12,000 bars the following equation can be used [81]:

$$\log [L^2 P/(L_0 - L)] = 5.8837 - 0.0004877T + 4.95 (L_0 - L)$$

where P is the pressure, bars; L is the length of the side of a cube containing a unit mass of mercury at pressure P, cm; L_0 is the value of L at atmospheric pressure, cm.

$$L = \left(v_P^{t_1}/v_0^t\right)^{1/3} \text{ and } L_0 = \left(v_0^{t_1}/v_0^t\right)^{1/3}$$

where v is the volume of a given mass of liquid, cm^3; t_1 is the experimental temperature, °C; t is the temperature taken as the initial temperature (usually 0°C). $\log [L^2 P/(L_0 - L)]$ is a linear function of $(L_0 - L)$.

---

* The viscosity of silicone oil No. 3 (VTU MKhP 2127-4a) is presented in the literature [77]. [This is a temporary technical specification.]

†Table 22 presents the logarithms of the relative viscosity at various pressures and temperatures. The viscosity at 30°C and at atmospheric pressure is taken as unity. Since the viscosity of a material changes very greatly with pressure, it is more convenient to have a curve of the dependence of the logarithm of viscosity on pressure, which will be smoother and more convenient for interpolation.

‡1 bar = 1.02 atm = 10^5 n/m^2.

Solid Pressure-Transmitting Media. Such media include polyfluorethylene resin [82], silver chloride [83], pyrophyllite, and others. These materials are plastic and have low coefficients of friction. Filling the space between compression surfaces, they transmit quasi-hydrostatic stress to the sample included between them. Gradually flowing out (under pressure) through the clearances between the piston and cylinder, these materials (with the exception of AgCl) simultaneously form an excellent seal in the high-pressure cavity.

Pyrophyllite (the aluminosilicate $Al_2O_3 \cdot 2SiO_2 \cdot 4H_2O$) has a molecular weight of approximately 360, density of approximately 3.15 g/cm^3. Katlenite, a variety of pyrophyllite, is a clay-like material. Both pyrophyllite and katlenite have a high melting point (approximately 1800°C), are electrically nonconductive, have high viscosity, and are used as media for the generation of quasi-hydrostatic pressures on the order of 200,000 atm at temperatures up to 2000-3000°C. Operation at such high temperatures is possible, apparently, due to the increased melting point of these minerals under pressure. Ductile solid bodies are also used as working media in the cylinders of powerful hydraulic presses [84].

A long list of materials which can be used for work under pressure, as well as a description of some of the properties of these materials, can be found in the literature [85].

BIBLIOGRAPHY

1. Mikhailov-Mikheev, P. B., Handbook on Metal Materials; Mashgiz, 1961; Zhuravlev, V. N., and Nikolaeva, O. I., Designer's Handbook, GNTI, Machine Literature, 1962.
2. Liberman, L. Ya., and Peisikhis, M. I., Handbook on Properties of Steels Used in Boiler and Turbine Construction, Mashgiz, 1958.
3. Nitrogen Worker's Handbook, Goskhimizdat, 1944.
4. Slavin, D. O., and Shteinman, E. B., Metals and Alloys in Chemical Machine Building and Apparatus Construction, Mashgiz, 1951.
5. Vyaznikov, N. F., Alloy Steel and Its Heat Treatment, Metallurgizdat, 1951.
6. Lancaster, I. F., Petrol. Refiner., 43(10):161 (1964).
7. Rakovskii, V. S., Smirnov, F. F., Rozhdestvenskii, L. A., and Kryukov, I. M., Hard Alloys in Machine Building; Mashgiz, 1955; Zelikman, A. N., Metallurgy of Tungsten and Molybdenum, Metallurgizdat, 1949.
8. Peters, E., and Byerly, I. I., Rev. Sci. Instrum., 34(7):819 (1963).
9. Sarrak, V. I. Uspeki Fiz. Nauk, 67(2):339 (1959); Garber, R. I., and Gindin, I. A., Uspeki Fiz. Nauk, 70(1):57 (1960).
10. Chem. News, 38:54 (1960); H. Eiring, Ind. Eng. Chem., 51:5 (1959).
11. Kurdyumov, G. V., and Osipyan, Yu. A., Vestn. Akad. Nauk SSSR, 5:7 (1963).

12.   Sakhin, S. I., and Sokolov, O. G., Metalloved. i Term. Obrabotka Meta.
       1:14 (1962).
13.   Ershova, T. P., Ponyatovskii, E. G., Dokl. Akad. Nauk SSSR, 151:1364
       (1963); Ershova, T. P., Fiz. Metal. i Metalloved., 16:91 (1963), 17:144
       (1964).
14.   Gulyaev, A. P., Metals Engineering, Oborongiz, 1948.
15.   Kondrat'ev, V. N., Zav. Lab., 15:472 (1949).
16.   Glikman, L. A., and Maksimov, N. M., Zav. Lab., 11:1091 (1945).
17.   Drozd, M. S.,Zav. Lab., 24:74, 1002 (1958).
18.   Kennedy, G. G., Am. J. Sci., 248:540 (1950).
19.   Shvarts, G. L., Shevelkin, B. N., and Toropov, V. N., Zh. Vses. Khim.
       Obshchestva, 8(3):317 (1963).
20.   Kornilov, I. I., and Polyakova, R. S., Zh. Vses. Khim. Obshchestva,
       8(3):305 (1963); Khimushin, F. F., Legirovaniye, Termoobrabotka
       i Svoystva Zharoprochnykh Staley i Splevov, Oborongiz, 1962;
       Kornilov, I. I., Sovremennyye Vysokoprochnyye Metal-Splavy,
       Experimental Investigations in the Area of Deep Processes, Izd. Akad.
       Nauk SSSR 1962; Hinde, J., Chem. Proc. Eng., 44(4):179 (1963).
21.   Shapiro, M. B., Moskvin, N. N., Kristal', M. M., and Makarov, V. V.,
       Tr. Nauchn.-Issled. Inst. Khim. Mash., No. 40, 1962.
22.   Kantarovich, Z. B., Principles of Design of Chemical Machines and
       Apparatus, Mashgiz, 1960.
23.   Zaikov, M. A., Zh. Tekh. Fiz., 19:684 (1949).
24.   Uzhik, G. V., Strength and Plasticity of Metals at Low Temperatures;
       Izd. Akad. Nauk SSSR, 1957; Fradkov, A. B., Kislorod, 5:34 (1945);
       Parker, C. M., and Sullivan, J. W. W., Ind. Eng. Chem., 55(5):18
       (1963).
25.   Basset, J., J. Phys. Radium, 1(18):121 (1940).
26.   Bridgman, P. W., Investigation of Great Plastic Deformations and
       Rupture, Izdatinlit, 1955.
27.   Ryabinin, Yu. N., Livshits, I. D., and Vereshchagin, L. F., Fiz. Tverd.
       Tela, 1(3):476 (1959); 1(6):960 (1959).
28.   Vereshchagin, L. F., and Shapochkin, V. A., Fiz. Metal. i Metalloved.,
       7:479 (1959); Vereshchagin, L. F., and Zubova, E. V., Dokl. Akad.
       Nauk SSSR, 134:787 (1960); Shapochkin, V. A., Fiz. Metal. i Metal-
       loved., 9:303 (1960).
29.   Ratner, S. I., Zh. Tekhn. Fiz., 19:408 (1949).
30.   Livshits, L. D., Ryabinin, Yu. N., Bersenev, B. I., and Martynov,
       E. D., Dokl. Akad. Nauk SSSR, 154:86 (1964).
31.   Bersenev, B. I., Vereshchagin, L. F., Ryabinin, Yu. N., and Livshits,
       L. D., Some Problems of Great Plastic Deformations of Metals at
       High Pressures, Izd. Akad. Nauk SSSR, 1960.
32.   Solids Under Pressure (eds. Paul., W., and Warschauer, D. M.) Aca-
       demic Press, New York, 1963.
33.   Krichevskii, I. R., and Khazanova, N. E., Zh. Fiz. Khim., 21:719 (1947).

34. Dodge, B. F., Chem. & Ind., 87(6):169 (1962); Thygeson, J. R., and Molstad, M. C., J. Chem. Eng. Data, 9(2):309 (1964).
35. Ness, H. C., and Dodge, B. F., Chem. Eng. Progr., 51(6):266 (1955).
36. Ashmarin, N. V., Amer. Tekhn. i Promyshlennost, No. 12 (1946).
37. Class, J., Werkstoff Korrosion, 5(8-9):281 (1954).
38. Comings, E. W., High-Pressure Technology, New York, 1956.
39. Bridgman, P. W., The Physics of High Pressure, ONTI, 1935; Bridgman, P. W., The Physics of High Pressure, London, 1958.
40. Gudtsov, N. T., and Govze, M. N., Izv. Akad. Nauk SSSR, Otdel. Tekhn. Nauk, 1:67 (1952).
41. Rozhanskii, V. N., Pertsov, N. V., Shchukin, E. D., and Rebinder, P. A., Dokl. Akad. Nauk SSSR, 116(5):769 (1957).
42. Stewart, J. W., Phys. Rev., 97(3):578 (1955).
43. Short Handbook on Processing of Nonferrous Metals, Metallurgizdat, 1945; Molyneax, F., Chem. Process. Eng., 43:204 (1962).
44. Black, G. N., Chem. & Ind., 30:727 (1952).
45. Afanas'ev, P. A., The Usage of Plastics in Machine Building, Mashgiz, 1961.
46. Swenson, C., Rev. Sci. Instrum., 25(8):834 (1954); Browman, H., and others, Rev. Sci. Instrum., 27:550 (1956).
47. Merkel, E., Chem. Ingr. Techn., 27(5):279, 284 (1955).
48. Polyethylene (Ed. by M. I. Garbar) Goshkimizdat, 1955.
49. Bockhoff, F. J., and Roth, R. F., Chem. Eng. Progr., 51(6):252 (1955); Gourlag, J. S., and Jones, M., Brit. Plastics, 29:446 (1956).
50. Dyment, J., and Ziebland, H., J. Appl. Chem., 8:203 (1958).
51. Saechtling, H., Chem. Ind. Techn., 27(10):602 (1955).
52. Knowles, J. K., and Diets, A. G. H., Trans. ASME, 77:177 (1955).
53. Aibinder, S. B., Lakes, M. G., and Maiors, N. Yu., Dokl. Akad. Nauk SSSR, 159(6):1244 (1964).
54. McGeer, P. L., and Duns, H. S., J. Chem. Phys., 20:1813 (1952).
55. Furucawa, G., McCoskey, R., and King, G., J. Res. Nat. Bur. Standards, 49:273 (1952); Chegodaev, D. D.., Polyfluoroethylene Resins, Goskhimizdat, 1956.
56. Zukowsky, R., and Gase, R., Nature, 183:37 (1959).
57. Welbergen, H. G., J. Sci. Instrum., 10:247 (1933).
58. Leipunskii, O. I., Zh. Tekhn. Fiz., 10:596 (1940).
59. Hiller, Osterr. Chem.-Ztg.; 45, 111 (1942); Zh. Khim. Prom., Vol. 17, No. 7 (1940).
60. Bol'shakov, P. E., Trudy GIAP, No. 1, Goskhimizdat, 1953.
61. Morey, Properties of Glass, New York, 1938.
62. Mendeleev, D. I., Zh. Russ. Fiz. Khim. Obshchestva, VI, No. 5, 1, 1 (1874).
63. Khalilov, Kh. M., Dokl. Akad. Nauk Azerb. SSR, 8:221 (1952).
64. Key, W. B., Chem. Eng. Progr. (Symp. Ser.), 48(3):71 (1952).

65. Bartenev, G. M., Dokl. Akad. Nauk SSSR, 91:523 (1953).
66. Morley, J. G., Nature, 189:1560 (1959).
67. Pugh, H. L. D., Hodgson, G., and Gunn, D. A., J. Sci. Instrum. 40:221 (1963).
68. Pryanishnikov, V. P., Khim. Prom., 1:15 (1954).
69. Canad. Chem. Proc. Ind., 30, No. 3 (1946).
70. Hyde, G. R., Friction at Very High Pressure (M. S. Thesis), 1957.
71. Reeves, L. E., Scott, G. I., and Babb, S. E., J. Chem. Phys., 40(12): 3662 (1964).
72. Kiyama, R., Teranishi, H., and Inoue, K., Rev. Phys. Chem. Japan, 23(1):20 (1953).
73. Speransov, N. N., Handbook of Petroleum Products Consumer, Gostoptekhizdat, 1940; Nevyazhskaya, L., and Novikov, L., Non-metallic Materials (a handbook), Mashgiz, 1948.
74. Volarovich, M. P., Izv. Akad. Nauk SSSR, Otdel. Tekhn. Nauk, 3:27 (1940).
75. Zolotykh, E. V., Izmerit. Tekhn., 3:2 (1955).
76. Zolotykh, E. V., Tr. Inst. Komiteta Standartov, Mer Izmerit. Prib. SSSR, 75(135):123 (1964).
77. Zolotykh, E. V., Tr. Inst. Komiteta Standartov, Mer Izmerit. Prib. SSSR, 46(106):81 (1960).
78. Babb, E. S., and Scott, G. I., J. Chem. Phys., 40(12):3666 (1964).
79. Perevertkin, S. M., Khrapovitskii, Yu. S., and Tsiklis, D. S., Trudy GIAP, No. 7, Goskhimizdat, 1957.
80. Bett, K. E., Hayes, P. H., and Newitt, D. M., Phil. Trans. Roy. Soc., A247(923):59, London (1954).
81. Bett, K. E., Weale, K. E., and Newitt, D. M., Brit. J. Appl. Phys., 5:243 (1954).
82. Andreatch, P., and Andersen, O. L., Rev. Sci. Instrum., 28:288 (1957).
83. Ryabinin, Yu. N., and Livshits, L. D., Zh. Tekhn. Fiz., 29:1167 (1959).
84. Vereshchagin, L. F., Fedorovskii, A. E., Isaikov, V. K., Slesarev, V. N., and Semerchan, A. A., Inzh. Fiz. Zh., 3:132 (1960).
85. Ind. Eng. Chem., 48, No. 5 (1956); 50, No. 9 (1958).

Chapter II

Design and Construction of High-
and Ultrahigh-Pressure Apparatus

DESIGN OF APPARATUS*

Stresses are present in a material under load, causing deformation. The design of apparatus operating at low pressures entails primarily determining wall thicknesses sufficient for the apparatus to withstand the load to be applied without permanent deformation. For equipment working at very high pressures, plastic deformation is included in calculation of wall thicknesses. The occurrence of this type of deformation is utilized in construction, as will be shown below.

Behavior of Material under Load

Before beginning our discussion of methods of design, it would be useful to recall just what stress and deformation are and what are the relations between them.

Stress. Stress is force (acting over a given cross section) per unit of area. Stress can be divided into two components: normal stress σ_n perpendicular to the cross section, and shear stress σ_τ acting in the plane of the cross section. The stress state which arises in a body under load is characterized by 9 stress components [1] parallel to the three coordinate axes x, y, and z (Fig. 6). Of these, three are normal stresses σ_x, σ_y, and σ_z, and six are the shear components of the three [total] shear stresses acting in the planes of the elementary cube selected for analysis.

The entire system of stresses can be reduced to three principal (normal) stresses by selecting the proper coordinate axes.

---

*This portion of the chapter written in cooperation with A. N. Rozen.

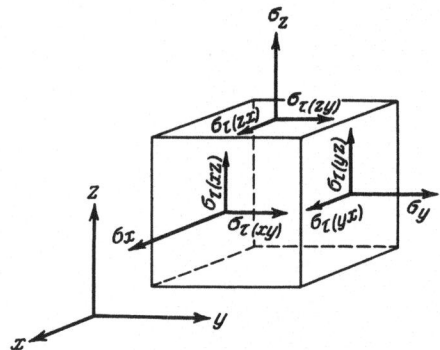

Fig. 6. Components of stresses acting on faces of an elementary cube.

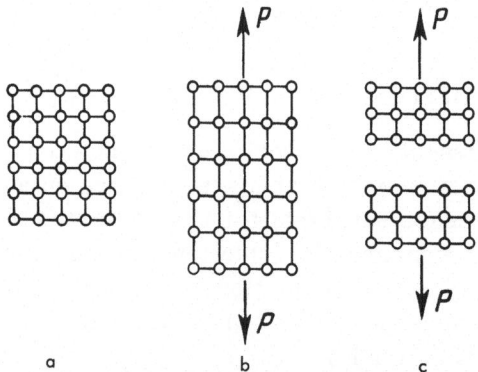

a   b   c

Fig. 7. Schematic representation of the deformation of a crystalline lattice under the influence of normal stresses: a) normal state; b) elongated; c) ruptured.

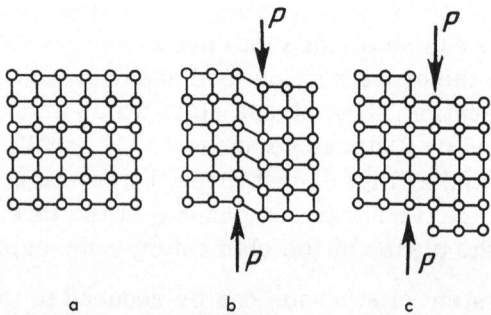

a   b   c

Fig. 8. Schematic representation of deformation of crystalline lattice under influence of shear stresses: a) normal state; b) elastic distortion; c) shear.

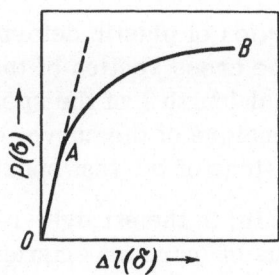

Fig. 9. Elongation as a function of tensile load (extension diagram).

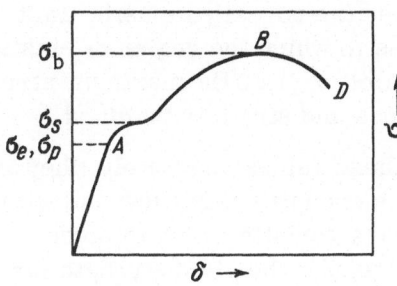

Fig. 10. Diagram of extension of metals with yield region. σ_e) Elastic limit; σ_p) proportional limit; σ_s) yield strength; σ_b) ultimate tensile strength.

Deformation. We differentiate between elastic deformations, those which disappear after the load is removed, and permanent (residual) deformations or plastic deformations, in which the displacement of one portion of the body relative to another does not disappear after removal of the load.

Figure 7 shows schematically the deformation and rupture of an ideal crystal lattice. It can be shown [2] that normal [tensile] stresses increase the separation between atoms (Fig. 7b), causing first elastic deformation then (when the distance between atoms is so great that the attractive forces are overcome) disrupting the crystal lattice by separation of one portion from another (Fig. 7c).

Shear stress (Fig. 8) also causes first elastic distortion of the crystal lattice and then displacement of one portion of the crystal relative to another without disruption of the integrity of the crystal; then plastic deformation takes place.

The vicinity of the shear plane and the neighboring volumes are, due to the distortion of the crystal lattice, stronger than those regions where shear has not taken place. Therefore, increasing the load causes new shear in planes which have not undergone shear previously and, conversely, shear does not reoccur where it has already taken place. Thus, the larger the number of individual shears which have occurred, the less a metal is suited to subsequent plastic deformation. When all planes and directions of shear have been exhausted, further increase in the load results in rupture of the material.

It is practically and theoretically important to know the maximum values of normal $\sigma_{n,\,max}$ and shear $\sigma_{\tau,\,max}$ stresses before rupture of a material takes place.

If a specimen is stretched and a graph is made of the dependence of absolute elongation Δl of the specimen on tensile load P, the result is a curve (Fig. 9) which will have a straight line sector OA, followed by a curved sector AB indicating transition to the region of plastic deformation. Placing the stress σ (force P per unit area of the cross section of the specimen) and unit elongation δ (Δl divided by initial length l of the specimen) on coordinate axes, we can, while retaining the nature of the curve, exclude dependence of the quantities on the dimensions of the test specimen.

The ratio of stress to unit elongation [strain] in the straight-line sector of the curve is a constant, called the modulus of (normal) elasticity or Young's modulus E (kg/cm^2). After transition through point A, the constancy of this ratio is no longer retained. Deformation begins to increase more rapidly than stress and in many cases permanent or plastic deformation appears (the dimensions and form of the specimen differ from their initial values after the stress is removed). Point A is a very important point on the diagram (Fig. 10). It determines the so-called elastic limit of the material σ_e, i.e., the maximum stress to which the region of elastic deformation extends, and the proportional limit σ_p, i.e., the maximum stress to which direct proportionality between stress and strain is retained.

Since it is very difficult to measure these values accurately (they are measured at very slight deviations from the proportionality law and with infinitesimal permanent deformations), a characteristic which is much easier to measure, and no less accurate, is used — the yield strength (or nominal yield point) σ_s (also represented as $\sigma_{0.2}$), i.e., the stress at which the permanent (plastic) deformation reaches 0.2% of the gauge length of the sample. This quantity is usually given in tables of properties.

With metals such as soft steel, a horizontal sector will appear above point A (see Fig. 10) showing that at this particular load elongation increases without an increase in load. This is the yield point of the material.

With further increase in load, first work-hardening and then rupture of the material occurs. Certain materials are ruptured at stresses corresponding to point B. These stresses are determined by dividing the maximum tensile load by the area of the cross section of the specimen and are called the ultimate tensile strength σ_b.

For most ductile materials, rupture under tensile load takes place not at the maximum load, but with a somewhat lower load (point D), since before rupture of a specimen of plastic material, it "necks down" so that less load is required to rupture it.

The stress can be determined by relating the force either to the initial cross section, or to the cross section at each given moment taking into account the reduction in area of the specimen during deformation. In the first case, so-called nominal stresses are produced; in the second case, true stresses* are produced. If the stresses are determined in the area before point A, i.e., in the area of elastic deformation, where the cross section changes very little, the true stresses are almost identical to the nominal stresses [3].

In the area of plastic deformation (to the right of point A) the area of the cross section changes so much that this change influences the values of true stress. The greater the plastic deformation of the material the greater is the difference between the nominal and true stresses. For highly ductile steels, the true stress may exceed the nominal stress by a factor of 2.

A very important characteristic of material [in the elastic region] is the ratio of stress to unit elongation (strain) — Young's modulus:

$$E = \frac{\sigma}{\delta}$$

The ratio of the unit reduction in cross-sectional linear dimension to the unit elongation [in the elastic region] is called Poisson's ratio:

$$\mu = \frac{\frac{\Delta d}{d}}{\frac{\Delta l}{l}}$$

The ratio [in the elastic region] of shear stress to the unit shear (shear strain) is called the shear modulus:

$$G = \frac{\sigma_\tau}{\theta}$$

where θ is the unit shear, i.e., the deformation caused by shear stress in the material.

---

* Editor's comment: "True stress" should be considered as a compound noun (a label). That does not necessarily imply that "true stress" is a value which actually exists or is more real than any of other value of stress in the specimen after necking occurs. Bridgman [17] has shown that stress on the midplane of the neck of a tensile specimen is composed of a uniform uniaxial stress, termed the flow stress, superposed on a hydrostatic tension. The maximum normal stress on this plane exceeds the average of normal stress on the plane (true stress) which exceeds the flow stress which exceeds the nominal stress.

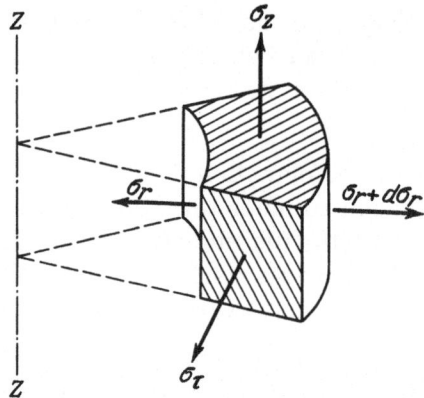

Fig. 11.   Stress components acting on element
of cylinder wall volume.

All of these values are interconnected by the following equation:*

$$\frac{1}{E}(1 + \mu) = \frac{1}{2G}$$

Stress Present in Thick-Walled Vessels

under Pressure [Elastic Case]

The relatively simple interrelations between stress and strain which hold in a body under uniaxial extension are replaced by a complex relation for the distribution of stress in the walls of an apparatus under internal-pressure loading.

As has been noted above, in the general case the stress state is characterized by nine components of stress (of which only six are independent). For analysis of the behavior of a cylinder wall (with external radius R and internal radius r_0) under external or internal pressure, we cut out an elementary volume dV (Fig. 11) from this wall. This volume element will be loaded only with normal stresses − the principal stresses. The principal stresses are the axial stress σ_z, the radial stress σ_r, and the tangential (circumferential) stress σ_τ; due to the conditions of symmetry, the shear stresses [on the faces of the volume element] will be equal to zero.

The dependence between these three mutually perpendicular stresses and the forces causing deformation is given by Lamé's formula. Using

* All of these relations are correct when direct proportionality between stress and strain is retained.

symbols introduced by A. M. Rozen and I. M. Naidich, this formula can be converted to a form convenient for calculation:

$$\left.\begin{array}{l} \sigma_r = -P_0 \cdot \dfrac{\alpha^2 - 1}{\alpha_0^2 - 1} - P_{\text{н}} \cdot \dfrac{\alpha_0^2 - \alpha^2}{\alpha_0^2 - 1} \\[2mm] \sigma_\tau = P_0 \cdot \dfrac{\alpha^2 + 1}{\alpha_0^2 - 1} - P_{\text{н}} \cdot \dfrac{\alpha_0^2 + \alpha^2}{\alpha_0^2 - 1} \cdot \\[2mm] \sigma_z = P_0 \cdot \dfrac{1}{\alpha_0^2 - 1} \, * \end{array}\right\} \tag{1}$$

where P_0 is the internal pressure, P_{H} is the external pressure, $\alpha = R/r$, $\alpha_0 = R/r_0$,† r is the radius at the point in question, R is the outer radius, r_0 is the inner radius.

Setting P_{H} equal to zero and letting $\alpha = \alpha_0$, we get the stress at the internal surface of the wall:

$$\left.\begin{array}{l} \sigma_r = -P_0 \\[2mm] \sigma_\tau = P_0 \cdot \dfrac{\alpha_0^2 + 1}{\alpha_0^2 - 1} \\[2mm] \sigma_z = P_0 \cdot \dfrac{1}{\alpha_0^2 - 1} \end{array}\right\} \tag{2}$$

Where $\alpha = 1$ and ($P_{\text{H}} = 0$) we get the following expression for the stress at the external surface of the wall:

$$\left.\begin{array}{l} \sigma_r = 0 \\[2mm] \sigma_\tau = P_0 \cdot \dfrac{2}{\alpha_0^2 - 1} \\[2mm] \sigma_z = P_0 \cdot \dfrac{1}{\alpha_0^2 - 1} \end{array}\right\} \tag{3}$$

Expressions (1), (2), and (3) show that the radial stresses in the wall of a vessel under internal pressure are compressive. In absolute value, they decrease from the value of the pressure P_0 at the internal wall to the zero pressure at the external wall.

The tangential stresses are always tensile and decrease toward the external surface, since $\alpha^2 + 1$ is always greater than 2. The tangential stresses are always greater than the radial stresses for any value of α_0.

---

*For a cylinder with a bottom, when the external pressure at the bottom is equal to zero.

†The quantities α and α_0 are also represented by q and q_0 (cf. [4], page 268).

The axial stresses are uniform at all points in the wall sufficiently distant from the ends and are always tensile.

Thus, the wall of a vessel is not uniformly stressed, the greatest stresses being developed at the internal layers. Even with an increase in wall thickness, its strength may still remain insufficient. It can be shown that a vessel with infinitely thick walls will not withstand pressure any greater than a given limiting value. In this case, $\alpha_0 = R/r_0 \to \infty$, and from formula (2) it follows that $\sigma_\tau = P_0$ at the internal surface of the vessel, i.e., the tangential stress at the internal surface is equal to the pressure, no matter what the wall thickness.

Equivalent Stress

The problem of determining the onset of plastic flow when a complex stressed state exists in the material has not yet been finally solved. This onset can be determined reliably only by direct experimentation [5].

The problem can be solved approximately by theory of strength of materials, replacing the given complex stressed state by an equivalent uni-axial tensile state. In this case, the actual stresses σ_r, σ_τ, σ_z are also replaced by an equivalent uniaxial tensile stress σ_e.

Using the theory of the energy of form change (distortion-energy theory)* we can derive the following expressions for equivalent stress :†

$$\sigma_{e.i.} = \frac{P_0 \sqrt{3}\,\alpha_0^2}{\alpha_0^2 - 1} \tag{4}$$

$$\sigma_{e.o.} = \frac{P_0 \sqrt{3}}{\alpha_0^2 - 1} \tag{5}$$

where $\sigma_{e.i.}$ and $\sigma_{e.o.}$ are the equivalent stresses at the internal and external surfaces of the wall.

---

*If we use the maximum shear-stress theory, in formula (4) in place of $\sqrt{3}$ we get the coefficient 2. The problem of the essence of the theory of strength has been rather fully discussed in the literature [1]. Up to the present, however, there is no single criterion for onset of plastic flow. Apparently, the coefficient in formula (6) takes on various values for various metals. However, with the amount of strength reserve [factor of safety, factor of ignorance] currently used, the deviation in values of this coefficient in the various theories of strength has no practical value.

†Here and below, we assume $P_H = 0$.

Assuming $\sigma_e = \dfrac{\sigma_s}{m}$ (where σ_s is the yield strength of the material, and m is the strength reserve [factor of safety]),* from equation (4) we get:

$$\left. \begin{aligned} \alpha_0 &= \frac{R}{r_0} = \sqrt{\frac{\sigma_s}{\sigma_s - 1.73mP_0}} \\[4pt] \text{from which the thickness of the wall is} & \\[4pt] S &= R - r_0 = r\left(\sqrt{\frac{\sigma_s}{\sigma_s - 1.73mP_0}} - 1\right) \end{aligned} \right\} \qquad (6)$$

Formulas (4–6) are used in the design of apparatus operating at pressures that cause only elastic deformation. With fixed vessel dimensions, formulas (4) and (5) can be used to calculate the equivalent stresses, due to internal pressure P_0, at the internal and external surfaces of the wall, and to compare these stresses with the yield point or strength of the material from which the apparatus is to be made. Equation (6) is used to determine the necessary wall thickness of an apparatus made of material with any given properties, intended to operate at pressure P_0 and with strength reserve m.

If a vessel is made with an infinitely thick wall ($S = \infty$) and m is assumed to equal 1, then from equation (6)

$$P_{0,\,\mathrm{lim}} = \frac{\sigma_s}{1.73} = 0.577\sigma_s$$

This expression is similar to that presented earlier for σ_τ and shows that the maximum pressure which will not cause plastic deformation is approximately 0.6† of the yield point or strength of the steel of which the apparatus is made.

It follows from the above that a high-pressure apparatus cannot be designed solely on the basis of the conditions of operation in the region of elastic deformation. In this case, even if we substitute the ultimate strength

---

* Editor's comment: A "factor of safety" in mechanical design is ordinarily composed of factors concerned with fatigue, corrosion, abrupt or impact loading, etc., as well as "true" factors of safety. The "strength reserve" mentioned here is simply a ratio of yield strength to equivalent static stress and hence the term "factor of safety" cannot be used as a substitute for "strength reserve."

† In various theories of strength [4] this value varies within small limits.

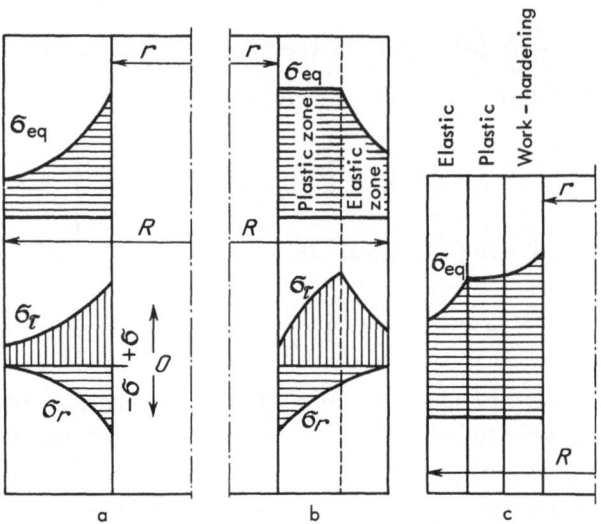

Fig. 12. Distribution of stress in the walls of a pressure vessel:
a) wall operating in the region of elastic deformations; b) wall
operating in the region of elastic-plastic deformations; c) wall
with a workhardening zone.

for the yield strength in the formulas, it is impossible to design an apparatus, even made of the best types of steel, which could retain a pressure higher than 15,000-20,000 atm. However, experience has shown [6, 7] that thick-walled vessels can retain considerably higher pressures: a vessel made of untempered high-carbon tool steel burst at 40,000 atm, whereas the ultimate strength of this steel was only 10,000 kg/cm^2.

This can be explained as follows: when the stresses in the metal exceed the yield strength, the metal goes over into the plastic state. The internal layer of the wall of the cylinder, where the pressure exceeds the yield strength, will be in the plastic state, and the equivalent stresses in it will have a certain value, uniform throughout the entire plastic layer. With a further increase in pressure, the stress in this layer will not increase, but the thickness of the plastic layer will increase. This agrees with the concepts of the nature of plastic deformation presented above.

The high-pressure cylinder is thus, as it were, divided into two concentric layers; the internal (plastic) layer, and the external (elastic) layer which protects the plastic layer from rupture. As the pressure is increased, the thickness of the plastic layer increases, and the thickness of the elastic layer consequently decreases until rupture finally occurs. Experiments have shown that rupture of a high-pressure cylinder begins from the outside [Ed. note: for ductile materials].

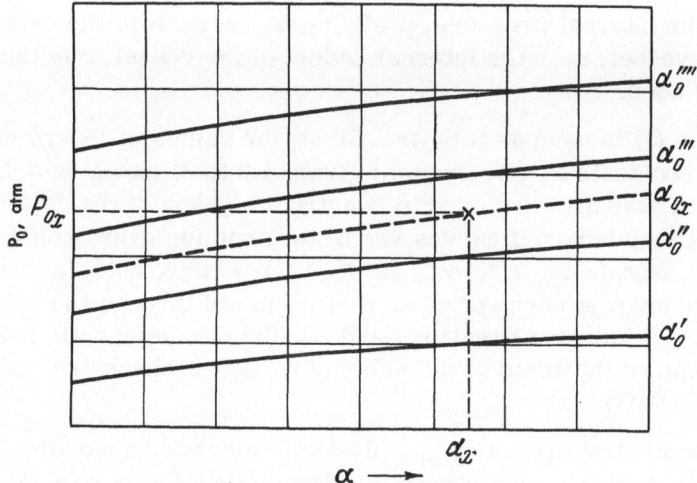

Fig. 13. Method of graphic calculation of operating pressure in a cylinder
with a plastic layer.

The diagrams of stress distribution in the walls of a vessel operating in the regions of elastic and elastic-plastic deformation (Figs. 12a and b) show that in the plastic-deformation zone the value of equivalent stress does not depend on the pressure, i.e., a horizontal sector [whose height is independent of pressure] is produced on the diagram. In this case, the material of the vessel wall is utilized much better than in the region of elastic deformation.

As a result of the strengthening caused by plastic deformation, the material is able to withstand a higher pressure. Here, three zones may exist simultaneously in the cylinder wall: the work-hardening, plastic, and elastic zones (Fig. 12c).

Finally, if we recall that the ultimate tensile strength, which for highly ductile steels may be only half the true stress at rupture (S_k),* is used in calculations, the difference between theory and practice becomes understandable.

We have already analyzed the design of apparatus working in the region of elastic deformation. The influence of a plastic layer may be considered using the expression [1, 9]:

$$P_0 = \frac{\sigma_s}{a^2 \sqrt{3}} \left(2a^2 \ln a_0 + a^2 - a_0^2\right) \tag{7}$$

---

* For methods of determining S_k, see, for example, [8].

where P_0 is the internal pressure, $\alpha = R/r_0$, $\alpha_0 = r/r_0$, R is the external radius of the vessel, r_0 is the internal radius of the vessel, r is the radius of the plastic zone.

Formula (7) is used as follows. First, the values of P_0 are calculated with various constants α_0 (for example, from 1.2 to 4) changing α from, say, 5 to 12. These data are used to construct a graph of the dependence of P_0 on α, and a number of curves are drawn, one for every constant value of α_0 (Fig. 13, curves α_0', α_0'', α_0''', etc.). Then a curve $\alpha_{0,x}$ is drawn parallel to the nearest curve of α_0 so that it passes through the point with coordinates $P_{0,x}$ (working pressure) and α_x (the diameter ratio selected). Having determined the scale of the value of $\alpha_{0,x}$, we check the value of P_0 using formula (7).

If the equivalent stress $\sigma_{e.el.}$ (elastic), calculated according to formula (4) exceeds the yield strength of the material only slightly ($\sigma_{e.el.} \leq 1.2\, \sigma_s$), the thickness of the plastic layer can be determined from the relation:

$$\frac{\delta}{r_0} = \frac{1}{2}\left(\frac{\sigma_{e.el.}}{\sigma_s} - 1\right) \tag{8}$$

If we set the radius of the plastic zone in Formula (7) equal to the external radius of the cylinder, we can obtain an expression for the pressure for complete flow of the cylinder, at which the plastic layer will reach the entire thickness of the wall:

$$P_s = \frac{2}{\sqrt{3}}\, \sigma_s \ln \alpha_0 \tag{9}$$

It is interesting to calculate, even approximately, the value of the rupture pressure for such a cylinder. This value is determined by the expression below, presented by A. M. Rozen and M. N. Efremov [10]:

$$P_b = \frac{2}{\sqrt{3}}\, \bar{\sigma} \ln \alpha_0 \tag{10}$$

where $\bar{\sigma}$ is the mean integral value of the equivalent stress in the cross section, $\sigma_s \leq \bar{\sigma} \leq \sigma_b$.

With a certain degree of approximation, we can replace $\bar{\sigma}$ in equation (10) by the ultimate tensile strength of the material σ_b; it is then easy to calculate the reserve of strength available in an apparatus of any given dimensions.

S. N. Sokolov [11], using the method of polygonal approximation of the tensile stress-strain curve constructed in "true stress-strain" coordinates, produced a general expression for determination of the rupture pressure in tubes. The author noted good correspondence between experimental data and theoretical data produced for the equation

$$P_{rup} = \sigma_b \ln \alpha_0$$

which differs little in essence from equation (9).

An overall solution to the problem is also given in reference [12], a graphic method of solution in reference [13].

Calculations of the rupture pressure of steel cylinders made according to equation (10) correspond satisfactorily with the results of experiments [14]. Thus, a tube of 30KhGSA steel with a diameter ratio of 3 was ruptured at a pressure of 12,600 atm, whereas the calculation according to equation (10) gave a value of 11,000 atm. A tube of type 1Kh18N9T steel was ruptured at a pressure of 7700 atm, whereas the calculated pressure was 8300 atm.

Determination of Thermal Stresses [15]

If the temperatures within and without the apparatus are different, stresses are present in the walls of the apparatus, often reaching considerable values. Let us assume that the temperature along the axis of a vessel is constant, but changes logarithmically through the thickness of the wall:

$$t = \frac{t_0 \ln \frac{R}{r} + t_1 \ln \frac{r}{r_0}}{\ln \frac{R}{r_0}} = t_1 + \Delta t \frac{\ln \frac{R}{r}}{\ln \frac{R}{r_0}} \tag{11}$$

where t_1 is the temperature of the outer surface of the vessel; t_0 is the temperature of the inner surface of the vessel; $\Delta t = (t_0 - t_1)$ is the temperature difference; t is the temperature at any radius r.*

We can then convert the expression for determination of stresses presented by Rozen and Naidich to a form convenient for calculations [definitions of α and α_0 for this section are given on page 45]:

---

*This assumption corresponds to heat transfer through a thick wall of a vessel with equilibrium thermal regime.

$$
\left.\begin{aligned}
\sigma_r &= q\left(\frac{\alpha^2-1}{\alpha_0^2-1}-\frac{\ln\alpha}{\ln\alpha_0}\right)=qK_r^t \\
\sigma_\tau &= q\left(-\frac{\alpha^2+1}{\alpha_0^2-1}+\frac{1-\ln\alpha}{\ln\alpha_0}\right)=qK_\tau^t \\
\sigma_z &= q\left(-\frac{2}{\alpha_0^2-1}+\frac{1-2\ln\alpha}{\ln\alpha_0}\right)=qK_z^t
\end{aligned}\right\}
\tag{12}
$$

A quantity q, which can be called the thermal pressure, is determined from

$$
q=\frac{E\alpha_t\Delta t}{2(1-\mu)}
\tag{13}
$$

where E is the Young's modulus, kg/cm^2; α_t is the thermal coefficient of linear expansion of the metal, cm/degree; μ is Poisson's ratio.

The thermal pressure q depends on the modulus of elasticity, co-efficient of linear expansion, and Poisson's ratio. The way in which Poisson's ratio changes with temperature is unknown. The modulus of elasticity decreases with increasing temperature, and the coefficient of linear expansion increases. Therefore it can be considered that, to a first approximation, the value $E\alpha_t/2(1-\mu)$ remains almost unchanged, so that

$$
q=k\Delta t
$$

where k = 15 for high-carbon nonalloyed steel; k = 16 for low-carbon (soft) nonalloyed steels; k = 18 for chrome-moly, molybdenum, and chrome-nickel steels.

We can see from formula (12) that where q > 0 (i.e., with internal heating) the radial stress throughout the entire mass of the wall is negative (compressive stress), since $(\ln\alpha/\ln\alpha_0)>(\alpha^2-1)/(\alpha_0^2-1)$ for any values of α, where $1<\alpha<\alpha_0$, and $\ln\alpha/\ln\alpha_0=(\alpha^2-1)/(\alpha_0^2-1)$ where $\alpha=1$ and $\alpha=\alpha_0$; consequently, the radial stresses on the internal and external surfaces of the wall are equal to 0.

The tangential (circumferential) stresses (with internal heating) are negative at the internal wall surface, equal to 0 at the point where $(\alpha^2+1)/(\alpha_0^2-1)=(1-\ln\alpha)/\ln\alpha_0$; and are positive at the external surface.

The axial stress is negative at the internal surface, equal to 0 at the point where $2/(\alpha_0^2-1)=(1-2\ln\alpha)/\ln\alpha_0$, and positive thereafter. With external heating, the signs of the thermal stresses change. The nature of change of thermal stresses is presented in Fig. 14, where we also see values for thermal stress coefficient. The values of thermal stresses are of the same order of magnitude as those of stresses caused by pressure.

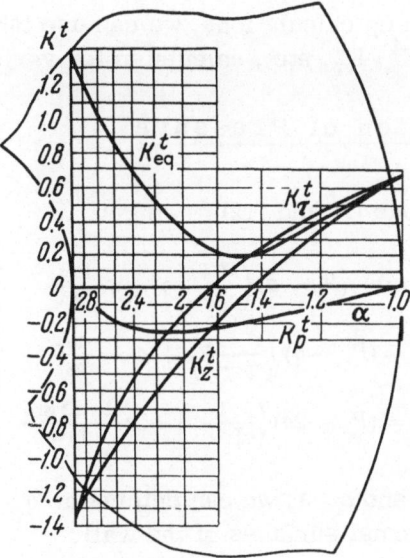

Fig. 14. Variation of coefficients of
thermal stress through thickness of
wall of hollow cylinder.

With a temperature difference between internal and external surfaces
of the wall $\Delta t = 30°C$ for chrome−nickel steel, formula (13) indicates $q = 540$
kg/cm^2.

Since σ_{T_0} (on the internal surface of the wall) varies within the limits $q \leq \sigma_{T_0} < 2q$, for a cylinder of chrome− nickel steel with radii $R = 60$
mm and $r_0 = 20$ mm, then when $\Delta t = 30°C$, $\sigma_{T_0} = 726$ kg/cm^2. If the temperature difference in the same cylinder $\Delta t = 100°C$, then $\sigma_{T_0} = 2420$ kg/cm^2.

An equation expressing the dependence among coefficients of thermal stresses can be used for calculation of the equivalent temperature
stresses:

$$K'_3 = \sqrt{(K'_r)^2 - K'_r K'_\tau + (K'_\tau)^2} \tag{14}$$

At the external and internal surfaces of the wall, $K_r^t = 0$; then

$$K'_{e.i.} = -K'_{\tau, i}$$
$$K'_{e.o.} = K'_{\tau, o.}$$

The sign is selected in consideration of the fact that the equivalent
stress (by definition) is a tensile stress. Calculating the value of K_T^t,
formula (14) is used to determine K_e^t. Then, calculating the thermal pressure q for a vessel of the given dimensions, the equivalent stress is
calculated.

In order to speed up calculations, we can use tables presented by I. M. Naidich [15] for K_e^t, K_r^t, etc., calculated for various diameter ratios.

Simultaneous Action of Pressure
and Temperature

Adding expressions (1) and (12),* we get:

$$
\left.
\begin{aligned}
\sigma_{r,\Sigma} &= -(P_0 - q)\left(\frac{a^2 + 1}{a_0^2 - 1}\right) - q \cdot \frac{\ln a}{\ln a_0} \\
\sigma_{\tau,\Sigma} &= (P_0 - q)\left(\frac{a^2 + 1}{a_0^2 - 1}\right) + q \cdot \frac{1 - \ln a}{\ln a_0} \\
\sigma_{z,\Sigma} &= (P_0 - 2q)\left(\frac{1}{a_0^2 - 1}\right) + q \cdot \frac{1 - 2\ln a}{\ln a_0}
\end{aligned}
\right\}
\tag{15}
$$

Assuming $\alpha = \alpha_0$ and $\alpha = 1$, we can determine $\sigma_{r,\Sigma}$, $\alpha_{\tau,\Sigma}$, and $\sigma_{z,\Sigma}$ at the internal and external surfaces of the wall.

Replacing the complex stress with the equivalent stress, we get

$$
\sigma_{0,\Sigma} = \sqrt{\frac{3P_0^2 + 3P_0 qa + q^2 a^2}{a_0^2 - 1}}
\tag{16}
$$

where

$$
a = \frac{a_0^2 - 1 - 2\ln a_0}{\ln a_0}
\tag{17}
$$

$\sigma_{0,\Sigma}$ is the equivalent total stress on the external surface of the wall.

If it is desirable that this stress on the external surface not exceed the equivalent stress from pressure [alone], we proceed as follows. $\sigma_{0,\Sigma}$ is set equal to $\sigma_{e.i.}$ from expression (18) and the permissible limiting temperature drop† is calculated:

$$
\Delta t_{\lim} = \frac{[\sqrt{4/3\,(a^4 - 1) + 1} - 1]P_0}{12a}
\tag{18}
$$

with which we can calculate the dimensions of the vessel from the stress on an external "fiber" (without considering temperature drop). If the experimental Δt is greater than the calculated $\Delta t_{\lim}$, the external fiber will undergo the greatest stress, and

$$
\sigma_{9,\Sigma} = \sigma_p \leqslant \sqrt{\frac{3P_0^2 + 3aP_0 q + q^2 a^2}{a_0^2 - 1}}
\tag{19}
$$

where σ_p is the permissible stress.

---

*External pressure P_H is assumed equal to 0.
†For the case of heat transfer from within the vessel outward.

In the same way we can calculate the thermal stress as a multilayered cylinder in the absence of slipping between layers, if we assume that the dimensions of the basic cylinder are equal to the dimensions of a monoblock.*

CONSTRUCTION OF APPARATUS

Autofrettaged Vessels

If a vessel in which plastic deformation has occurred is unloaded, its walls will remain in a stressed state due to residual deformation.† This type of deformation is termed autofrettage. Stresses arise at the boundary between the plastic and elastic layers caused by the fact that the expanded plastic layer stretches the external elastic layer, which in turn applies pressure to the internal layer. As a result, even when there is no internal pressure, stresses exist in the wall. A diagram of these stresses is presented in Fig. 15.

An autofrettaged vessel can withstand a higher pressure than a similar one not autofrettaged, due to the compression of the internal layer and work-hardening of the material during plastic deformation. This method of strengthening vessels, suggested in the 1870s by A. S. Lavrov,‡ is now widely used in practice.

The design of autofrettaged vessels [18, 19, 20] consists of determination of the pressure required for autofrettaging and of the dimensions of the vessel involved. There are several methods of calculation, of which we will analyze only the method relating to the so-called semielastic state, i.e., a state in which there are two zones in the vessel — the plastic and elastic zones.**

$$P_0 = 2.23\sigma_e \log \frac{\rho}{R_0} + 325 \cdot \frac{\sigma_e}{E}\left(\frac{\rho^2}{R_0^2} - 1\right) + \frac{\sigma_e}{2}\left(1 - \frac{\rho^2}{R_1^2}\right) \qquad (20)$$

---

*The literature [16] also presents methods for calculating high-pressure vessels; the influence of circular elliptical apertures on stress concentration is analyzed in particular.

†A great deal of experimental material has been accumulated [17] allowing us to draw conclusions about the nature of the various deformations accompanying prestrengthening.

‡The method used by A. S. Lavrov was the hardening of the material by forcing a die through a weapon barrel.

**The more accurate method of G. A. Smirnov-Alyaev [18] is not presented here since its application involves difficult graphical calculations.

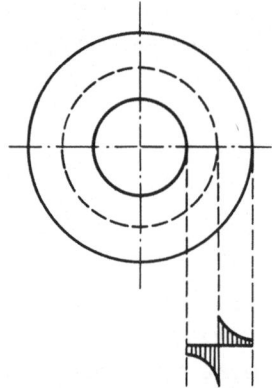

where P_0 is the pressure corresponding to the elastic limit of the material, σ_e is the elastic limit of the metal, E is the Young's modulus, R_0 is the internal radius, R_1 is the external radius of the prepared tube, ρ is the radius of the plastic layer $(R_0 < \rho < R_1)$.

The strength reserve

$$m = \frac{P_0}{P_{\max}}$$

Fig. 15. Distribution of equivalent stresses in an autofrettaged cylinder.

where $P_{\max}$ is the maximum internal pressure.

Since an autofrettaged cylinder is made by a final mechanical treatment of the autofrettaged piece, the dimensions of this piece must be determined. The following empirical equations are used for this purpose:

$$r' = R_1 + l_1$$

and

$$r_0' = R_0 - l_0$$

where r' is the external radius of the piece; r_0' is the internal radius of the piece; l_1 and l_0 are the radial allowances for internal and external radii. Usually l_1, expressed in millimeters, is numerically equal to the radius of the tube in meters, and l_0 is six millimeters greater than l_1.

The autofrettaging pressure P_a of the piece is determined from the formula

$$P_a = 2.23\sigma_e \log\frac{\rho}{r_0'} + 325\cdot\frac{\sigma_e}{E}\left(\frac{\rho^2}{(r_0')^2} - 1\right) + \frac{\sigma_e}{2}\left(1 - \frac{\rho^2}{(r_1')^2}\right) \qquad (21)$$

If the cylinder is subjected to pressure and then the pressure is reduced, yielding results and the dimensions of the cylinder change due to the fact that the external layer, whose radius is now greater than ρ, is extended, while the internal layer is compressed. The change in external and internal radii after autofrettaging can be determined from the formulas

$$\Delta r_1' = \frac{5}{6}\left(\frac{\sigma_e}{E}\cdot\frac{p^2}{(r')^2} - \frac{2P_a}{E}\cdot\frac{1}{(a')^2-1}\right)r_1' \qquad (22)$$

$$\Delta r_0' = (a')^2\cdot\frac{\Delta r'}{r_1'}r_0'$$

where

$$a' = \frac{r_1'}{r_0'}$$

Finally, the change in external radius under the autofrettaging pressure, i.e., before reduction of internal pressure, will be:

$$(\Delta r_1')_P = \frac{5}{6}\cdot\frac{\sigma_e}{E}\cdot\frac{p^2}{(r_1')^2}$$

Multilayered Vessels

The distribution of stresses in the walls of a cylinder as shown in Fig. 15 can be attained without creating the preliminary residual deformations. If tightly fitted cylindrical shells are inserted one inside the other (which is in the hot state), the result after they cool will be a cylinder consisting of individual layers with excess tightness [interference]. This tightness causes an external pressure to be applied to the internal shell and an internal pressure to be applied to the external shell, so that the wall of an unloaded cylinder is stressed (Fig. 16). This method [shrink-fitting] was developed and suggested by A. V. Gadolin (1856-1861).

The highest pressure which such a vessel can withstand when made of several layers of materials with various properties can be determined from the equations below, developed by A. M. Rozen:

$$P_0 = \frac{n}{\sqrt{3}}\left(\sigma_a - \frac{\sigma_g}{a_0^{2/n}}\right) \qquad (23)$$

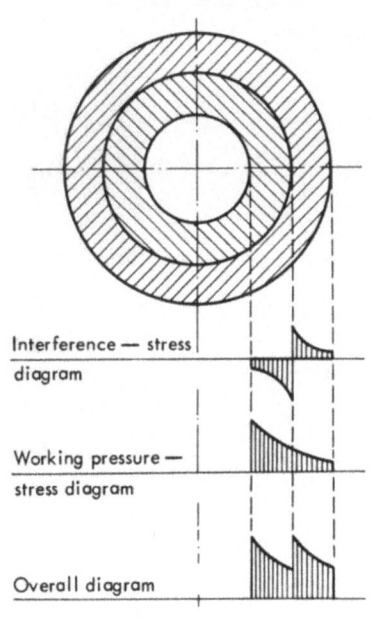

Interference — stress diagram

Working pressure — stress diagram

Overall diagram

Fig. 16. Distribution of equivalent stresses in walls of a two-layered vessel [under pressure].

$$a_0 = \frac{R}{r_0} = \sqrt{\left(\frac{\sigma_g}{\sigma_a - \sqrt{3} \cdot \frac{P_0}{n}}\right)^n}$$

$$\sigma_a = \frac{\Sigma \sigma_{0,i}}{n} = \frac{\sigma_{0,1} + \sigma_{0,2} + \sigma_{0,3} + \cdots + \sigma_{0,n}}{n}$$

$$\sigma_g = \sqrt[n]{\sigma_{0,1} \sigma_{0,2} \sigma_{0,3} \cdots \sigma_{0,n}}.$$

where n is the number of layers, σ_a and σ_g are the arithmetic and geometric means of permissible stresses of all layers, $\sigma_{0,i}$ is the equivalent stress in layer i.

The dimensions of each layer are determined by the expression

$$a_i = \sqrt{\frac{\sigma_{0,i}}{\sigma_a - \sqrt{3} \cdot \frac{P_0}{n}}} \tag{24}$$

where α_i is the ratio of radii of the i-th cylinder. If all the layers are made of the same material, and $\sigma_{0,i} = \sigma_g = \sigma_a = \sigma_0$, then:

$$P_0 = \frac{n}{\sqrt{3}} \sigma_0 \left(1 - \frac{1}{a_0^{2/n}}\right) \tag{25}$$

$$a_0 = \left(\frac{\sigma_0}{\sigma_0 - \sqrt{3} \cdot \frac{P_0}{n}}\right)^{n/2} \tag{26}$$

$$a_i = \sqrt{\frac{\sigma_0}{\sigma_0 - \sqrt{3} \cdot \frac{P_0}{n}}} \tag{27}$$

Representing equation (25) in the form

$$\frac{P_0}{\sigma_0} = \frac{n}{\sqrt{3}} \left(1 - \frac{1}{a_0^{2/n}}\right) \leqslant \frac{P_0}{\sigma_s}$$

we obtain, where n → ∞,

$$\frac{P_0}{\sigma_s} \longrightarrow \frac{2}{\sqrt{3}} \ln a_0 \tag{28}$$

From this it follows that, as the number of layers increases, the stress becomes distributed ever more evenly. This is characteristic [also] for plastic flow of the shell. However, the number of layers cannot be in-

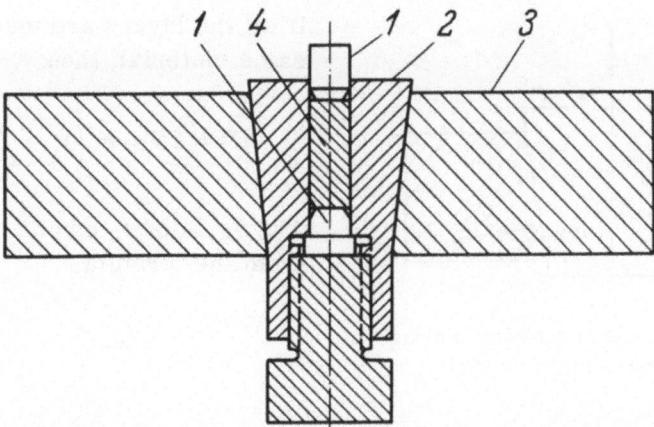

Fig. 17. Plan of method of automatic generation of external compression (support) of a vessel: 1) pistons; 2) vessel; 3) block; 4) material being compressed.

creased infinitely, since the multilayered vessel would be burst by the internal stresses alone (interference stress).

If it is necessary that the assembly stress be less than σ_s, the maximum pressure which can be withstood by a multilayered cylinder without plastic deformation is:

$$P_{0,\text{max}} = 1.07\sigma_s \tag{29}$$

Fulfillment of condition (29) should be checked during design calculations.

Before making a multilayered vessel, the interferences must be determined. Design values of interferences, i.e., the differences in the diameters of the i-th and (i − 1)-th layers manufactured but not assembled are calculated directly from the drawings:

$$S_i = \frac{\Delta r}{r_i} = \frac{r_{i,\,i} - r_{i-1,\,o}}{r_{i,\,i}} \tag{30}$$

The assembled interferences, which consist of the differences in diameters of layers during the process of assembly, are greater than the design interferences, since the deformation of shells from their preceding dimensions will be additive during assembly:

$$ES_i = \sigma_{\tau,i+1} - \sigma_{\tau,i} = \left(\sigma_{e,i+1} - \frac{\sigma_{e,i}}{a_i^2}\right) \cdot \frac{2}{\sqrt{3}} \tag{31}$$

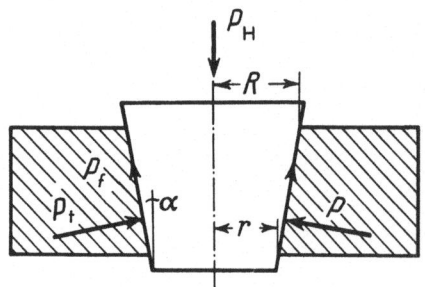

Fig. 18. Plan of forces in apparatus working
with external compression (support).

If all the layers are made of the same material, then

$$ES_i = ES = \sigma_e \left(1 - \frac{1}{\alpha_0^{2/n}}\right) \qquad (32)$$

The value of σ_e can be determined from the formula:

$$\sigma = \frac{\sigma_s}{m}$$

where m is the reserve of strength.

Assembly interferences δ_i are determined from the expression

$$\delta_i = \sum_1^{i-1} S_{i-1} \cdot \frac{\beta_{i-1}^2 - 1}{\beta_i^2 - 1} + S_i \qquad (33)$$

where $\beta_i = \frac{r_i}{r_0}$ is the ratio of the radius of the i-th cylinder to the internal radius of the first shell.

Multilayered vessels are manufactured by various methods. For example, thin steel sheets can be wound onto a thin-wall tube, each layer being welded.

The wound vessel is a variation of multilayered vessel made by winding shaped steel band or square section wire onto a tube. The band or wire is first heated and strongly stretched [21]. Finally, a multilayered vessel can be made by applying several heated discs around a thin-walled tube so as to cover its entire length.

A detailed description of all these construction methods as well as the design of wound vessels is presented in the literature [22, 23].

Vessels with Variable External Mechanical Support

As was noted above, the highest internal pressure can be withstood by multilayered vessels whose internal layers have a plastic zone. However, the operating pressure for such vessels must be limited, since they break under the action of their assembly stresses. Vessels made of the best modern steels can withstand pressures no higher than 20,000-25,000 atm. The simplest method for increasing the working pressure on the basis of the above is to strengthen the external support of the vessel in proportion to the increased internal pressure.

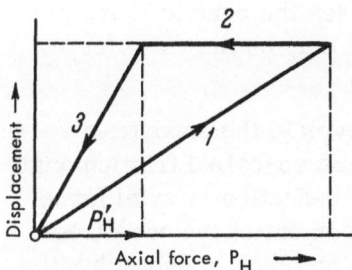

Fig. 19. Dependence of displacement of conical vessel on axial force.

In multilayered vessels, the maximum compression of the internal layer is attained when the internal pressure is equal to zero. If we make a vessel in a conical shape and place it in a conical support (Fig. 17), the external pressure will automatically increase with increasing internal pressure. Pistons 1, entering vessel 2, increase the pressure in the vessel and simultaneously press the vessel down into the body of block 3. When the cone is forced into the block, forces arise on the contiguous tangent surfaces of cone and support which act like the forces on the surface of a wedge.

The force P_H (total load on the piston) (Fig. 18) results in stress P acting normal to the side surface of the cone plus friction force P_f tangent to this surface and directed opposite the movement of the cone. The force P_t will act over the entire surface of the cone.

Projecting all forces vertically, we see that force P_H is composed of two component forces:

$$P_H = P_t \sin \alpha + P_f \cos \alpha \qquad (34)$$

Ignoring friction between support and cone, and taking angle α so small that the height of the cone will differ very little from the length of the generator of the cone, we get:

$$P_t = Ph\pi (R + r) \qquad (35)$$

where h is the height of the cone, R and r are the large and small radii of the cone.

Since angle α is very small, we can consider that

$$\sin \alpha = \frac{R - r}{h}$$

from which

$$P = \frac{P_H}{\pi (R^2 - r^2)} \qquad (36)$$

Using (36) we can approximately determine the support pressure, i.e., the stress arising on the surface of the cone as it is pressed into the support.

Force P_H, also called the axial force, causes the cone to move into the support. The amount of the shift depends not only on the axial force, but also on the force of friction.

A reduction in axial force as the piston moves in the opposite direction will allow the cone to move back only when the so-called friction angle becomes less than angle α. Up to this point, the reduction in axial force will not involve a change in the amount of displacement of the cone (see Fig. 19, line 2). The movement of the cone into the block is shown by line 1. The axial force for unloading P_H', i.e., the force at which the cone begins to move back out of the block, will be less than P_H, and will be equal to

$$P_{\text{н}}' = P_t \, \sin \alpha - P_f \cos \alpha \tag{37}$$

If the results of experimentation, i.e., the axial force and the displacement of the cone, are entered on the graph, the friction force can be determined:

$$P_{\text{н}} - P_{\text{н}}' = 2P_f \cos \alpha \tag{38}$$

Since the friction force $P_f = fP_t$,

$$\left. \begin{array}{l} P_{\text{н}} = P_t \sin \alpha + fP_t \cos \alpha \\ P_{\text{н}}' = P_t \sin \alpha - fP_t \cos \alpha \end{array} \right\} \tag{39}$$

where f is the coefficient of friction.

From this,

$$f = \frac{P_{\text{н}} - P_{\text{н}}'}{P_{\text{н}} + P_{\text{н}}'} \cdot \frac{\sin \alpha}{\cos \alpha} \tag{40}$$

Thus, the coefficient of friction between the support and the cone can be determined experimentally.

The support pressure [see (36)] is higher the less the angle α. However, reduction of angle α also causes the force of friction to increase. Consequently, the axial force applied to the conical vessel is expended on generating the support pressure and overcoming the force of friction. We call the portion of P_H used to create the support pressure the effective force $P_{H.\,ef.}$.

Adding up expressions (39), we get:

$$P_{\text{H.\,ef.}} = P_t \sin \alpha = \frac{P_{\text{н}} + P_{\text{н}}'}{2} \tag{41}$$

$P_{H. ef.}$ depends on the value of angle α (it is less, the less α). The stress P can be expressed by the following formula:

$$P = \frac{C}{\sin \alpha + f \cos \alpha} \qquad (42)$$

where

$$C = \frac{P_H}{2\pi R_{av} h}, \qquad R_{av} = \frac{R+r}{2}$$

Then

$$\frac{dP}{d\alpha} = \frac{C(\cos \alpha - f \sin \alpha)}{(\sin \alpha + f \cos \alpha)^2}$$

At the point of the extreme,

$$\tan \alpha = \frac{1}{f}$$

Therefore, in order to increase the support pressure, not only the angle of the cone but also the friction between support and cone should be reduced. For this, lead foil 0.1-0.3 mm thick is placed between them (lead is used because it has the least shear stress); the foil is lubricated with a paste consisting of glycerin, water, and graphite. However, even this does not allow any substantial increase in the support pressure, since the [component of] force tending to eject the vessel from the support also increases. It is therefore expedient to press the conical vessel into the support with a separate piston; in order to create pressures on the order of 50,000 atm, several stages of support pressure are used, i.e., a conical vessel with several supports, each moving within another, can be applied.*

Even higher pressure can be created by supporting the entire vessel in a liquid compressed to a pressure on the order of 30,000 atm. This reduces the pressure differential from inside to outside of the apparatus and

---

* The construction of such installations is described in more detail in Chapter III. [Editor's comment: Bridgman had designed, and he and others had constructed, a vessel with a single stage only of conical support for use to 50,000 atmospheres. A single-stage intensifier (about 1600 : 1 area ratio), fed by a low-pressure pump, actuated a cemented-carbide piston with chamfer-ring seal for pressurization of solids. The high-pressure piston was similar to Fig. 17 on page 59.]

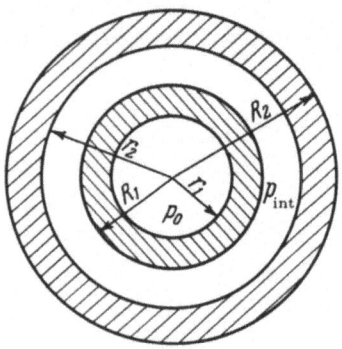

Fig. 20. Scheme of vessel with
hydraulic support.

greatly improves the properties of the material, especially cemented
carbides, used to make the apparatus (see Chapter I). This approach allows the internal pressure to be increased to 100,000 atm [24].

When a conical vessel is forced into a support, the support is expanded and the vessel is compressed. In order to evaluate the displacement of the cone, the change in internal diameter of the support must be
known; this value can be calculated by well-known methods [25]. The
radial displacement of a point located at distance ρ from the center of the
vessel is determined by the formula:

$$U = \frac{1-\mu}{E} \cdot \frac{r^2 P_i - R^2 P_o}{R^2 - r^2} \rho + \frac{1+\mu}{E} \left(\frac{P_i - P_o}{R^2 - r^2} \right) \cdot \frac{r^2 R^2}{\rho} \tag{43}$$

where R and r are the internal and external radii of the vessel; P_o is the
external pressure; P_i is the internal pressure.

In our case, the external pressure $P_o = 0$, and $\rho = r$. Then

$$U = \frac{P_i \cdot r}{E} \left[\frac{R^2 + r^2}{R^2 - r^2} + \mu \right] \tag{44}$$

The displacement of the vessel into the support causes compression
of the vessel and a reduction in the internal diameter. In order to prevent
binding of the packing and the piston in the bore of the conical vessel, the
amount of change of the diameter of this bore must be known. This value
can be determined from (43). Here, $P_i = 0$ and $r = \rho$. Then,

$$U = - \frac{2r R^2 P_o}{E(R^2 - r^2)} \tag{45}$$

In designing a conical-vessel piston, its diameter should be reduced by 2U.

Vessels with Hydraulic Support

An apparatus constructed on the hydraulic support principle (Fig. 20) consists of several vessels placed one inside the other. The clearances between the vessels are filled with liquid under hydrostatic pressure somewhat less than the pressure in the next vessel; each shell thus operates with a certain pressure differential across it. The limitations connected with the increase of assembly stresses in multilayered vessels are eliminated since the pressure in the clearances can be increased gradually with increasing internal pressure. The higher the pressure, the better the conditions under which the materials are operating. Placing the entire apparatus in a liquid compressed to 30,000 atm allows simultaneous usage of hydraulic support.

The design of vessels with hydraulic support is carried out using the formulas developed by Kagan [4]. The external radius of the internal cylinder is equal to

$$R_1 = \sqrt{\frac{\psi}{\alpha} r_1 R_2} \tag{46}$$

and the external radius of the external cylinder is

$$R_2 = \frac{2ar_1\sigma'}{\psi(\sigma' + \sigma'' + 2P_0)} \tag{47}$$

where r_1 is the internal radius of the internal cylinder, σ' is the permissible stress of the material of the internal cylinder, σ'' is the same for the external cylinder, P_0 is the maximum working pressure;

$$\psi = \sqrt{\sigma'/\sigma''}$$

If we represent the pressure in the clearance space between cylinders by P_{int}, then

$$P_{int} = \frac{\sigma'' - \sigma' + 2P_0}{4} \tag{48}$$

$$P_0 = \frac{R_2^2 - a^2 R_1^2}{2R_2^2} \sigma'' + \frac{R_1^2 - r_1^2}{2R_1^2} \sigma' \tag{49}$$

where $\alpha R_1 = r_2$ (internal radius of external cylinder).

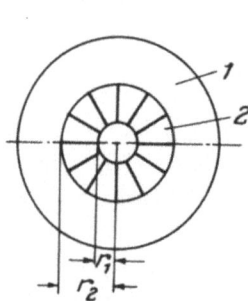

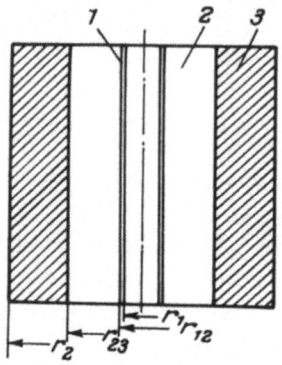

Fig. 21. Wedge vessel: 1) ring;
2) wedges.

Fig. 22. Longitudinal section
of wedge vessel: 1) shell; 2)
wedges; 3) ring.

In order to determine the maximum permissible pressure change
in the gap δP_{int}, we represent the permissible increase in stress in the
wall of the external cylinder as $\delta\sigma'$. Then, the permissible decrease in
pressure in the gap will be

$$\delta' P_{int} = \frac{2P_0 + \sigma' + \sigma''}{4} \cdot \frac{\delta\sigma'}{\sigma'}$$

and the permissible pressure increase will be

$$\delta'' P_{int} = P_{int} \cdot \frac{\delta\sigma''}{\sigma''} \tag{50}$$

Wedge Vessels

The important directions in the design of high-pressure apparatus
have been: replacement of tensile stress by compression stress, and usage of a method which might be called the stress-multiplication method in a
solid medium. These have resulted in the manufacture of high-pressure
vessels using such materials as tungsten carbide and highly hardened
steels, whose compressive strengths are three to four times greater than
their tensile strength. The use of stronger materials and the possibility
of reducing stresses in the crucial parts of installations have permitted
the creation of apparatus capable of maintaining pressures of 200,000 atm
and higher.

The principle of stress replacement and the method of multiplication
of stress in a solid body have been used, in particular, in the design of
wedge vessels [26].

A wedge vessel* consists of two layers (Fig. 21): the external layer is a ring 1 made of high-quality heat-treated steel, the internal layer is made of a large number of wedges 2 made of a harder material and carefully ground to fit each other. There are no tangential stresses in the wedges, and the radial stresses arising at the internal surface of such a wedge vessel when pressure is created inside it propagate through the bodies of the wedges and are decreased at the external surface of the wedge layer (at the limit when there is no friction between wedges) by r_2/r_1.† Therefore, ring 1 supports much less stress.‡

If we assume [28] that the material of the shell 1 (Fig. 22) (which provides hermetic sealing during compression of gases and liquids) is deformed as a rigid plastic body, ring 3 has the deformation diagram of an elastic-plastic body, and the compressive stress acting on the internal surface of wedges 2 is less than the yield point of the material of which the wedges are made, we can define pressures P_1 and P_2 which can be withstood by a wedge vessel and by an ordinary vessel of the same dimensions respectively.

With elastic deformation [29], P_1 and P_2 are the pressures which create identical equivalent stresses at the danger points of the vessels. With elastic-plastic deformations, P_1 and P_2 are the pressures at which plastic deformation extends to the entire cross section of the vessels.

Representing P_1/P_2 as m, we can show that

$$m_y = \frac{a_2^2}{a_2^2 - 1}\left[\frac{2\sigma_{s_1}}{\sigma_{per}}(K-1) + \frac{a_1^2 - 1}{a_1^3}\cdot\frac{a_2}{K}\right] \tag{51}$$

$$m_{pl} = \frac{a_3}{a_1 K}\cdot\frac{2\ln a_1 + a_1 K a_2\sqrt{3}(K-1)}{2\ln a_2}\cdot\frac{\sigma_{s_1}}{\sigma_{s_3}} \tag{52}$$

where m_y is the pressure ratio with elastic deformations, m_{pl} is the pressure ratio with plastic deformations,

$$a_1 = r_2/r_{2,3}$$
$$a_2 = r_2/r_1$$

---

*The idea of using wedges was first suggested by R. V. Mil'vitskii, but was not applied during his time [27].

†Editor's comment: F. J. Fuchs, Jr., additionally has tapered the lengths of the wedges. The length of the outer surface of the wedge can be, say, three times the length of the inner surface. Thus, the stress-multiplication factor for a wedge vessel with a radius ratio of r_2/r_1 and taper ratio of L_2/L_1 would be $r_2 L_2/r_1 L_1$.

‡This method is also called control of radial stresses [28].

σ_{S_1} and σ_{S_3} are the yield point of material of shell 1 and ring 3, σ_{per} is the permissible stress,

$$K = r_{1,2}/r_1$$

With α_2 and K fixed, the value of m is a function of α_1. The maximums of this function are equal to:

$$m_{y(max)} = \frac{2a_2^2}{3\sqrt{3}K(a_2^2-1)}\left[\alpha_2 + 3\sqrt{3}K(K-1)\cdot\frac{\sigma_{s_1}}{\sigma_{per}}\right] \text{ where } \alpha_1 = \sqrt{3} \qquad (53)$$

$$m_{pl(max)} = \frac{\alpha_2}{eK}\cdot\frac{2+\frac{eK}{\alpha_2}\sqrt{3}(K-1)}{2\ln\alpha_2}\sigma_{s_1} \text{ where } \alpha_1 = e \qquad (54)$$

from which the optimal value of $r_{2,3}$ is equal to $r_2/\sqrt{3}$ or r_2/e.

If a solid body is being compressed in the vessel, shell 1 becomes unnecessary. Then $K = 1$

$$m_{y(max)} = \frac{2}{3\sqrt{3}}\cdot\frac{a_2^3}{(a_2^2-1)} \qquad (55)$$

and

$$m_{pl(max)} = \frac{\alpha_2}{e\ln\alpha_2} \qquad (56)$$

Analysis of equilibrium in a unit cell [30] (see Fig. 11) leads to analogous conclusions. From the conditions of equilibrium, we can write:

$$\frac{d(\sigma_r,r)}{dr} = \sigma_\tau \qquad (57)$$

Representing the radial displacement of a unit cell by u, we can produce the following differential equation:

$$r^2\cdot\frac{d^2u}{dr^2} + r\cdot\frac{du}{dr} - u = 0 \qquad (58)$$

from which it follows that $\sigma_r = -P_1$, where $r = r_{2,3}$, and $\sigma_r = 0$, where $r = r_2$.

Then

$$\sigma_r = P \cdot \frac{r_1 r_{2,3}}{r_2^2 - r_{2,3}^2} \left(1 - \frac{r_2^2}{r^2} \right)$$

$$\sigma_\tau = P \cdot \frac{r_1 r_{2,3}}{r_2^2 - r_{2,3}^2} \left(1 + \frac{r_2^2}{r^2} \right) \tag{59}$$

$$\sigma_j = \sigma_\tau - \sigma_r = P \cdot \frac{r_1 r_{2,3}}{r_2^2 - r_{2,3}^2} \cdot \frac{r_2^2}{r}$$

where r is the radius of interest.

As is known, σ_j (difference between tangential and radial stresses) has its greatest value at the internal surface of the wedge vessel, i.e., where $r = r_{2,3}$. Then

$$\sigma_{j(r_{2,3})} = 2P \cdot \frac{r_1 r_{2,3}^2}{r_{2,3}(r_2^2 - r_{2,3}^2)} \tag{60}$$

Taking the derivative of expression (60) and equating it to zero, we get:

$$\frac{d\sigma_{j(r_{2,3})}}{dr_{2,3}} = -2P r_1 r_2^2 \cdot \frac{r_2^2 - 3r_{2,3}^2}{r_{2,3}^2(r_2^2 - r_{2,3}^2)} = 0 \tag{61}$$

from which the optimal radius

$$r_{2,3} = \frac{r_2}{\sqrt{3}} \tag{62}$$

Expression (62) is correct only where $r_1 < r_2/\sqrt{3}$.

Tetrahedral Press

The approach of replacing tensile stress by compressive stress is also used in the device called the tetrahedral press (anvil) [31]. This is a high-pressure apparatus (Fig. 23) made of four pistons consisting of a hard material. The end of each piston consists of a triangular area. The four pistons form a tetrahedral cavity which is a high-pressure vessel designed for compression of materials enclosed in an envelope of pyrophyllite. Four interconnected pressure intensifiers and pistons form the basis of a tetrahedral press.* This apparatus can be used to generate pressures on the order of 100,000 atm at temperatures up to 2000°C.

---

* Various designs of such presses will be discussed in the following chapter.

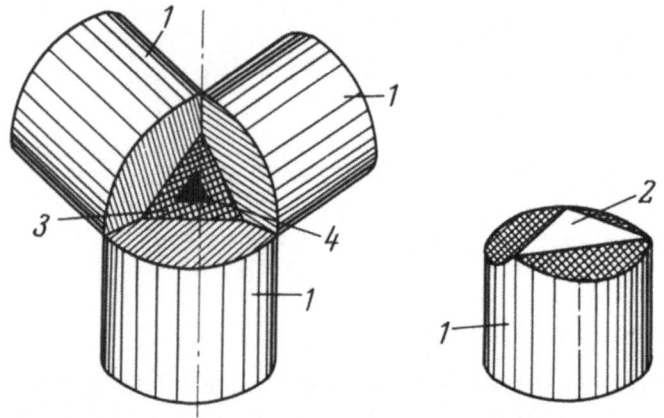

Fig. 23. Diagram of tetrahedral press: 1) piston; 2) triangular area (anvil);
3) pyrophyllite; 4) material to be compressed.

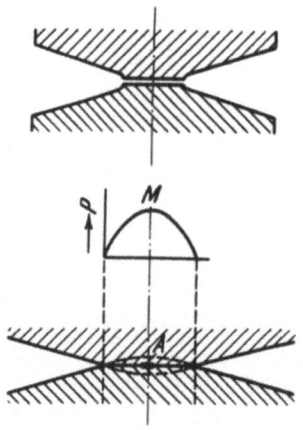

Fig. 24. Plan of compression of
plastic bodies to 425,000 atm.

Conical Dies (Bridgman Anvils)

An interesting tendency in design of high-pressure apparatus has been the compression of material over very small areas, surrounded and reinforced by less highly loaded masses of metal. We know [31] that the most suitable geometric form for specimens to be subjected to axial load is that of truncated cone. It will withstand higher pressures than a cylinder with the same cross-sectional area, since the atoms at the truncated surface have mechanical bonds which extend fan-wise toward the larger base of the cone.

One of the first designs using this principle was created by Bridgman [32]. Placing a ductile material under a carboloy cone (Fig. 24) pressed into a carboloy block submerged in a liquid compressed to 30,000 atm, he produced a pressure on the order of 400,000 atm. The highest pressure is achieved at point A, which corresponds to point M on the pressure diagram.

This method has developed in two directions. One has led to the invention of apparatus in which the compression of a thin layer of material is performed simultaneously with the application of a shear force.* The other direction has been determined by the attempt to overcome the

---

*This apparatus will be described in Chapter III.

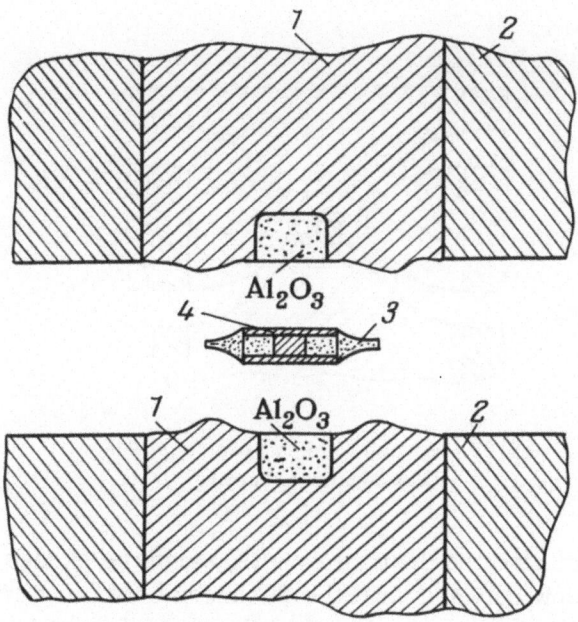

Fig. 25. "Anvil" with heat insulation: 1) Carboloy; 2) banding; 3) packing; 4) graphite.

primary shortcoming of the method, i.e., the impossibility of heating a
thin layer of material between the blocks to a high temperature and retaining this temperature without weakening the surface of the dies as a result
of their being heated.

First of all, an attempt was made to increase the thickness of the
central portion of the specimen [32] (Fig. 25). At the same time, thermal
insulation (Al_2O_3) was placed at the center of the anvil to protect the metal
from heating. However, the pressure which can be attained in this version
of the apparatus is limited since the compressing surfaces of the apparatus
touch each other after the specimen is compressed. Increasing the sample

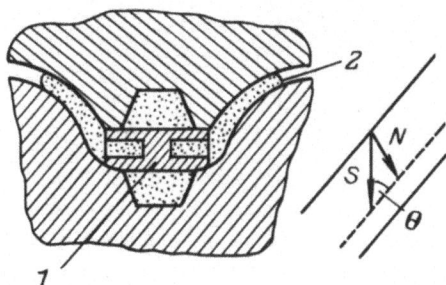

Fig. 26. Apparatus with conical packing: 1) material being compressed; 2) pyrophyllite packing.

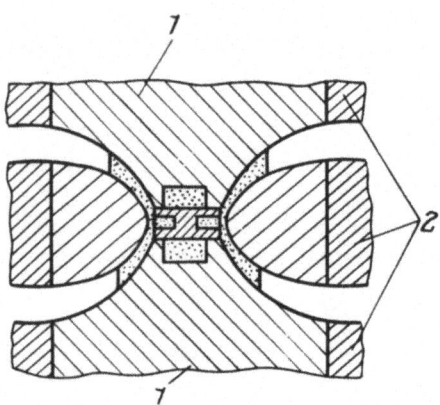

Fig. 27. Apparatus with two conical packings:
1) Carboloy; 2) banding.

thickness does not give the desired results, since the excess material plus
packing merely flow outward until the thickness becomes "critical," i.e.,
until the friction forces become equal to the shear forces in the material
being compressed.

Considerably higher pressure has been attained in the apparatus
shown in Fig. 26. With this form of sample being compressed, and movement of piston S, the packing is moved out only the distance N, where N =
S sin θ. Therefore, the central portion can be compressed to a higher pressure
than the packing.

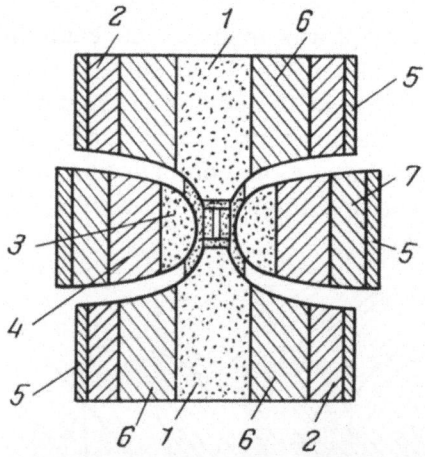

Fig. 28. Belt apparatus: 1) cemented-carbide
anvils; 2, 6) straps for strengthening dies; 3)
circular vessel of cemented carbide; 4, 7)
circular vessel straps; 5) soft steel ring.

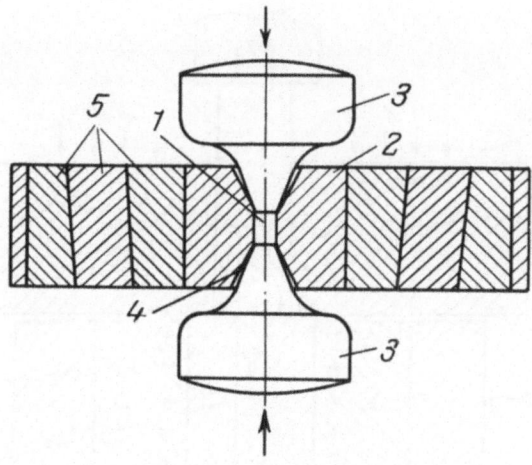

Fig. 29. Institute of High Pressure Physics' device with
conical anvils: 1) material being studied; 2) support;
3) anvils; 4) pyrophyllite packing; 5) supporting rings.

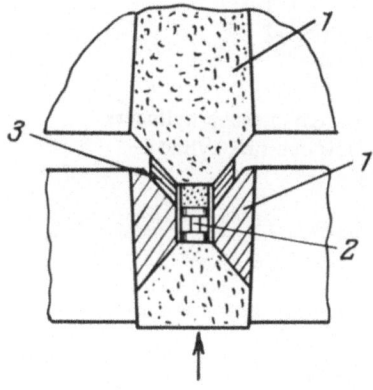

Fig. 30. Vessel with conical piston: 1)
Carboloy; 2) sample; 3) packing.

Figure 27 shows an apparatus
in which the upper and lower blocks
have identical form. Here, the sample thickness can be doubled. In this
case, there is a central ring-shaped
section which is made up prestressed
by bands 2 which protect the internal
ring-shaped vessel from rupture.

Belt Apparatus

An assembly consisting of two
conical pistons, a ring-shaped vessel,
and supporting belts has been called
the belt apparatus [33] (Fig. 28). The
sample is enclosed in pyrophyllite insulation, and the pistons are insulated by conical layers of packing which
flow into the clearance spaces between pistons and vessel during compression. Thus, sealing of the vessel as well as thermal and electrical insulation of the sample are attained.

If the supporting belts are made conical, the vessel can be considerably strengthened due to the mechanical support. Figure 29 shows an installation of this design created at the Institute of High Pressure Physics
of the Academy of Sciences USSR [34] independently from those described
above.

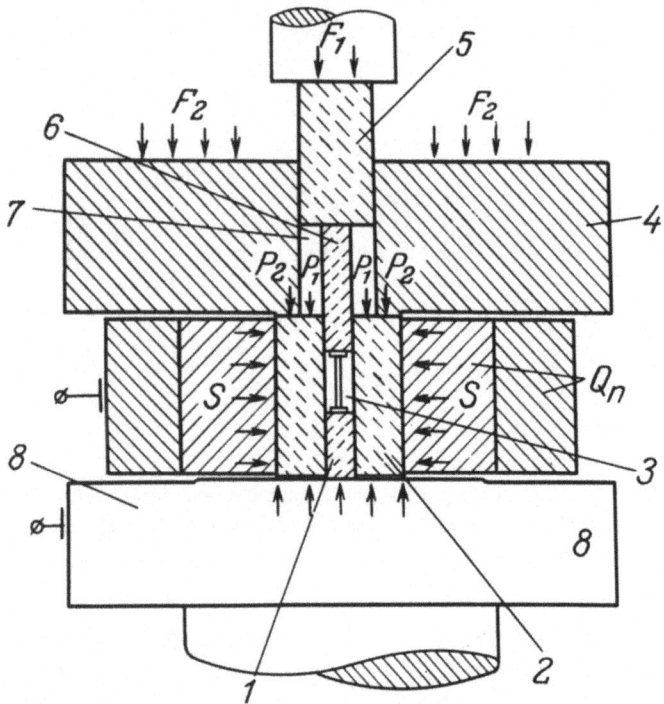

Fig. 31. Piston piezometer: 1, 5, 6) pistons; 2) high-pressure cylinder;
3) chamber holding material being studied; 4) low-pressure cylinder; 7)
silver chloride cylinder; 9) hydraulic press bed.

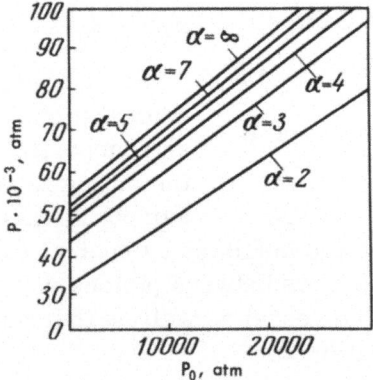

Fig. 32. Dependence of maximum pressure
P within piezometer (operating in the elastic
region) on external pressure of support P_0 for
cylinders with various ratios of external
diameter to internal diameter (α).

We know from experience of working with pressures of tens of thousands of atm that pistons without side support will be broken, and vessels, even those strengthened by belts or with mechanical support, will be burst without end support. In order to prevent rupture of pistons or vessels, the following design has been suggested [35] (Fig. 30). A conical piston is placed in a vessel whose ends are formed by an internal cone with the same angle (90°) as the piston. The ratio of length of cylindrical portion of the piston to its diameter is slightly over one. During compression of the sample, the packing material flows into the gap between piston and vessel and supports both the piston and the ends of the vessel.*

Piston Piezometer (Supported Piston and Cylinder) [36]

We know that the increase in ductility of brittle materials under high hydrostatic pressure is caused by a change in the nature of the stressed state. When using solid ductile materials for generation of quasi-hydrostatic pressure, we can considerably increase the strength of the apparatus, since the mechanical properties of such materials improve under high pressure. This phenomenon is used in the apparatus called the piston piezometer with quasi-hydrostatic support.

The apparatus (Fig. 31) consists of a cylinder 2 plus pistons 6 and 1 creating a stage of high pressure. The material being investigated is placed in cylindrical chamber 3. Cylinder 4 and piston 5 are used to create the supporting pressures. The circular chamber 7 is filled with a ductile material such as silver chloride.† Piston 5, moving under the influence of force F_1, compresses the ductile material and simultaneously moves piston 6, which creates the pressure in the material being investigated. As this happens, quasi-hydrostatic stresses arise in the ductile material supporting piston 6 and the ductility of the material of the piston is increased. Axial forces F_1 and F_2 are created by a hydraulic press. Radial forces S are caused by assembly tension [interference stresses]. These forces prevent brittle fracture of the cylinder. The maximum pressure [39] which can be generated by a piston of this type is equal to the proportional limit [under pressure] of the material of the piston (σ_p).

The walls of cylinder 2 are extended under the action of the internal pressure P. The greatest stresses arise at the internal surface of the cylinder. In this case σ_r, σ_T, P, the external pressure on the wall P_0 created by radial forces S, and the ratio of radii of the cylinder α are connected by the equation

---

*The so-called girdle apparatus [35] is constructed according to the same principle.
†Other ductile materials are used as well as silver chloride [37, 38].

$$P = -\sigma_r \; ; \; P = \frac{\sigma_\tau(a^2 - 1) + 2a^2 P_0}{a^2 + 1} \qquad (63)$$

The value of a and the value of σ_τ for the cemented carbide of which the cylinder is made are fixed. Since cemented carbides are brittle, on the basis of the first theory of strength, only the greatest stress is considered; in this case this means the tangential stresses on the internal surface of the vessel.

If $\sigma_\tau = \sigma_p$ (σ_p being the proportional limit in tension for the cemented carbide under high hydrostatic pressure), then, according to formula (63), we can calculate the maximum permissible pressure for the given cylinder.

Figure 32 shows the results of calculation where $\sigma_p = 54,000 \, \text{kg/cm}^2$ and the support pressure* is varied from one to 30,000 atm.

Let us note in conclusion that an apparatus operating under pressures from 10,000 to 100,000 atm, even one made of the best modern steels, will not last long, since the stresses in some of its individual parts so greatly exceed the strength of the material that these parts will be rapidly destroyed. †

BIBLIOGRAPHY

1.   Nadai, A., Plasticity (translated from English), Otdel. Nauchn. Tekhn. Inst., 1936.
2.   Gulyaev, A. P., Metals Engineering, Oborongiz, 1948.
3.   Likharev, K. K., Zavodsk. Lab., 15:1343 (1949).
4.   Kantarovich, Z. V., Principles of Design and Chemical Machines and Apparatus, Mashgiz, 1952.
5.   Belyaev, N. N., Design, Construction and Investigation of Chemical Apparatus and Machines, Tr. Nauchn. Issled.-Inst. Khim. Mash., 5, Mashgiz, 1950, p. 60.
6.   Bridgman, P. W., Physics of High Pressure, Otdel. Nauchn. Tekhn. Inst., 1935. [Bridgman, P. W., Physics of High Pressure, London, 1958].
7.   Mayers, A. F., Chem. & Ind., 72(4):643 (1954).
8.   Chipizhenko, A. P., Zavodsk. Lab., 15:1452 (1949).
9.   Belyaev, A. M., and Sinitskii, A. K. Izv. Akad. Nauk SSSR, Otdel. Tekhn. Nauk, Nos. 2, 3, 4, and 6 (1938).
10.   Krichevskii, I. R., and Tsiklis, D. S., Zh. Fiz. Khim., 17:115 (1943).
11.   Sokolov, S. N., Designing for Strength, Issue 2, Mashgiz, 1958.

---

*Detailed design of multistaged support is presented in the literature [40].
†Editor's comment: This is not true for a number of kinds of apparatus which have infinite lives at pressures considerably above 10,000 atm.

12. McGregor, G. W., Coffin, L. F., and Fisher, J. C., J. Appl. Phys., 19, March (1958); Lomakin, V. A., Inzh. Sb. Akad. Nauk SSSR, 21 (1955).

13. Makhonina, T. A., Designing for Strengths, Issue 2, Mashgiz, 1958.

14. Gladkovskii, V. A., Vereshchagin, L. F., and Ivanov, V. E., Fiz. Metal. i Metalloved., 6:1100 (1958).

15. Naidich, I. M., Tr. Nauk Issled.-Inst. Khim. Mash., Collection of articles, 10:125 (1951).

16. Faupil, J. H., Petrol. Refiner., 23(9):247 (1959); Ind. Eng. Chem., 49:1979 (1957).

17. Bridgman, P. W., Large Plastic Deformation and Fracture (translated from English, McGraw-Hill, 1952), IL, 1955.

18. Smirnov-Alyaev, G. A., The Theory of Autoreinforcement of Cylinders, Oborongiz, 1940.

19. De la Shez, Design of Self-Loading Weapons, II, GIZ, 1940.

20. Machine Building (Encyclopedic Handbook), Vol. 1, Book 2, Mashgiz, 1952, p. 377.

21. Korndorf, B. A., High Pressure Techniques in Chemistry, Goskhimizdat, 1952.

22. Class, J., and Mayers, A. F., Chem. Ing. Tech., 24:184 (1952); Siebel, E., and Schwaigerer, S., Chem. Ing. Tech., 24:199 (1952).

23. Comings, E. W., High Pressure Technology, New York, 1956; Neiman, E. Ya., and Pimshtein, P. G., Khim. Mash., 2:23 (1964).

24. Bridgman, P. W., Smithsonian Inst. Publ. Rep., 199 (1951).

25. Timoshenko, S. P., and Lessel's, D., Applied Theory of Elasticity, GONTI, 1931.

26. Gonikberg, M. G., Tsiklis, D. S., and Opekunov, A. A., Dokl. Akad. Nauk SSSR, 129:88 (1959).

27. Mil'vitskii, R. V., Khim. Mash., 1:31 (1938).

28. Mirinskii, D. S., Zh. Prikl. Mekhan. i Tekhn. Fiz., 2:165 (1960).

29. Butuzov, V. P., Mirinskii, D. S., and Kats, G. S., Symposium on Deep Processes, Izd. Akad. Nauk SSSR, 1962, p. 172.

30. Wentorf, R. H., (Editor), Modern Very High Pressure Technique, London, 1962 [Russian translation, Izd. "Mir," 1964].

31. Hall, H. T., Rev. Sci. Instrum., 29(4):267 (1958); Bovenkerk, H. P., Bundy, F. P., Hall, H. T., Strong, H. M., and Wentorf, R. H., Nature, 184:1094 (1959).

32. Bridgman, P. W., Most Recent Work in the Area of High Pressures, GIIL, 1948.

33. Hall, H. T., Rev. Sci. Instrum., 31(2):125 (1960).

34. Vereshchagin, L. F., Galaktionov, V. A., Semerchan, A. A., and Slesarev, V. N., Dokl. Akad. Nauk SSSR, 132: 1059 (1960).

35. Wilson, W. B., Rev. Sci. Instrum., 31(3):331 (1960).

36. Ryabinin, Yu. N., and Livshits, L. D., Zh. Tekhn. Fiz., 29:1167 (1959).

37. Boyd, F. R., and England, J. L., Yearbook Carnegie Inst., 57:170 (1958).
38. Daniels, W., and Jones, M. T., Rev. Sci. Instrum., 32(8):885 (1961).
39. Ryabinin, Yu. N., Fiz. Metal. i Metalloved., 6:393 (1958).
40. Chirstiansen, E. W., Kistler, S. S., and Gogarty, W. B., Rev. Sci. Instrum., 32(7):775 (1961); Andreatch, P., and Andersen, O. L., Rev. Sci. Instrum., 28(4):288 (1957).

Chapter III

Methods of Generating Pressure
at Normal and Elevated Temperatures

Generation of pressure in equipment is the most difficult part of high-pressure experimental work. We will analyze below some methods for generating pressures in gases, liquids, and solid bodies.

COMPRESSION OF GASES

Laboratory Compressors*

Laboratory compressors now used are four- or five-cylinder horizontal machines compressing gases respectively to 300 or 1000 atm, i.e., each cylinder has a compression ratio of about 3.5. The [volumetric] capacity of these compressors is approximately 10 m³/hr.

A compressor is supplied with a crankshaft-connecting rod mechanism and a flywheel driven by an electric motor (belt drive); the lubricating valve is in the first compression stage. The compressed air passes through an oil catcher, where most of the oil picked up by the gas from the compressor is removed. However, the compressed gas contains oil not only in the form of a fog or droplets, but also in the dissolved state [2]. Also, if the gas is stored in a water gas holder, it is saturated with water vapor. Moisture can be removed relatively easily by placing a vessel filled with silica gel and calcium chloride before the inlet tube to the first stage of the compressor. Also, the compressed gas is passed through a series of high-pressure vessels filled with silica gel saturated with calcium chloride, caustic potash, activated carbon, and anhydrone.†

The oil and moisture contained in a gas can be completely removed by absorption of the oil and moisture by an adsorbent or by cooling to a temperature at which the oil and moisture condense.

---

* These machines are described in detail in literature [1].

†Anhydrone (magnesium perchlorate) is a strong oxidizer, so that it cannot be used for drying of hydrocarbons.

79

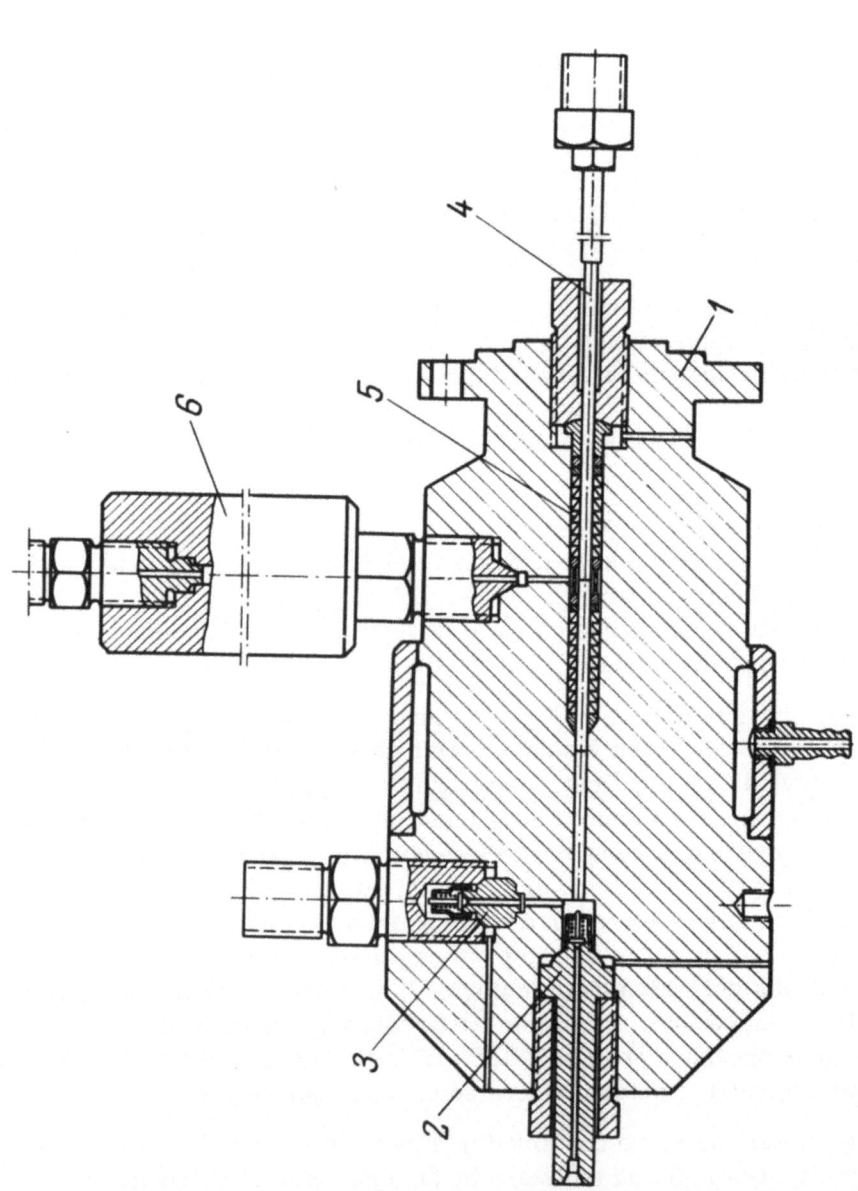

Fig. 33. Cross section of cylinder of single-stage compressor: 1) cylinder; 2) feed valve; 3) delivery valve; 4) shaft; 5) gland; 6) oil reservoir.

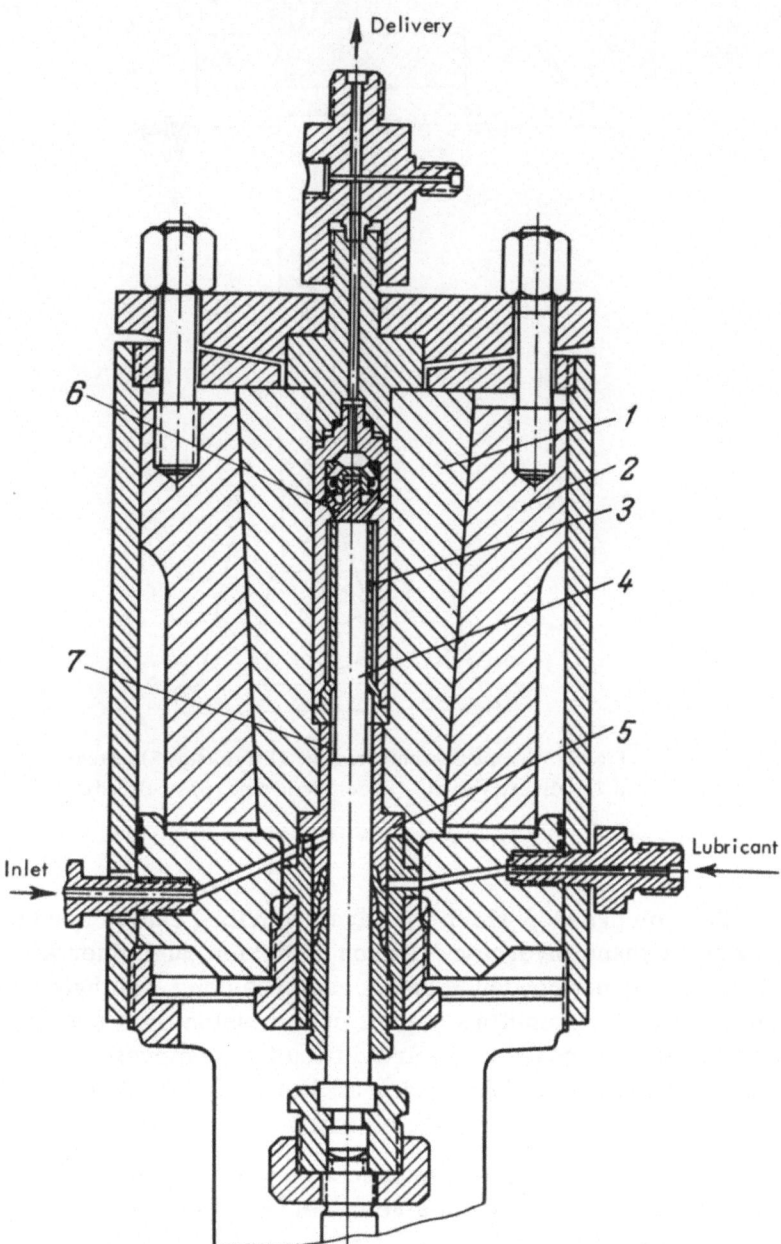

Fig. 34. Vereshchagin and Ivanov gas compressor: 1) Cylinder; 2) support; 3) sleeve seal; 4) shaft; 5, 6) valve; 7) space beneath piston.

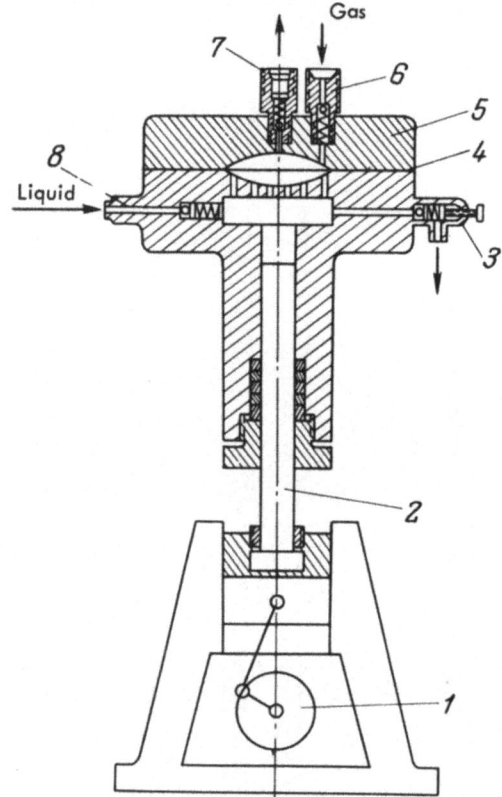

Fig. 35. Membrane compressor: 1) flywheel; 2) piston;
3,7) delivery valves; 4) steel membrane; 5) steel body;
6, 8) valves.

The compressor which we have described can be used to compress
nitrogen, methane, hydrogen, carbon monoxide, and other so-called constant
gases which do not condense under compression. The hydrocarbons (even
if their critical parameters permit compression or if special measures are
used to prevent condensation) are difficult to compress with this type of
compressor, since the lubricating oil is almost totally carried off by the
hydrocarbons due to its high solubility.*

This type of compressor cannot be used to compress oxygen, since
the oxygen and oil form explosive mixtures. For the same reason, air can
be compressed with this type of compressor only to 200 atm. Oxygen is

---

*It is also possible that emulsification of oil is considerably increased by
the compressed gases (especially hydrocarbons) due to reduction in surface tension at the phase boundary.

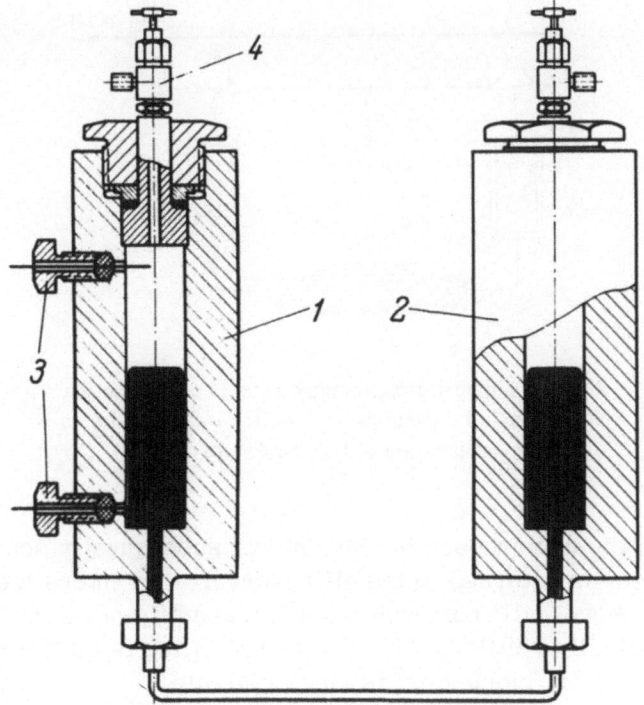

Fig. 36. Compression by mercury (mercury transfer): 1, 2) connected vessels; 3) contacts; 4) valve.

compressed in special compressors using glycerin lubrication.* However,
under laboratory conditions different methods can be used to compress oxygen; these methods will be described below.

Additional Compression Stages

For compression of considerable quantities of gases to pressures over
1000 atm, a device referred to as a final compression stage is often used.

B. A. Korndorf Single-Stage Compressor [1]. This
compressor has an increased compression ratio and is supplied by compressed gas from an ordinary laboratory compressor. The principal parts
of the compressor (Fig. 33) are the working cylinder 1, water cooled and
designed for a pressure of 5000 atm, the inlet valve 2 and delivery valve 3,
and shaft 4 with gland 5. The gland consists of conical babbitt rings directed in opposite directions. They are lubricated from an oil reservoir 6
which is periodically filled with oil. The upper portion of the oil reservoir
is connected to the lower valve of the compressor oil separator. The oil

---

*The literature [4] contains a description of a new type of compressor
 whose pistons use Teflon sealing rings without lubrication.

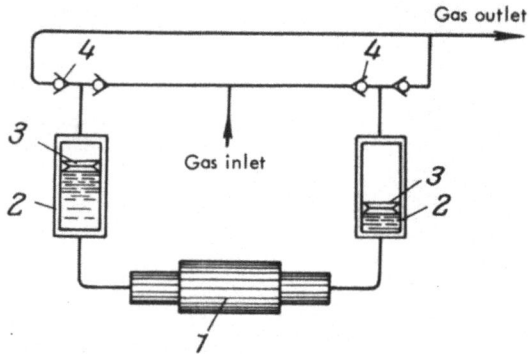

Fig. 37. Liquid-transfer compressor: 1) double-action
intensifier; 2) high-pressure vessels; 3) floating piston
to reduce evaporation of liquids; 4) valves.

removed from the compressor by the gas and subsequently separated in
the oil separator is returned to the oil reservoir and enters the gland once
more. Since the gas still removes a certain quantity of oil, the oil reservoir
must be periodically refilled. This method of oiling is inefficient. Lubrication by a spray oiler (lubricator) is more reliable.

When the compressor is supplied with a gas compressed to 1000 atm,
the output pressure at the oil separator reaches 5000 atm.

L. F. Vereshchagin and V. E. Ivanov Compressor.
This compressor [5] is a single-stage machine with a high compression
ratio, designed to work with an inlet pressure of approximatly 100 atm (i.e.,
the pressure in a compressed-gas cylinder). The compressor (Fig. 34)
consists of a conical cylinder 1 retained in a support 2. The bore of the
cylinder contains polished sealing ring 3, in which stepped piston 4 moves.
As the piston moves downward, gas enters through valve 5. As the gas is
compressed, the lower section of the piston closes the valve. Delivery
valve 6 does not have a valve body and consists of an inverted conical bell
with elastic walls and a very slight rise over the seat.

Sealing of the piston is achieved by sleeve 3,* which presses against
the piston, the degree of compression [pressure] being proportional to the
increase in pressure in the cylinder. Also, the lower portion of the piston
creates a gas blanket in the space beneath the piston 7. Since the clearances between piston, ring, and valve 5 contain lubricant (solidol or nigrol,
State Standard GOST 542-41), its high viscosity in the gaps creates a pressure gradient which prevents leakage of gas. In spite of the high compression ratio (approximately 100 : 1), the parts of the compressor are not heated,

---

*The principle of operation of the sleeve seal is described in detail below.

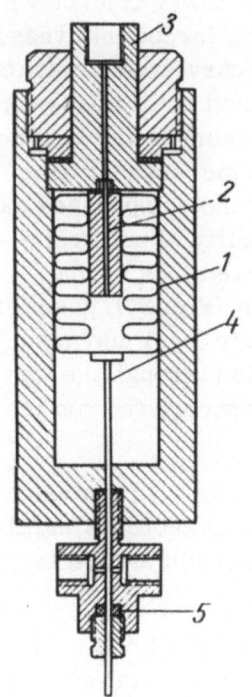

Fig. 38. Compression by
metal bellows: 1) bellows;
2) insert; 3) head; 4) rod;
5) gland.

since the hot gas is rapidly removed from the
cylinder and the gas remaining in the dead
space of the cylinder (approximately 0.01 times
the working volume of the cylinder) is cooled
by expansion. This compressor can be used to
produce pressures of up to 6000 atm.

Compressing Devices

When a small quantity of gas is sufficient for investigations, or when especially
pure gas is required, as well as when oxygen
or a rare gas must be compressed, several
kinds of compressing devices are used.

Membrane [Diaphragm] Com-
pressor [6]. The compressor consists
of a steel body 5 (Fig. 35) which contains a
lens-shaped cavity divided by a steel membrane 4 held between the body and a head. As
the flywheel 1 is rotated, piston 2 compresses
liquid located in the block, which forces the
membrane toward the head. During this cycle,
gas leaves the lens-shaped cavity through valve
7. During the return stroke, the membrane is
relieved and draws more gas in through valve
6. Since the displacement of the piston is somewhat greater than the capacity of the lens-
shaped cavity, a small portion of liquid is
pumped through inlet valve 8 and out through controlled outlet valve 3 with
each cycle. Since the membrane is fully supported* by the spherical surfaces of the block, it can withstand high pressure ratios.

Compression by Mercury. Two connected vessels 1 and 2
are half filled with mercury (Fig. 36). One vessel has two contacts 3. The
space over the mercury in vessel 2, which is connected to a pressure generating device, is filled with a liquid which is to be used to generate the
pressure (oil, water, glycerine, etc.).

Connecting a pressure cylinder filled with compressed gas to valve
4, some gas is drawn off [the cylinder into vessel 1]. The level of mercury
in vessel 1 is reduced, and rises in vessel 2 (the liquid generating the pres-

---

* The principle of the fully supported membrane [diaphragm] is also used
in the design of the membrane [diaphragm] valve (Chapter VI) and the membrane ["null"] instrument (Chapter IV).

sure is released from valve 2). When the circuit of the lower contact 3 is opened, valve 4 closes and the oil (or other liquid) being forced into vessel 2 begins to compress the gas. The position of the mercury (which must be known in order to prevent loss of mercury) is determined as it closes contact 3. Since it is possible to practice to raise the mercury to the very top of vessel 1, this method can be used to produce high compression ratios fairly easily. By using a collecting vessel and feeding the compressed gas into it, it is possible to accumulate a considerable quantity of pure compressed gas rather quickly. If both vessels are connected to a double-action intensifier and valves are included in the system (Fig. 37), gases can be compressed to very high pressures [7]. The mercury level may also be determined by the resistance of a carbon rod inserted through the top of the vessel: as the mercury rises, the current path through the rod is shortened and the resistance is reduced.

This method of gas compression has its deficiencies. It requires a considerable quantity of mercury, which is in itself undesirable. Also, the mercury is contaminated by the pressure-transmitting liquid and begins to contaminate the gas after a certain period of time.

Compression with Metallic Bellows (Sylphon). Bellows 1 with insert 2 serving to reduce the internal space is installed in a high-pressure column (Fig. 38). The bellows is connected through the insert to head 3 in turn connected to a gas line. The bellows is compressed from without by oil fed through the lower inlet of the column by an oil pump.

Before gas is fed into the apparatus, the bellows is compressed by oil. Then, as the oil is gradually drained from the column, gas is fed into the bellows under pressure. The expansion of the bellows is allowed to continue until the bellows reaches its maximum expanded position, after which the inlet to the bellows is closed and the compression of the gas is begun by pumping oil into the apparatus.

In order to determine the position of the bellows in the column, rod 4 (diameter 4 mm) is attached to the bottom of the bellows. The rod moves through gland 5 made of polyfluoroethylene rings. Due to the low coefficient of friction of the polyfluoroethylene and the low rate of movement of the rod, the resistance in the gland is slight and a small pressure differential is sufficient to move the rod. The position of the bellows can be determined by observing the portion of the rod which extends outside the apparatus.

Compression of Oxygen. In compressing oxygen, the mercury or oil is replaced by a mixture of glycerin and distilled water. The plan of an apparatus for compressing oxygen to 1500 atm was described by L. F. Vereshchagin and V. A. Preobrazhenskii [8]. We compressed oxygen to 500 atm with a bellows compressor operating with a mixture of glycerin and water.

In order to compress oxygen with a gas-pressure booster, the packing of the piston is made of polyfluoroethylene rings. In this method, oxygen can be compressed to very high pressures (on the order of 10,000 atm). An apparatus designed for preliminary compression of oxygen using metal bellows as described above was tested by the author [9].

The pressure in the reaction column connected to the gas-pressure booster was measured by manganin manometer. The coil of the manometer was not protected from the oxygen, yet no disruptions in the operation of the manometer at pressures up to 10,000 atm were noted over a long period of service.

In compressing oxygen, all parts of the installation used must be carefully cleaned of oil and other organic materials; otherwise they may ignite and large steel parts may begin burning.* If cleanliness is assured, the operation of compressing oxygen is safe; the pressure can be increased, as was noted above, to 10,000 atm.

Other Methods of Compressing Gases

Considerable pressure can be created by evaporation of a compressed gas in a closed space [10-12] as well as by heating a compressed gas.

S. S. Boksha [13] constructed a thermo-compressor for compression of gases. This apparatus consists of 4 thick-walled vessels with capacities of 2130, 1360, 500, and 480 ml. Each vessel contains an electric element with capacities of 9, 7, 6, and 5 kw respectively. The entire system is filled with gas from a pressure cylinder, after which a relay is closed which first connects the first heating element. The gas in the outermost vessel is heated, and the pressure in the entire system increases. Then, the relay disconnects the first heating element and connects the second heating element. The pressure in the second vessel and the system connected to it begins to increase. Then, the third and finally the fourth heating elements are connected, and after the system is further filled with more gas, the first heating element is connected once more. Check valves are installed between the vessels to prevent gas from entering a vessel in which the heating element has been turned off.

In eleven minutes, a thermo-compressor increased the pressure in a 200 ml capacity vessel from 150 to 2000 atm.

Very high pressure can be produced by using the kinetic energy of a flying projectile for adiabatic compression of a gas [14]. In this case, the compressed gas is heated to 5000-7000°C. This method is at the present time the only method for simultaneous creation of high pressures and very high temperatures.

---

*See Chapter VIII.

The compressed gas expands very rapidly; the reaction products of any reaction which occurs between the components of the gas mixture at high temperatures and pressures are instantly cooled and undergo quenching. This method makes it possible to eliminate the influence of the walls of the apparatus on the reacting materials; it has therefore been used for investigation of certain reactions [15-17].

An installation for adiabatic compression of gases consists of a column (closed at one end) in which a piston moves to compress the gas. The piston must be carefully fitted to the column in order that leakage will be minimal. The piston, while located at one end of the column, is struck by compressed gas and begins moving in the column, compressing the gas in front of it. After it gives up all its energy to the gas, the piston stops, then moves back under the action of the gas which it has compressed. After a few cycles, the piston stops. An entire cycle last a few hundredths of a second, so that the walls of the column are not heated.

The piston is manufactured very carefully and fitted to the column so that the clearance between column and piston does not exceed 5 μ. In spite of this, gas losses occur as the gas slips through the space around the piston, and the accelerating gas leaks into the gas to be compressed at the beginning of each cycle.

Yu. N. Ryabinin [18] has developed a single-cycle unit in which the piston compresses the gas and then returns to its initial position only once. For this, a piston-valve is placed in the end of the column; this piston begins to move as the compressing piston moves. When the compressing piston reaches its end position and stops, the valve piston flies out of the column, opening an exit for the accelerated gas. Therefore the compressing piston, flying backward, extracts all the accelerated gas from the column and stops in its initial position.

COMPRESSION OF LIQUIDS AND SOLIDS

Liquid Pumps and Compressors

Vereshchagin Hydraulic Compressor [19]. The compressor (Fig. 39) consists of a block 1 of alloy steel, in which piston 2 moves. The block contains inlet valve 3 and delivery valve 4. The piston, made of ball-bearing steel type ShKh15 annealed and tempered to a Rockwell hardness of 40-45, is moved by link 5, held to the block by tension element (rods) 6.

An interesting feature of this design is the gland.* Usually, the operation of a gland is improved by increasing its length. However, this re-

---

* Editor's comment: This gland is the key to reliable operation of the higher-pressure compressors.

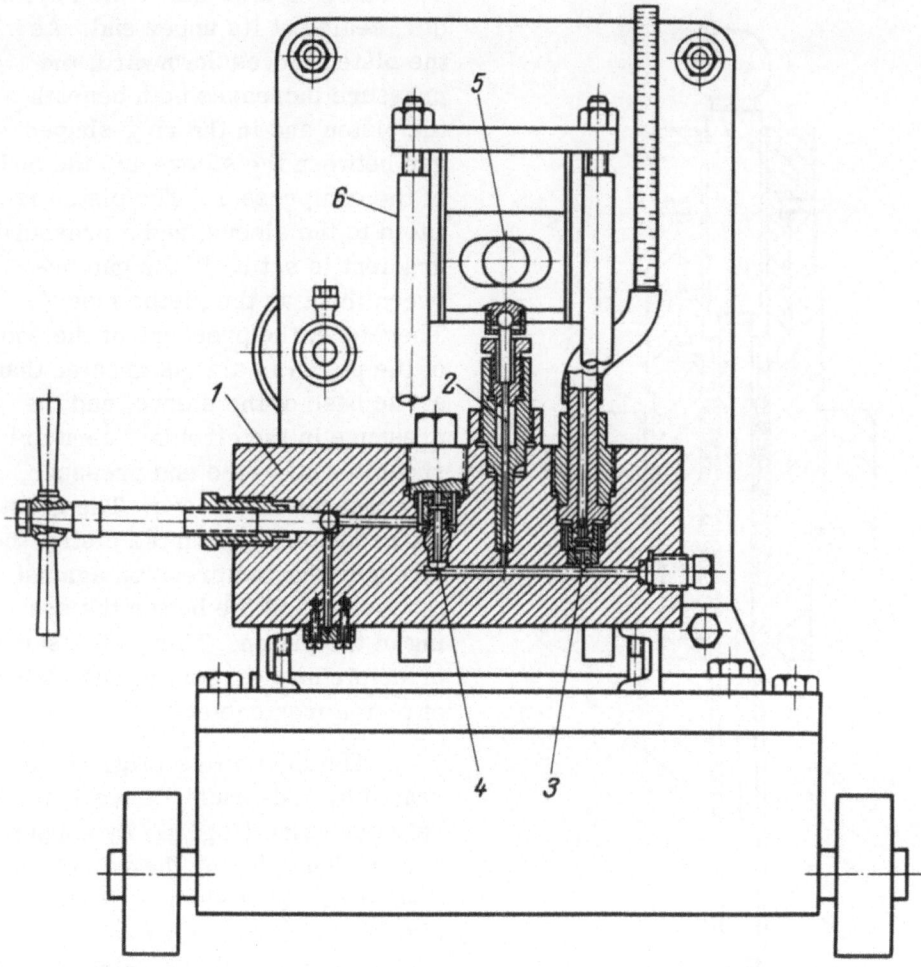

Fig. 39. Vereshchagin hydraulic compressor: 1) block; 2) piston; 3) inlet valve; 4) delivery valve; 5) link; 6) tension members.

quires an increase in the length of the piston as well. This causes an increase in the friction and a reduction in the resistance of the piston to longitudinal bending.* As a result, the gland undergoes wear and may fail. The instability of a long piston makes it necessary to decrease the stroke, which causes an increase in the relative volume of the dead space.

In the compressor being described here, the gland consists of two sets of packing (Fig. 40). The system of leather and steel rings 1 creates the initial pressure gradient at which the sleeve seal begins to operate.

---

*A piston for this type of compressor must have a small diameter in order to reduce the forces acting on it.

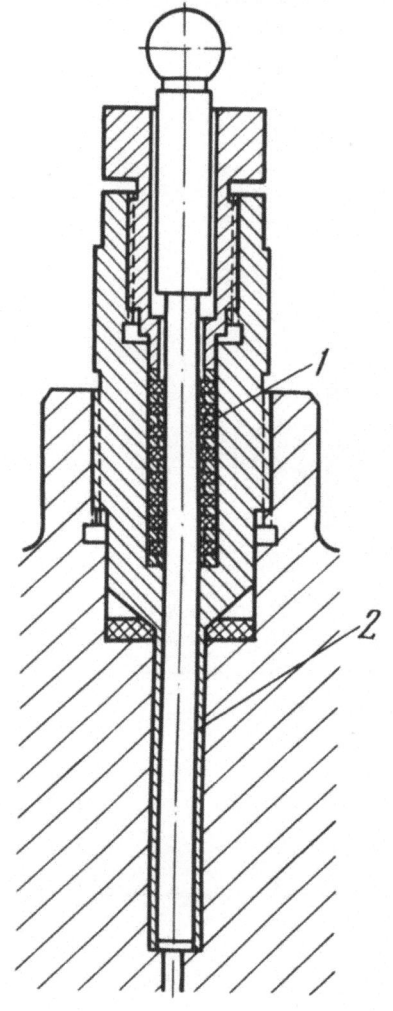

Fig. 40. Packing of piston: 1) steel rings; 2) sleeve.

The sleeve 2 is a thin-walled cylinder, sealed at its upper end. As the piston moves downward, the pressure increases both beneath the piston and in the ring-shaped gap between the sleeve and the body of the compressor. The piston is fitted to the sleeve, and a pressure gradient is set up in the gap between them as the piston moves. Therefore, the pressure at the end of the piston is always greater than at the base of the sleeve, and the pressure in the circular [annular] space, equal to the end pressure, compresses the sleeve. The greater the pressure beneath the piston, the stronger the compression against the sleeve and the better the sealing of the piston. This is the same principle used in sealing the pistons of piston manometers.

The inlet valve (Fig. 41) is sealed by red-brass sleeve 1, the delivery valve (Fig. 42) by copper ring 1. The plug of the valve* is also sealed by a sleeve. Plates 2 of the valves are very carefully fit (to two or three Newton rings).† This compressor can be used to compress both liquids and gases, which have been preliminarily compressed to 500-600 atm, to a final delivery pressure of 5000-6000 atm.

The Ivanov Hydraulic Compressor [20].‡ The liquid to be compressed is fed in through nipple 14 (Fig. 43) under a pressure of

---

*The valves are the most vulnerable part of any pump. This problem will be analyzed further below.

†Editor's comment: This kind of valve operates reliably at the higher pressures, whereas ball valves are subject to attrition at similar pressures.

‡Editor's comment: This design of compressor is very reliable in perform-

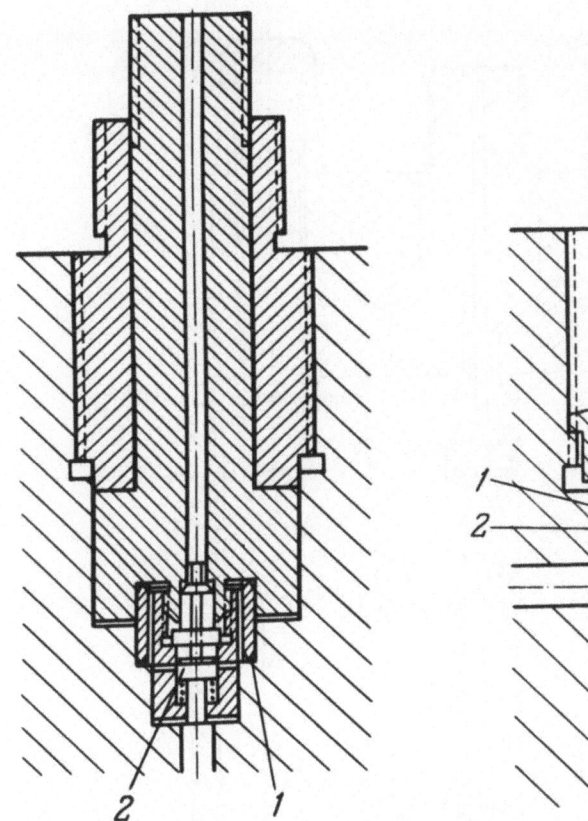

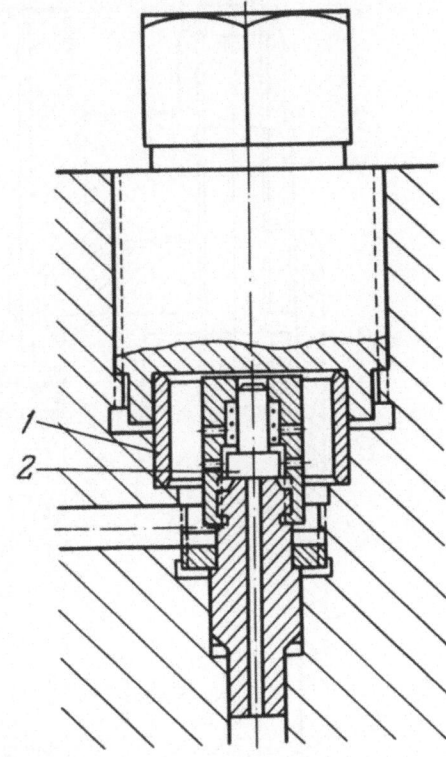

Fig. 41. Inlet valve: 1) sleeve; 2) plate.   Fig. 42. Delivery valve: 1) ring; 2) plate.

25-30 atm into the annular space between the body 13 and gland 12 from
which it passes through three apertures to the annular space between the
shaft 16 and the gland 12. As the piston moves downward, the liquid passes
through lateral apertures in the shaft to the bore of the piston 11 and, opening inlet valve 3, fills the compression chamber. As the piston moves upward, valve 3 closes and the pressure in the compression chamber rises.
When the pressure in the chamber exceeds the pressure beyond the delivery
valve 4, the valve opens and the compressed liquid enters the receiving
vessel.

ance. It is entirely capable of supplying a manifold that is piped to several
locations in a laboratory; in this way liquid at 150,000 psi is on tap, much
as steam, compressed air, vacuum, and other services are piped into modern laboratories. A pressure of 200,000 psi is reliably available for mod-
erate-duty use from certain models of this type of compressor.

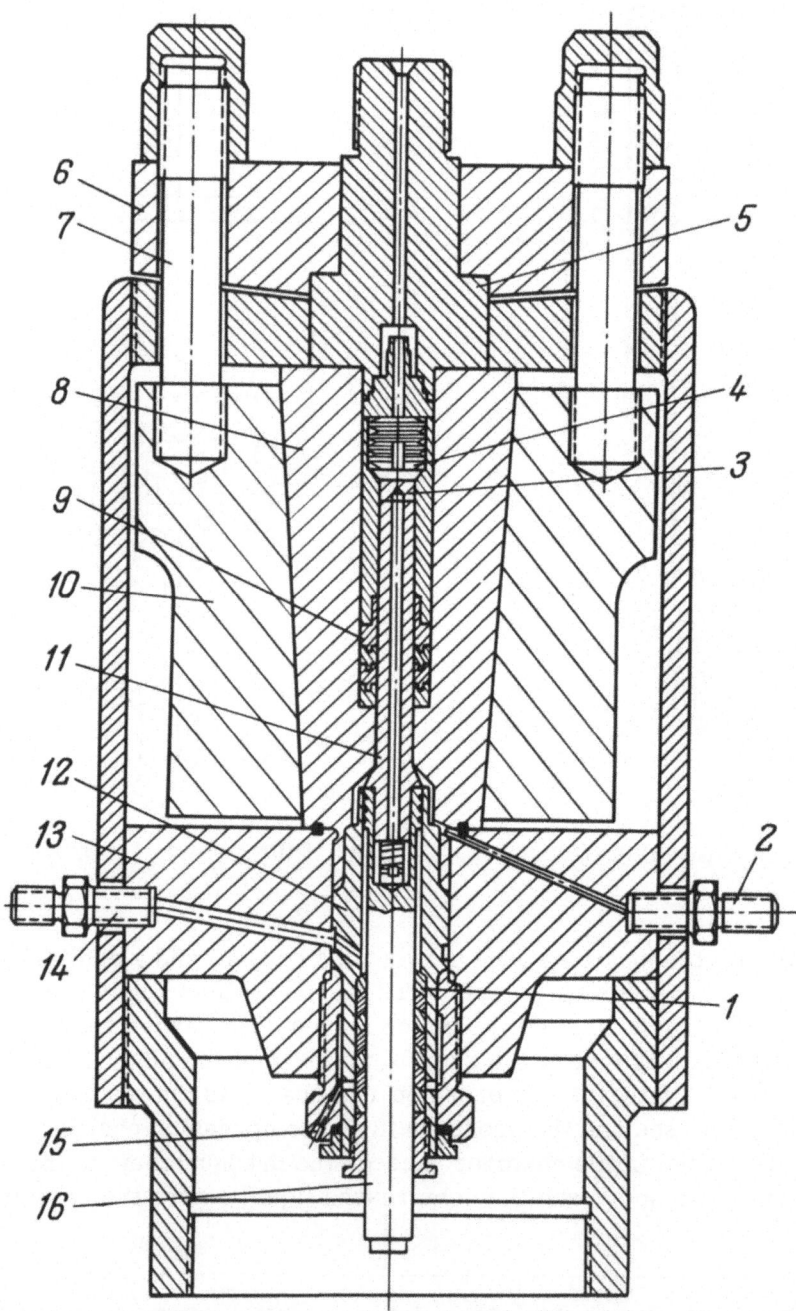

Fig. 43. Hydraulic compressor with separate supply of liquid and lubricant: 1, 9) glands; 2, 14, 15) nipples; 3) inlet valve; 4) delivery valve; 5) plug; 6) flange; 7) tension member; 8) cone; 10) support; 11) piston; 12) gland; 13) body; 16) shaft.

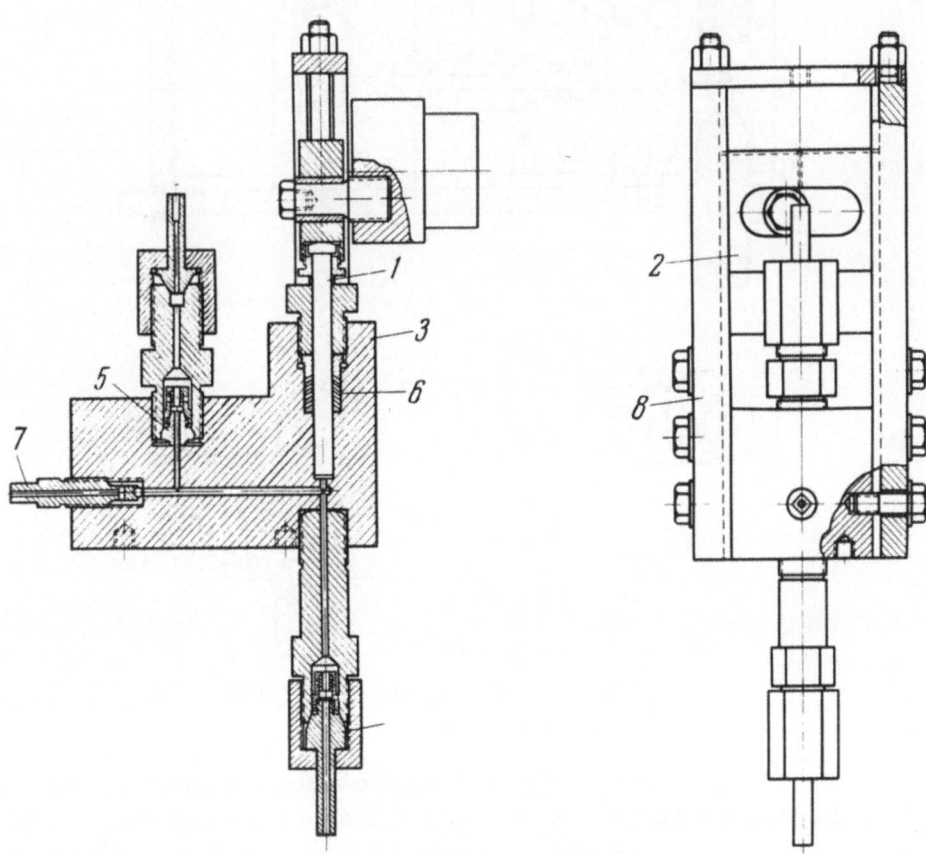

Fig. 44. Tsiklis hydraulic pump: 1) piston; 2) link; 3) pump body; 4) inlet valve; 5) delivery valve; 6) gland; 7) valve; 8) slide rails.

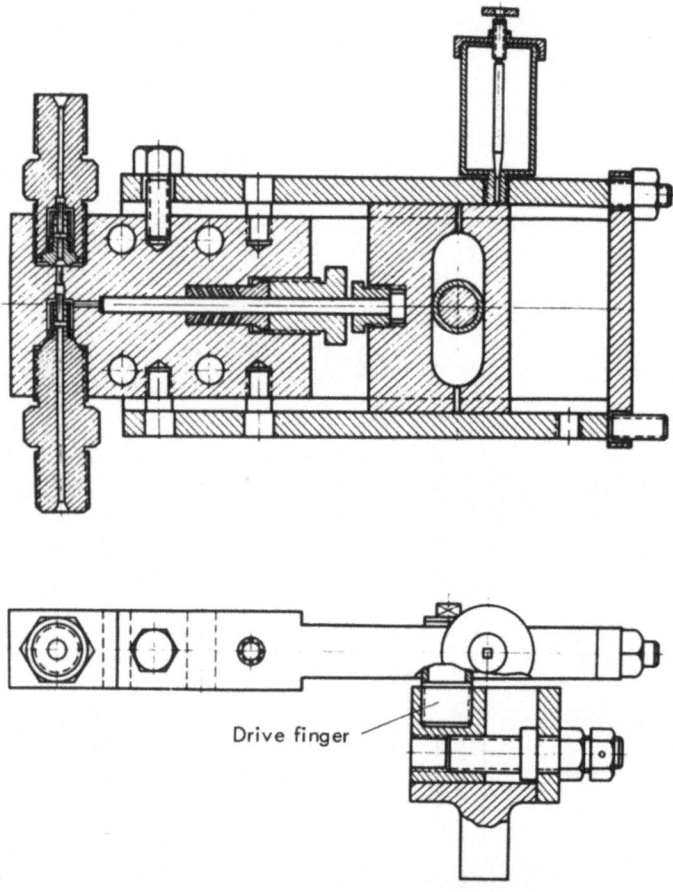

Fig. 45. Hydraulic pump with variable output.

The compressor automatically controls the mechanical support of the compression chamber. Cone 8 which contains the chamber is held in support 10. As the pressure increases, plug 5 tends to move away from cone 8. When this happens, support 10 is forced even more tightly around the cone by flange 6 and tension members 7, compressing it further.

The sealing around the moving piston is a result of the hydraulic resistance of the liquid flowing through the clearance space [21]. Piston 11 and sealing sleeves 1 and 9 are lubricated by a hypoid grease, fed in under a pressure of 25-30 atm through nipples 2 and 15. The output of the compressor is 120 liters of liquid per hour at pressures up to 5000 atm, or 60 liters per hour at a pressure up to 8000 atm. Other models of similar design yield 12 liters per hour at 12,000 atm and 2000 liters per hour at 2000 atm [22].

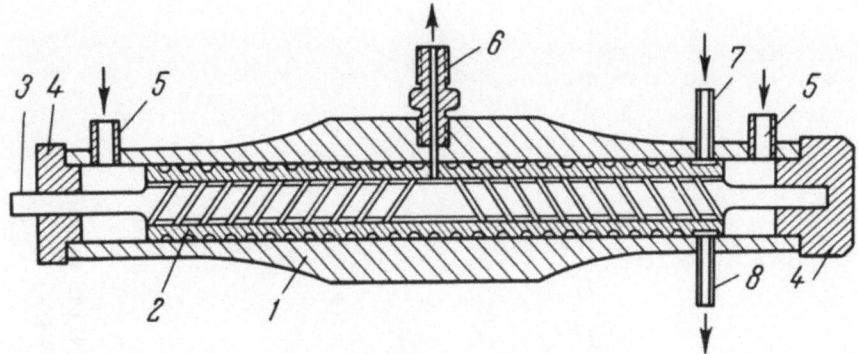

Fig. 46. High-pressure screw pump for viscous liquids: 1) body; 2) insert; 3) drive; 4) ends; 5) tubes; 6) nipple; 7, 8) tubes.

Fluid Pump. The pump design which we use (Fig. 44) can be recommended for generation of pressures up to 1000–1500 atm.

Piston 1, moved by link 2, moves through the hump of the L-shaped body 3 in which inlet valve 4 and delivery valve 5 are located. These disc valves are sealed on the spherical and conical surfaces of the valves themselves. Gland 6 consists of alternating brass and leather or plastic rings. Valve 7 is used for flushing the pump. Slide rails 8 along which the link moves are fastened directly to the body of the pump.

The pump is driven by a speed reducer. With a piston diameter of 12 mm and an operating rate of 30 cycles per minute, its yield at 1000–1500 atm is 6 liters per hour.

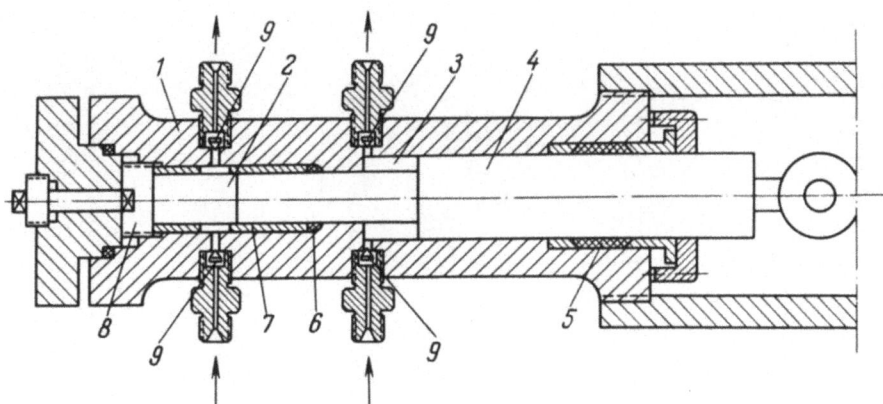

Fig. 47. Pump for transferring corrosive media: 1) body; 2, 3) chambers; 4) plunger; 5, 6) glands; 7) sleeve; 8) ring; 9) valves.

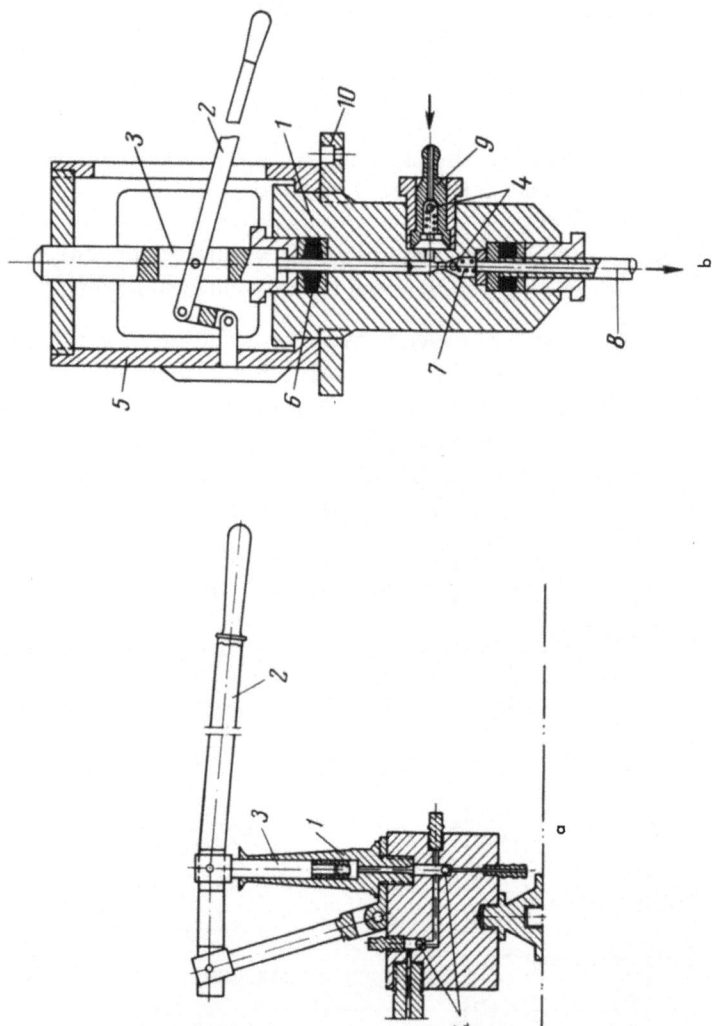

Fig. 48. Hand pumps of various designs (a and b [26]): 1) cylinders; 2) handles; 3) pistons; 4) ball valves; 5) frame; 6) gland; 7) delivery valve; 8) tube; 9) inlet valve; 10) mounting flange.

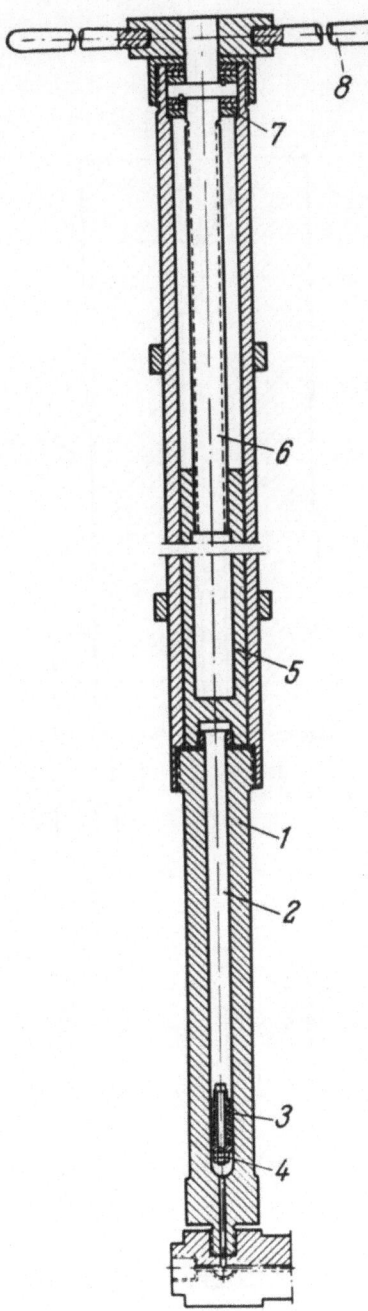

Fig. 49. Sidorov mechanical press: 1) cylinder;
2) piston; 3) sealing rings; 4) nut; 5) traveling
nut; 6) drive screw; 7) ballbearings; 8) arms.

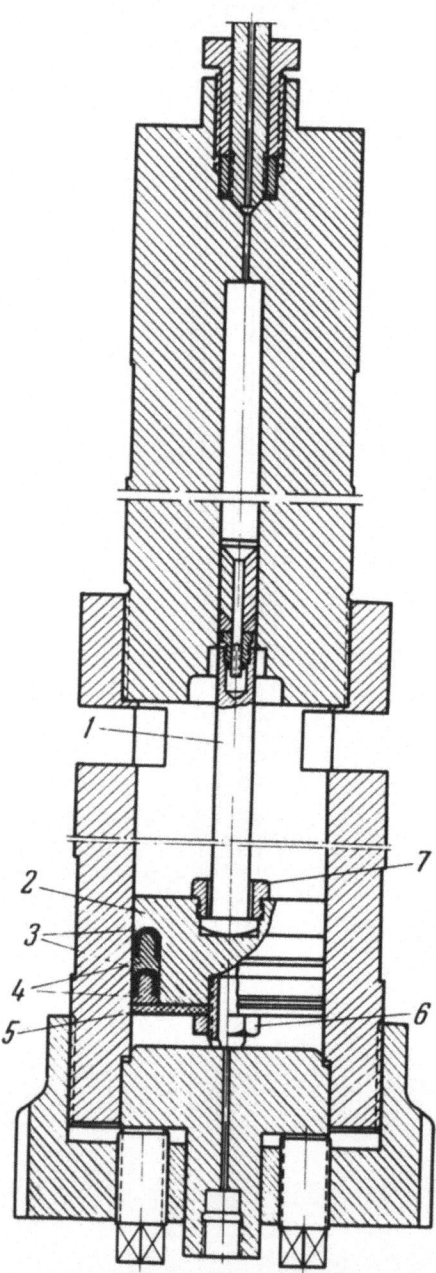

Fig. 50. Sidorov intensifier: 1) small piston;
2) large piston; 3) leather collars; 4) rings; 5)
plate; 6) nut; 7) retaining nut.

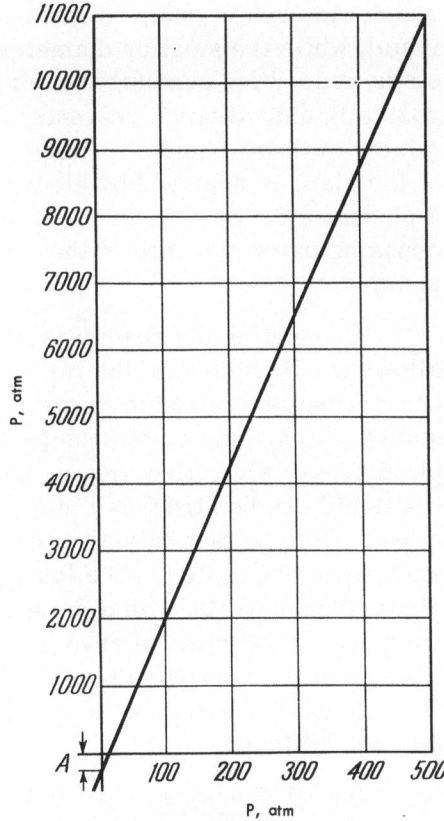

Fig. 51. Calibration of intensifier.

Figure 45 shows a similar pump of another design* in which the output can be regulated by changing the stroke of the piston by moving the finger [pin] of the sliding drive relative to the axis of the reducing gear drive. The horizontal position of the piston allows a more favorable location of the valves.†

A pneumatic drive is used in place of the mechanical drive for compression of liquids to high pressures. The diameter of the working shaft of one such compressor, designed for creation of pressures up to 10,000 atm, is 5.5 mm. The shaft is connected to the piston of an air compressor with a diameter of 220 mm. Due to the high ratio of areas of the pistons, an air pressure of 15 atm is sufficient for creation of a pressure of 10,000 atm [23].

Screw Pump for Viscous Liquids [24]. The pump (Fig.46) consists of body 1 with bronze insert 2 impressed to cool the liquid. Screw drive 3 with spiral channels rotates within the insert. The drive rotates within glands located in the ends 4 of the body. The gap between drive and insert is .01 mm. The liquid is fed to the pump through tubes 5. The drive, rotating at several thousand rpm, picks up the liquid and forces it through nipple 6. The opposition of the screw threads on the drive equalizes the axial forces and removes the loads from the glands. The cooling liquid flows through tubes 7 and 8.

Sidorov Pump. Compression and pumping of corrosive liquids at high pressures can be easily done using the Sidorov pump [25]. A body 1 (Fig. 47) made of stainless steel is divided into chambers 2 and 3; a stepped plunger 4 moves within these chambers. The large-diameter stage plus

---

*Engineer I. S. Radle took part in its development.

†The NZhR pump (controllable liquid pump) manufactured by the school production shops of MIFI [Moscow Engineering and Physics Institute] is similarly constructed. It is capable of producing pressures up to 1000 atm. Depending on the operating pressure and diameter of the plunger, the output of this pump varies from 1 to 44 liters per hour.

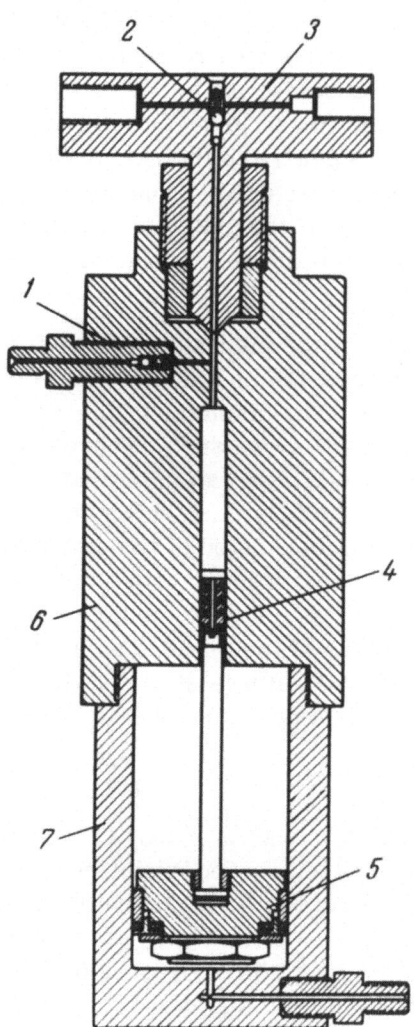

Fig. 52. Pressure booster with valves: 1) inlet valve; 2) delivery valve; 3) valve body; 4) small piston; 5) large piston; 6) high-pressure cylinder; 7) low-pressure cylinder.

gland 5 is used to pump an inert liquid, while the smaller diameter stage is used for pumping the corrosive liquid. Gland 6 prevents mixing of these liquids. Sealing of the gland is achieved by sleeve 7 plus ring 8. Conical valves in opposing pairs are used in the pump.

Connecting the chambers allows the liquids from the two chambers to be mixed in the required proportions or to be supplied separately. Also, the inert liquid can be circulated in a closed cycle so that only a very small quantity of its will be lost. Since gland 6 works with a pressure drop of practically zero, leakage of the corrosive liquid is excluded. The sealing of gland 5 is not difficult.

Hand Pumps. A hand pump (Fig. 48a, b) consists of cylinder 1 in which lever 2 moves piston 3. The pump shown in Fig. 48a has two ball valves 4. It is easy to use and simple to make. If the piston diameter is 1.2 cm and the arm length is 50 cm, this pump can be used to produce pressures of 600–700 atm. The pump shown in Fig. 48b can produce even higher pressures.

Sidorov Mechanical Press

The mechanical press (Fig. 49) is convenient in cases when pressure must be changed smoothly in a system by adding very small quantities of liquid to the system. Cylinder 1 contains piston 2 which is sealed at its tip by alternating leather and brass rings 3, compressed by nut 4. The piston is moved by nut 5 driven by thread 6 mounted in ball-bearing races 7. The screw is rotated by long handles 8 or by a chain drive from a motor. The productivity of the press depends on the diameter and travel of the piston, and may be varied over a wide range.

A deficiency of this design is that the reverse movement of the piston accomplishes no work. A device similar to this one but with a gland at the top of the cylinder can be used for introduction of mercury under pressure to high-pressure apparatus (see Chapter IX).

Pressure Intensifier

A pressure intensifier (booster), or hydraulic press, is a system consisting of two pistons in two cylinders. Figure 50 shows a Sidorov intensifier. Piston 1 has a considerably smaller diameter than piston 2. When a pressure is created beneath piston 2, the force applied to the piston is equal to the pressure times the area of the piston. The pressure over the upper piston will then be equal to this force divided by the area of the upper piston.

In other words, theoretically the pressure over the small piston is as many times greater than the pressure beneath the large piston as the area of the large piston is in the ratio to the area of the small piston. In practice, this ratio is not obtained since the pistons move in cylinder bores with friction.

In order to calibrate the intensifier, the values of pressures beneath the large piston and above the small piston are plotted against each other on a graph. The experimental line will not pass through the coordinate origin, but will be shifted by a certain value A (Fig. 51). Pressure P above the small piston is determined by the expression

$$P = pB + A$$

where p is the pressure beneath the large piston (low-pressure piston); B is the ratio of areas of large and small pistons; A is the pressure used in overcoming friction.

When the pistons are raised, the value A is negative; when they are lowered it is positive. Once the value of A is determined by experiment, the pressure in the apparatus can be measured approximately.

This formula is unsuitable for intensifiers with self-energizing seals, in which friction increases with increase of pressure.

At pressures on the order of 20,000 atm, the compressibility of fluids (not gases) reaches 30%, and it is difficult to reach the required pressure with one cycle of piston movements; certain additional methods must be used. One such method is to supply the pressure intensifier with two valves, thus converting it to a low-speed compressor. One such intensifier [27] is shown in Fig. 52.

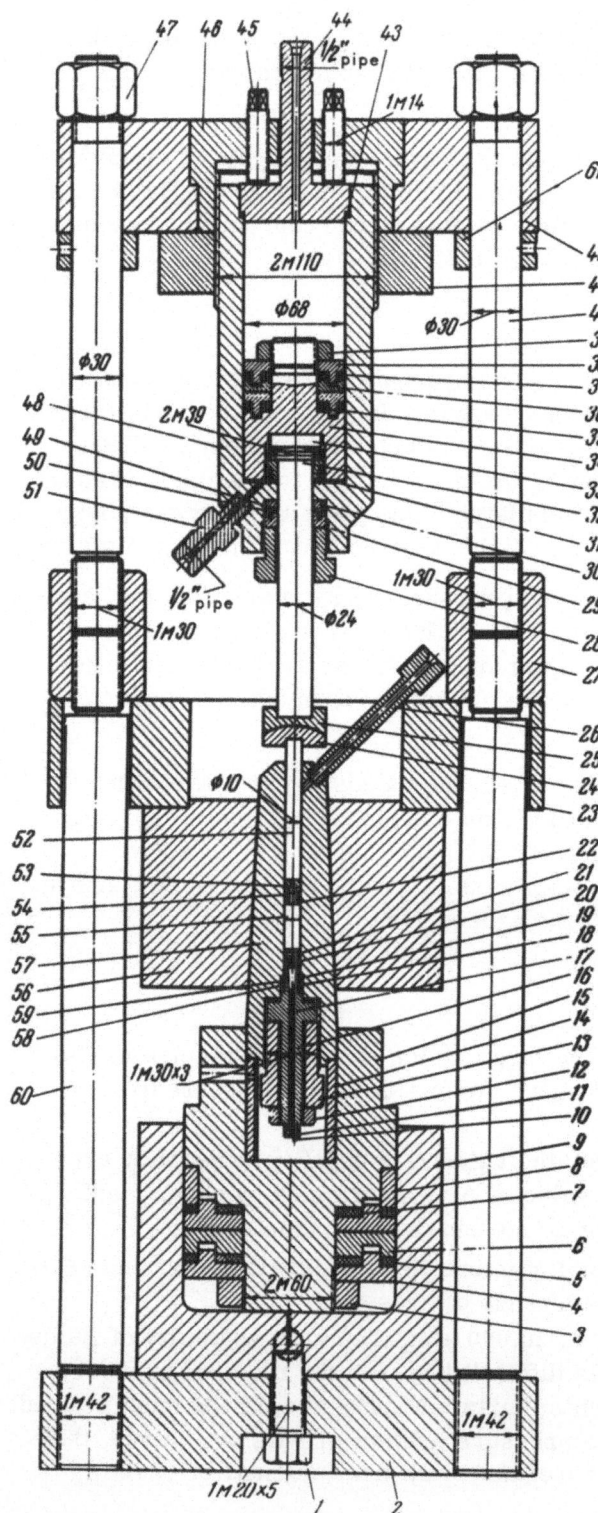

Fig. 53. Pressure booster with support: 1, 45) bolts; 2) bottom platen; 3, 11, 12, 27, 31, 39, 47) nuts; 4, 6) pressure rings; 5, 7, 22, 35, 37, 43, 49, 50, 58, 59) packing; 8) rings; 9, 48) cylinder; 10) screw; 13, 28) compression nuts; 14, 53) inserts; 15, 32, 34, 52) pistons; 16) plug; 17) electrodes; 18) insulating cone; 19, 29, 30, 36, 38) pressure rings; 20) coil; 21) shell; 23) middle platen; 24, 25) reinforcing rings; 26) tube; 33) plate; 40) upper tension member; 41) support; 42) upper platen; 44) head (plug); 46) top; 51) nipple; 54) conical [chamber] ring; 55) plug; 56) support; 57) conical vessel; 60) lower tension members; 61) support rings. The M notes are thread designations.

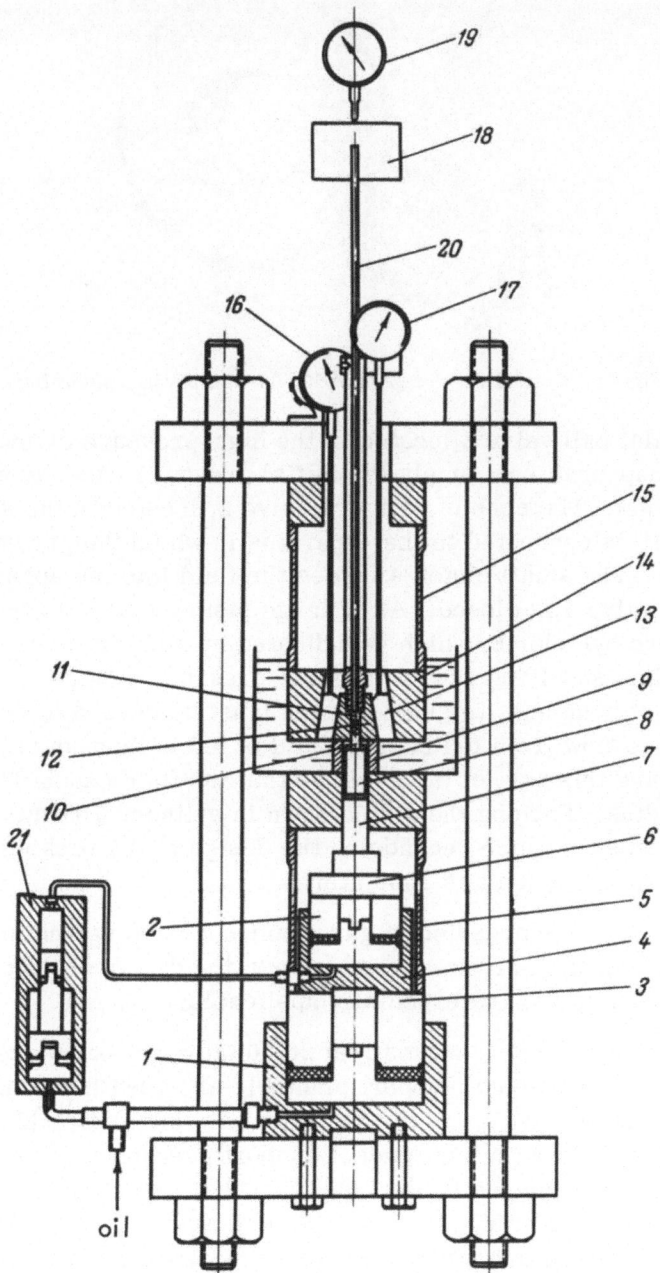

Fig. 54. Pressure intensifier with double support: 1, 2, 21) pressure intensifiers; 3, 6, 10, 11) pistons; 4, 15) cylinders; 5, 9) rings; 7) plug; 8, 18) blocks; 12) conical vessel; 13, 14) supports; 16, 17, 19) indicators; 20) rod; 21) control intensifier.

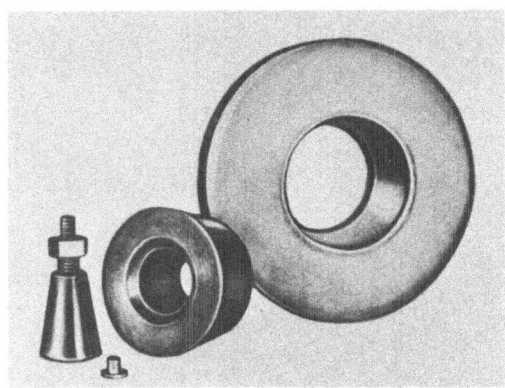

Fig. 55. Overall view of parts of 50,000-atm pressure intensifier.

The inlet ball valve 1 located in the high-pressure cylinder is connected to an apparatus to supply the initial pressure (such as with a mechanical press). The cone of delivery valve 2, located in the so-called
valve body 3* is connected to the apparatus in which the pressure must
be produced. The small piston 4 forces the fluid into the apparatus as it
rises, when valve 1 is closed. After large piston 5 rises to the level of the
high-pressure cylinder 6, which is indicated by a sharp increase in pressure in the line supplying fluid to the low-pressure cylinder 7 (indicating
termination of pumping), the large piston must be caused to drop. Allowing the fluid to flow from cylinder 7, fluid is fed in through valve 1. Valve
2 closes during this part of the cycle so that the fluid cannot flow back out
of the apparatus. Forcing the piston down in cylinder 7 requires a certain
amount of pressure which depends on the design of the packing in the small
piston and may be as high as 1000 atm.

With direct compression of gas, hand-operated valves may be used
in place of automatic valves. Press valves, the design of which is described in Chapter VI, are especially applicable.

For generation of pressures over 20,000 atm, pressure intensifiers constructed on the mechanical support principle are used (see Chapter II).
Such a pressure intensifier is the Vereshchagin intensifier [28]. Figure 53
shows an intensifier with independent support used for creation of pressures
up to 30,000 atm (designed by the author). The conical vessel 57, made of
heat treated type 45KhNMF steel (ultimate tensile strength $\sigma_b = 18,500$ kg
per cm^2) is placed in support 56, made of type 40Kh steel.

According to data in the literature [28, 29], the support can be made
of three steel discs. An aperture is drilled in each disc, after which, with

---

*The part in which the manganin manometer and pressure-reduction valve
  are mounted.

the disc mounted on a rod, the edge of the support is peened so that the internal surface is hardened by cold working. After this, the discs are heat treated to $R_c = 53$ and ground on the ends and internal surfaces with a specific tolerance. Then, each disc is hardened by pressing over a conical die of the same dimensions as the conical vessel, producing a residual strain of about 5%. The die is pressed into the support using a lubricant made of water, glycerine, and graphite. The pressure on the walls of the support reaches 15,000 atm, and the cone is displaced along its axis by 10 mm. After the hardening, the three rings are fastened together and the aperture is ground.

Our experience has shown that the manufacture of the support can be simplified if it is made of one piece of annealed and normalized metal. The external surface of the conical vessel and the internal surface of the support are ground on a machine tool and lapped in with emery paper and boron carbide powder. In processing these surfaces, it is important not only to produce polished surfaces, but also to avoid a rippled surface, which might cause the mount to break when the conical vessel shifts within it.*

The supporting force is created by piston 15 moving in cylinder 9. The piston seal is achieved by the unsupported-area principle. Insert 14 transmits the force of the piston to the conical vessel. It is made of 30KhGS steel and is hardened to a high hardness. The manganin manometer lead goes through this insert.

The pressure in the conical vessel is produced with the pressure intensifier consisting of cylinder 48 and piston 34. The cylinder is sealed by plug 44 which compresses copper spacer 43 under the pressure of bolts 45 threaded into cover 46. The body of the piston carries hardened plate 33, on which driving piston 32 bears. Piston 32 travels in a gland with self-energizing seal. Reverse travel of the booster piston can be produced by feeding oil through nipple 51.

Piston 32 is made of ShKh 12 steel, hardened to high hardness ($R_c = 62-65$) and polished. Breakage of the piston usually begins with light scratches around the edge. In order to increase the life of the piston, reinforcement rings 24 and 25, made of hard steel, are wrapped around its end. Working piston 52 is made of KhVG steel, which changes its form and volume very little during heat treatment.†

---

*The technology of making of the parts of the installation was borrowed to a great extent from L. F. Vereshchagin.

†It is most expedient to make the piston of tungsten carbide. Young's modulus and Poisson's ratio for this material are: $E = (6.44-6.94) \cdot 10^6$ kg/cm^2 and $\mu = 0.267-0.359$.

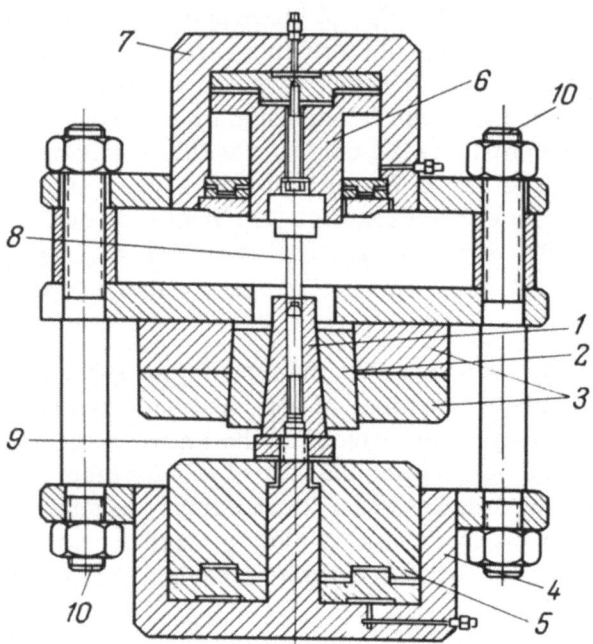

Fig. 56. Butuzov, Shakhovskii, and Gonikberg pressure intensifier
with double support: 1) conical vessel; 2, 3) supports; 4, 7)
cylinder; 5, 6, 8) pistons; 9) electrical lead; 10) tension members.

The piston is hardened, tempered to R_c = 62-65 and polished. The
piston diameter is made 0.1 mm smaller than the bore diameter. This is
necessary to prevent seizure of the piston in the bore of the conical vessel
when the piston presses the vessel into the support (see calculation in
Chapter II). Rings 24 and 25 have spherical surfaces,* which aid in the
self-centering of the piston. Piston 52 compresses the packing consisting
of plug 55, made of KhVG steel, against vinyl chloride packing members 22,
conical [chamfer] ring 54, and washers 53. The conical [chamfer] rings 54,
made of type 18 KhNVA steel, prevent packing 22 from entering the gap between the seal and the cylinder.

Tube 26 is used to produce the initial liquid pressure. After the
piston moves the packing past the aperture of the tube, the channel through
which the liquid is supplied is cut off, and the compression of the liquid
continues in the channel of the cylinder. With a maximum pressure of
30,000 atm at room temperature, only isopentane remains sufficiently fluid.
All other liquids are hardened [solidify or are made too viscous].†

---

*Advised by L. F. Vereshchagin.
†Editor's comment: Other liquids are also suitable, viz., certain gasolines.

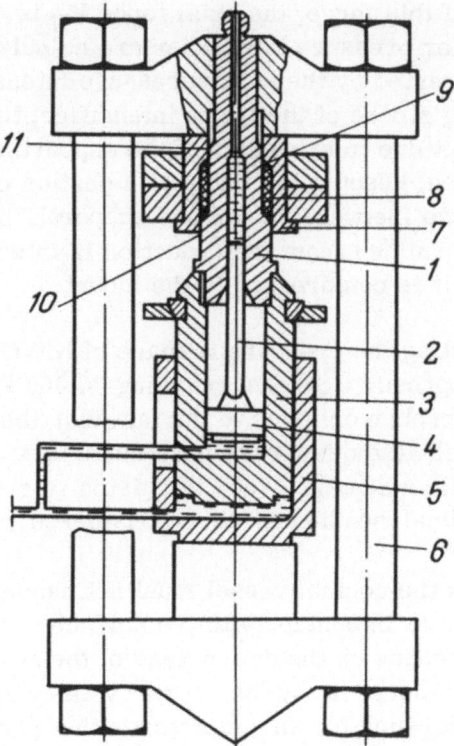

Fig. 57. Pressure intensifier with external support of cylinder by compression of packing: 1, 3, 5) cylinders; 2, 4) pistons; 6) tension members; 7) 2-layered cylinder; 8) rubber rings; 9) steel rings; 10, 11) shoulders.

The installation is assembled on three fastening platens, 2, 23, and 42, held with four tension members 40 and 60. In operating with fluids, the pressure is measured with a manganin [coil] manometer. In working with solid bodies, the manometer is replaced by a steel plug and the pressure is determined from the area ratio of the pistons.

Before the conical vessel is placed in the mount, it is wrapped in lead foil 0.1 mm thick, lubricated with a paste consisting of graphite and glycerine.

The method of designing vessels with mechanical support is presented in Chapter II. Design requires, based on properties of the material being used and the nominal dimensions of the conical vessel chosen, determination of the pressure to cause yield and selection of the proper support pressure.

The effective force is calculated from the dimensions of the conical vessel; on the basis of this force, the axial force P_H is determined. The dimensions of the lower pressure intensifier are calculated from the value of P_H and the force created by the upper pressure intensifier. In order to determine the working stroke of the lower intensifier, the shift of the conical vessel in the mount due to expansion of the support and compression of the vessel is calculated, keeping in mind that a portion of the lead foil will flow out through the gap [between vessel and support]. The diameter of the working piston is determined from the reduction in internal diameter of the conical vessel as it is compressed in the mount.

The piston creating the pressure is made of KhVG steel. It undergoes tremendous compressive stress, reaching 30,000 kg/cm^2 in this installation, and is not broken only due to the fact that the compressive strength of KhVG steel, like other very hard materials, is considerably greater than the tensile strength. When the piston is manufactured, the possibility of longitudinal bending must be considered.

The pressure in the conical vessel must not cause an excessive stress in the support where it is in contact with the surface of the conical vessel. In turn, this stress depends on the dimensions of the apparatus and the amount of working pressure. It can be shown by calculation (see Chapter II) that the limiting pressure for an apparatus with a piston diameter of 10 mm and a conical vessel with an [semivertical] angle of 2.5° will be 30,000 atm. At higher pressures, the support stresses exceed the yield strength and even the ultimate strength of the steel of the support and the conical vessel, which leads to rupture of these parts.* Therefore, so-called double support is used to create higher pressures, increasing the support pressure by stages. For this, a second stage of support must be created

---

* Editor's comment: The 30,000-atm "pressure barrier" for single-stage support equipment appears to be in a class with the "sound barrier" and the "thermal barrier" of aerodynamic terminology, both of which proved to be of smaller significances then originally posited. The editor and others employ a supported cone for routine use to 38,000 atm and for occasional use above 40,000 atm (at room temperature, with noncorrosive liquids). A barrelling of the chamber is observed (as occurs also with pressures of 30,000 atm and lower), however, presumably because restraint by unpressurized metal at the ends of the chamber strengthens the ends. This barrelling does not interfere with operation of the chamber in general. Similarly, 30,000 atm pressure with both pressure chamber and external support at 500°F does not appear to be harmful. Pinch-off is present too.

around the initial support by placing the first support mount in a second mount. Double support, in spite of its greater complexity in comparison with single support, makes a larger selection of cone angles possible.

An installation [30] constructed on the double-support principle is shown in Fig. 54. It consists of a primary conical vessel 12 placed in supports 13 and 14. The piston 10 (diameter 6.3 mm) which creates the high pressure is made of Carboloy and presses on Carboloy block 8, diameter 12.5 mm. The block presses on plug 7, made of hard steel. Piston 10 is pressed into the cylinder by piston 6.

The high-pressure vessel 12 moves into support 13, which in turn moves into support 14 under pressure from piston 3 moving in cylinder 1. The external surface of vessel 12 has a solid angle of 18°. The solid angle of the external surface of support 13 is 13°. The maximum pressure on the internal surface of the support 13 is equal to 26,300 atm, on the external surface, 11,400 atm.

Displacement of vessel 12 in support 13, and support 13 in support 14, are measured by the lever indicator 16 and 17. These data can be used to get an idea of the friction of the moving parts and to calculate the support pressure.

Displacement of piston 10 in vessel 12 is measured by special rod 20, made of hardened steel. The rod passes in through the plug closing the pressure vessel and contacts Carboloy piston 11, located above the sample being investigated. The displacement of the entire vessel relative to the piston is read on indicator 19.

Block 18 is made of bronze and designed for absorption of shock in case the vessel ruptures.

It is very important that the pressure in intensifiers 1 and 2 increase smoothly and in the proper relationship; otherwise the relationship between the support pressures will be disrupted. This is the job of pressure intensifier 21, which is connected to both working intensifiers and assures that the required pressure ratios are maintained as oil is fed to both intensifiers. The capacity of this intensifier must be sufficient to achieve the maximum pressure.

The pressure under piston 6 is measured by a weight manometer.

An overall view of the conical vessel, small piston, and supports is presented in Fig. 55.

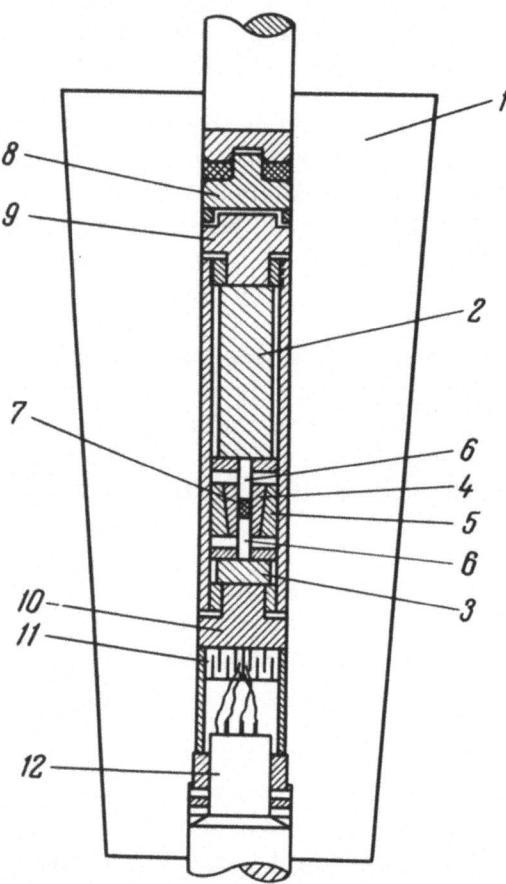

Fig. 58. 100,000-atm pressure intensifier: 1) conical vessel;
2, 3) blocks; 4) small conical vessel; 5) support; 6) pistons;
7) [specimen] material; 8) plug; 9, 10) plates; 11) compressometer; 12) plug with electrical lead inputs.

A double-support pressure intensifier of a considerably more convenient design can be used to create pressures up to 40,000 atm; it was
described by V. P. Butuzov, G. P. Shakhovskii, and M. G. Gonikberg [31]
(Fig. 56). Conical vessel 1 is located in support 2 which, in turn, is located in supports 3. When pressure is created in cylinder 4, piston 5 forces
the conical vessel into support 2, and support 2 into support 3. As this occurs, a pressure on the order of 14,000 atm is created on the surface of
the vessel, while a pressure on the order of 12,000 atm is created at the
surface of support 2. Upper piston 6, moving in cylinder 7, is used to
generate pressure in vessel 1. This piston presses on piston 8, 16 mm in
diameter. Piston 8 rides on a sealing plug which serves to generate pres-

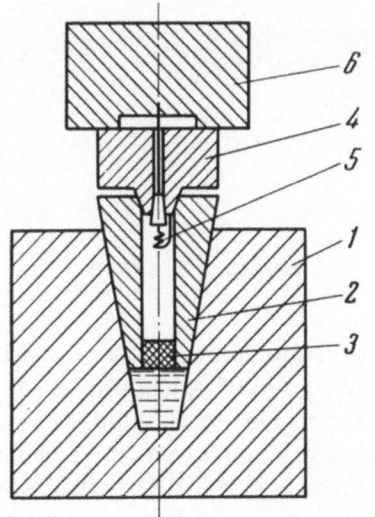

Fig. 59. Compression by conical piston:
1) support; 2) cone; 3) lead plug; 4)
sealing plug; 5) manometer; 6) plate.

sure in the cavity of the conical vessel.
The pressure is measured by a manganin manometer, the end of whose
coil protrudes through electrical lead
9. The cylinders of the pressure intensifier are held down by 8 tension
members 10.

Figure 57 shows a pressure intensifier designed by M. K. Zhokhovskii [26]; the mechanical support in
this booster is replaced by hydraulic
support. High-pressure cylinder 1
and piston 2 and low-pressure cylinder
3 and piston 4 form an ordinary pressure intensifier. Cylinder 3 is also
the piston for hydraulic-press cylinder
5, which creates the support pressure.
The pressure at the lower side of the
intensifier and in the hydraulic press
is created by one pump. The bottom
of cylinder 5 bears, through a cushion,
on the lower plate of the frame. Cylinder 7 and its support fit on the
shoulders of the external surface of cylinder 1. Packing insert 8 made of
rubber rings and two anti-extrusion rings 9, which exclude leakage of the
rubber through the gaps, are located in the cavities between shoulders 10
and 11.

As pressure is generated at the lower end of the intensifier and in
the hydraulic press, the external support pressure increases simultaneously. If friction is ignored, then

$$p = p_1 \cdot \frac{S_1}{S_2}$$

where p is the pressure in cylinder 1; p_1 is the pressure in the low-pressure cylinder 3; S_1 is the area of piston 4; S_2 is the area of piston 2.

In turn,

$$p_2 = p_1 \cdot \frac{S_3}{S_4}$$

where p_2 is the pressure in packing 8; S_3 is the area of cylinder 5; S_4 is
the area of packing ring 8.

Then,

$$p = kp_2$$

where

$$k = \frac{S_1 S_4}{S_2 S_3}$$

The stress in the rubber approximates hydrostatic pressure [distribution].

In order to achieve even higher pressures, the conical vessel and its mount can be submerged in a fluid compressed to 30,000 atm [32]. An installation for compression [pressurization] of material to 100,000 atm constructed on this principle is presented in Fig. 58. Conical vessel 1 of an ordinary 30,000 atm installation with independent support, filled with isopentane, contains a casing in which a small conical vessel 4 with support 5 and piston 6 is compressed between blocks 2 and 3 by two nuts. The material being investigated 7 is placed within the small conical vessel.

The quantity of liquid in vessel 1 is calculated such that as it is compressed plug 8 contacts plate 9 at the moment when the pressure in the liquid reaches 25,000-30,000 atm. Then, further movement of the plug causes motion of plate 9 which transmits the force of piston 6 by pressing on block 2. The piston, moving into cone 4, compresses material 7 and simultaneously moves the cone into mount 5. Thus, reinforcement of materials by surrounding external pressure is used in this design.

The pressure in the apparatus is measued by compressometer 11, the design of which is described in Chapter IV.

L. F. Vereshchagin, A. I. Likhter, and V. I. Ivanov [33] used a conical vessel as a piston to create high pressures. Steel support 1 (Fig. 59) contains conical vessel 2 with an aperture straight through, which is soldered to lead plug 3. Glycerine is poured into the bottom of support 1. The cavity of the conical vessel contains a compressible fluid. The vessel is closed with plug 4 containing the electric leads and manganin manometer 5. Part 6 transmits the pressure from the hydraulic press. As cone 2 is pressed into support 1, it is compressed and moves down into the support. The volume of the internal cavity is thus decreased and the pressure within the cavity increases.

Anvil Apparatus

The usage of the principle of replacement of tensile stress by compressive stress (see Chapter II) has allowed the creation of apparatus cap-

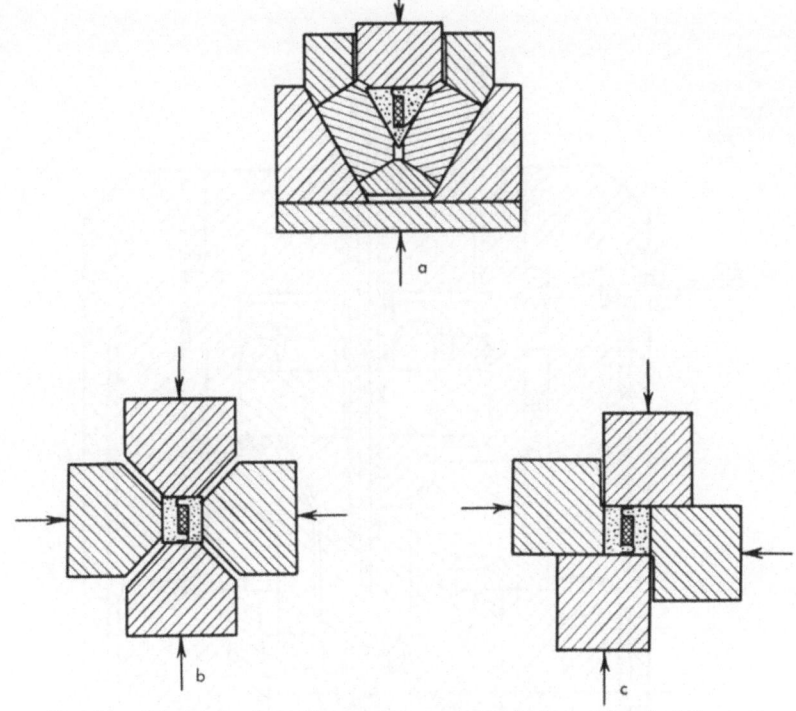

Fig. 60. Diagrams of anvil apparatus: a) tetrahedral; b) and c) 6-anvil [cubic, if of equal sides].

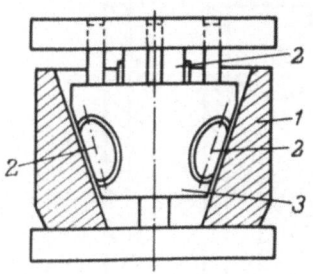

Fig. 61. Tetrahedral installation with a single source of power: 1) support; 2) anvils; 3) guiding frame.

able of producing pressures on the order of 100,000 atm and higher with considerable volumes of sample material. Figure 60 shows diagrams of anvil apparatus of designs improved over the design of the Hall tetrahedral press described in Chapter II.

In the apparatus constructed according to plan a, the force creating the pressure is applied to one piston only.* This eliminates one of the primary deficiences of the Hall installation, in which the forces of four pressure intensifiers must be equalized. A pressure intensifier of this design [34, 35] with one source of power is presented in Fig. 61. The installation consists of conical sup-

_____

* Editor's comment: In the USA, this scheme is termed the NBS (National Bureau of Standards) technique after reference [34].

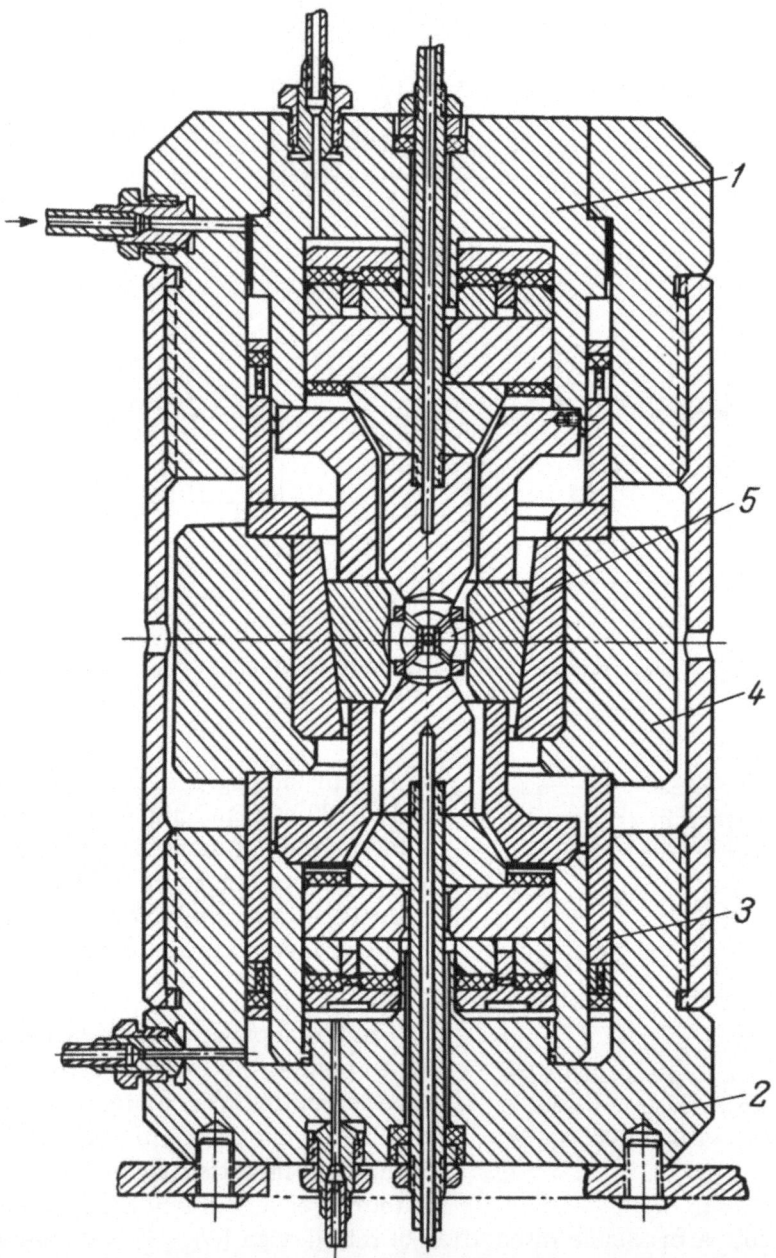

Fig. 62. Cubic clamp intensifiers: 1) upper press; 2) lower press; 3) ring press;
4) wedge mount; 5) high-pressure chamber with 6 anvils.

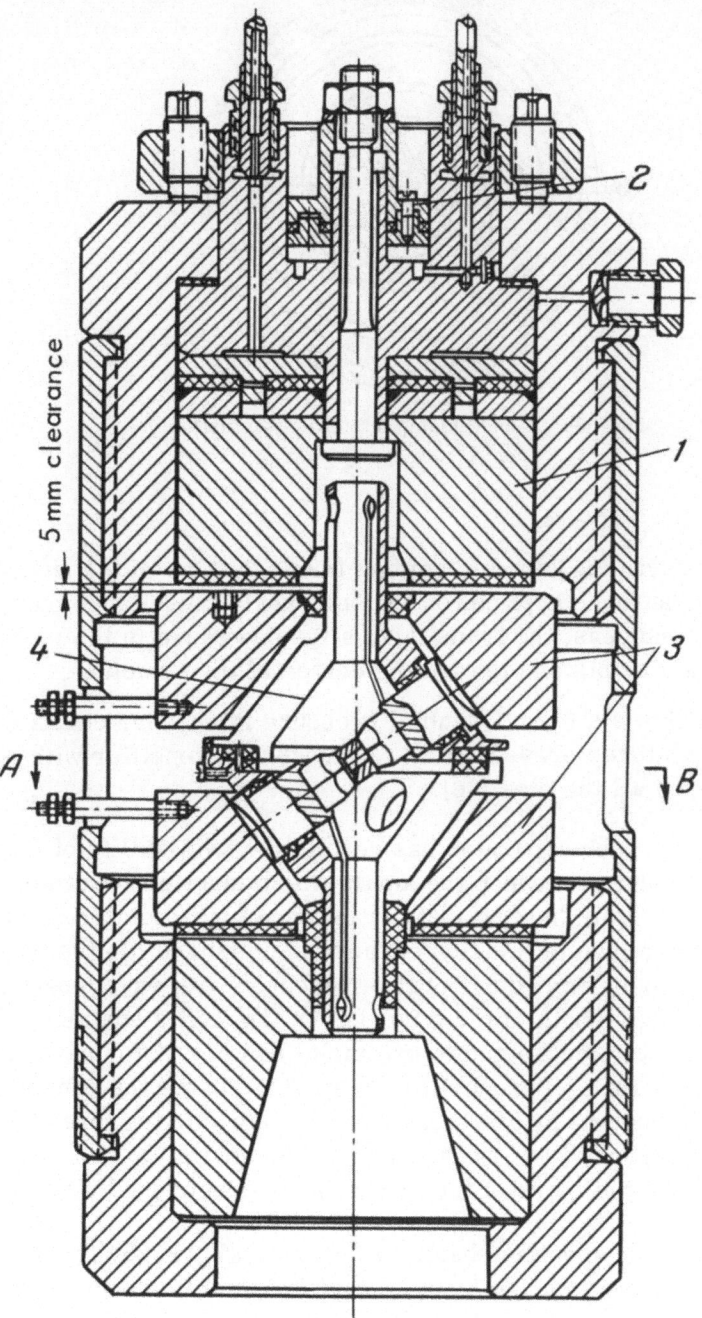

5 mm clearance

Fig. 63a. 6-anvil clamp intensifier with a single source of power: 1) press; 2) reverse travel [retraction] press; 3) wedge supports; 4) anvil frame. AB Section on following page.

Section through AB

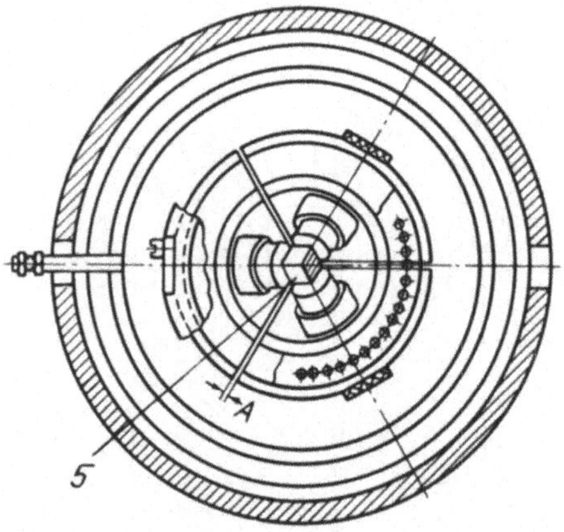

Fig. 63b.   6-anvil clamp intensifier (section)
compressing sample.

port 1 plus four anvils 2, joined in guiding frame 3. One of the anvils is located above. The force applied to the upper anvil causes it to move downward. When this happens, the frame also slips downward in the mount, moving the remaining pistons toward the center of the structure.

Figure 62 shows a 6-anvil [cubic] pressure intensifier constructed in cubic form with three presses; Fig. 63 shows an intensifier with one press which moves all the dies [36].

The design of such apparatus has resulted in availability of a comparatively small volume between the anvils which must contain the pyrophyllite packing, sample, heating installation, and measuring transducers. Figure 64 shows the plan of location of these elements within the tetrahedron. Here, pyrophyllite equilateral tetrahedron with edge length 25% greater than the edge length of the triangles on the anvils contains metal tube 1 with the sample, connected to two metal contacts 2 for passage of current through the tube. Pyrophyllite prism 3 insulates the contact. The temperature is measured by thermocouples 4. In order to increase friction between the pyrophyllite tetrahedron and the anvils (so as to prevent leakage [extrusion]) the faces of the tetrahedron are covered with ocher.

Observations of specimen harm in this type of pyrophyllite tetrahedron [37] have shown that the least amount of distortion occurs when the sample is placed oriented from one side to an angle peak. The generation of a pressure gradient in the pyrophyllite can vary the results of measurements from experiment to experiment by 40%.

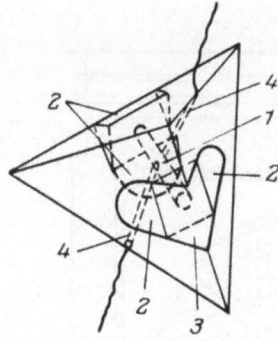

Fig. 64. Arrangement of sample and thermocouples within tetrahedron: 1) sample tube; 2) contacts; 3) insulation; 4) thermocouples.

Wedge Apparatus

Figure 65 shows the diagram of an apparatus [38] constructed on the principle of the wedge vessel. The high-pressure cylinder consists of four carefully fitted wedges 1 with spherical [sic] surfaces, placed in steel band 2. The apparatus is closed with a plug 3 from below, and a cylindrical steel piston 4 is placed above in the bore formed by the wedges. The pressure on this piston is created by conical piston 5. The bore contains pyrophyllite cylinder 6 in turn containing the sample being tested 7. As the piston moves, the pressure in the bore rises sharply and reaches a maximum at the end of the piston stroke, when the piston cone closes the wedge-shaped vessel. A compact unit is created which is capable of withstanding a pressure over 50,000 atm at a temperature up to 1500°C.

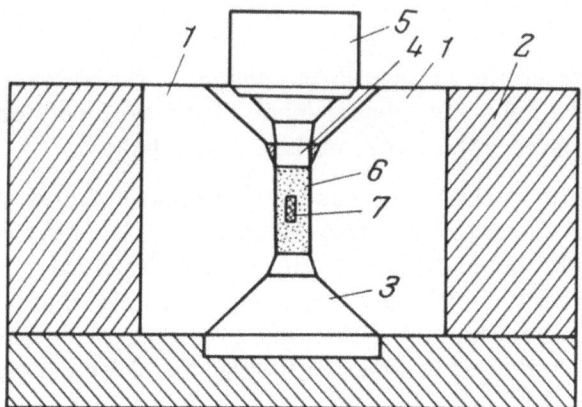

Fig. 65. Wedge pressure intensifier: 1) wedges; 2) ring; 3) plug; 4) cylindrical piston; 5) conical piston; 6) pyrophyllite cylinder; 7) sample.

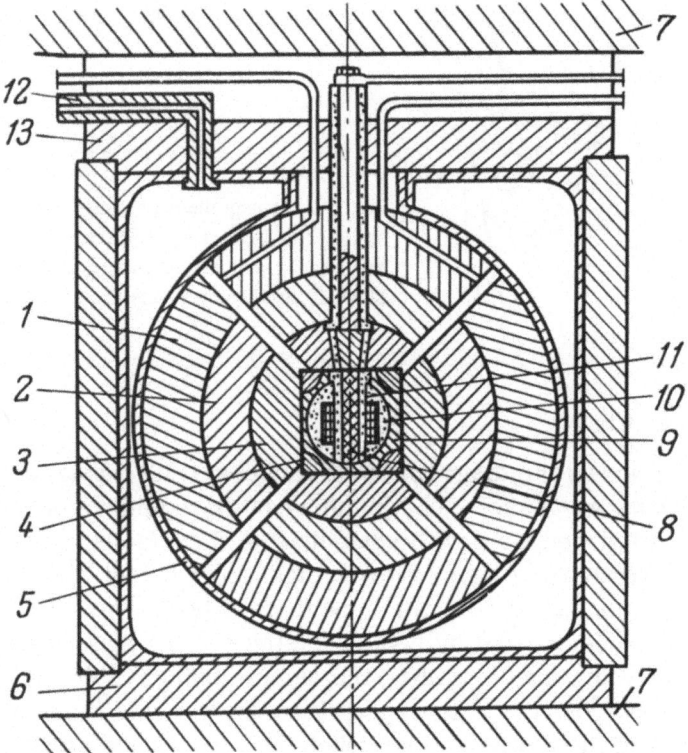

Fig. 66a. Spherical vessel *a*, longitudinal section. 1) Top of segment; 2) middle of segment; 3) base of segment; 4) hard steel rib; 5) spherical shell; 6, 13) pistons; 7) platens; 8) graphite resistor; 9) cubic soft copper shell; 10) thermal insulation; 11) electric insulating shell; 12) tube.

This same principle is used in the design of the apparatus shown in Fig. 66 [39]. It is a spherical vessel made of six segments each of which consists of three parts — the top 1, middle 2, and base 3.

The internal portions of the bases have square planes which form a cubic volume when the apparatus is assembled. This cavity contains cubic shell 9 made of soft copper, which itself contains a spherical volume that holds a graphite resistance 8 which is used to heat the sample, electric insulation shell 11, and the material being investigated compressed into a tube. This tube is placed in thermal insulation 10 (soapstone) [or lava].

The spherical vessel is located in spherical shell 5, 52 cm in diameter, placed in a tube covered by pistons 13 and 6. Water is forced into the tube through tubing 12 under a pressure of up to 6000 atm, causing the segments to move together and compress the cubic cavity. It can be seen from

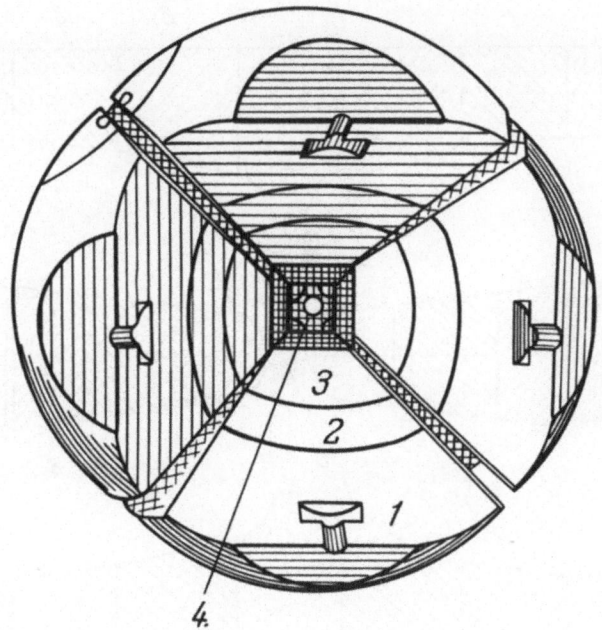

Fig. 66b. Spherical vessel, view from above: 1) Top of segment; 2) middle of segment; 3) base of segment; 4) cubic soft
copper shell.

Fig. 66 that the areas of the bases of the segments contacting the copper
cube are less than the outer projected areas of the segments. Due to this,
the pressure created within the cubic volume is greater than the pressure
of the water by a factor equal to the number of times that the area of the
projection of the surface exceeds the area of a side of the cubic container.

As the pressure is increased within the copper shell, the shell flows
into the clearances between parts 2 and 3. In order to prevent puncturing,
ribs 4 are made of hard steel. With a pressure in the tube of about 6000
atm, pistons 13 and 6 withstand a force of 13,000 tons. This force is applied to a shell consisting of plates 7 placed in steel support and wrapped
with stressed steel bands (not shown on the figure). The steel bands form
the body of the apparatus, similar to the body of a wound vessel. Synthesis
of diamonds has been performed in Sweden in this type of apparatus.*

---

* A wedge vessel was also used in this work [40].

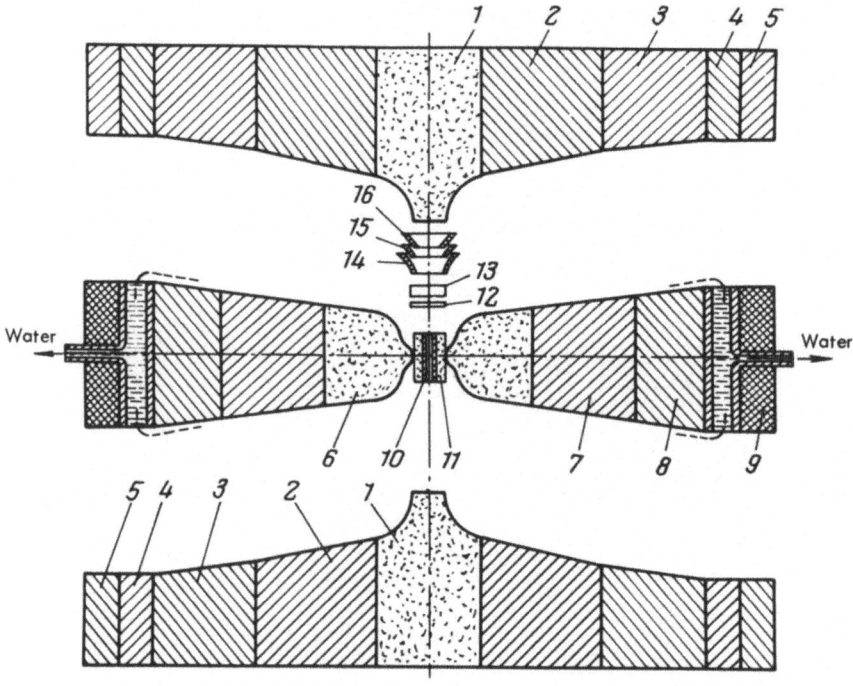

Fig. 67. Belt apparatus (apparatus with conical dies): 1) Carboloy pistons; 2, 3, 7, 8) belts; 4) steel ring; 5) copper ring; 6) Carboloy chamber; 9) rubber belt; 10) tube; 11) pyrophyllite shell; 12) nickel packing; 13) steel packing; 14, 16) pyrophyllite cones; 15) steel cone.

Belt Apparatus

The highest pressure (on the order of 200,000 atm) has been produced in so-called belt apparatus.†

A belt apparatus (or apparatus with conical anvils) is shown in Fig. 67. It consists of three parts – two pistons and a high-pressure chamber in the form of a ring. The pistons consist of Carboloy anvils 1, of the most suitable form (conical), reinforced by belts 2 and 3 made of hardened self-loaded (press-fitted) steel. These rings are surrounded by rings 4 and 5 made of soft steel and copper, which protect the surrounding area from flying fragments in case the inner rings burst.

---

†These apparatus were created almost simultaneously in the USSR [40] and the USA [41]. They can be used to produce this order of pressure with simultaneous heating of the material being investigated to 2000-4000°C.

Chamber 6 is a Carboloy ring, also reinforced by rings 7 and 8. If the experiments are performed at high temperature, the chamber can be cooled with water. The water jacket is surrounded with rubber cushioning ring 9.

The material being investigated is placed in tube 10 with pyrophyllite shell 11, which is placed in Carboloy chamber 6. Then, current-conducting nickel packing 12 and steel packing 13 are placed on the pyrophyllite, and a layered packing consisting of pyrophyllite 14 and 16 and steel 15 conical rings is applied to the piston. The apparatus is then placed between the platens of a powerful hydraulic press, which is used to compress the entire assembly. By passing a current through the pistons, which are insulated from the body by packing 14 and 16, tube 10 and the sample being investigated can be heated to several thousand degrees.

Figure 68 shows the pyrophyllite container [42] in which the material being investigated and the thermocouple are inserted; these containers are used for studying phase transformations, and measurement of the dependence of electrical resistance and melting point of material on pressure.

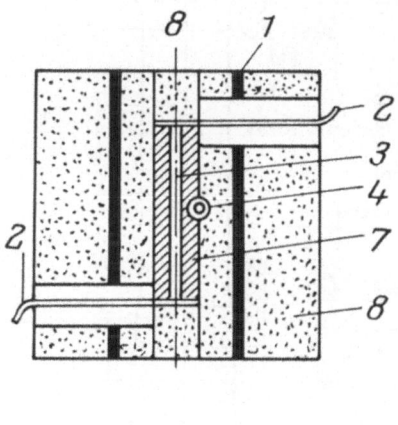

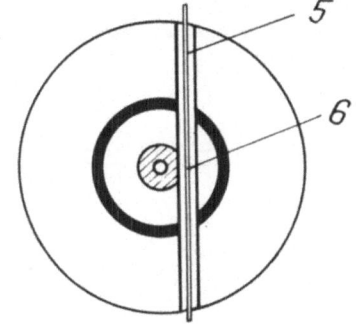

Fig. 68. Container: 1) graphite furnace; 2) stainless steel wire 2.5 mm in diameter; 3) material being investigated; 4) insulated thermocouple lead; 5) thermocouple; 6) solder; 7) silver chloride; 8) insulation.

At the present time, investigations are performed in belt apparatus at pressures up to 200,000 atm and temperatures up to 5000°C [43].

Small Hydraulic Presses

The usage of belt and anvil apparatus, in which the sample being investigated may be 10 to 12 mm in diameter, requires the construction of powerful intensifiers capable of creating forces on the order of several thousand tons. It is rather difficult to manufacture such intensifiers. Industrial hydraulic presses cannot be used for these purposes, since they

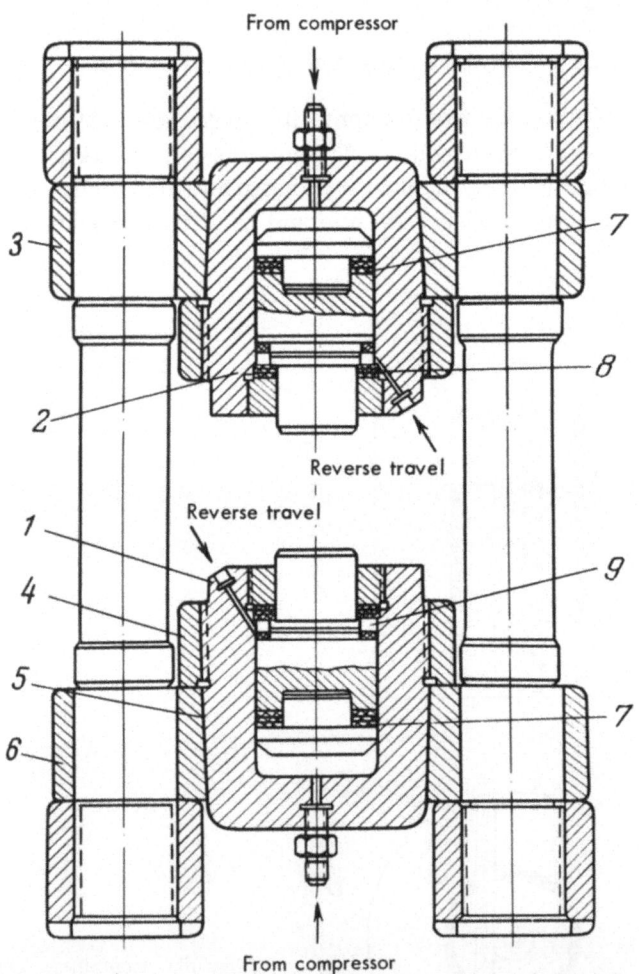

Fig. 69. Small laboratory hydraulic press: 1, 2) cylinders; 3, 6) platens; 4) nut; 5) conical surface; 7, 8) packing; 9) reverse travel cavity.

cannot long maintain the forces involved, are complex in design, are expensive, and are extremely large and heavy.*

Figure 69 shows a small press [44], in which the force required is achieved by increasing the operating pressure of the fluid in the cylinders of a press. The press has a lower cylinder 1 and an upper cylinder 2 with conical external surfaces 5, and mechanical support platens 3 and 6. Nut 4 transmits the load from the press to the platens. The pistons of the cylinders are packed with vinyl chloride and textolite rings 7, operating on the unsupported-area principle. Reverse travel of the piston is produced by forcing fluid into cavity 9. The platens of the press are connected by four columns.

When the pressure in the cylinders is 5000 atm, the press can create a force of up to 1000 tons, corresponding to a pressure in anvil or belt apparatus (with a piston diameter of 10 mm) equal to 700,000 atm.

GENERATION OF HIGH PRESSURE WITH SIMULTANEOUS APPLICATION OF SHEAR FORCE

The method of compression of material in a thin layer (see Chapter II) is highly effective. Bridgman [45] discovered that the combination of high pressures generated in such apparatus and the application of shear forces aids in the progress of certain physical and chemical conversions of material or mixtures of materials. These conversions can be studied by means of the behavior of curves of dependence of shear force on applied pressure.

L. F. Vereshchagin, E. V. Zubova, and V. A. Shapochkin [46, 47] improved this method and created an installation in which shear force can be applied at pressures up to 500,000 atm. The installation (Fig. 70) consists of four anvils 1, made of a cerametallic alloy, pressed into belts 2. The material being investigated is located between two anvils. The anvils are placed in a hydraulic press and compressed. Then, block 3 is rotated approximately 60°, thus applying shear force to the compressed material.

The apparatus for compression of material in a thin layer is shown in Fig. 71. The apparatus is located between base plate 1 and piston 2 of

---

*Editor's comment: There are now available, from several sources, simple presses with low ratios of weight-to-tonnage capacity, intended specifically for high-pressure usage. Such presses are designed to have small bed areas and small daylight, and ordinarily are fed by hydraulic fluid at relatively high pressures, much as described here. Presses with capacities up to 10,000 tons are in use.

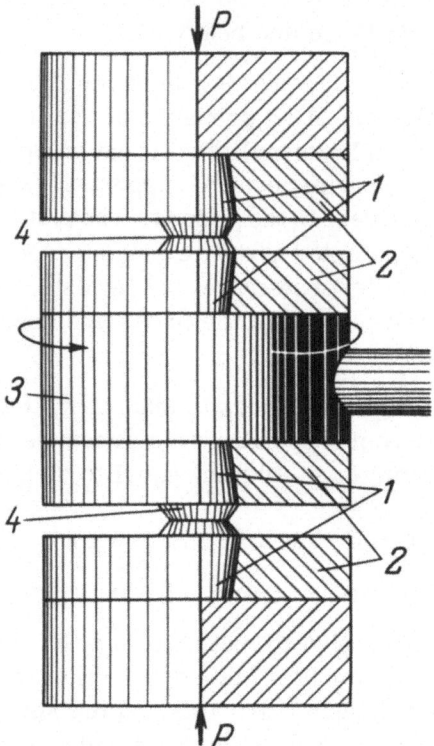

Fig. 70. Installation for simultaneous creation of
high pressures and shear: 1) anvils; 2) belts; 3)
block; 4) material being investigated.

a hydraulic press which bears on the apparatus through thrust bearing 3.
Pobedit conical anvils 4 are enclosed in supporting belts 5. The entire system is enclosed in aligning vessel 6 which contains a slot for lever 7. By
applying a force to this lever (measured by a dynamometer), mobile shaft 8
is moved, also rotating belts 5 with the anvils. If we know the force required for rotation of shaft 8, we can calculate the shear force and establish the dependence of this force on the pressure being applied to the material being investigated.

OTHER METHODS OF PRODUCING HIGH PRESSURES

High pressures can be created using the property that materials
change their volume upon melting.

V. G. Lazarev and L. S. Kan created a pressure of 1700 atm by freezing water in a closed space [48]. According to the calculations of Kan,
ethyl alcohol frozen at normal pressure and heated to 20°C in a closed
vessel would create a pressure of 6000 atm. If the alcohol is frozen with

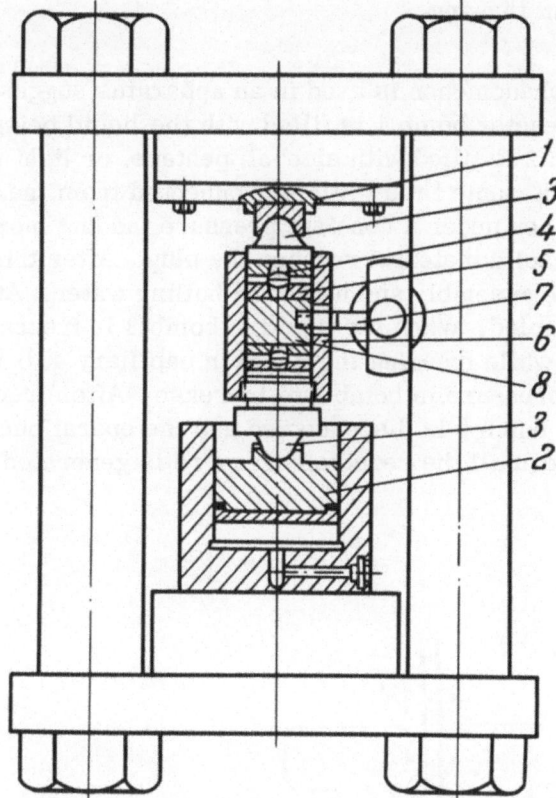

Fig. 71. Apparatus for compression of material in a thin layer: 1) base plate; 2) piston; 3) thrust bearing; 4) anvils; 5) belts; 6) housing; 7) lever; 8) roll.

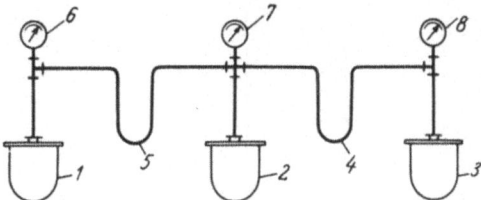

Fig. 72. Plan of Kan installation for compression of fluids by freezing: 1, 2, 3) bombs; 4, 5) capillary tubes; 6, 7, 8) manometers.

an initial pressure of 5000 atm, a pressure of 10,000 atm can be produced
by subsequent thawing.

This phenomenon is used in an apparatus suggested by L. S. Kan [49]
(Fig. 72). Reactor bomb 1 is filled with the liquid being investigated; generator bomb 2 is filled with alcohol, pentane, or light gasoline. Bomb 3 is
filled with the same fluid while disconnected from the apparatus and the
liquid is frozen under a constant pressure, adding more liquid until the
level of the solid material reaches the plug. After this, bomb 3 is connected to the assembly and heated in boiling water. At the same time,
bomb 2 is cooled. When the liquid in bomb 2 is frozen, bomb 3 is disconnected and, while freezing the liquid in capillary 4, bomb 2 is heated. This
causes the pressure in bomb 1 to increase. After freezing the liquid in
capillary 5, bomb 2 is disconnected and the operations with bombs 2 and 3
are repeated until the required pressure is generated in bomb 1.

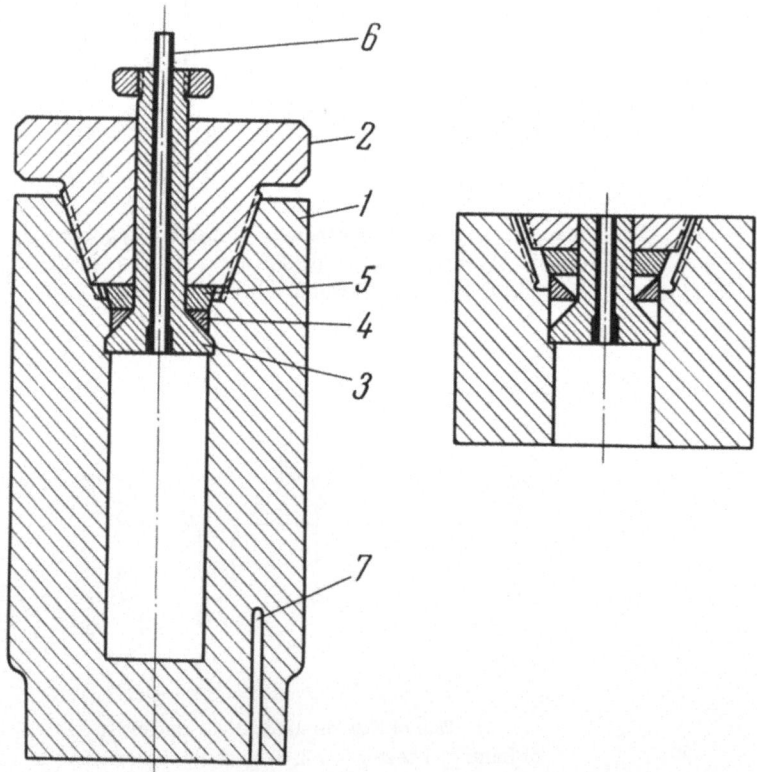

Fig. 73. High-pressure vessel with conical thread: 1) body; 2) nut; 3) head; 4)
packing; 5) pressure ring; 6) capillary; 7) thermocouple well.

E. A. Itskevich, et al., used the same method to create the pressure in the low-pressure cylinder of an intensifier. For this purpose, water is pumped in beneath the low-pressure cylinder and frozen. The expansion of the water causes the piston to move. The small piston can be used to compress a material cooled to the temperature of liquid helium. The loss of heat through the piston is considerably reduced.

N. B. Brandt and A. K. Tomashchik [50] discovered that samples frozen in water under pressure are subjected to heterogeneous [nonhydrostatic] stresses; therefore they used a mixture of water with ethyl alcohol to create high pressures. It was shown that the ductility of the solid phase of such solutions is considerably greater than the ductility of ice, and increases with increasing concentration of alcohol in the solution.*

HEATING AT HIGH PRESSURES

A material compressed to high pressures can be heated by two methods: 1) the entire high-pressure apparatus can be heated in a furnace to the required temperature (external heating); 2) the high temperature can be created within the apparatus (internal heating).

External Heating. With external heating, the temperatures of the parts of the apparatus are identical, so that the pressure which the apparatus can withstand is determined by the strength of the steel of which it is made, at the temperature to which it is heated.

At high pressures and temperatures, packing and threads operate under very difficult conditions; therefore, the high-pressure apparatus must be made of special steels and special designs must be used.

Figure 73 shows an apparatus [52] for investigation of the compressibility of water at temperatures up to 1000°C and pressures up to 2550 atm, made of a steel whose composition is presented on page 12. This steel has high tensile strength at a temperature of 1000°C and withstands corrosion well.

The body [bore] 1 of the bomb is covered [sealed] by head 3 with copper packing 4 and pressure ring 5. Parts 3, 4, and 5 amount to an unsupported-area plug. The packing is compressed by nut 2. Experience has shown that at high temperatures and pressures, it is the threads on the nut that fail most frequently. In this design a tapered (conical) thread with large pitch [few threads per unit length] is used, allowing the nut to be rapidly removed even if it sticks [seizes].

---

*A mixture of water and methyl alcohol can also be used for these investigations [51].

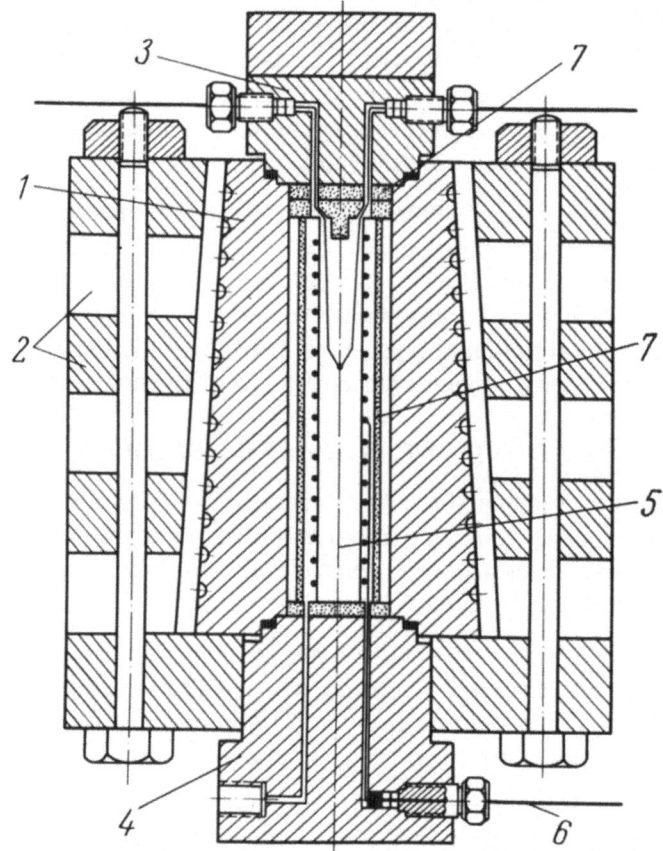

Fig. 74. High-pressure apparatus with internal furnace (for investigations at pressures up to 12,000 atm and temperatures up to 1200°C):
1) conical vessel; 2) support; 3, 4) covers [heads]; 5) furnace; 6)
electric lead; 7) screens.

Connecting capillary 6, made of stainless steel, is threaded into head
3. In order to assure proper sealing at the thread, the head and capillary
are heated red-hot and the thread is expanded by placing a mandrel in the
capillary. After this, the thread will not leak.

With external heating, there is no temperature gradient in the walls
of the high-pressure apparatus and no thermal stress which, as we have
seen in Chapter II, may be considerable in other types of apparatus. The
absence of thermal stresses and the relative simplicity of design of the
furnaces required are advantages of this method of heating.

<u>Internal Heating</u>. With internal heating, the heat source is placed inside the vessel. The wall temperature within the vessel may be considerably reduced by good thermal insulation (refractory and thermal-insulating [radiation] screens, as well as artificial cooling of apparatus walls). However, considerable thermal stresses may arise in the walls of the apparatus.

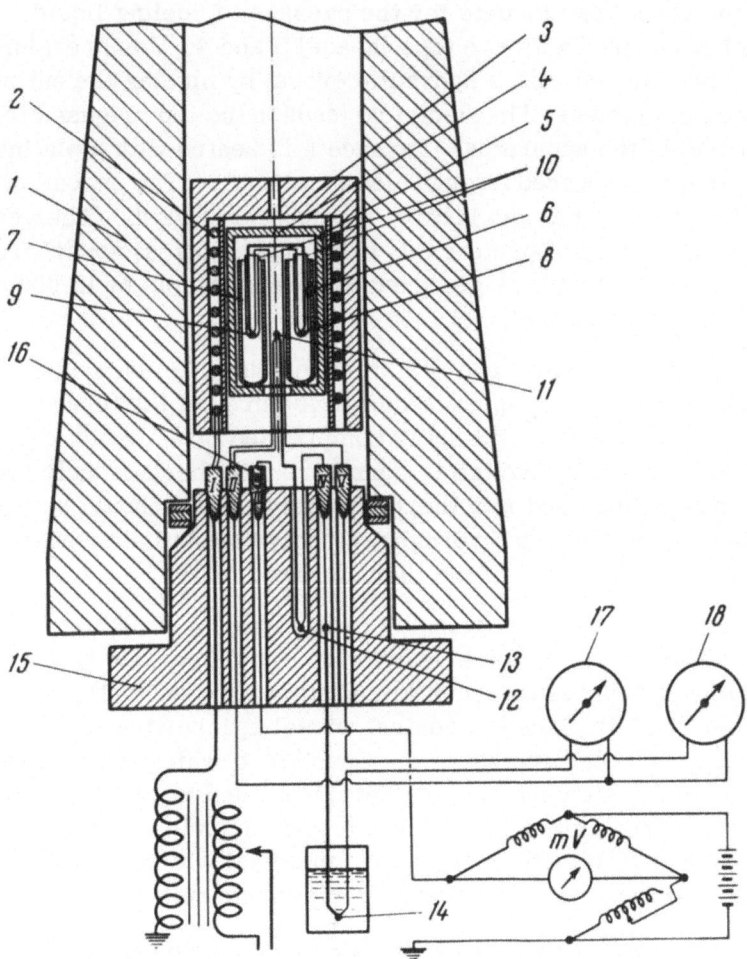

Fig. 75. Installation for investigation of polymorphic transformations: 1) high-pressure vessel; 2) heater; 3) thermal insulation; 4) crucible [enclosure]; 5) pack-ing; 6, 7) porcelain ampules; 8, 9, 11, 12, 13, 14) thermocouple junctions; 10) quartz capillary tubes; 15) plug; 16) manganin manometer; 17, 18) galvanometers.

Heat sources may be explosive materials [53], thermite [54], etc. The most widely used method of internal heating is the passage of an electric current through a heating element or the striking of an electric arc [55].

The electric leads into this type of furnace must be designed to pass a high-power current. Various designs of such apparatus used for mineralogical investigations at high pressures and temperatures are described in the literature [56].

Figure 74 shows the design of an apparatus with internal heating [57]. Conical vessel 1 is surrounded by support mounts 2. There is a channel between the vessel and mounts for the passage of cooling liquid. The conical vessel is covered with two tops [heads] 3 and 4. Compression [chiefly restraint] is achieved with a hydraulic press, by placing the entire apparatus between the platens. The electrical leads in the top are used to place a thermocouple in the apparatus. Furnace 5 is heated with a platinum-rhodium winding, powered through electrical lead 6. The second conductor for the electric power is the body of the apparatus itself. The vessel contains screens 7, used to protect the walls of the conical vessel from heat radiation. The apparatus is designed for pressures up to 12,000 atm and temperatures up to 1200°C.

Sometimes, graphite rod resistors are used in place of the electrical winding for heating. The specific design features of the furnace are determined by the type of material being compressed (gas, liquid, or solid). If the material being investigated is corrosive, the thermocouple placed in the apparatus is enclosed in a thermocouple well. It should be recalled that certain gases, such as hydrogen, dissolve in platinum under pressure and change the indications of the thermocouple.

Figure 75 shows the apparatus designed by V. P. Butuzov, S. S. Boksha, and M. G. Gonikberg [58] for investigation of polymorphic transformations at extremely high pressures and temperatures above 1000°C. Electrical heating element 2 is placed in conical vessel 1, insulated from the walls by the thermal-insulating enclosure 3. Crucible 4, which is insulated from the electrical heating element by mica covering 5 and has a hole in its bottom, contains two porcelain ampules 6 and 7. One ampule contains the material being investigated, the other, any other material which does not undergo polymorphic transformation under the [given] experimental conditions.

The differential thermocouple, whose junctions 8 and 9 are placed in the investigated and standard material, can fairly precisely fix the temperature situation during the polymorphic transformations, which may be accompanied by absorption or production of heat. The junctions of the differential thermocouple are protected from corrosive materials by quartz capillaries 10.

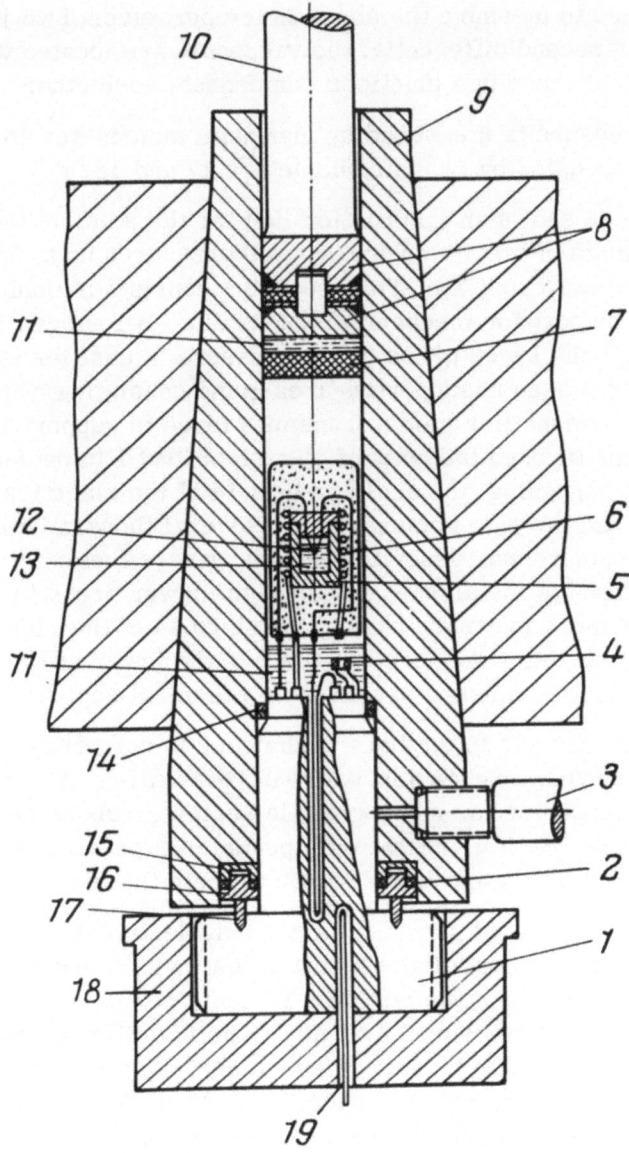

Fig. 76. Assembly for studying melting points of metals: 1) electrical inlet lead; 2) plastic seal; 3) nipple; 4) manganin manometer; 5) crucible; 6) electric furnace; 7) rubber plug; 8) piston seal; 9) conical vessel; 10) piston; 11) liquid; 12, 19) thermocouples; 13) material being studied; 14) seal around electrode carrier; 15, 16) steel ring; 17) compression ring; 18) base.

A second differential thermocouple, with four junctions, 11, 12, 13, and 14, is used to measure the absolute temperature. Two junctions 12 and 13 of this second differential thermocouple are located within the plug 15; the emf's of these two junctions compensate each other.

The pressure is measured by manganin manometer 16, and the emf of the thermocouples by two galvanometers 17 and 18.

Figure 76 shows an installation [59] for the study of the dependences of melting points of metals on pressure (at pressures up to 30,000 atm and temperatures up to 2000°C). The apparatus consists of conical vessel 9 located in a support for mechanical constraint. Before gas is passed in through joint 3 the lower press of the pressure intensifier is used to compress packing 2 in an unsupported-area gland consisting of rings 15 and 16. The pressure-intensifier piston transmits force to support 18, which transmits this force through the body of electrical lead 1 to pressure ring 17. While this is happening, the sealing rings 14 of the electrical lead are not compressed and the gas, compressed to several thousand atm (for example, by a compressor described above), passes freely into the bore of the cone. After the installation is filled, the lower press is connected again and, applying more pressure to ring 17, rings 14 of the electrical lead are compressed. At this time, base 18 touches the lower end of cone 9 and creates the force necessary to achieve mechanical support of the vessel.

Further, the gas in the bore of the cone is compressed by piston 10 which, as it moves, touches unsupported-area seal 8. A hydraulic plug is used in this apparatus for more reliable sealing: rubber plug 7, onto which liquid 11 (a mixture of glycerine with pentane) is poured, is installed before the piston. The same liquid is poured onto the electrical lead plug.

After the required pressure is created, electrical heater 6 is connected (the heating element consists of a coil of platinum or tungsten wire or a graphite resistance) and the material 13 contained in vessel 5 is heated. The temperature is measured by a dual-junction differential thermocouple 12. The hot junction is placed in the medium being investigated, the cold junction in a well in the body of the electrical lead. The temperature of the cold junction is measured by a second thermocouple 19. The pressure is measured by a manganin manometer 4.

One specific feature of internal heating of compressed gases is the sharp increase in required power at the furnace as pressure is increased. According to the data of the author, an internal heater used to heat a column approximately 30 mm in diameter to 520°C required 5 amps at 8 volts at atmospheric pressure. At a pressure of 600 atm nitrogen, the same temperature could be maintained only by a current of 20 amps at 30 volts. A qualitatively identical picture was observed by J. Basset [60], L. F. Vereshchagin and Ya. A. Kalashnikov [61], and J. Robin and S. Robin [62]. This phenomenon is apparently explained by the great increase in density of the medium and by convection to the walls of the apparatus.

L. F. Vereshchagin and Ya. A. Kalashnikov investigated another phenomenon accompanying internal heating of a compressed gas in detail; namely the displacement of the high-temperature zone toward the lower portion of the high-pressure apparatus: they established that the displacement of the heated zone, as well as the instability of indications of thermocouples at high pressures [63], are caused by changes in the laminar [flow] state of the compressed gas by turbulent convection currents within the vessel.

All these phenomena must be considered when investigations are performed in high-pressure vessels with internal heating.

BIBLIOGRAPHY

1. Korndorf, B. A., High Pressure Techniques in Chemistry, Goskhimizdat, 1952; Chem. Eng., 56: 175 (1956); Comings, E. W., High Pressure Technology, London, 1956.
2. Gamburg, D. Yu., Neft. Khoz., Vol. 46, No. 9 (1947).
3. Tsiklis, D. S., Zh. Khim. Prom., No. 7 (1958).
4. Chem. Eng. News, 41:28, 56 (1963).
5. Vereshchagin, L. F., and Ivanov, V. E., Pribory i Tekhn. Éksperim., 4:73 (1957).
6. Wolf, R. C., and Bowen, J. C., Ind. Eng. Chem., 49(12):1962 (1957).
7. Newholl, D. H., Ind. Eng. Chem., 49(12):1949 (1957).
8. Vereshchagin, L. F., and Preobrazhenskii, V. A., Izv. Akad. Nauk SSSR, Ser. Khim., 4:359 (1945).
9. Tsiklis, D. S., and Kulikova, A. I., 30:1670 (1956); Zh. Fiz. Khim., 39, No. 7 (1965).
10. Simon, F., Ruheman, and Edwards, W. A. M., Phys. Z. Chem., B6: 331 (1929).
11. Rudenko, N. S., and Zlunitsyn, S. A., Zh. Éksperim. i Teor. Fiz., 16:776 (1946).
12. J. Sci. Instrum., 14:34 (1937).
13. Boksha, S. S. Kristallografiya, 2(1):199 (1957).
14. Ramsauer, S., Phys. Z., 34:890 (1933); Chem. Fabr., 10:391 (1937).
15. Khariton, Yu. B., Izv. Akad. Nauk SSSR, Otdel. Khim. Nauk, 6:461 (1944).
16. Kousmine, E., Ind. Chim., Vol. 36, No. 380 (1949).
17. Furman, M. S., and Tsiklis, D. S., Dokl. Akad. Nauk SSSR, 91:597 (1953); Tsiklis, D. S., Dokl. Akad. Nauk SSSR, 91:327 (1953); Tsiklis, D. S., and Borodina, M. D., Dokl. Akad. Nauk SSSR, 140:1376 (1961); 147:860 (1962).

18. Ryabinin, Yu. N., Zh. Éksperim. i Teor. Fiz., 23:461 (1952).

19. Vereshchagin, L. F., Zh. Tekhn. Fiz.,16:669 (1946).

20. Ivanov, V. E., Vereshchagin, L. F., and Demyashkevich, B. P., Pribory i Tekhn. Éksperim. No. 1, 126 (1960).

21. Ivanov, V. E., Sb. Tr. Inst. Mash., Akad. Nauk SSSR, 1958.

22. Bersenev, B. I., Vereshchagin, L. F., and Livshits, L. D., Some Questions of Large Plastic Deformations of Metals at High Pressures, Izd. Akad. Nauk SSSR, 1958.

23. Whalley, E., and Lavergene, A., Rev. Sci. Instrum., 32(9):1062 (1961).

24. Straub, R. A., Mach. Design, 25(7):149 (1953).

25. Sidorov, I. P., USSR Patent No. 101990, Class 59a, 4 and 5.

26. Borzunov, V. A., and Semin, V. P., Tr. Inst. Kom. Standartov, Mer i Izmerit. Prib., 46(106):107 (1961).

27. Krichevskii, I. R., and Tsiklis, D. S., Zh. Fiz. Khim., 17:126 (1943).

28. Vereshchagin, L. F., Freindlin, L. Kh., Rubinshtein, A. M., and Numanov, N. K., Izv. Akad. Nauk SSSR, Otdel. Khim. Nauk, No. 6, 809 (1951).

29. Bridgman, P. W., Proc. Am. Acad. Arts Sci., 72:45 (1937).

30. Bridgman, P. W., Proc. Am. Acad. Arts Sci., 74:21 (1940).

31. Butuzov, V. P., Shakhovskoi, G. P., and Gonikberg, M. G., Tr. Inst. Kristallogr. Akad. Nauk SSSR, 11:233 (1955).

32. Bridgman, P. W., Proc. Roy. Soc., 203(1072):1 (1950).

33. Vereshchagin, L. F., Likhter, A. I., and Ivanov, V. I., Zh. Tekhn. Fiz., 26:874 (1956).

34. Lloyd, E., Hitton, H., and Johnson, D., J. Res. Nat. Bur. Stand., 630, CNI (1959).

35. Galaktionov, V. A., Vest. Akad. Nauk SSSR, 7:74 (1962).

36. Butuzov, V. P., Mirinskii, D. S., and Kats, G. S., Tr. Vses. Nauk Issled. Inst. Pyezoopt. Syr'ya, 3, No. 2 (1960).

37. Ashcroft, K., and Lees, J., Nature, 198:957 (1963).

38. Gonikberg, M. G., Tsiklis, D. S., and Opekunov, A. A., Dokl. Akad. Nauk SSSR, 129:88 (1959).

39. Wentorf, R. H., Ed., Modern very high pressure techniques, London, 1962, p. 118, B. Platen.

40. Vereshchagin, L. F., Galaktionov, V. A., Semerchan, A. A., and Slesarev, V. N., Dokl. Akad. Nauk SSSR, 132:1059 (1960); Vereshchagin, L. F., Semerchan, A. A., Zubkov, V. M., and Kuzin, N. N., Dokl. Akad. Nauk SSSR, 145:71 (1962).

41. Hall, H. T., Rev. Sci. Instrum., 31(2):125 (1960).

42. Bundy, F. P., Phys. Rev., 110(2):314 (1958).

43. Bundy, F. P., J. Chem. Phys., 38(3):618 (1963); Bundy, F. P., and Wentorf, R. H., Op. Cit., 1144.

44. Vereshchagin, L. F., Semerchan, A. A., Isaikov, V. K., and Ryabinin, Yu. N., Pribory i Tekhn. Éksperim., 5:93 (1960).

45. Bridgman, P. W., Studies in Large Plastic Flow and Fracture, Izdatinlit, 1955.

46. Vereshchagin, L. F., Zubova, E. V., and Shapochkin, V. A., Pribory i Tekhn. Éksperim., 5:89 (1960).

47. Vereshchagin, L. F., Shapochkin, V. A., and Zubova, E. V., Fiz. Metal. i Metalloved., 10:135 (1960).

48. Lazarev, V. G., and Kan, L. S., Zh. Éksperim. i Teor. Fiz., 14:439 (1944).

49. Kan, Ya. S., Zh. Tekhn. Fiz., 18:1156 (1948).

50. Brandt, N. B., and Tomashchik, A. K., Pribory i Tekhn. Éksperim., 2:118 (1958).

51. Bewilogua, L., and Roöner, R., Monatsber. Deut. Acad. Wiss., Berlin, 1(2):85 (1959).

52. Kennedy, G. C., Am. J. Sci., 248(8):540 (1950).

53. Crookes, W., Proc. Roy. Soc., London, A 76:458 (1905).

54. Bridgman, P. W., J. Chem. Phys., 15:292 (1947).

55. Basset, J., Compt. Rend., 208:267 (1939); Vereshchagin, L. F., and Fateeva, K. S., Pribory i Tekhn. Éksperim. 1:133 (1960).

56. Eitel, W., Die Experimentallen Hilfsmittel zur Mineralsyntese u. s. w., 1925; Griggs, D. T., and Kennedy, G. S., Am. J. Sci., 254:722 (1956); Slutskii, A. B., Experimental Investigations in the Area of Deep Process, Akad. Nauk SSSR, 1962.

57. Goranson, R.W., Sci. Monthly, 51:524 (1940).

58. Butuzov, V. P., Boksha, S. S., and Gonikberg, M. G., Dokl. Akad. Nauk SSSR. 108:337 (1956).

59. Boksha, S. S., Shakhovskoi, M. G., Pribory i Tekhn. Éksperim., 3:86 (1958).

60. Basset, J., Compt. Rend., 203:1338 (1936).

61. Vereshchagin, L. F., and Kalashnikov, Ya. A., Zh. Tekhn. Fiz., 25:1458 (1955).

62. Robin, J., and Robin, S., Phys. et radium, 17:499 (1956).

63. Goranson, R. W., Am. J. Sci., 22:481 (1931).

Chapter IV

Methods of Measuring High Pressures

PRESSURE UNITS

Pressure is force applied to a unit of area. The most widely used pressure unit is the atmosphere (normal [standard] or physical atmosphere), equal to the pressure of a column of mercury 760 mm high on an area of 1 cm^2 with mercury density 13.5951 g/cm^3 (at 0°C) and an acceleration of gravity equal to 980.665 cm/sec^2. The weight of such a mercury column is 1.033 kg. Thus, a normal atmosphere (n. atm.) corresponds to 1.033 kg/cm^2.

For convenience in calculations, the so-called technical atmosphere (atm.), equal to exactly 1 kg/cm^2 has been introduced and is used in Soviet technical writing.* In the SGS [CGS] system, the unit of pressure is one dyne per square centimeter, also called a "barye."

In 1961, the International Unit System SI (State Standard GOST 9867-61) was introduced in the USSR; in this system, pressure is expressed in newtons per square meter: 1 n/m^2 = 1.019716 × 10^{-5} kg/cm^2; 1 kg/cm^2 = 0.980665 × 10^5 n/m^2.

Pressure is also measured in kilograms per square meter, in pounds per square inch, in tons per square inch, in meters of water, in mm of water and mercury, and in inches of water and mercury. Appendix III presents conversion factors for these pressure units.

*   *   *

Pressure is measured by manometers, which are divided into so-called absolute and relative types.

---

* Ed. note: This is the "atmosphere" used throughout this book.

ABSOLUTE MANOMETERS

Absolute manometers are instruments which permit direct measurement of pressure by equilibration of its force, the value of which can be accurately determined. These manometers include liquid and piston (weight) manometers.

Liquid Manometers

By equilibrating a pressure to be measured with the weight of a column of liquid, the magnitude of the pressure can be determined. The heavier [more dense] the liquid, the shorter the column of liquid required to equalize the given pressure. Therefore, high pressures are measured most frequently by using mercury. The weight of a column of mercury 76 cm high with a cross-section of 1 cm^2, as is known, corresponds to 1 atm. Therefore, measurement of pressures on the order of hundreds of atm requires a mercury column hundreds of meters high. Such manometers have been used, but they are cumbersome and inconvenient in use. Recently, a change has been introduced in the design of these manometers which has considerably simplified operation with them.

An improved mercury manometer (Fig. 77) was used by D. I. Mendeleev [1]. The manometer consists of a series of interconnected columns. Each column is filled with mercury. The intervals between columns of mercury are filled with water. The pressure is measured as the sum of the heights of the mercury columns (h_1, h_2, h_3) minus the sum of the heights of the water columns, multiplied by the specific gravities of mercury and water respectively.

In spite of certain deficiencies (difficulty in sealing the joints, requirement of correcting factors, etc.), mercury manometers of this design are still used [2, 3, 4]. One such is the precision multi-stage mercury manometer with steel tubes [5], consisting of nine columns, each approximately 17.5 m high. The steel tubes have an internal diameter of 1.6 mm.

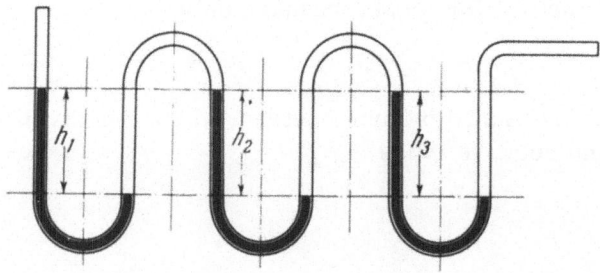

Fig. 77. Mendeleev manometer.

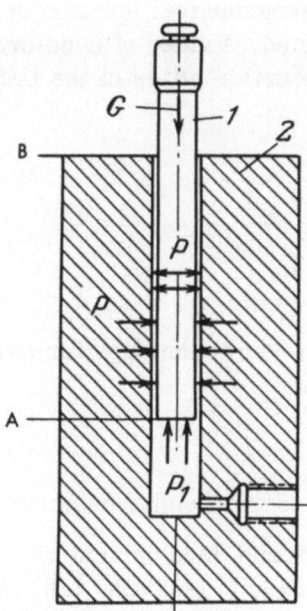

Fig. 78. Piston manometer:
1) piston; 2) cylinder.

Toluene is the fluid between the mercury columns. The mercury level in each column is determined by a cathetometer through a transparent plastic window, the distance between windows by the resistance of a measuring strip of Invar.

For accurate readings, especially where there are a large number of columns, correction factors must be introduced (for temperature differences between various portions of the column, compression of mercury at the pressure being measured, force of gravity, etc.). Thus, the mercury manometer, strictly speaking, is not absolute.

If the manometer is placed in a thermostat, the necessity for temperature correction can be eliminated. Corrections for the force of gravity are still necessary. The pressure is equalized by the weight of a column of mercury, i.e., by the force acting on the area of the tube cross-section. This force is equal to

$$F = mg$$

where m is the mass of the mercury; g is the acceleration of gravity at the point where the measurement is performed.

For so-called normal acceleration (980.655 cm/sec^2), this force will be equal to

$$F = mg_n$$

Expressing the mass of the mercury column by the area of its cross-section, height **h**, and density of the mercury, and equalizing the right portions of these equations and simplifying, we get

$$h_n = h \frac{g}{g_n}$$

where h_n is the height of the column for normal acceleration; h is the measured height of the column.

The value of the correction depends on the geographical location of the place where the measurement is being performed. Values of acceleration of gravity (cm/sec^2) are presented below for certain cities of the USSR:

Tbilisi 980.32
Odessa 980.74
Kiev 981.08
Moscow 981.52
Leningrad 981.93
Arkhangel'sk . . . 982.13

More detailed information on the corrections and techniques for calculation can be found in [6], by M. K. Zhokhovskii.

Piston Manometers

Another type of absolute manometer is the piston (weight) manometer.

Detailed investigations of piston manometers and development of their theory were performed by M. K. Zhokhovskii [6, 7] (see also [8]). He divides piston manometers into four basic types:

1) a simple, unpacked piston;
2) a simple piston in a cylinder with lateral pressure;
3) a single differential piston;
4) a dual differential piston.

Operation of a Piston Manometer. The pressure P_1 to be measured acts on piston 1 (Fig. 78), which moves freely in cylinder 2. The piston is carefully fitted to the cylinder and has no additional packing.* The pressure in the cylinder attempts to eject the piston. By equalizing this pressure with known load G, if the area of the piston is known, the pressure under the piston can be determined.

There is a clearance between the piston and the cylinder, which is filled with a liquid. Sealing of the piston is achieved by liquid [viscous] friction between the walls of the piston and the cylinder as the liquid moves through the clearance under the action of the pressure.

Pressure P_1 and load G applied to the ends of the piston compress the piston. A pressure which varies from the hydrostatic pressure in the cylinder (at base of piston A) to atmospheric pressure (at level B) exists in the clearance between the lateral surfaces of the piston and the cylinder; this pressure acts normal to the lateral surfaces. These forces cause deformation of the piston and cylinder.

---

* Manometers with pistons sealed by a gland will not be analyzed here.

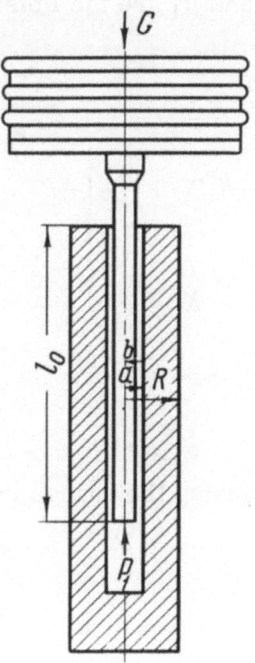

Fig.79. Simple piston.

Load G applied to the piston creates pressure P_1 beneath the piston. The liquid fills the clearance and flows out of the clearance under the action of this pressure. Under stable conditions, the liquid in the clearance flows under the action of the constant pressure difference between the beginning and end of the piston. The liquid flow is compensated by a drop in this piston. The rate of descent of the piston must be small; therefore, the clearance must vary within a range of a few microns.

Rate of Descent of Piston. It is important to know the dependence of the rate of descent of the piston on its dimensions, the clearance width, etc.

With a small pressure P_1, such that deformation of the piston and the change in viscosity of the pressure-transmitting fluid can be ignored, the rate of drop of a simple piston, as shown on Fig. 79, is equal to:

$$V = \frac{P_1 h^3}{6 \eta_0 b l_0}$$

where V is the speed [rate of descent], cm/sec; P_1 is the pressure, dyn/cm^2; $h = a - b$ is the clearance, cm (a is the radius of the cylinder bore, b is the radius of the piston); η_0 is the viscosity of the fluid, poise; l_0 is the length of the portion of the piston within the bore, cm.

This equation allows the dimensions of a planned instrument to be estimated or the rate of descent of a piston to be determined if the dimensions of the piston are known.

However, at high pressures, the cylinder and piston are deformed and the clearance changes. Since the pressure along the length of the clearance is variable, the viscosity of liquid in the clearance also changes along the length of the piston. In order to compute the descending rate of a piston under these conditions, the change in clearance width must be estimated. The introduction of deformation coefficients is very convenient in this case. For a simple piston (Fig. 79) these coefficients are equal to:

$$K = \frac{a}{E}\left(\frac{R^2 + a^2}{R^2 - a^2} + \mu_1\right) + \frac{b}{E_1}(1 - \mu_1), \qquad K_1 = \frac{b}{E}\mu_1$$

where R is the external radius of the cylinder; E and E_1 are the moduli of elasticity of the materials of cylinder and piston; μ and μ_1 are the Poisson's ratios for the materials of the cylinder and piston.

Then, the rate of descent of the piston is:

$$V = \frac{1}{6\eta_0 bl_0 c}\left\{\left[(h - K_1 P_1)^3 + \frac{3K}{c}(h - K_1 P_1)^2 + \frac{6K}{c^2}(h - K_1 P_1) + \frac{6K^3}{c^3}\right] - \right.$$

$$+ \frac{1}{e^{cP_1}}\left[\left(h + P_1(K - K_1)\right)^3 + \frac{3K}{c}\left(h + P_1(K - K_1)\right)^2 + \right.$$

$$\left.\left.\frac{6K}{c^2} + \left(h + P_1(K - K_1)\right) + \frac{6K^3}{c^3}\right]\right\}$$

where c is the isothermal piezo-coefficient characterizing the change in viscosity of the liquid with pressure.

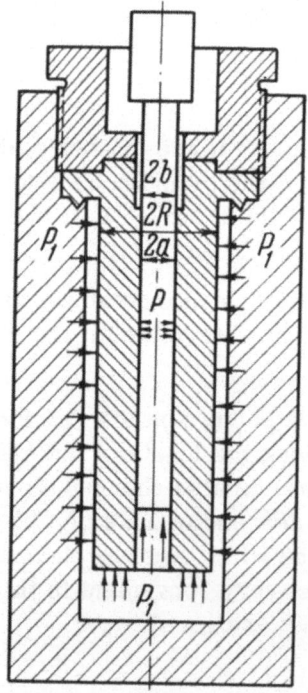

Fig. 80. Simple piston and cylinder with lateral pressure.

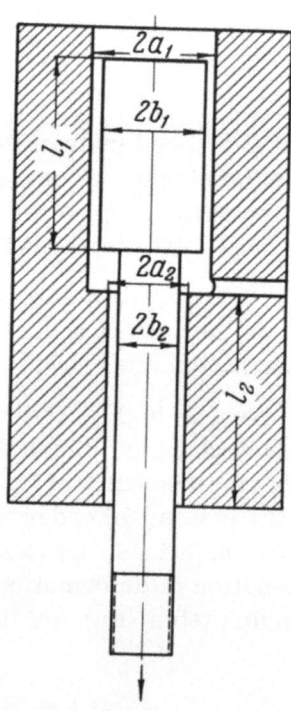

Fig. 81. Single differential piston.

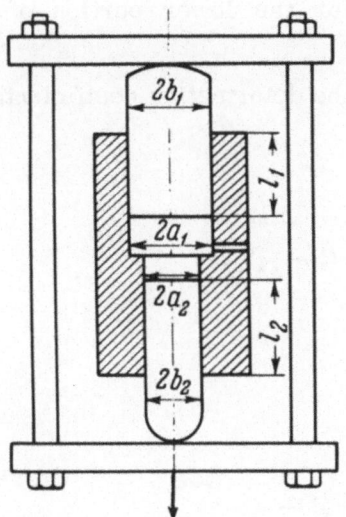

Fig. 82. Dual differential piston.

It can be considered, on the basis of experimental data, that the viscosity of the liquids changes with pressures as follows :*

$$\eta = \eta_0 e^{cP}$$

where η is the dynamic viscosity under pressure; η_0 is the viscosity at atmospheric pressure.

This formula can be used to determine the rate of descent of a simple piston at high pressure. The expressions for deformation coefficients of manometers of other types are presented below, and the expressions for the rate of descent of the pistons of these manometers can be found in literature [6, 7].

The deformation coefficients K_5 and K_6 for a simple piston in a cylinder with lateral pressure (Fig. 80) are equal to:

$$K_5 = \frac{a}{E}\left(\frac{R^2 + a^2}{R^2 - a^2} + \mu\right) + \frac{b}{E_1}(1 - \mu_1)$$
$$K_6 = \frac{a}{E}\left(\frac{2R}{R^2 - a^2} - \mu\right) + \frac{b}{E_1}\mu_1$$

For a single differential piston (Fig. 81), the deformation coefficients are:

for the upper portion,

$$K_2 = \frac{a_1}{E}\left[\frac{R^2 + a_1^2}{R^2 - a_1^2} + \mu\right] + \frac{b_1}{E_1}(1 - \mu_1)$$

for the lower portion,

$$K_3 = \frac{a_2}{E}\left[\frac{R^2 + a_2^2}{R^2 - a_2^2} + \mu\right] + \frac{b_2}{E_1}(1 - \mu_1)$$
$$K_4 = \frac{b_1^2 - b_2^2}{b_2 E}\mu_1$$

where a_1, b_1, and h_1 are the radii of the cylinder and piston, and the magnitude of the clearance for the upper portion of the piston respectively; a_2,

---

*See Chapter I.

b_2, and h_2 are the corresponding values for the lower portion of the piston.

For a dual differential piston (Fig. 82) the deformation coefficients are respectively:

for the upper portion

$$K' = \frac{a_1}{E}\left[\frac{R^2 + a_1^2}{R^2 - a_1^2} + \mu\right] + \frac{b_1}{E_1}(1 - \mu_1)$$

$$K_1' = \frac{b_1}{E_1}\mu_1$$

for the lower portion

$$K'' = \frac{a_2}{E}\left[\frac{R^2 + a_2^2}{R^2 - a_2^2} + \mu\right] + \frac{b_2}{E_1}(1 - \mu_1)$$

$$K_1'' = \frac{b_2}{E}\mu_1$$

Rate of Rotation of Piston. No matter how well a piston manometer is made, defects in centering of the system may cause friction between the piston and cylinder. In order to eliminate this, the piston is rotated. The piston may be given a certain angular velocity by hand, after which it will rotate with decreasing speed until it stops. The piston may also be rotated with constant speed by a motor.

When there is no non-liquid friction, under low-pressure conditions, and when the clearance and viscosity of the liquid are constant, the rate of rotation of a simple piston is:

$$\omega = \omega_0 e^{-\alpha t}$$

$$\alpha = \frac{4\pi l_0 \eta b^3}{mr^2 h}$$

where ω is the instantaneous value of angular velocity of the piston; ω_0 is the initial angular velocity of the piston; r is the radius [of gyration] of the load; m is the mass of the load.

If friction force T is applied to the piston (for example friction in the shaft bearing), then:

$$\omega = \omega_0 e^{-\alpha t} - \frac{\beta}{\alpha}(1 - e^{-\alpha t}), \qquad \beta = \frac{2T b_1}{mr^2}$$

where b_1 is the radius of the bearing shaft.

At high pressures, the value of the torque applied to the piston by liquid friction changes greatly due to deformation of the clearance space and variation in viscosity of the liquid. In this case, coefficient α can be calculated from the deformation coefficients presented above.

Effective Area of Piston. As was noted above, during the process of measurement of pressure, the area of the piston changes due to its deformation under load. If the area determined from measurement of the diameter is compared with the area determined by calibration of the manometer with a known pressure, the latter value will be greater even at [low] pressures where the deformation of the piston can be ignored. This is determined by the fact that the piston diameter is effectively increased by the liquid film flowing through the clearance space between cylinder and piston under the influence of the pressure.

The area determined by calibration is called the effective area of the piston. This value must be known in order to determine the pressure being measured by the manometer with respect to a given load.

If the friction arising when the liquid and piston move in the cylinder is taken into consideration, the pressure measured by the manometer will be equal to

$$P = \frac{G - T}{S}$$

where G is the weight of piston plus load; T is the friction; S is the area of cross-section of the piston.

We know that the liquid friction

$$T = P\pi b\,(a - b)$$

Then,

$$P = \frac{G}{S + \pi bh}$$

The expression in the denominator is the effective area. Thus,

$$S_{\text{ef}} = \pi b^2 + \pi b\,(a - b)$$

i.e., the effective area is greater than the true area by approximately one half the area of the ring-shaped gap. This approximate expression for determination of the effective area is often used in calculations. However, this expression, like the expression for the accurate determination of effective area

$$S_{ef} = \frac{\pi}{2}\,(a^2 + b^2)$$

is correct only at low pressures. At high pressures, when deformation of the piston and cylinder takes place, the effective area changes. This change can be calculated from the following formulas:

for a simple piston in a normal cylinder

$$S_{P_1} = S_{ef}\left\{1 + P_1\left[\frac{3\mu_1 - 1}{E_1} + \frac{1}{b}\left(\frac{K}{2} - K'\right)\right]\right\}$$

for a piston in a cylinder with lateral pressure

$$S_{P_1} = S_{ef}\left\{1 + P_1\left[\frac{3\mu_1 - 1}{E_1} - \frac{1}{b}\left(K_6 - \frac{K_5}{2}\right)\right]\right\}$$

for a dual differential piston

$$S_{P_1} = S_{ef}\left\{1 + P_1\left[\frac{3\mu_1 - 1}{E_1} + \frac{b\left(\frac{K'}{2} - K'_1\right) - b_2\left(\frac{K''}{2} - K''_1\right)}{b_1^2 - b_2^2}\right]\right\}$$

for a single differential piston

$$S_{P_1} = S_{ef}\left\{1 + P_1\left[\frac{3\mu_1 - 1}{E_1} + \frac{b_1\frac{K_2}{2} - b_2\left(\frac{K_3}{2} + K_4\right)}{b_1^2 - b_2^2}\right]\right\}$$

Here S_{P_1} is the effective area at pressure P_1; S_{ef} is the effective area calculated or determined at low pressures; K_1, K_2, etc., are the deformation coefficients.

Representing the expressions in brackets as λ, we get

$$S_{P_1} = S_{ef}\,(1 + P_1\lambda)$$

where λ is a coefficient characterizing the relative change in area per unit of pressure.

It can be shown that the simplest method for determining the effective area is measurement of the diameters of the piston and cylinder bore.

However, with very small bore diameters, it is very difficult to measure this diameter with high accuracy. Even if the diameter is measured indirectly, such as by weighing the liquid filling the bore, the required accuracy is still not achieved.

The effective area can be calculated from the diameter of the piston and the size of the gap, determined hydrodynamically, using a method suggested by M. K. Zhokhovskii. The value of clearance h is determined from the descent rate of piston V, which is found experimentally:

$$h = \sqrt[3]{\frac{6\eta b l_0 V}{P_1}}$$

Another method for determining the effective area is comparison of two pistons, where the area of one piston is known.

Also, a known pressure can be created beneath the piston, as by the vapor of some material at known temperature, for example the vapor of pure carbon dioxide, the vapor pressure of which at 0°C is 26114.7 ± 1 mm Hg (g = 980.665) or 34.4009 ± 0.0013 atm. At high pressures, reference points which will be given below can be used.

The most accurate method of determining the effective area is measurement of this area from the indication of some accurate manometer, such as a mercury manometer.

The effective area can be measured at pressures of several thousand atm by the following method [9]: The piston manometer is connected to a mercury manometer, consisting of a steel column, approximately 900 cm high, filled with mercury. The column is terminated by steel plugs. Two tubes filled with nonviscous oil are used to transmit the pressure to the top and bottom of the column. The position of the division boundary [meniscus] between mercury and oil is maintained at a constant level by electrical contacts. The height of the mercury column is determined from the distance between these contacts with an accuracy of 0.013 cm, using a calibrated Invar tape. The mercury column and the oil line leading to the top of the column are held constant at 25°C.

The top of the column is connected to the atmosphere, the bottom to the piston manometer to be calibrated. The pressure which equalizes the mercury manometer can be easily determined if the dimensions of the manometer are known. A pressure is created at the upper terminus of the column through a system of valves by a press; the manometer to be calibrated is also connected to the top of the column. A second piston manometer is connected to the bottom of the column. The pressure at which it is equalized is equal to the pressure at the bottom of the column during the first measurement plus the pressure of the mercury column. As the two piston manometers are switched, the pressure is increased each time by 11.73 atm, finally reaching a pressure of 2550 atm.

If the pressure of mercury and oil are known, the value of the pressure generated and the effective area of the manometer at this pressure can be calculated. The accuracy of this method at 510 atm is ±0.015 atm; at 2550 atm, the accuracy is ±0.14 atm. This method gives good results, but it is complex and requires the construction of expensive apparatus.

A simpler method [10] consists of connecting two piston manometers of identical design and dimensions, made of different materials. The modulus of elasticity of these materials should be known. The dimensions of the clearance under the conditions defined will change in proportion to the moduli of elasticity.

Ignoring second-order quantities, the effective areas A and B of the pistons of the two manometers are determined at the given pressure:*

$$A = A_0 [1 + \alpha f(P)]$$
$$B = B_0 [1 + \beta f(P)]$$

where A_0 and B_0 are the effective areas at atmospheric pressure; α and β are constants inversely proportional to the moduli of elasticity of the materials; $f(P)$ is the unknown pressure function.

Then:†

$$\frac{A}{B} = \frac{A_0}{B_0} [1 + (\alpha - \beta) f(P)]$$

Equalizing the two manometers at identical pressure, we determine $(\alpha - \beta)f(P)$; knowing α/β, we can determined the absolute value of effective area at the known pressure. This method is applicable only if the ratios of moduli of elasticity and Poisson's ratios are identical.

When these conditions are observed, the effective area changes in direct proportion to the change in pressure, which permits reliable extrapolation to be performed for higher pressures.

Experiments have shown that this method makes it possible to measure the change in effective area with an accuracy to $10^{-4}\%$ of the total area. The effective area at atmospheric pressure may be determined from the indications of a mercury manometer.

---

*These expressions are identical to equations presented above for determination of effective area determined by M. K. Zhokhovskii.
†This expression has been derived with certain assumptions.

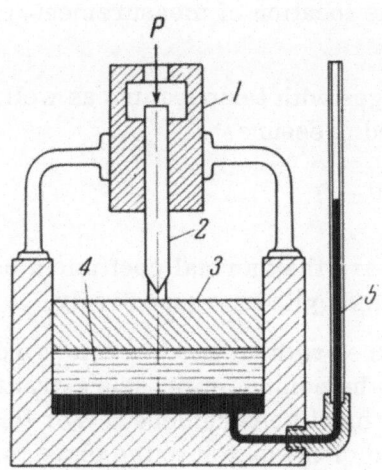

Fig. 83. Amagat manometer: 1) high-pressure cylinder; 2) high-pressure piston; 3) low-pressure piston; 4) oil; 5) mercury manometer.

By measuring the capacitance of the condenser formed by the cylinder, piston, and oil layer in the gap, we can determine the width of the gap and, consequently, the effective area of the piston [11].

Corrections to Reading of Piston Manometers. It is clear from the above that accurate pressure measurement with a piston manometer requires introduction of certain corrections to the readings of the piston manometer. The most important correction is that for the change in effective area with pressure.

The pressure, measured by the piston manometer

$$P_1 = \frac{G}{S_0}$$

where S_0 is the effective area; G is the weight of the load plus piston.

If the effective area is changed by ΔS, the change in pressure

$$\Delta P_1 = -P_1 \frac{\Delta S}{S_0}$$

The total change in effective area of the piston with pressure

$$\Delta S = S_0 \lambda P_1$$

from which

$$\Delta P_1 = -\lambda P_1^2$$

where λ is a coefficient characterizing the relative change in area per unit of pressure. The value of λ for various types of manometers can be found in the literature [7].

It is further necessary to introduce a correction for the acceleration of gravity:

$$\Delta P_2 = P_1 \frac{g - g_n}{g_n}$$

where g is the acceleration of gravity at the location of measurement; g_n is the normal acceleration.

The effective area of the piston changes with temperature as well. In this case, the correction to the measured pressure

$$\Delta P_3 = P_1 (\alpha + \beta)(20 - t)$$

where P_1 is the pressure at 20°C; α and β are the thermal coefficient of linear expansion of the material of piston and cylinder respectively.

In cases when the pressure is measured with an accuracy to hunddredths of an atmosphere, the pressure of the column of pressure-transmitting liquid in the connecting capillaries must be considered. For this, the height of the liquid column from the end of the piston to the place to which the measured pressure is related must be known.

In practice, the location of all high-pressure capillaries connecting the manometer with the apparatus can be drawn to scale and the difference in the columns of liquid between the levels of the end of the piston and, for example, the level of the mercury in the mercury equalizer or [diaphragm] membrane null-instrument can be calculated. Then, the correction

$$\Delta P_4 = \pm H \gamma$$

where H is the calculated height of the liquid column, cm; γ is the specific gravity of the liquid at the given temperature, g/cm^3.

The correction is added or subtracted, depending on the location of the manometer and the instrument in which the pressure is being measured.

Manometers with Single Piston.*   The greater the diameter of the piston, the more slowly it will descend into the cylinder with the same clearance width. However, an increase in diameter leads to an increase in equalizing load required.

Amagat Manometer [12].   In order to reduce the load, Amagat used two pistons of different diameters (Fig. 83).† The small piston receives the load to be measured, while the large piston, connected directly to the small piston, equalizes the pressure of the mercury column. With a sufficiently high ratio of areas of pistons, this manometer can be used to

---

*We will not describe the simplest manometer — a Rucholz press, which the reader can find in the literature [6].

†Another method for reducing the load being equalized is reduction of the piston diameter.

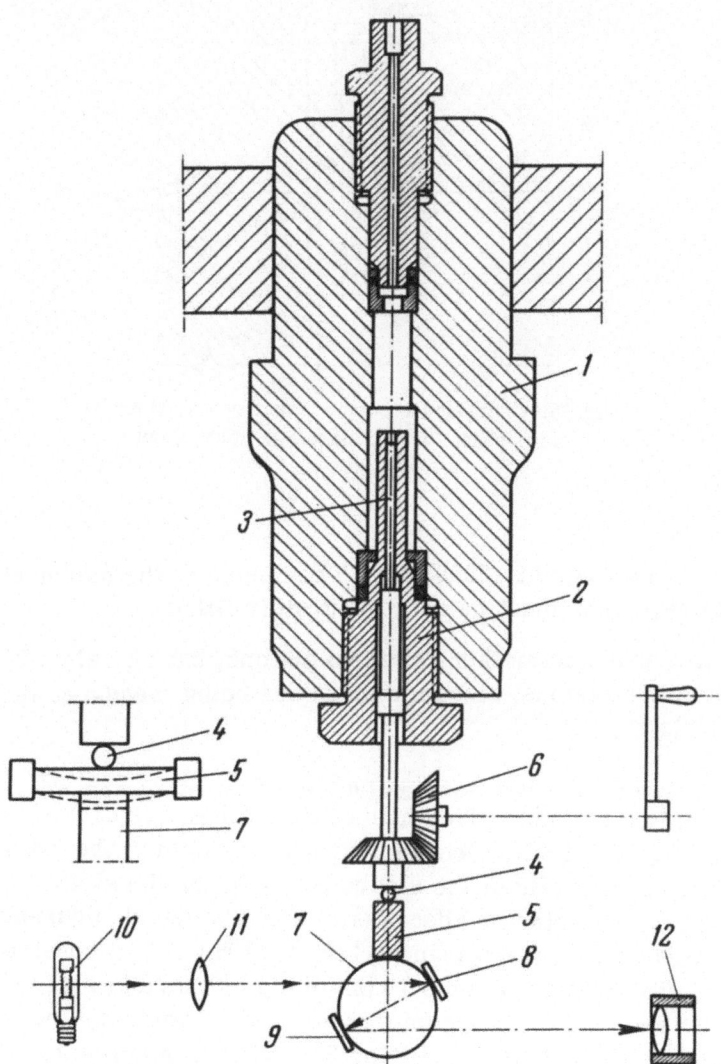

Fig. 84. Vereshchagin-Aleksandrov piston manometer: 1) bomb; 2) sleeve; 3) piston; 4) ball; 5) beam; 6) bevel gear; 7) steel ring; 8, 9) mirrors; 10) lamp; 11) lens; 12) microscrope.

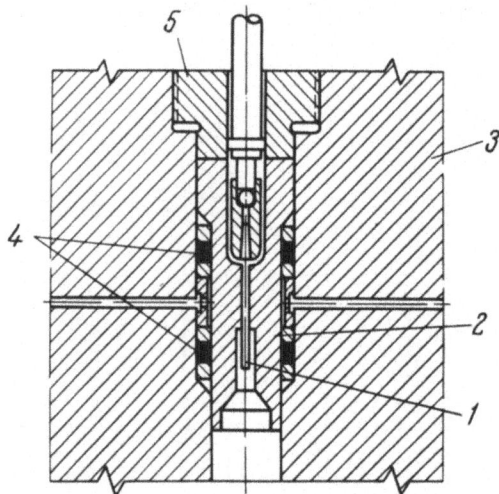

Fig. 85. Controlled-clearance manometer: 1) measuring piston; 2, 3) cylinders; 4) seal; 5) nut.

measure pressures up to 3000 atm. The design of the manometer is complex and the pistons must be very accurately fitted.

In this manometer there are two pistons, but we relate it to manometers with single pistons, since the pressure being measured is borne by
only a single piston.

Manometer with Cylinder with Lateral Pressure.   Good sealing of the piston
is achieved in a cylinder which has a thin-wall sleeve at its lower portion.
Since the pressure in the clearance drops continually, the external pressure acting on the walls of the sleeve compresses the sleeve, improving
the sealing of the piston and decreasing the diametral clearance. The
Vereshchagin-Aleksandrov manometer [13] (Fig. 84) is constructed on this
principle. It consists of a bomb 1 containing cylindrical sleeve 2 made of
hardened steel. Sleeve 2 is compressed in the bomb, and the pressure
presses it around the piston. Piston 3, 1.52 mm in diameter, made of ball-
bearing steel, is contained in the sleeve. The lower end of piston 3 bears
on steel beam 5 through steel ball 4; beam 5 is supported on two
mounts and can bend between them. As a working fluid, the authors used
a viscous mixture of glycerine and glucose.

In order to reduce friction, the piston is rotated by bevel gear 6. The
movement of the piston under pressure is measured by the bending of steel
beam 5. The amount of bending is determined from the compression of
light steel ring 7, which carries mirrors 8 and 9.

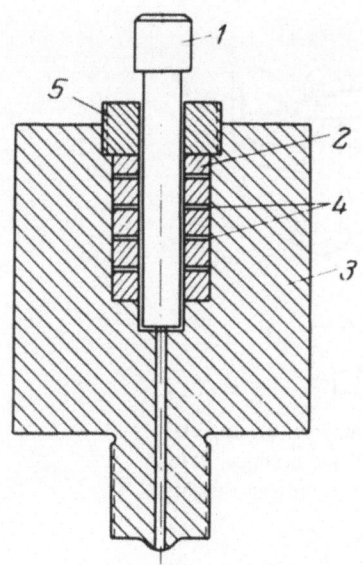

Fig. 86. Piston manometer with constant clearance: 1) piston; 2) insert; 3) manometer; 4) apertures; 5) nut.

Light from lamp 10 passes through lens 11, is reflected from these mirrors and focused in the lens of measuring microscope 12. The movement of the dot of light on the scale of the microscope is used to measure the change in ring diameter under the influence of bending of the beam. At 10,000 atm, the bending reaches 20 μ.

As can be seen from the description of the manometer, the authors do not use a load for measurement of pressure at all. They have actually used the method of relative pressure measurement.* There are similar designs in which the displacement of the piston is measured from the bending of powerful springs made of disks of sheet steel 2 mm thick.

A somewhat different design [14] is the piston manometer with controlled clearance (Fig. 85). The measuring piston 1 is located in cylinder 2, which in turn is located in cylinder 3. A definite, calculated pressure is created between cylinders 2 and 3, the amount of the pressure depending on the amount of pressure to be measured. The pressure in the gap between cylinders 2 and 3 deforms cylinder 2, changing the clearance between piston 1 and cylinder 2. In order to prevent leakage of the liquid, unsupported-area seal 4 is used, compressed by nut 5.

The dependence of pressure P_1, which must be created in the clearance between cylinders 2 and 3, on the measured pressure P_2 is expressed by the formula

$$P_1 = K + LP_2$$

where K is the pressure required for reduction of the clearance between the cylinders at atmospheric pressure; L is a coefficient dependent on the position of the so-called "minimal-clearance ring."

---

*This method of measuring the position of a piston is inefficient and eliminates the advantage of a piston manometer as an absolute manometer.

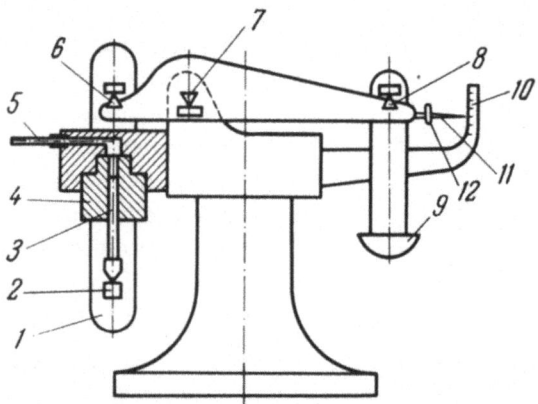

Fig. 87. Shtyukrat balance: 1) fork support; 2) support;
3) piston; 4) cylinder; 5) tube; 6, 7, 8) knife edges; 9) load
container; 10) scale; 11) indicator; 12) adjusting weight.

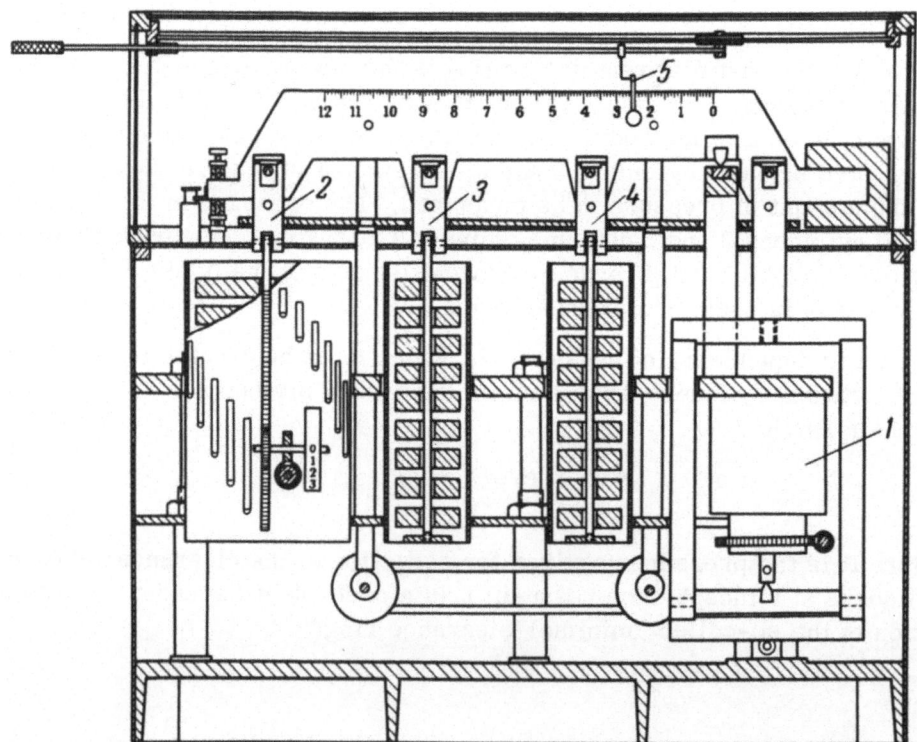

Fig. 88. Decimal manometric balance: 1) manometer; 2, 3, 4) equalizing weights; 5) rider.

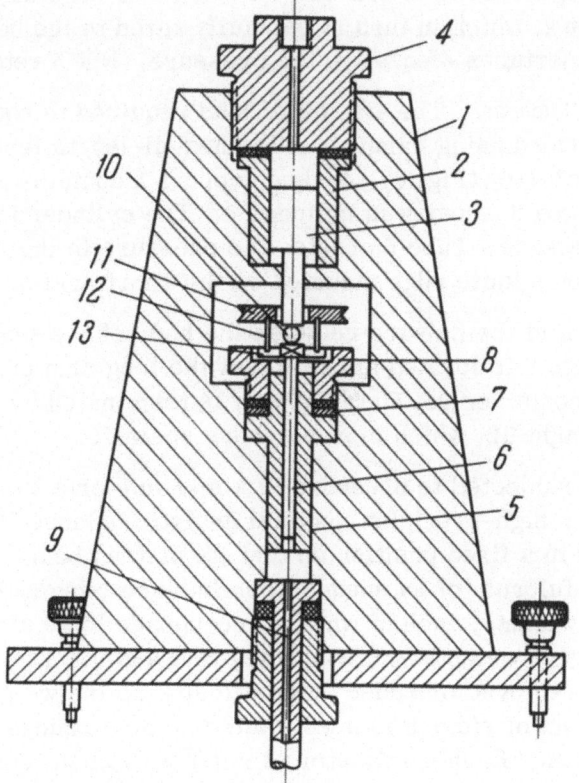

Fig. 89. Zhokhovskii manometer: 1) body; 2, 5) cylinders;
3, 6) pistons; 4, 8) nuts; 7) packing; 9) tubing; 10) ball; 11)
pulley; 12) fingers; 13) plate.

If the minimal-clearance ring is located near the top end of the cylin-
der, then $L \approx 0.7$; if the ring is located near the middle, then $L \approx 0.35$. Depending on the initial size of the clearance, the support pressure P_1 may be
greater or less than the pressure being measured.

The attempt is made to move the sealing between the support cylinder 3 and the measurement cylinder 2 as far as possible from the ends of
the piston, in order to avoid local deformations resulting from end effects.

A Carboloy piston is used in this manometer, which allows measurements to be performed with high accuracy.

A piston manometer with constant clearance (independent of pressure)
[15] is shown on Fig. 86. In this manometer, the cylinder is symmetrically
loaded from within and without, so that the clearance dimension and the

force of liquid friction remain constant. Piston 1 is carefully fitted to cylindrical insert 2, which in turn is carefully fitted to the body of the manometer 3. Apertures 4 equalize the pressure. Nut 5 retains insert 2.

Manometric Balances.   The amount of load required to equalize the piston can be reduced using balances with unequal-length levers. In the Shtyukrat balance* [16] (Fig. 87), forked support 1 contains support 2, which is pressed by piston 3, moving in cylinder 4. The cylinder is fastened to the body of the balance. Tube 5 creates the pressure in the manometer. Support 1 hangs on a knife edge mounted on the short arm of the balance.

The fulcrum of the balance rests on the body of the scale at knife edge 7. Knife edge 8 is located at the end of the long arm of the balance, carrying weight-container 9. The long arm is terminated by indicator 11 with adjusting-weight 12, which moves across scale 10.

The piston subjected to pressure in a manometeric balance (Fig. 88) is weighed using a high-precision decimal analytic balance [17]. Manometer 1 is installed in a fixed position so that its piston† bears on a frame supported by the fulcrum of an unequal arm balance (stages 2, 3, and 4). Stage 2 has nine weights, each of which corresponds to 70 atm. Stages 3 and 4 consist of nine weights corresponding to 7 and 0.7 atm respectively. The weights are removed or added automatically as the weight ring rises or falls. Movement of rider 5 is used to determine a change in pressure of less than 0.7 atm. Each centimeter of rider movement corresponds to 0.03 atm.

The balance described was constructed in 1940, and has been periodically checked over the last 14 years [18]. In this time, the calibration coefficient A in the expression

$$P = As + B$$

has changed by 0.0004. Here P is the pressure being measured; s is the indication of the balance; B is a constant.

Manometers with Differential Piston [6, 7, 19, 20]. In order to reduce the load on the piston, manometers with so-called differential pistons are used. The pressure in these manometers is borne by the area of a ring with diameters equal to the larger and smaller pistons. This area may be comparatively slight, which allows a reduction in the load equalized by the pressure being measured. The diameters of the pistons

_____

*In this manometer, the piston moves in a gland.

†In this design, the body of the manometer is rotated rather than the piston.

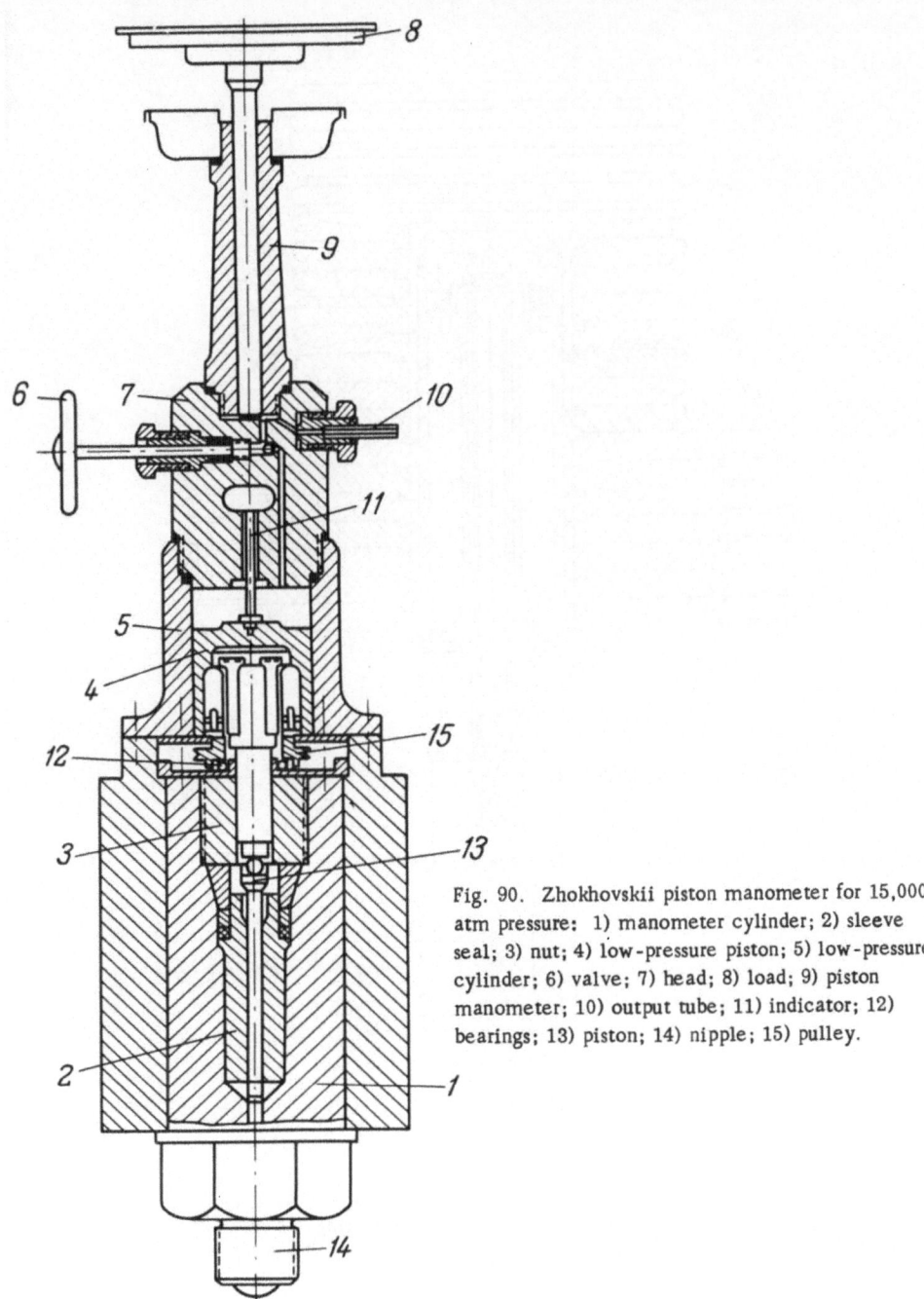

Fig. 90. Zhokhovskii piston manometer for 15,000 atm pressure: 1) manometer cylinder; 2) sleeve seal; 3) nut; 4) low-pressure piston; 5) low-pressure cylinder; 6) valve; 7) head; 8) load; 9) piston manometer; 10) output tube; 11) indicator; 12) bearings; 13) piston; 14) nipple; 15) pulley.

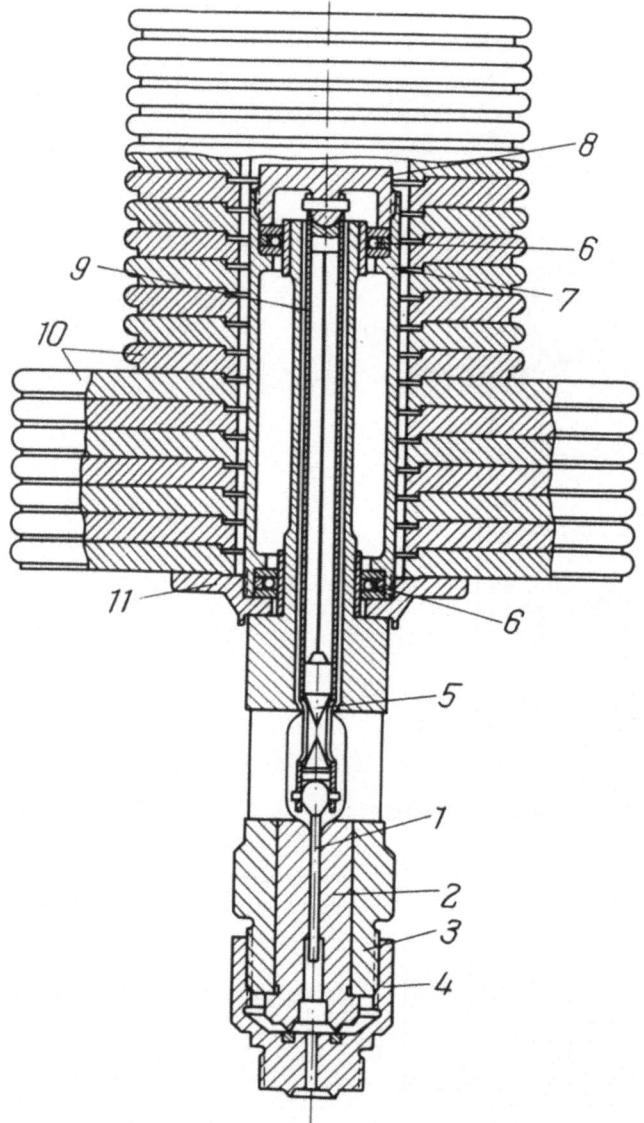

Fig. 91. Manometer with displaced center of gravity: 1) piston;
2) cylinder; 3) nut; 4) nipple; 5) plumb; 6) bearing; 7) bushing;
8) head; 9) tube; 10) load; 11) plate.

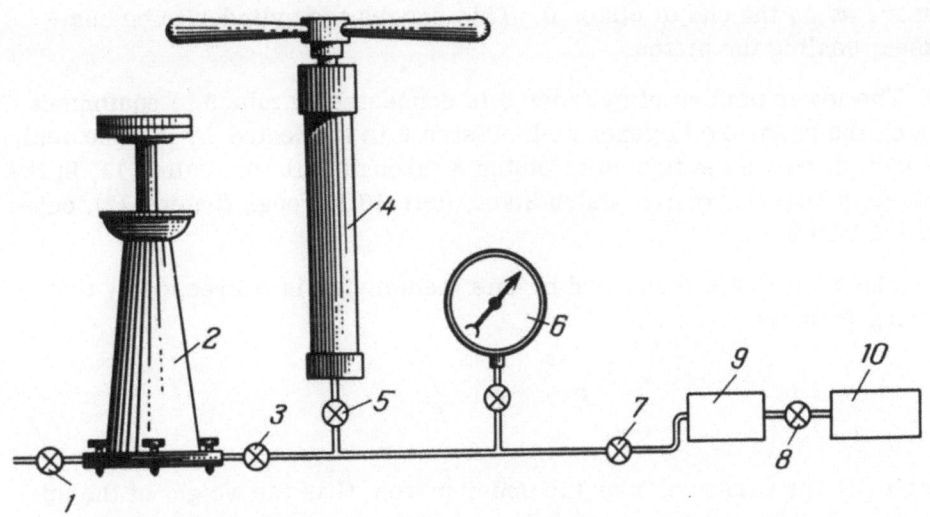

Fig. 92. Measuring installation plan: 1, 3, 5, 7, 8) valves; 2) piston manometer; 4) oil press; 6) manometer [pressure gauge]; 9) dividing (null) instrument; 10) system in which pressure is to measured.

may be rather great, so that the pistons will drop slowly. Also, the load is not placed on the piston, but is rather suspended from it, which also improves the usage conditions of such manometers.

In addition to these advantages, differential piston manometers have deficiencies: it is more difficult to manufacture the piston; the piston drops more rapidly, since leakage takes place on two sides; the accuracy of a differential manometer is somewhat less than the accuracy of a normal piston manometer. For measurement of high pressures, a dual manometer with a differential piston is used, i.e., a second manometer is connected in series to a first one so that the ring space of the first is connected to the large piston of the second. In essence, the second manometer acts as a pressure intensifier. It is also connected to the space in which the pressure must be measured.

Zhokhovskii Manometers. The design of a manometer with a measurement booster for determination of pressures up to 2000 atm has been created by M. K. Zhokhovskii (Fig. 89). Body 1, which is a truncated cone with a window cut out, contains cylinder 2, in which piston 3 is fitted. The cylinder is sealed by nut 4, through which a tube connecting the chamber of cylinder 2 with a piston manometer used to create an accurately known pressure is connected.

The lower portion of body 1 contains cylinder 5 with fitted piston 6. The cylinder is sealed by packing 7 and nut 8. There is clearance between the body and cylinder 5, into which liquid can penetrate under the same pressure as on the end of piston 6. This causes the cylinder to be compressed, sealing the piston.

The lower portion of cylinder 5 is connected by tube 9 to equipment in which the pressure is generated. Piston 6 is connected by a spherical head (not shown on the figure) to piston 3 through ball 10. Pulley 11, in the aperture in body 1, rotates piston 3 and plate 13 (through fingers 12), connected to piston 6.

The pressure P measured by this manometer is expressed by the following formula:

$$P = P_1 K + \frac{G}{S}$$

where P_1 is the pressure over the upper piston; G is the weight of the upper piston; S is the cross-sectional area of the upper piston.

$$K = \frac{S}{s}$$

where s is the cross-sectional area of the lower piston.

With a large area ratio of the pistons, a small accurately-known pressure can be used to equalize and therefore measure a high pressure.

For measurement of very high pressures (up to 25,000 atm), other measuring intensifiers [21, 22] are used. Figure 90 shows a manometer designed for a pressure of 15,000 atm. The apparatus in which the pressure must be measured is connected through nipple 14 to cylinder 1 of the manometer, in which carefully fitted piston 13 moves. The force from this piston is transmitted through the ball to the low-pressure piston 4, which moves in cylinder 5. The space over piston 4 is filled with oil, the oil pressure being less than the pressure in the apparatus by a factor equal to the ratio of areas of piston 4 to piston 13.

Cylinder 5 is covered by head 7 and valve 6, through which oil can be transmitted from the space over piston 4 to the low-pressure piston-manometer bore 9 with load 8, connected to head 7. Indicator 11, which is observed through the viewing window, is used to determine the position of the piston system 13-4. The position of the piston of manometer 9 can be controlled by feeding oil through tube 10.

Bearing 12 and pulley 15 rotate the pistons of the manometer. The pressure being measured is

$$P = P_0 + Gn$$

where P_0 and n are constants of the instrument; G is the weight of load 8 and the piston in manometer 9;

$$n = \frac{s_2}{s_3 s_1}, \qquad P_0 = G_0/s_3$$

where s_1 is the effective area of the piston in manometer 9; s_2 is the effective area of piston 4; s_3 is the effective area of piston 13; G_0 is the weight of piston 4 and the parts connected to it.

MP-600 Manometer.   In order to reduce the load necessary for equalization of a high pressure, a very small diameter piston can be used. In this case, however, in addition to the difficulties of manufacture of the piston and cylinder, the danger of bending of the piston-load system as it is rotated arises. This deficiency is eliminated in manometers in which the center of gravity is moved downward. One such manometer [23] is shown in Fig. 91. Piston 1 is placed in cylinder 2, which is pressed against nipple 4 by nut 3. The upper portion of nut 3 carries bushing 7 through bearing 6. Tube 9 is mounted on the head of the piston; within the tube there is plumb 5, used to set the measuring system in a strictly vertical position. The tube is covered by head 8 screwed into bushing 7. Plate 11 carrying the load 10 is screwed into the bottom part of the bushing. As the piston moves, tube 9, connected to it, raises the bushing and load weights. The load weights are rotated by hand. Manometers of this design can be used to measure pressures up to 2500 atm with accuracy to hundredths of an atmosphere.

Methods of Measurement.   For measurement of pressure with a piston manometer, an installation (Fig. 92), consisting of the piston manometer 2, oil press 4, and null instrument 9 (equilibrium indicator) dividing the medium in the cylinder of the piston manometer from system 10, in which the pressure is measured, must be assembled.

The oil press is necessary to replace oil lost flowing through the clearance between cylinder and piston. Also, it is used to set the manometer piston in the position at which the measurement is to be performed. This position is generally adjusted for manometers of various designs.

The instrument dividing the oil from the medium being investigated is extremely important: it is the indicator of equilibrium between load and measured pressure.

Before measuring the pressure with a piston manometer, the pressure is measured approximately with Bourdon gauge 6. To do this, valves

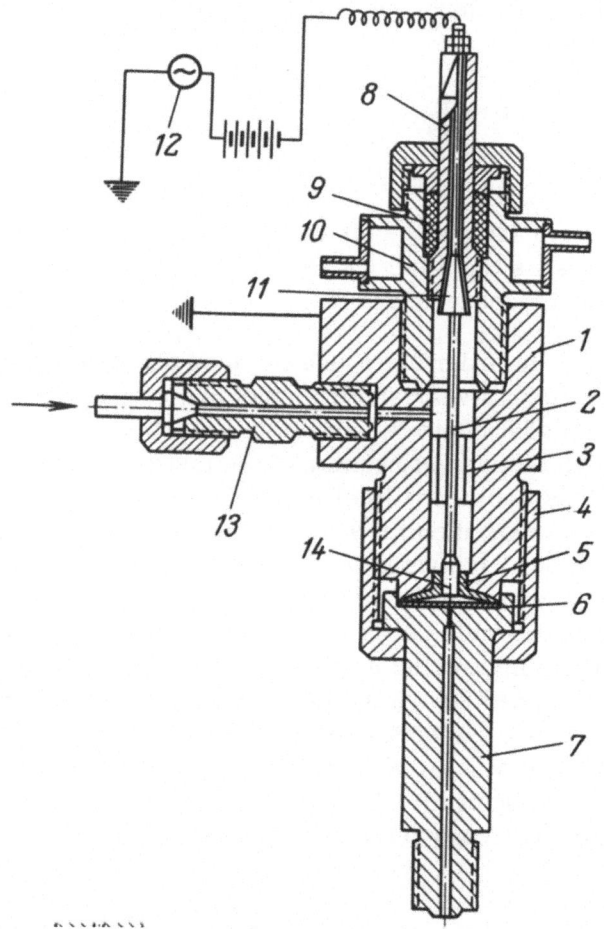

Fig. 93. Membrane [diaphragm] null instrument: 1) body; 2)
electrodes; 3, 5) packing; 4) nut; 6) membrane; 7, 10, 13)
nipples; 8) spindle; 9) gland; 11) electric lead; 12) lamp; 14)
contact terminal.

3, 7, and 8 are closed and an oil pressure approximately equal to the pressure in system 10 is produced by the press. Then, valves 7 and 8 are gradually opened, and the indications of instrument 9 (which may be a U-shaped
metal manometer filled with mercury and supplied with contacts used to set
the mercury at the same level in both tubes) are used to produce a pressure
in manometer 6 equal to the pressure in system 10. Then, valve 7 is closed
and valve 3 is opened, the piston of manometer 2 is raised to its working
position and is loaded to match the indication of gauge 6. Valve 5 is then
closed, the piston plus load is rotated and valve 7 is opened. By adding or
removing loads from the plate, a position is attained when the indicator of

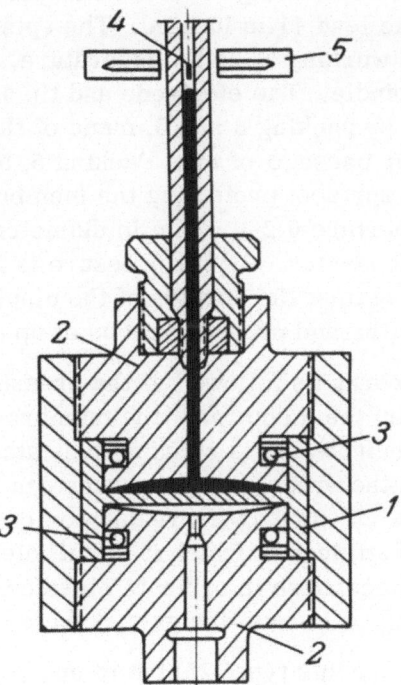

Fig. 94. Membrane mercury null instrument:
1) membrane; 2) pressure nuts; 3) sealing rings;
4) float; 5) coil.

instrument 9 reacts to a load change minimally for the given type of manometer. The amount of this load is used in calculating the pressure in system 10.

The equilibrium indicator between load and pressure may be a so-called membrane [diaphragm] divider* or a null instrument [24] as shown on Fig. 93. This instrument is constructed on the principle of the fully supported membrane.

Membrane 6, 0.1–0.2 mm thick, is clamped between body 1 and nipple 7 by nut 4. The membrane is made of brass or stainless steel. Since the membrane rests on the surface of nipple 7 and the fluoroplast packing 5, it can withstand pressure drops of dozens of atmospheres without rupture. The sensitivity of the instrument permits measurement of pressures with an accuracy of thousandths of an atmosphere.

———

*This instrument may be used for measurement of pressure by a Bourdon gauge if the medium being measured cannot be permitted to touch the curved tube of the gauge.

The upper portion of the body 1 contains nipple 10 in which spindle 8 with insulated electric lead 11 is located. The spindle can move in gland 9 which is cooled when working at high temperature. Electrode 2 with tip 14 is attached to the spindle. The electrode and tip are insulated from the body of the instrument by packing 3 and 5, made of fluoroplast in which aperatures are made for passage of oil. Packing 5, as was noted above, is also the upper support surface, protecting the membrane from rupture. Nipple 7 contains an aperture 0.2-0.5 mm in diameter, connecting the membrane cavity with the apparatus. As the pressure is increased above the membrane, it presses against the surface of the nipple; since the aperture is very small, it can withstand pressure drops of up to 100 atm.

The instrument works as follows: if the pressure in the apparatus is greater than that over the membrane, the membrane touches tip 14 and closes the electric circuit, which is noted on indicator 12. By feeding oil through nipple 10 from the press, the membrane can be moved away from the contact, opening the circuit. By increasing or decreasing the oil pressure, a position can be attained at which it is sufficient to change the pressure by 100ths of an atmosphere in order to close or open the circuit. A small, low-voltage lamp or a sensitive galvanometer is used as an indicator.

A null instrument of this type [25] but of another construction is shown on Fig. 94. The instrument consists of machined membrane 1, 25 mm in diameter and 0.2 mm thick, made of stainless steel. The membrane is compressed between two nuts 2 and sealed by plastic O-rings 3. Pressure nuts 2 have lens-shaped cavities allowing the membrane to move to either side by 0.1 mm. Mercury is poured in above the membrane to a height of 40-100 mm; float 4, made of soft iron, floats on the top of the mercury. The position of the float is determined inductively by the position of coil 5, connected into the measuring circuit. The instrument permits measurement of changes in the position of the membrane by 0.005 mm, i.e., allows measurement of a pressure of 1500 atm with an accuracy of 0.07 atm.

RELATIVE MANOMETERS

Manometers Using Elastic Properties of Materials

It is difficult to work with mercury or piston manometers. Therefore, when there is no need to measure the pressure with extreme accuracy, secondary (relative) instruments which can be calibrated using a piston manometer are used.

Manometer with Tubular Spring.
One of the most widely used relative manometers is the tubular-spring manometer [Bourdon gauge]. Its design and construction are well known [6], therefore, they will not be analyzed here.

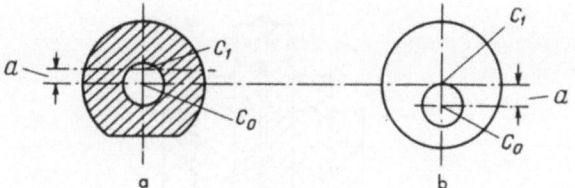

Fig. 95. Section of tubular springs: a) with flat; b) with
displaced aperture [eccentric bore].

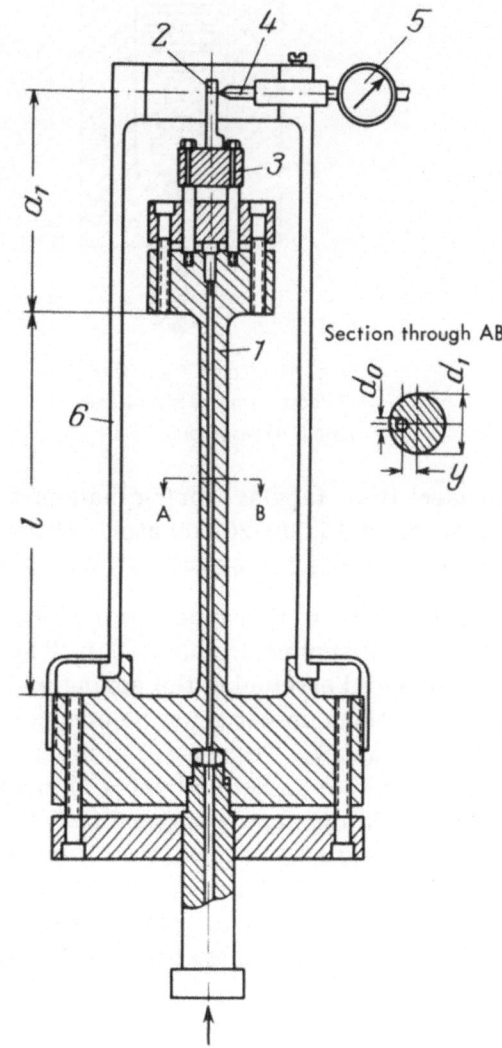

Section through AB

Fig. 96. Straight-tube manometer: 1) tube; 2) rod;
3) stand; 4) lever; 5) indicator; 6) protective cover.

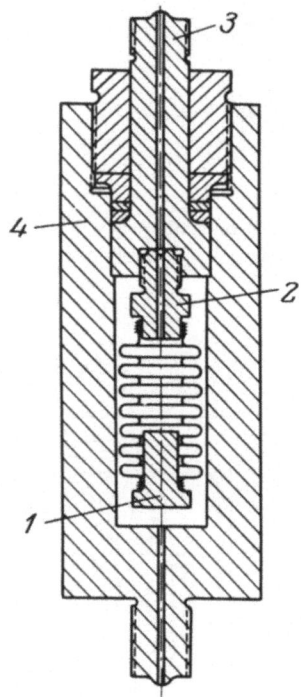

Fig. 97. Connection of bellows:
1) bottom plug; 2) passage; 3)
head; 4) reservoir.

At the present time, tubular-spring manometers are available for measuring pressures of 15,000-20,000 atm. Although their measurement accuracy is low (error 3-4%),* they are convenient as indicating instruments.

Standardized gauges of this type are made for measurement of pressures up to 5000 atm (instrument class 0.35). They require extreme care in use. In order to avoid stretching the spring and destroying the accuracy of the instrument, the manometer must be connected only at the moment measurement is to be made. Therefore, in measurement practice this type of standardized manometer is set up beside an ordinary device constantly showing the pressure in the apparatus, and the standardized manometer is only used at the time of measurement, then immediately disconnected and drained of liquid or gas. Tubular springs are subject to the phenomenon of hysteresis; therefore, these manometers, especially standardized instruments and manometers designed for high pressure, require frequent checking.

_____

* Editor's comment: There are available in the USA Bourdon gauges for use at pressures up to at least 7000 atm with a claimed full-scale accuracy of 0.1-0.2%.

The limiting measurable pressure is increased by using tubular springs [26] of cross-sections as indicated on Fig. 95. The movement of the free end of such a spring under pressure is caused not by deformation of the profile, as in ordinary tubular manometers, but by displacement of the neutral axis (C_1) passing through the center of gravity of the cross-section by a certain amount relative to the axis of the bore (C_0). The force acting on the bore axis creates a bending torque relative to the cross-sectional axis, which causes the tubular spring to bend toward the thicker wall.

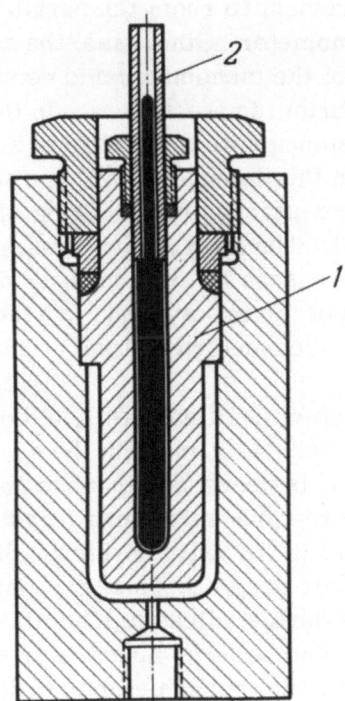

Fig. 98. Tait manometer: 1) cylinder; 2) capillary.

Straight-Tube Manometer. A tube with a channel displaced relative to the axis is also used in the manometer shown on Fig. 96 [27]. The manometer is a thick-walled steel tube 1 with two flanges. The aperture in the tube is displaced from the center by distance y; therefore, when a pressure is created in the tube, it bends toward the thicker wall. In this case rod 2, fastened to stand 3, presses level 4 of indicator 5. The deviation f of the rod is proportional to the pressure and can be calculated from the equation:

$$f = \frac{l^2}{d_1}\left(1 - \frac{2}{\mu}\right)\left(1 + \frac{2a_1}{\mu}\right)\frac{P \cdot 4\xi\alpha^2}{E\left(1+\alpha\right)\left(1-\alpha^4\right) - \left(1-\alpha\right)4\alpha^2\xi^2}$$

where μ is Poisson's ratio; E is the Young's modulus;

$$\xi = y\left(r_1 - r_0\right)$$
$$\alpha = r_0/r_1$$

l, a_1, r_0, r_1, and y are dimensions of the instrument in mm (see Fig. 96). According to the author of [27], $l = 332.4$, $a_1 = 130.2$, $r_0 = 2.48$, $r_1 = 8.21$, and $y = 2.85$.

The tubular manometer described above can be used for measurement of pressures of noncorrosive media and mixtures whose composition does not change (due to condensation of the components) as the temperature is changed from the temperature of measurement to room temperature, which is the temperature of the tubular manometer. Otherwise, the condensing material will adhere to the spring of the manometer and remain in the apparatus, which will disrupt the equilibrium in the system. In this case it is useful to isolate the manometer spring with a metallic bellows from the material being investigated. When this is done, plug 1 and passage 2, threaded for screwing into head 3, are attached by soldering or contact (roller) welding to the bellows (Fig. 97). The head closes reservoir 4, connected with the apparatus in which the pressure is to be measured. The reservoir is maintained at the temperature of the experiment. A steel capillary leading to the tubular manometer is connected to head 3.

The bellows-capillary-manometer system is filled with liquid oil (vaseline or transformer oil) so that no air bubbles remain in the system. In order to prevent crushing of the bellows it is necessary that the change in volume of the bellows under the action of the pressure being measured be greater than the compressibility of the oil in the entire system. Since upon heating of the oil, its pressure increases due to thermal expansion, the temperature cannot be increased in this device unless a preliminary pressure of 10 to 15 atm is produced in it. The bellows-manometer system must be first calibrated using another tubular manometer in order to determine the correction required for the rigidity of the bellows. For stainless steel bellows 10 to 15 mm in diameter, the correction will be a few atmospheres.

A great deficiency of tubular manometers is the frequent breakage of the springs. Operation with these manometers, especially if the compressed medium is a gas, requires special safety measures such as the installation of protective screens or plates of shatter-proof glass in front of the dials.

The accuracy of this type of manometer can be increased by using a helical spring with suppressed zero. This type of spring makes it possible to measure only a portion of the pressure interval near its upper limit. For example [28], a device for measurement of pressures up to 665 atm is supplied with a scale of 630-700 atm.

Tait Manometer. Manometers of other designs exist in which the pressure is measured by the elastic deformation of a material. One such is the Tait manometer [29] (Fig. 98). Cylinder 1, filled with mercury and terminating in glass capillary 2 is subjected to external pressure. The compression of the cylinder causes the mercury to rise in the capillary. This instrument has considerable hysteresis and requires large temperature corrections since it is in essence a mercury thermometer.

Manometers Using Change in Volume of Liquid or Gas Under Pressure. If we know the compressibility of any gas or liquid, we can determine the pressure by measuring the volume of the material under pressure. In particular, Michels [30] used a hydrogen manometer for calibration of his piston manometer.

A manometer based on the compressibility of water is a glass sphere filled with water and subjected to biaxial pressure. A glass capillary filled with mercury and subjected to the measured pressure is connected to the sphere. The mercury enters the sphere of water, moving down the capillary. Its position is determined by the resistance of a platinum wire embedded in the capillary. This manometer also is inconvenient to use. It allows pressures to be measured for which data on the compressibility of water are available. The indications of this manometer must be corrected for the compression of the glass. Also, the accuracy of the instrument is reduced due to drop formation of the mercury on the wire.

Electrical Manometers

The inconvenience of relative manometers based on the measurement of the elastic properties of materials has caused investigators to turn to the so-called electrical effect of pressure. Several designs of manometers have been suggested; these designs can be divided into two types: 1) electrical manometers, i.e., those in which the pressure changes the electrical properties of a material, and 2) so-called elastic-electrical manometers, in which the deformation of an elastic element is measured by the change in some electrical parameter of a transducer.

Manganin Manometer. Purely electrical manometers include manometers which measure a change in electrical resistance of a metal wire due to the pressure. Since this effect is not great, these manometers are convenient to use only for measurement of pressures on the order of several thousand atmospheres.

This type of manometer was first used [31] in 1903. Since that time, almost all the metals plus many alloys have been tested as wire materials; however, the material which has proven the best has been manganin, the material first used for such purposes.*

---

*The literature [32] contains an indication that an alloy containing 15.9% manganese and 84.1% silver has a greater rate of change of resistance with pressure and a lower temperature coefficient than manganin; information is also to be found concerning a new pressure-sensitive material [33] which is a product of the reaction between zirconium tetrachloride and the rare earth metals.

The advantage of manganin (a copper alloy containing 11% manganese and 2.5-3% nickel in addition to the copper) is that its resistance is a linear function of pressure. At the present time, it has been experimentally established [34] that the linear dependence is retained up to 10,000 atm with an error of 0.7%, and up to 30,000 atm with an error of a few percent. The accuracy of pressure measurement is a function of the accuracy with which the electrical resistance of the manganin wire is measured and the constancy of the properties of the manganin itself.

The change in resistance of manganin with pressure $\Delta R/R_0 \Delta P$ varies for different samples within limits of $2 \cdot 10^{-6}$-$2.5 \cdot 10^{-6}$ cm^2/kg. Let us take a measurement accuracy of resistance of 0.01 Ω, which is not unreasonable for measurements performed in a laboratory with ordinary instruments (bridges or potentiometers). The accuracy of pressure measurement then depends on the initial resistance of the coil used. For a number of reasons, this resistance is taken as 200 Ω. Then ΔP, with a change in resistance of the coil of 0.01 Ω will be:

$$\Delta P = \frac{0.01}{2.0 \cdot 10^{-6} \cdot 200} = 25 \text{ atm}$$

The value of ΔP, however, cannot be used to judge the accuracy of measurements which must be evaluated by relating ΔP to the value of the pressure measured. If the manometer for which a value of ΔP is 25 atm is used to measure a pressure of 10,000 atm, the measurement error is 0.25%, which is the accuracy which can be given by a manometer of a high degree of accuracy with a tubular spring (at considerably lower pressures). But pressures of 1000 and 500 atm with the same value of ΔP will be measured with errors of 2.5 and 5%. Thus, manganin manometers can best be used at pressures over 2000-3000 atm.

In order to determine the pressure more accurately, the resistance of the coil should be measured with an accuracy of no less than ± 0.001 Ω. For this, the manometer coil must be kept at a constant temperature since the coefficient of change in resistance of manganin with pressure depends on temperature [35], and changes with a temperature variation of 1°C (between 0 and 95°C) by an average of $2.2 \cdot 10^{-4}$ times its initial value.* This means that a change in surrounding temperature of 5°C during measurement or in comparison with the calibration temperature will produce an error of 0.1%.

An alloy of gold and chromium [36] has a considerably smaller variation in resistance with temperature. The best results are produced with a wire made of an alloy of chromium and gold containing 2.1% chromium.

---

* Change in temperature of the wire also leads to a change in its resistance.

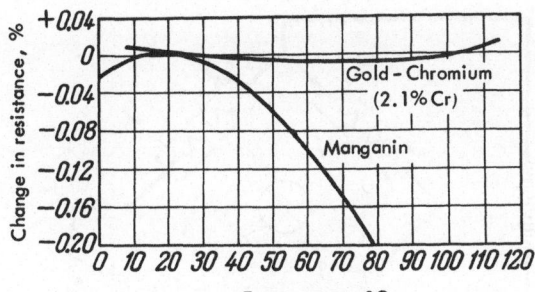

Fig. 99. Change in resistance of gold—chromium alloy
and manganin with temperature.

The resistance of gold—chromium alloy wire changes almost linearly
with pressure. The coefficient of resistance with pressure for this alloy
is $0.96 \cdot 10^{-6}$, i.e., gold—chromium alloys are three times less sensitive
to changes in pressure than manganin. However, the independence of the
indications with respect to temperature is an extremely valuable advantage.
The dependence of resistance on temperature for this alloy and for manganin
are presented on Fig. 99. The gold—chromium alloys change their resistance very little over a wide temperature interval. However, this alloy is
extremely sensitive to mechanical action, as well as to cooling. The stress
formed in the wire as it is wound into a coil disappears after heating to
150°C. It should be noted, however, that when heated to 175°C the wire
takes on a drastically increased sensitivity to temperature change.

Depending on the required accuracy, various methods of measurement of coil resistance are used.

Balanced Bridge Circuit.   A Wheatstone bridge is used as a measuring
arm, permitting measurement of coil resistance with an accuracy of $\pm 0.01\,\Omega$.
The bridge is supplied by a six-volt battery. The higher the voltage, the
more sensitive the circuit. However, too strong a current may cause overheating of the thin wire in the coil, which distorts the measurement results.

Unbalanced Bridge Circuit.   This circuit is more simple and reliable, but
less accurate. It uses a galvanometer rather than a null instrument. When
there is no pressure, the bridge is so balanced that the galvanometer indicates zero. When the bridge is unbalanced due to a change in resistance
of the manganin coil, the arrow of the galvanometer is deflected. If the
galvanometer is properly graduated, pressure can be measured directly in
atmospheres. The bridge operates on DC, the voltage being stabilized with
high accuracy. The circuit is convenient in that it includes an indicator
(in contrast to the preceding circuit).

More Accurate Circuits.   In order to increase the accuracy of determination of pressure, the electrical resistance of the coil of the manometer must

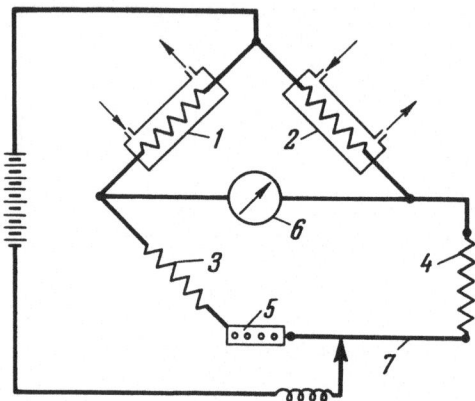

Fig. 100. Katsnel'son-Ikhlov circuit for measurement of resistances using a manganin coil:
1) coil under measurement pressure; 2) coil
at atmospheric pressure; 3, 4) bridge coils; 5)
resistance box; 6) mirror galvanometer; 7)
rheostat.

be measured with an accuracy of no less than $\pm 0.001\ \Omega$. This cannot be
done with ordinary instruments with coil resistances of 200 Ω. Therefore,
Katsnel'son and Ikhlov suggested a circuit (Fig. 100) in which only the
change in resistance with pressure is measured. The resistance R_x of
coil 1, which is subjected to the pressure to be measured, is unknown in
the balanced Wheatstone bridge shown in the figure. The initial resistance
of the coil is accurately measured. The second arm of the bridge is coil
2, which has the same resistance (with an accuracy to 0.2-0.3 Ω), but which
is not under pressure. Both coils are temperature-controlled. Coils 3 and
4, with the same impedance as coils 1 and 2, are connected in the bridge
as well. They are placed side by side, in order that the temperature conditions are identical. Resistance box 5 with selections 0.5, 1, 2, 3, and 4 Ω
is connected in series with coil 3. The circuit is closed by rheostat 7, made
of manganin wire 1 m long, with a resistance of about 1 Ω. Moving the rheostat mobile contact 1 mm to the right corresponds to connection of a resistance of about 0.001 Ω to the arm of the bridge containing coil 3 and removal
of an identical resistance from the arm containing coil 4. Mirror galvanometer 6 is used as a null instrument.

After the circuit is connected, electrical calibration of the rheostat
must be performed, i.e., the values of its division must be determined.
This requires that the resistance box be connected in the circuit in place
of the manganin coil and, while changing resistances, the position of the
rheostat mobile contact at bridge equilibrium is noted. The results of the
measurement are used to construct a graph showing R_x directly from the

indications of the rheostat. This dependence cannot be calculated, since movement of the contacts on the rheostat changes the resistance of two arms of the bridge at the same time.*

A variant of this plan with a potentiometer is described in work [39].

Calibration of Manganin Manometer. The manganin manometer is a secondary instrument requiring calibration. It can be calibrated most precisely by using a weight manometer. Since the indications of a manganin manometer are linear over a large pressure interval, the manometer can be calibrated using only two reference points, one of which may be atmospheric pressure, the other the melting pressure of mercury at a given temperature. M. K. Zhokhovskii [40, 41] presents an empirical formula for the dependence of melting pressure of mercury on temperature up to 25,000 atm:

$$\log(P + 37663) = 1.21458 \log T + 1.69765$$

where P is the pressure, kg/cm^2; T is the temperature, °K.

The freezing point or melting point of mercury under pressure has been determined numerous times. The authors [14] performed a very careful investigation and produced a mercury solidification pressure at 0°C of 7716 ± 51.5 kg/cm^2; according to the data of M. K. Zhokhovskii, this pressure is 7715 kg/cm^2; these data differ from those of Bridgman [31], which are still in use (7640 kg/cm^2).

In calibration, the movement of a piston compressing the mercury or another material is measured as a function of pressure. The curve drawn in the coordinates "movement of piston vs. pressure" will show a bend [change in slope] at the mercury solidification point.

A special installation is required for this type of calibration. The author has suggested a simpler and sufficiently reliable method of calibration using the melting point of mercury [42]. A metal vessel containing 400 g of carefully purified mercury is placed in a high-pressure column. The column is closed and the zero point of the manganin manometer is tested. Then, the pressure in the insulation is increased to approximately 7000 atm. The pressure can be determined approximately from the reading of the manometer on the low-pressure line to the pressure intensifier or from the coefficient of change in resistance of manganin with pressure. Before cali-

---

*Other precise circuits are also presented in the literature [14, 37, 38].
For example, the sensitivity of the indicator of an AC balanced bridge (frequency 1000 Hz) allows a change in resistance equal to $1:10^7$ to be measured.

bration, this coefficient can be accepted as equal to $2.5 \cdot 10^{-6}$ cm^2/kg. Then, the temperature is lowered to 0°C and calibration is begun. Benzine is fed into the column, with a waiting period of 5 to 10 minutes after each dose of benzine is fed in, in order to allow the heat of compression to dissipate and to allow the new dose of liquid to take on the temperature of measurement. During calibration, the resistance of the manometer coil is measured as it changes, depending on the quantity of liquid fed into the apparatus, generating pressure in the mercury. Since the mercury freezes only after a considerable excess pressure is produced (a phenomenon like supercooling), it is easy to note that after a smooth increase in resistance in the coil, at a certain point the resistance will drop sharply and will remain at a certain value, which can be accepted as the resistance at 7715 atm. On the other hand, when the pressure is reduced, the resistance of the coil, after first dropping, increases back to the same value. These changes are so clear and so easily reproducible that they allow an accurate reading of the resistance value corresponding to the freezing pressure of mercury at 0°C.*

Polymorphic conversion temperatures, determined from the sharp break in the curve of volume vs. pressure, can be used as reference points for the calibration of a manometer at higher pressures. The most reliable scale of such conversions at the present time is that produced by Kennedy and La Mory [44] using an installation in which the piston compressing the material being investigated was rotated to reduce friction.

The polymorphic conversion pressures of certain materials are presented below (in bars):

$Bi_I - Bi_{II}$	$Bi_{II} - Bi_{III}$	$Tl_{II} - Tl_{III}$	$Cs_{II} - Cs_{III}$
$25\,410 \pm 95$	$26\,975 \pm 190$	$36\,690 \pm 100$	$41\,800 \pm 1000$

For the measurement of even higher pressures, data on the change in melting point of germanium under pressure can be used [45]. The dependence of melting point on pressure is linear up to 180,000 atm. The reduction in temperature equals $(3.27 \pm 0.7) \cdot 10^{-3}$ deg/atm.

The points of sharp change in electrical resistance of certain metals [46] with pressure can be used for this same purpose (Table 28).

It should be noted that in calibration of apparatus, the ratio of load to anvil area is sometimes defined as a function of pressure. In reference [47] it was shown that the generation of a stress gradient in the pyrophyllite used as a plastic medium caused the results of this sort of calibration to differ by up to 40% from experiment to experiment, depending on the location of the sample.

---

*The calibration of instruments for measuring pressure is described in the literature [43].

TABLE 28. Reference Points for Change of Electrical Resistance

Metals	Pressure, kbar	Nature of change
Bismuth	83	Sharp reduction in resistance
Iron	133	Sharp increase in resistance
Barium	144	" " " "
Lead	161	" " " "
Rubidium	193	" " " "
Calcium	375	Maximum point in graph

M. K. Zhokhovskii [48] (see also [19]) suggested a thermodynamic method for reproducing a scale of extremely high pressures. According to this method, the dependence of melting point of the material being investigated on pressure is determined, and expressed empirically by an equation; the curve is extrapolated to higher-pressure areas. After this, the piston manometer is used to calibrate a manganin manometer, the dependence produced being also expressed by an empirical equation and extrapolated to even higher pressures. Then, these measurements are combined into one experiment. The curves produced should match. Then, the melting of other materials is investigated, selecting the materials so that their melting curves intersect with the melting curve of the first material. The points of intersection of the extrapolated sectors should be confirmed experimentally.

Later, the author [49-51] showed that if a generalized melting-process parameter, such as the specific melting energy $\lambda/\Delta V$ (where λ is the heat of melting, and ΔV is the change in volume upon melting) is introduced, we find that $\ln(\lambda/\Delta V)$ is a linear function of $\ln T$ along the melting curve, and $\lambda/\Delta V$ is a linear function of pressure. The slope of these two lines is identical, and their value equals the constant C in the Simon [52] equation:

$$\frac{a+P}{a} = \left(\frac{T}{T_0}\right)^C$$

This empirical equation, as we know, indicates the dependence of melting point of the material on pressure, the constants A and C for each material generally being determined experimentally.

The Simon equation can be represented in a new form, including the parameter $\lambda/\Delta V_0$, i.e., the specific melting energy of the material, in place of the empirical constant a at the triple point:

$$P = \frac{1}{C} \cdot \frac{\lambda_0}{\Delta V_0}\left[\left(\frac{T}{T_0}\right)^C - 1\right] \tag{1}$$

$$C = \frac{\ln \frac{\lambda}{\Delta V} \cdot \frac{\Delta V_0}{\lambda_0}}{\ln \frac{T}{T_0}} \qquad (2)$$

The constancy of constant C can be established by experiments which do not require pressure measurement. It is sufficient for this purpose to measure the latent heat of melting λ and the change in volume ΔV, as well as the temperature, along the melting curve. If the value of constant C calculated according to equation (2) does not change, equation (1) is an analytic expression of the thermodynamic scale of pressure in the pressure interval for which the values of C are constant.

If we know the values of $\left(\frac{\partial \Delta V}{\partial T}\right)_P = \Delta \beta$ and $\left(\frac{\partial \Delta V}{\partial P}\right)_T = \Delta \alpha$, the change in temperature with pressure can be expressed by the following differential equation:

$$\frac{dP}{dT} = -\frac{1}{\Delta \alpha}\left[\frac{b}{T_0}\Delta V_0 e^{-b\,(T/T_0-1)} + \Delta \beta\right]$$

where b is an empirical constant.

This equation is correct for materials in which

$$\Delta V = \Delta V_0 e^{-b(T/T_0-1)}$$

This condition, calculations have shown [50], is fulfilled for the majority of materials investigated.

A method of constructing a high-pressure scale using points of intersection of melting curves of various materials on the plane PT has also been suggested [8].

*Design of Manganin Manometers.**   Designs of manganin manometers differ primarily in the method of introducing the electrical lead and the method of manufacturing the coil.

In order to reduce the dimensions of the coil, a conductor diameter of 0.05-0.03 mm is used. The use of such a thin conductor in enamel and double silk insulation requires skill and dexterity on the part of the experimenter. Practice has shown that the conductor should be wound bifilar on a shell of wax paper and glued with cellulose nitrate varnish.

---

*The first work performed in the USSR on the creation of a manganin manometer for measurement of pressure up to 10,000 atm was published by B. S. Aleksandrov and L. F. Vereshchagin [53].

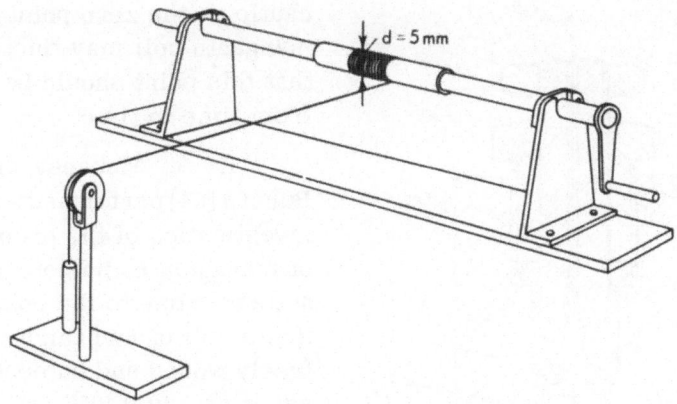

Fig. 101. Device for winding manganin coils.

We can recommend the device shown on Fig. 101 for winding coils. To use the device, aluminum foil is wrapped around the center shaft, then covered with wax paper. Then, the center of the conductor to be wound around the shaft is fastened to the waxed paper using a silk thread. After this, both ends of the conductor are put under tension using a light load suspended over the pulley. After the wire is straight, and the experimenter has satisfied himself that there are no defects in the wire, the wire is slowly wound around the center shaft, with each turn touching the next, leaving the last 1-2 cm of the conductor unwound. This free end is fastened with silk thread and the entire coil is covered with a thin layer of cellulose nitrate varnish. After drying of the varnish, the coil is wound with silk thread and varnished once more. During winding, we recommend that the conductor be handled with waxed paper in order to ensure that no oil or moisture from the skin gets on the conductor.

When a coil is ready, the foil covering is removed from the shaft, and the shaft is pulled out of the coil. If the center hole in the coil is 5 mm and the wire diameter is 0.03 mm, the coil height will be 3 to 4 mm. The prepared coil is aged, in order to remove stresses from the conductor and thereby eliminate fluctuation in the zero point of the manometer. For this, the coil is heated in a current of dry nitrogen for 2 hours at 130°C, after which it is placed in a vessel with dry ice for 10-15 minutes. These operations are repeated for 30-40 hrs. It is then useful to place an ebonite or Plexiglas insert containing copper conductors inside the coil, soldering the ends of the coil to the copper conductors. The coil is then subjected to a pressure somewhat higher than its working pressure, after which it can be calibrated. All coils should be calibrated, even those made from the same spool of wire, since the resistance of various segments of wire may differ.

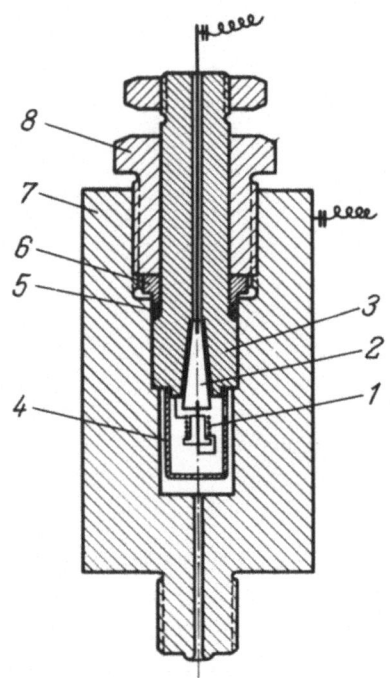

Fig. 102. Manganin manometer: 1) coil; 2) insulated cone; 3) head; 4) cap; 5) packing; 6) bottom box; 7) body; 8) nut.

Even with all these precautions, the zero point of a manganin coil may fluctuate, so that this point should be checked from time to time.

K. A. Alekseev and L. L. Burova [54] performed a detailed investigation of the properties of manganin manometers. These authors came to the conclusion that a coil of manganin wire, freely wound and subjected to aging at 130-140°C, still has a nonlinear dependence of resistance on pressure. The nonlinearity, i.e., the change in pressure coefficient [$\Delta R/(R_0 \Delta P)$] with pressure, reaches approximately 1.5% at 10,000 atm, according to these authors. For wire of the same type but different diameters, the coefficient fluctuates within limits of ~ 4%, whereas for wire of the same diameter but taken from various points on the spool the fluctuation is 0.2%.

K. A. Alekseev et al. [55] suggested another design for manganin coils and a new method of aging manganin. A coil is prepared in the form of a thin spiral, placed in longitudinal grooves in a steatite frame. Four such spirals are connected, joining the ends with copper foil petals. After the transducer is assembled, the coil is annealed with a current pulse from an 800 μf condenser. With this method of annealing, the pressure coefficient of all transducers tested differed by less than 1%.

Figure 102 shows a manometer used by the author for many years. One end of coil 1 is soldered to lead 2, which consists of an insulated cone, while the other end is soldered to a conductor connected to the body of head 3. Then, cap 4 is placed over the coil. The head is completely assembled, i.e., packing 5 and bottom box 6 are installed. Pure, dry benzine is poured into body 7, in the vertical position.

The way in which the pressure is transmitted to the coil is very important. If the pressure created by benzine, kerosene, or another nonelectric-conducting liquid (in the absence of moisture) is measured, direct trans-

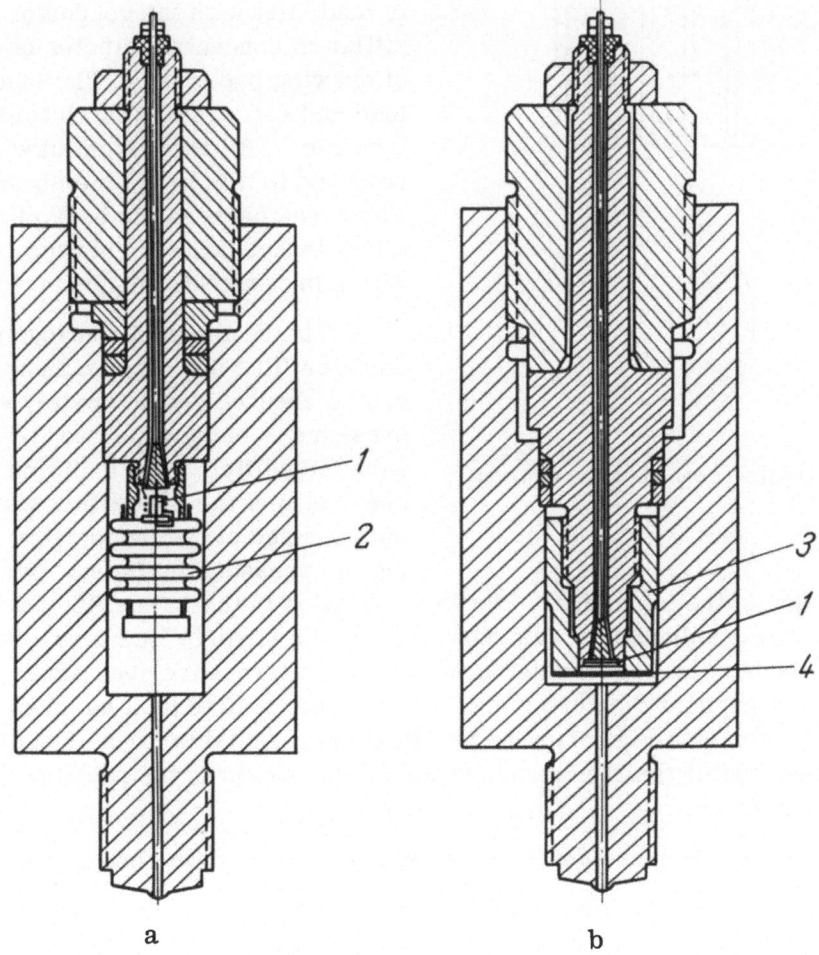

Fig. 103. Section of bellows (a) and membrane (b) coil devices: 1) coil; 2) bellows; 3) sleeve; 4) membrane.

mission of the pressure to the coil is possible. This method can be used only up to 10,000 atm, due to the great increase in viscosity of the liquid and resulting appearance of shear force in the coil, which distorts the readings of the manometer at higher pressures.

The pressure of noncorrosive gases and oxygen can be measured directly, without filling the manometer with a pressure-transmitting fluid. In all other cases, the coil of the manometer must be separated from the medium in which the pressure is being measured. This can be done using metal bellows, placed over the coil and filled with a fluid such as pentane (Fig. 103a), or by a membrane (Fig. 103b).

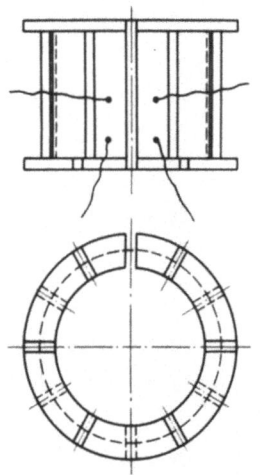

Fig. 104. Plan of compressometer.

If the membrane is used, the coil is made flat with the conductor wound bifilar in concentric circles on a base of drawing paper [56]. The electrical lead and coil are screwed into sleeve 3 (see Fig. 103), the bottom of which is soldered to thin brass membrane 4. The space between the body of the electrical lead and the membrane is filled with a pressure-transmitting fluid.

The readings of a manganin manometer (like any other manometer) can be checked approximately, where pressure is being generated by a pressure intensifier, by calculation of the area ratio of the intensifier pistons and the readings of the manometer indicating the pressure in the low-pressure cylinder of the intensifier. However, friction in the intensifier packing must be considered; as was shown above, the straight line drawn in the coordinates "pressure under large piston vs. pressure over small piston" does not begin at the zero point, but is rather shifted by a value equal to the friction (see Fig. 51). This method is not suitable for self-sealing [self-energizing] piston packings, since the stress in the packing changes with pressure.

Manganin manometers of other designs are described in the literature [6, 20].

Compressometer [57]. Pressures over 30,000 atm can be measured with a device (Fig. 104) consisting of a thin-walled steel sleeve with a number of slots. One of the slots goes from top to bottom, while the rest do not, ending alternately short of top and bottom. This positioning of the slots lengthens the path of an electric current passing through the sleeve. As the sleeve is compressed, its electrical resistance changes. By measuring this resistance, the force acting on the sleeve can be determined and the pressure in the apparatus can be calculated.

The sensitivity of the device is higher the less the cross-sectional area of the sleeve. It is therefore important that the material of which it is made have great strength and be able to withstand considerable uniaxial compression without rupture [see Fig. 58 for clarification of presence of uniaxial compression].

The resistance of a compressometer changes both as it is compressed [uniaxially] by the piston of the apparatus in which the pressure is being gen-

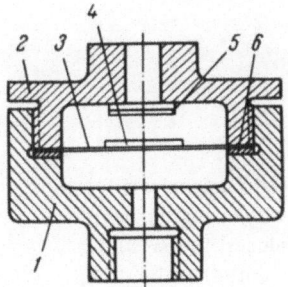

Fig. 105. Tensimetric ma-
nometer: 1) body; 2) cover;
3) diaphragm; 4, 5) transducer; 6) packing.

erated and under the influence of hydrostatic pressure. Therefore when
the compressometer is calibrated, its resistance is determined as a function of the compressive load at atmospheric pressure and of the hydrostatic pressure; the change in compression coefficient of the steel with
pressure is also determined. The change in resistance of the compressometer is measured potentiometrically. This requires four electrical leads.
Two of these act as current-carrying leads (one grounded), the two others
leading to the potentiometer.

The hydraulic support pressure in the apparatus is measured with a
manganin manometer.

Compressometers are used to measure pressure from 30,000 to
100,000 atm in an apparatus with variable mechanical and hydraulic support (see Chapter III). Pressures over 10,000 atm are measured by the
load applied to the anvil [or piston] of the apparatus and checked from reference points by placing checking materials in the apparatus together with
the material being investigated.

Tensimeters. If a manganin or constantan wire is wound around
a steel tube and the tube is subjected to internal pressure, the change in resistance of the extended wire can be used to measure the internal pressure.

The tube should be so designed that its walls will undergo only elastic
deformation.

This method can be used to measure pulsating pressures [58] such
as in internal combustion engines. In this case, the change in resistance
of the tube can be registered on an oscilloscope. Even more rapidly changing pressures (such as during adiabatic compression of gases, when the process lasts only hundredths or thousandths of a second) are measured with special instruments.

Tensimetric Manometer [59]. A manometer of this type
(Fig. 105) is a chamber 1 in which cover 2 clamps diaphragm 3 to packing
6. A tensimetric transducer 4 is applied to the diaphragm. Transducer 5
is used to compensate for temperature deformation of transducer 4.

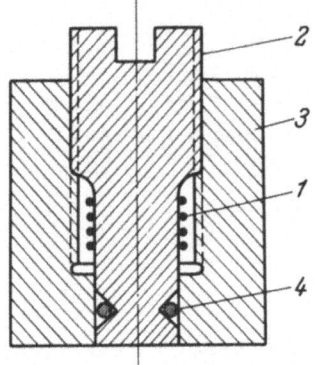

Fig. 106. Electric manometer for measuring rapidly changing pressures: 1) transducer; 2) rod; 3) body of instrument; 4) sealing ring.

Under pressure the diaphragm bends, and stresses arise in the diaphragm which are measured by the tensimetric transducer. By using diaphragms of various thicknesses, pressures over a wide pressure range can be measured. Following is a dependence between diaphragm thickness δ and measured P:

δ, mm	0.1	0.2	0.5	1.0
P, atm	0.004—0.02	0.016—0.7	0.1—4.5	0.4—18
δ, mm	2.0	5.0	7.0	
P, atm	1.6—75	10—450	19.6—900	

The same principle is used in the design of an apparatus described in work [60]. This apparatus carries four wire tensimetric transducers with resistances of $\sim 200 \, \Omega$ each. Two are used for initial balancing of the measuring bridge, a third for compensation of the influence of temperature. The measuring tensimetric transducer is fastened to the high-pressure vessel; the transducer for compensation of change in temperature is placed beside the measuring transducer. The two remaining transducers are fastened to a steel plate (on opposite sides). The bridge circuit is balanced by bending the steel plate.

A change in pressure in the vessel can be read directly in scale divisions of a galvanometer acting as a null instrument.

The literature [61] contains a description of a method for measuring pressure from the volumetric compressibility of a metal probe under hydrostatic pressure.

Capacitive Manometers. A membrane manometer in which the bending of the membrane is measured using an electric circuit is called a capacitive manometer. The membrane serves as one side of a

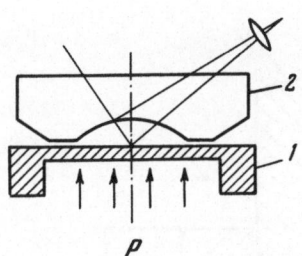

Fig. 107. Plan of interference manometer: 1) diaphragm; 2) quartz plate.

condenser whose capacitance is measured as the membrane moves. These manometers are described more fully in the literature [62-65].

Descriptions are available of other types of electric manometers for measurement and recording of rapidly changing pressures [66-68]. One such is the instrument shown on Fig. 106 [69]. Transducer 1, made of constantan wire, is applied to rod 2, screwed into the body of the instrument 3. Sealing ring 4 is used to protect the wire from the action of the gases whose pressure is being measured. The pressure being measured causes elastic deformation (longitudinal compression) of the rod, causing its cross-section to change. This causes the constantan wire wrapped around the rod to be extended and changes its resistance. The change in resistance is measured by a sensitive circuit. This instrument allows the entire picture of pressure change with time to be traced.

Interference Manometer. Of the many manometer designs suggested recently, the interference manometer is one of the most interesting [70]. This instrument (see Fig. 107) consists of a steel diaphragm 1, which receives the pressure. The diaphragm is carefully polished and covered with quartz plate 2. A ray from a special light source strikes the diaphragm and interferes [optically]. A change in the form of the diaphragm with pressure changes the number of interference bands, which is recorded on photographic film. This instrument can be used to measure rapidly changing pressures.

Piezoelectric Manometer [62-65, 71]. The action of this manometer is based on the fact that mechanical oscillation of a quartz plate is transformed into an electric potential.

This effect, discovered in 1817 by Ayui, is explained by the fact that when a quartz plate is compressed on certain planes (perpendicular to the so-called electrical axis of the crystal) a potential is formed proportional to the force applied to the plates. The sign of the potential depends on the pair of planes to which the force is applied. In addition to quartz, other materials also have this property (such as Seignette salt [potassium tartrate-hydrate]).

By measuring the charge potential, the pressure applied to the plate can be determined.

In view of the fact that electrostatic charges arise in quartz instantaneously and disappear rapidly, they are recorded by transforming them,

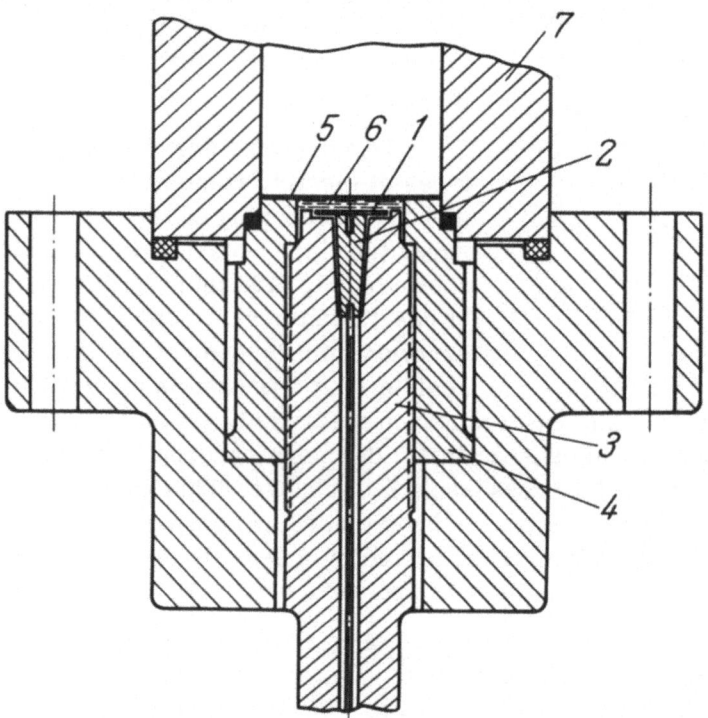

Fig. 108. Manganin transducer for measuring rapidly changing pressures:
1) flat coil; 2) electric lead; 3) nipple; 4) sleeve; 5) membrane; 6) pressure-transmitting fluid; 7) adiabatic-installation column.

using a special circuit, to electric current oscillations, which are then
recorded by an oscillograph. One specific feature of this type of manometer is its ability to measure rapidly changing pressures, due to which this
type of manometer is often used for measurement of pressures in the bar-
rels of artillery weapons, in internal-combustion-engine cylinders, etc.
Although the instrument itself is simple, the measurement circuit required is complex, requiring high-quality insulation of all current-conducting leads; also, the indications of the instrument depend on many external conditions (moisture content of the air, etc.).

Rapidly changing pressures may be measured using the manganin
transducer [56] whose design is shown on Fig. 108.

Ryabinin Crusher [Upsetting] Manometer [72]

Another type of instrument for measurement of dynamic pressure is
the crusher [upsetting] manometer which can be used only for measurement of maximum pressure.

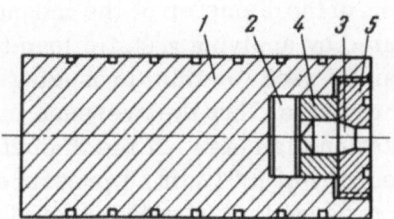

Fig. 109. Crusher [upsetting] manometer:
1) piston; 2) crusher [material being in-
dented]; 3) block; 4) ring; 5) nut.

The Ryabinin crusher [up-
setting] manometer* differs from
ordinary crusher manometers.
Piston 1 (Fig. 109) contains crusher
2 consisting of a plate of well-an-
nealed red copper, containing hardened steel block 3 with tip angle
120°. The block moves in guide ring
4, and its movement is limited by
nut 5, which presses ring 4 through
the crusher. Thus, all parts except the block are immobile.

The piston in the column of an adiabatic installation moves, compressing the gas toward the closed end of the column, then stops and flies
back. When the piston stops, the block, freely moving within the piston,
continues moving in its previous direction due to inertia, and presses into
the copper plate. Doing this, it leaves an imprint in the plate. The force
f with which the block is impressed into the copper plate is equal to

$$f = mw$$

where m is the mass of the block; w is the acceleration (in this case deceleration) experienced by the block.

The same deceleration is experienced by the piston. Therefore,

$$w = \frac{f}{m} = \frac{F}{M} = \frac{PSg}{Q}$$

where F is the force decelerating the piston; P is the pressure; S is the
area of the cross-section of the piston; M is the mass of the piston; g is
the acceleration of the force of gravity; Q is the weight of the piston.

Representing the weight of the block as q, we get the relation

$$P = \frac{Qf}{qS}$$

If we know the weight of the piston, the area of its cross-section,
and force f, we can determine the pressure. The force f is measured
from the diameter of the indentation made in the brass by the block.

---

*The Ryabinin crusher [upsetting] manometer is used for measurement
of pressures in an adiabatic installation.

In order to determine the dependence of the diameter of the indentation on the force f, the crusher is calibrated by applying a static load to
the plug. A Rockwell or Brinell device can be used for this, producing a
so-called calibrating graph which shows not force f but the pressure in
atmospheres corresponding to each indentation diameter. If the diameter
of the indentation is measured with an accuracy of 0.01 mm, a pressure
measurement accuracy of ±20 atm can be produced. It should be noted that
the accuracy of this manometer is usually lower than this. The error may
reach 10%, since the mechanical properties of the copper change from
sample to sample. Also, the change in plasticity of the crusher depending
on the rate of compression introduces an element of error.

It is clear from all the above concerning manometers that a high-
pressure laboratory should have piston and manganin manometers for extremely precise measurement of pressure and Bourdon gauges for approximate measurements.

DIFFERENTIAL MANOMETERS

In the process of investigation, it is often necessary to measure small
pressure drops under very high total pressures. Instruments which can be
used to make these measurements are called differential manometers.

The methods of measurement of pressure drops are highly varied in
their degree of accuracy and in the design of the measuring instruments
used. When the pressure drop to be measured does not exceed a few dozen
cm of a liquid column, and the total pressure is not too great, thick-walled
glass capillaries are used and the fluid level is observed directly. Often
(for higher pressures) special types of glass are used. Although very
narrow glass capillaries are capable of withstanding high pressures, they
cannot be used for manufacture of manometer columns, since capillary
phenomena distort the indications. Wider tubes are not strong. Also, it
is extremely difficult to join glass with metal, to seal the joints, etc.
Therefore, indirect methods of determining the level of the liquid in manometer columns are more satisfactory, though not as simple. If the liquid used does not conduct electricity, its level in a closed vessel can be
determined by measuring the resistance of a platinum wire heated by a
current to a certain temperature and connected into a bridge circuit. As
the wire touches the liquid, it is cooled and its resistance decreases sharply,
unbalancing the bridge.

Usually in high-pressure work, the manometer liquid used is mercury. This facilitates the problem and allows the mercury level to be determined by its ability to close electrical contacts located at various levels
in the differential manometer. This method is insufficiently accurate, allowing only a rough determination of the mercury level with an accuracy

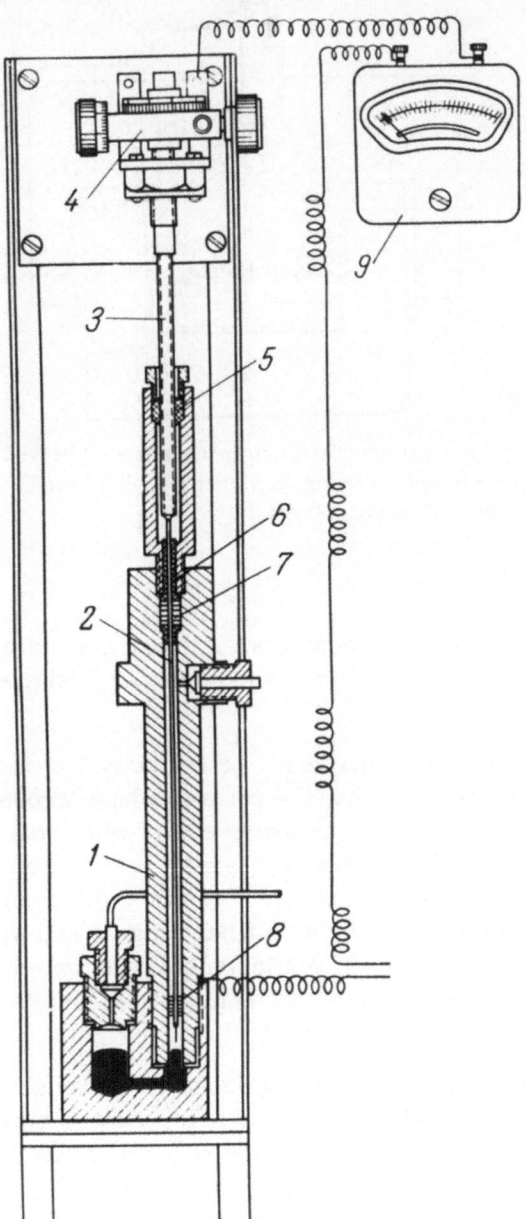

Fig. 110. Bol'shakov differential manometer: 1) body;
2) pin; 3) screw; 4) micrometer; 5) fiber rings; 6) fiber
sleeves; 7) leather gland; 8) directing ring; 9) galvanometer

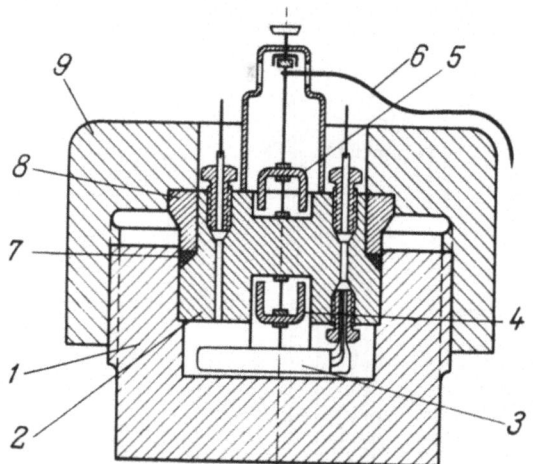

Fig. 111. Aristov contactless manometer: 1) body; 2)
cover; 3) tubular spring; 4, 5) magnets; 6) arrow; 7)
packing; 8) bushing; 9) nut.

equal to the distance between contacts. A more accurate method is the
moving-contact technique used in the manometer designed by P. E. Bol'-
shakov [73].

Bol'shakov Manometer. Steel body 1 of the instrument (Fig.
110) has two connected columns, the ratio of whose cross-sections is 1 : 25.
The narrow column has pin 2, the lower end of which carries a platinum
contact. The pin is held with screw 3, which has a fine thread. The screw
is firmly connected to micrometer 4. The pin and thread are insulated
from the body of the manometer with fiber rings 5 and sleeves 6, leather
gland 7, and guiding rings 8. The micrometer has two drive gears and a
disk with a vernier drive, which can be used to measure movement of the
screw to an accuracy of 0.001 mm.

The body and pin of the differential manometer are connected to an
electric circuit, supplied by a thermocouple. As the mercury rises, it
touches the platinum contact in the right column of the manometer, which
causes the arrow of galvanometer 9 to move. By raising or lowering the
pin, it is easy to determine the level of the mercury.

A deficiency of this manometer is the possibility of play in the gear
teeth of the micrometer and some adhesion of the mercury to the contact.

Contactless Manometers. So-called contactless instruments include the Aristov manometer [74] (Fig. 111). The body 1, filled

with oil, is closed by cover 2, which is connected to tubular spring 3. The spring is also filled with oil and connected to magnet 4, rotating with the shaft to which it is attached. Another magnet 5 is attached to a shaft on the outside of the nonmagnetic steel cover. Its shaft is connected to arrow 6 which reads on a scale on the side of the device.

The tubular spring is under differential pressure and shows the pressure difference. Depending on the rotation of the spring, magnet 4 rotates, turning magnet 5, connected to the arrow.

This manometer allows measurement of pressure drops over a rather wide range of pressures. Since the tubular spring is under differential pressure, its thickness need not be too great. A standard spring, designed for measurement of pressures up to 2 atm, will give a measurement accuracy of no less than 0.05 atm.

Another design for a similar manometer is the Shelaputin and Voitovich instrument [75], in which the flexible element is an 80/55 mm bellows with 13 convolutions. This manometer operates under pressures up to 160 atm and measures pressure drops up to 1000 mm Hg.

An electropneumatic differential manometer for measurement of pressure drops from 5 to 80 atm with a total pressure of 700 atm has been described by D. V. Kiselev [76].

Gamburg-Katsnel'son Manometer [77]. The device consists of two interconnected columns filled with mercury. One has considerably greater cross-section, which increases the accuracy of the manometer. The narrow column is terminated with a transducer (Fig. 112) located within the metal portion of the apparatus. The transducer consists of glass tube 1 with trap* 2. The tube is surrounded by coil 3 containing a resistance winding of approximately 40 Ω. Hollow iron core 4 (float) containing 3 notches at its ends,

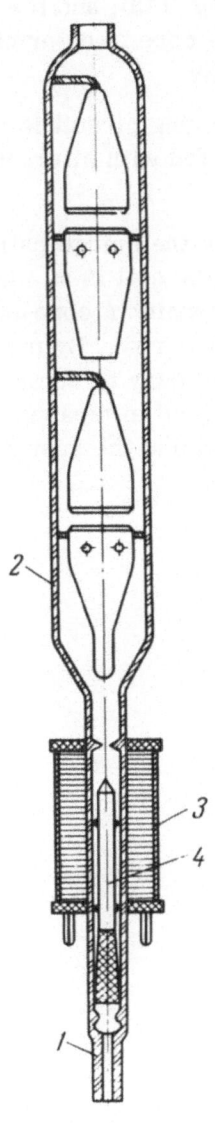

Fig. 112. Contactless Gamburg-Katsnel'son manometer transducer: 1) glass tube; 2) trap; 3) coil; 4) float.

_____

*The trap prevents the mercury from entering the high-pressure apparatus.

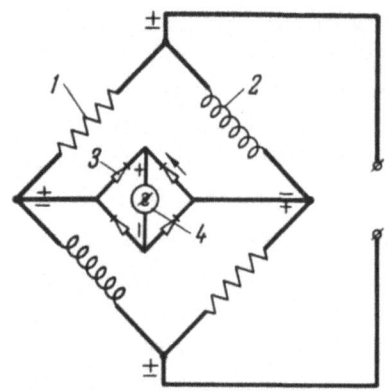

Fig. 113. Electrical circuit of contactless manometer: 1) resistance; 2) transducer coil; 3) copper-oxide rectifier; 4) galvanometer.

which center the float in the tube, moves within the tube. As the mercury rises or drops, the core moves within the tube in the coil. The coil is connected to an electrical measuring circuit (Fig. 113), and the movement of the core is determined galvanometrically.

The measuring circuit is an ordinary bridge fed with alternating current.

The arm of the bridge paired with the measuring coil is an identical coil, within which a core moves on a micrometer screw. By moving the core, it is easy to determine the true position of the float. This device is designed to measure extremely small pressure drops (approximately 5 mm Hg) and the accuracy is 0.02 mm Hg.

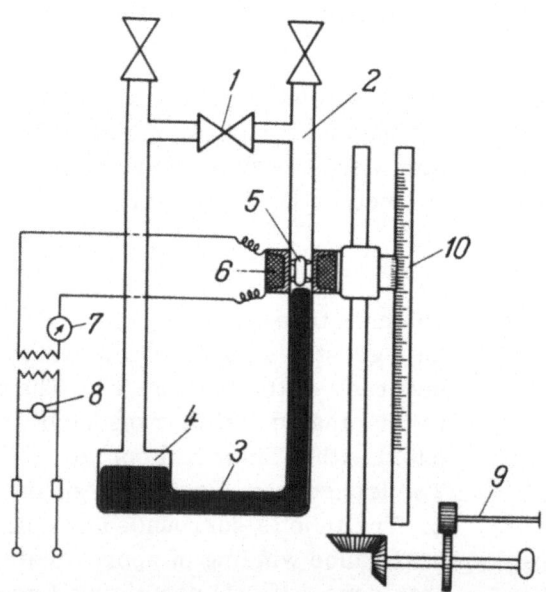

Fig. 114. Manometer with electromagnetic transducer: 1) valve block; 2, 4) columns; 3) mercury; 5) float; 6) induction coil; 7) milliammeter; 8) lamp; 9) precise control; 10) scale with vernier gauge.

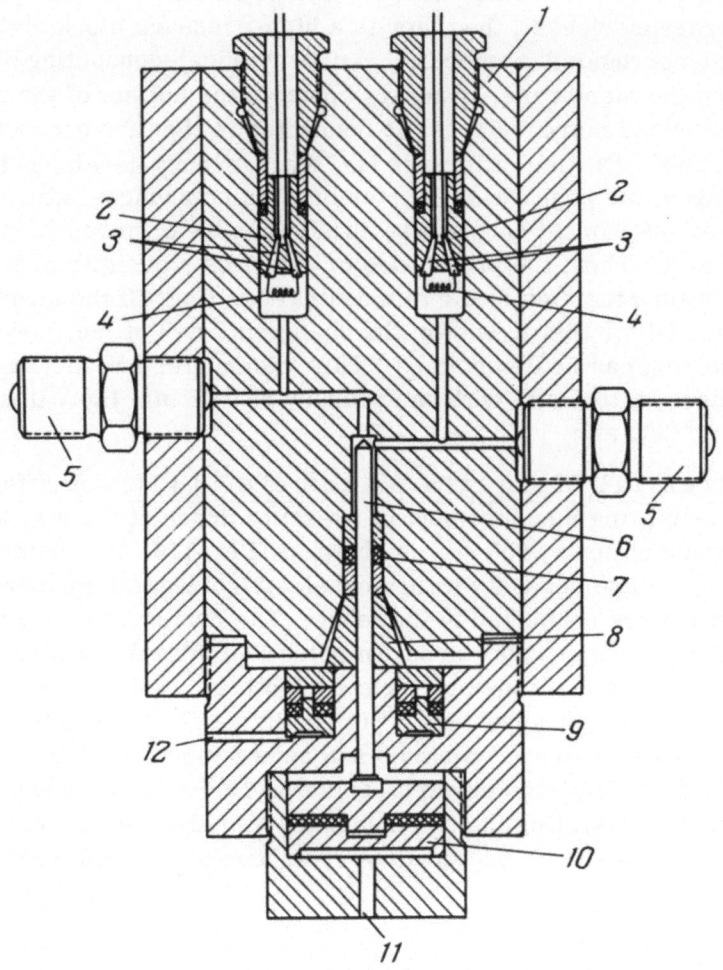

Fig. 115. Differential manganin manometer: 1) body; 2) heads; 3) electrical leads; 4) coils; 5) nipples; 6) needle valve; 7) gland; 8) cone; 9, 10) pistons; 11, 12) oil inlets.

This device is used in investigations not as a differential manometer, but rather as a device allowing pressure to be maintained within two interconnected vessels with an accuracy to 0.02 mm Hg with an overall pressure on the order of 1000 atm.

Another design exists for a manometer with electromagnetic transducer [78]. Valve block 1 (Fig. 114) connects two columns of a manometer 2 and 4, filled with mercury. Float 5 floats on top of the mercury column. The presence of the float in the induction coil 6 is determined from milliammeter 7, the position of the float by reading the scale.

In a manometer of another design [79] one column is a tube placed in the high-pressure vessel, the other is a high-pressure block divided into two parts by a corrugated membrane. A nonelectrical-conducting liquid is poured in atop the membrane. Mercury contacts the bottom of the membrane. A calibrated mercury press is connected to the tube connecting the two columns. As the difference in the mercury levels in the columns changes, the position of the membrane also changes, which is determined by an instrument connected with electrical leads mounted in the body of the block. Then, the press adds or removes a quantity of mercury sufficient to return the membrane to a neutral position. If the quantity of mercury required for this is known, the change in level of mercury in the measuring column can be determined. This manometer can measure pressure differences up to 1 atm with an accuracy of 0.05 mm Hg with an overall pressure of 500 atm.

Differential Manganin Manometer [80]. A differential manometer permitting measurement of pressure drops of 0.2 atm with an overall pressure of up to 7000 atm is shown on Fig. 115. It consists of body 1, into which two heads 2 with conical electrical leads 3 are screwed. Manganin coils 4 are connected to the leads. The manometer is connected to the high-pressure apparatus through nipples 5. Hydraulic needle valve 6 allows the coils to be disconnected from each other. Piston 9 seals the valve gland, using the unsupported-area principle. When oil is forced by a hand press into aperture 12, piston 9 and cone 8 move, sealing packing ring 7. When oil is fed into aperture 11, piston 10 begins to move, also moving needle 6 separating the two manganin transducers. If needle 6 is closed and nipples 5 are connected to vessels containing different pressures P_1 and P_2, then

$$\frac{R_1 + \Delta R_1}{R + \Delta R} = \frac{R_2 + \Delta R_2}{R_3}$$

where R_1 and R_2 are the initial resistances of the manganin coils; ΔR_1 and ΔR_2 are the changes in resistances of R_1 and R_2 under the influence of the pressure; R is the initial resistance of the manganin at which the bridge circuit was balanced at atmospheric pressure; ΔR is the change in resistance of the manganin compensating the unbalancing of the bridge circuit; R_3 is the constant standard resistance, and

$$\Delta R = R_3 \frac{R_1 + \Delta R_1}{R_2 + \Delta R_2} - R$$

For a bridge with equal arms, $R_1 = R_2 = R_3 = R$,

$$\Delta R = \frac{R(\Delta R_1 - \Delta R_2)}{R + \Delta R_2}$$

Since ΔR_2 is very slight in comparison with R, then

$$\Delta R = \Delta R_1 - \Delta R_2$$

We can express the change in resistances ΔR_1 and ΔR_2 by pressures

$$\Delta R_1 = k_1 R P_1 \quad \text{and} \quad \Delta R_2 = k_2 R P_2$$

where k_1 and k_2 are the coefficients of resistance change of the coils with pressure.

If the coils are made of identical wire, $k_1 = k_2 = k$. Then

$$\Delta R = k R (P_1 - P_2)$$

The term $\Delta R/kR$ is equivalent to a certain pressure drop which is measured by the instrument.

BIBLIOGRAPHY

1. Mendeleev, D. I., Zh. Russ. Fiz. Khim. Obshchestva, 4:309 (1872).
2. Jellinek, K., Lehrbuch der physikalische Chemie, Vol. 1, Berlin (1928).
3. Handbuch der experimental Physik, 8:2 (1929).
4. Tiller, F. M., Anal. Chem., 26:1252 (1929).
5. Roebuck, I. R., and Ibser, H. W., Rev. Sci. Instrum., 25:46 (1954).
6. Zhokhovskii, M. K., Techniques for Measurement of Pressures and Rarefactions, Mashgiz, 1952; ASME Publications, High Pressure Measurements, Bibliography on Bourdon Tubes and Bourdon Gages, Paper N53-YKD-1, 1953.
7. Zhokhovskii, M. K., Theory and Design of Instruments with Unsealed Pistons, Mashgiz, 1959.
8. Voronel', A. V., Fiz. Metal. i Metalloved., 9:174 (1960).
9. Bett, K., Hayes, P. F., and Newitt, D. M., Phil. Trans. Roy. Soc., London, Ser. A., 247(923):59 (1954).
10. Dadson, R. S., Nature, 176(4474):189 (1955).
11. Butterman, H. I., and Schuster, M., Z. Angew. Phys., 9(1):29 (1957).
12. Amagat, E. H., Ann. Chim. Phys., 29:68 (1893).

13. Vereshchagin, L. F., and Aleksandrov, B. S., Zh. Tekhn. Fiz., 9:348 (1939).
14. Jonson, D. P., and Newhall, D. H., Trans. ASME, 75:301 (1953).
15. Zhokhovskii, M. K., Izmerit. Tekhn., 8:14 (1959).
16. Zhokhovskii, M. K., Izmerit. Tekhn., 2:1 (1940); 5:1 (1940).
17. Sage, B. N., and Lacey, R., Trans. Am. Inst. Min. Met. Eng. Pert. Div., 136 (1940).
18. Reamer, H. H., and Sage, B. H., Rev. Sci. Instrum., 26:592 (1955).
19. Razumikhin, V. N., and Borzov, V. A., Tr. Inst. Kom. Standartov Mer i Izmerit. Prib. SSSR, 46 (106):55 (1960).
20. Comings, E. W., High Pressure Technology, New York, 1960.
21. Zhokhovskii, M. K., Konyaev, Yu. S., and Levchenko, V. G., Pribory i Tekhn. Éksperim., 3:118 (1959).
22. Konyaev, Yu. S., Pribory i Tekhn. Éksperim., 4:107 (1963).
23. Indrik, P. V., Vestn. Mashinostr., 33(7):74 (1953).
24. Tsiklis, D. S., Pribory i Tekhn. Éksperim., 2:142 (1961).
25. Coffin, D. O., Rev. Sci. Instrum., 29(10):896 (1958).
26. Nogatkin, A. G., Priborostroenie, 5:13 (1956).
27. Gielessen, J., Z. Angew. Phys., 8(4):193 (1956).
28. Metallurgia, 52:44 (1955).
29. Tait, P.G., Report of the voyage of H.M.S. Challenger, II, App. A (1881).
30. Michels, A., Ann. Physik, 72:285 (1923); 73:377 (1924).
31. Bridgman, P. V., High Pressure Physics, ONTI, 1935.
32. Schultze, A., Chemiker. Ztg., 67:228 (1943).
33. Bomberg, H. I., Rev. Sci. Instrum., 30(1):43 (1959).
34. Zolotykh, E. V., and Burova, L. L., Tr. Inst. Kom. Standardtov, Mer i Izmerit. Prib. SSSR, 46(106):62 (1960).
35. Adams, L. H., Goranson, R. W., and Gibson, R. F., Rec. Sci. Instrum., 8:230 (1937).
36. Darling, H. E., and Newhall, D. H., Trans. ASME, 75:311 (1953).
37. Weir, C. E., J. Res., 45:468 (1950).
38. Merret, F., and Norrish, R., Proc. Roy. Soc., A206, 309 (1951).
39. Warschauer, D. M., and Paul, W., Rev. Sci. Instrum., 29(8):675 (1958).
40. Zhokhovskii, M. K., and Razumikhin, V. N., Tr. Inst. Kom. Standartov Mer i Izmert. Prib. SSSR, 46(106):68 (1960).
41. Zolotykh, E. V., and Burova, L. L., Izmerit. Tekhn., No. 1 (1959).
42. Tsiklis, D. S., Zh. Tekhn. Fiz., 15:960 (1945).
43. Lloyd, E. K., and Johnson, L. P., Static and Dynamic Calibration of Instruments for Measurement of Pressure at the National Bureau of Standards, Izd. Akad. Nauk SSSR, 1960.
44. Kennedy, G. G., and La Mory, P. N., in Collection: "Progress in Very High Pressure Research," Proc. of Confer. at Lake George in 1960, New York, 1961; La Mory, P. N., Am. Soc. Mech. Engs., N. WA-340, 1962.

45.   Hall, H. T., J. Phys. Chem., 59:1144 (1955).

46.   Dricamer, H. G., and Balchan, A. S., in Collection: Modern Very High Pressure Techniques, London, 1962.

47.   Ashcroft, K., and Lees, I., Nature, 198(4884):957 (1963); Myers, M. B., Dachille, F., and Roy, R., Rev. Sci. Instrum., 34(4):401 (1963).

48.   Zhokhovskii, M. K., Izmerit. Tekhn., 2:3 (1957).

49.   Zhokhovskii, M. K., Zh. Fiz. Khim., 37:2635 (1963); Izmerit. Tekhn., 7:11 (1959).

50.   Zhokhovskii, M. K., Zh. Fiz. Khim., 38:33 (1964).

51.   Zhokhovskii, M. K., Izmerit. Tekhn., 4:29 (1964).

52.   Simon, F., Trans. Faraday Soc., 33:65 (1937).

53.   Aleksandrov, B. S., and Vereshchagin, L. F., Zh. Tekhn. Fiz., 13:843 (1939).

54.   Alekseev, K. A., and Burova, L. L., Tr. Inst. Kom. Standartov Mer i Izmerit. Prib. SSSR, 75(135):36 (1964).

55.   Alekseev, K. A., Tr. Inst. Kom. Standartov Mer i Izmerit. Prib. SSST, 75(135):44 (1964).

56.   Tsiklis, D. S., and Borodina, M. D., Pribory i Tekhn. Éksperim., 2:200 (1965).

57.   Bridgman, P. W., Investigation of High Plastic Deformations and Rupture [in Russian], Izdatinlit, 1955.

58.   Gross, M., Mech, M. J., and Lane, P. H., J. Sci. Instrum., 30:1 (1953).

59.   Zelyaev, A. F., Shumov, K. M., and Alekseev, E. N., Zav. Lab., 22:1368 (1958).

60.   Leonidov, G. G., and Polandov, I. N., Pribory i Tekhn. Éksperim., 2:159 (1960).

61.   Newhall, D. H., and Abbot, L. H., Chem. Eng. Progr., 56(3):112 (1960).

62.   Newitt, D. M., The Design of High Pressure Plant, London, 1940.

63.   Aarons, A. B., and Cole, R. H., Rev. Sci. Instrum., 21:31 (1950).

64.   Jolly, Legallis, and Cherry, J. Appl. Phys., 18:613 (1947).

65.   Higgs, P. M., Rev. Sci. Instrum., 20:23 (1949).

66.   Thomson, M., and Consins, E., Instruments, 20:330 (1947).

67.   Willey, B., J. Sci. Instrum., 23:264 (1946).

68.   Montgomery, L. H., and Ward, J. W., J. Sci. Instrum., 18:289 (1947).

69.   Diment, J., Carson, I. A., and Charters, A. C., Rev. Sci. Instrum., 26:879 (1955).

70.   Bush, W., and Barcas, W., Rev. Sci. Instrum., 19:678 (1948).

71.   Kedi, U., Piezoelectricity and Its Practical Techniques, Izdatinlit, 1946.

72.   Ryabinin, Yu. N., Zh. Éksperim. i Teor. Fiz., 23:461 (1952).

73.   Bol'shakov, P. E., Tr. GIAP, No. 1, Goskhimizdat, 1953, p. 30.

74.   Aristov, G. E., Zh. Khim. Prom., 16:45 (1939).

75.   Shelaputin, K. N., and Voitovich, M. V., Tochnaya Industriya, 11:7 (1940).

76. Kiselev, D. V., Byull. Inf. Nauchn. Tekhn. Byuro NIIlaborpribor, 3:33 (1953).
77. Gamburg, D. Yu., Zh. Fiz. Khim., 24:272 (1950).
78. Brennstoff-Chem., 37(12):9 (1956).
79. Reamer, H. H., and Sage, B. H., Rev. Sci. Instrum., 31(3):337 (1961).
80. Zhokhovskii, M. K., and Bakhvalova, V. V., Tr. Inst. Kom. Standartov Mer i Izmerit. Prib. SSSR, 3:12 (1960).

Chapter V

Measurement of the Flow and Temperature
of a Compressed Medium

MEASUREMENT OF FLOW OF A COMPRESSED MEDIUM

High-Pressure Rheometers

When designing rheometers, it is first necessary to choose a means of measuring the level of a liquid under pressure in the column of the rheometer. This can be done by one of the methods previously described. Secondly, the capillaries must be properly selected.

The use of a metal capillary involves considerable difficulty. Design of a rheometer for air has shown that even if water is used as the liquid filling the rheometer, the generation of a liquid-level difference in the columns of 15 cm with a total air flow rate of 50 cm^3/min at a pressure of 300 atm requires a capillary with an internal diameter of 0.1 cm and a length of 150 cm. A reduction in capillary diameter to 0.05 cm allows its length to be reduced to approximately 10 cm. However, it is very difficult to manufacture a metal capillary with a bore of 0.05 cm. It is consequently more convenient to use a glass capillary, mounted as shown in Fig. 116.

Capillary 1 is sealed at one end by gland 2 which must be able to withstand a pressure equal to the height of the fluid column in the tube of the rheometer. This design is convenient in that it allows easy replacement of capillaries. The sleeve containing the capillary is mounted in the rheometer assembly, which in turn is no different than an ordinary rheometer operating at atmospheric pressure. Details of design depend on the method of liquid-level measurement to be used by the investigator [1, 2].

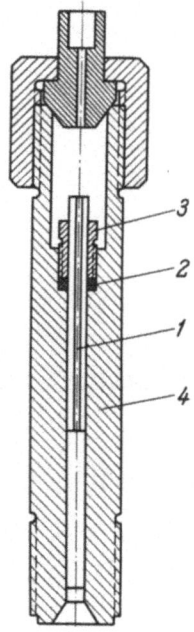

Fig. 116. Reinforce-
ment of high-pressure
rheometer capillary:
1) capillary; 2) gland;
3) nut; 4) nipple.

Another high-pressure rheometer [3] (shown
in Fig. 117) consists of two columns. The left
column contains a high-pressure vessel 1. The
right column contains a high-pressure chamber 2
which is movable vertically and which contains a
window for observation of mercury level. The two
columns are connected by a flexible steel capillary
4. The rheometer is filled with mercury to the
line I–I. Water covers the mercury, and there are
manometer connections to the inlet of each column
(not shown on the figure), so designed as to cool
the water during measurement of the pressure drop
in a hot medium.

When no flow passes through the tube, the
height of the movable chamber is set so that the
level of the mercury surface corresponds to
guide line 3 on the glass. When the medium flows
through the tube and through the measuring element, the mercury level in the mobile chamber
rises. An elevating screw is then turned, raising
the chamber until the mercury level corresponds
once again with the guide line. Then the mercury
level in the left column is returned to its initial
position. The amount of elevation of the chamber
required to return the mercury to its initial position is proportional to the pressure difference in
the rheometer column.

When the liquid flows through the tube, $h_1 = h_2$, $y_1 + y_2 = R$. Then:

$$P_1 + \gamma_a y_1 - \gamma_b (y_1 + y_2) + \gamma_a y_2 = P_2$$

where h_1 and h_2 are distances from the guide lines to the surfaces of the
liquid over the mercury; y_1 and y_2 are deviations in mercury level in the
left and right columns of the instrument; P_1 and P_2 are pressures ahead
of and behind the measuring element; γ_a is the specific gravity of the
fluid over the mercury; γ_b is the specific gravity of the mercury in the
movable metal tube.

The volume occupied by the manometric liquid between marks A and
B remains constant if the mercury levels correspond to the guide lines
and if we assume that the volume of tube 6 is constant. Then $y_1 = 0$ and

$$P_1 - P_2 = y_2 (\gamma_b - \gamma_a)$$

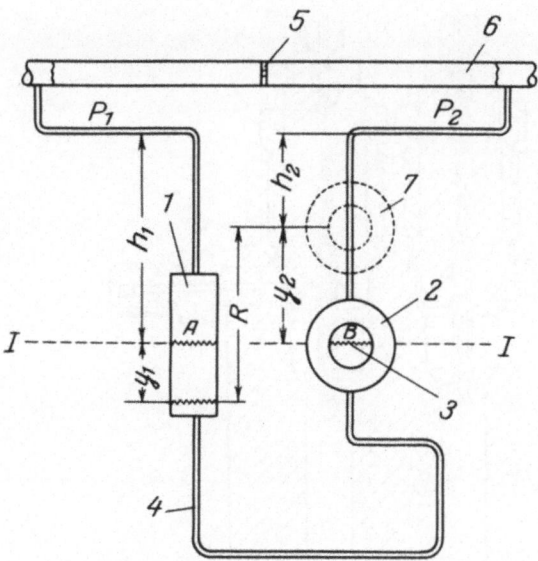

Fig. 117. High-pressure rheometer: 1) high-pressure vessel;
2) chamber with window; 3) guide line; 4) steel capillary; 5)
measuring diaphragm; 6) tube; 7) position of chamber when
$P_2 < P_1$.

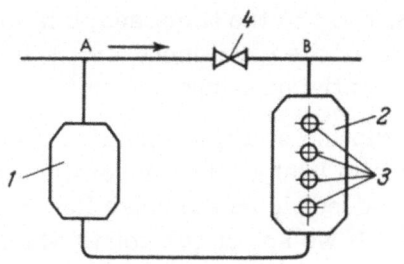

Fig. 118. Plan of flow meter: 1) high-pressure vessel; 2) collector; 3) windows; 4) valve.

The mobile chamber is moved by a micrometer screw with an accuracy of 0.0025 mm, and the level is established using a cathetometer.

This device can be used to measure pressure drops of a few millimeters of mercury with a total pressure of up to 1500 atm.

Gas Flow Meters for Gas Under Pressure.
V. V. Kamzolkin and A. N. Bashkirov [4] have suggested the following device for measurement of the rate of gas flow under high pressure (Fig. 118). The gas moves in direction A → B. Valve 4 is closed and the time required to fill collector 2 is determined by observing the volume of liquid through windows 3.

The gas flow rate (in liters/h) is

$$V = \frac{V_1}{t}\left(1 + \frac{h\gamma}{P}\right)\frac{3600}{1000}$$

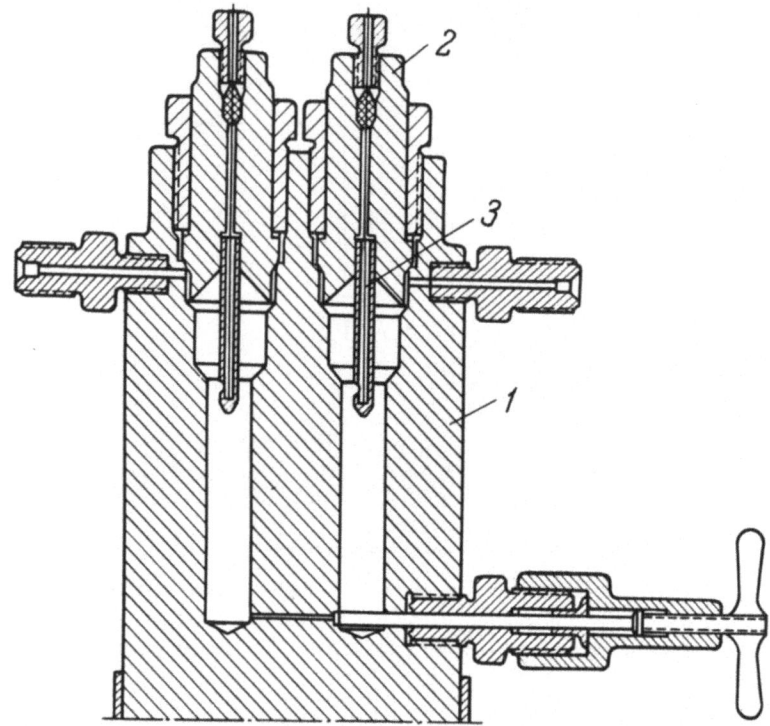

Fig. 119. Aristov-Sidorov flow meter: 1) high-pressure vessel; 2) heads; 3) contacts.

where V_1 is the standard volume of gas, ml; t is the time, sec; h is the difference in liquid levels in the columns of the instrument, cm; γ is the specific gravity of the liquid, kg/cm^3; P is the pressure, kg/cm^2.

The literature [5] contains a description of a flow meter of another design. When gas passing through a tube is heated, the temperature distribution in the gas and walls of the tube depends on the mass flow rate rather than on the volumetric flow rate. If we select two points which are at identical temperatures in a nonmoving gas, the temperature difference between these two points in a moving gas stream can be used to measure gas flow rates on the order of a few liters per minute (at atmospheric pressure).

The author of [5] describes a flow meter with external heating and two thermocouples placed on the axis of the gas stream for measurement of hydrogen flow rate under pressures up to 900 atm.

Another device [6] (Fig. 119) consists of vessel 1 with two connected chambers, filled over half full with mercury. The two heads of the chamber 2 contain platinum contacts 3 connected through an automatic switch to the solenoid-operated valves. The chambers of the instrument are

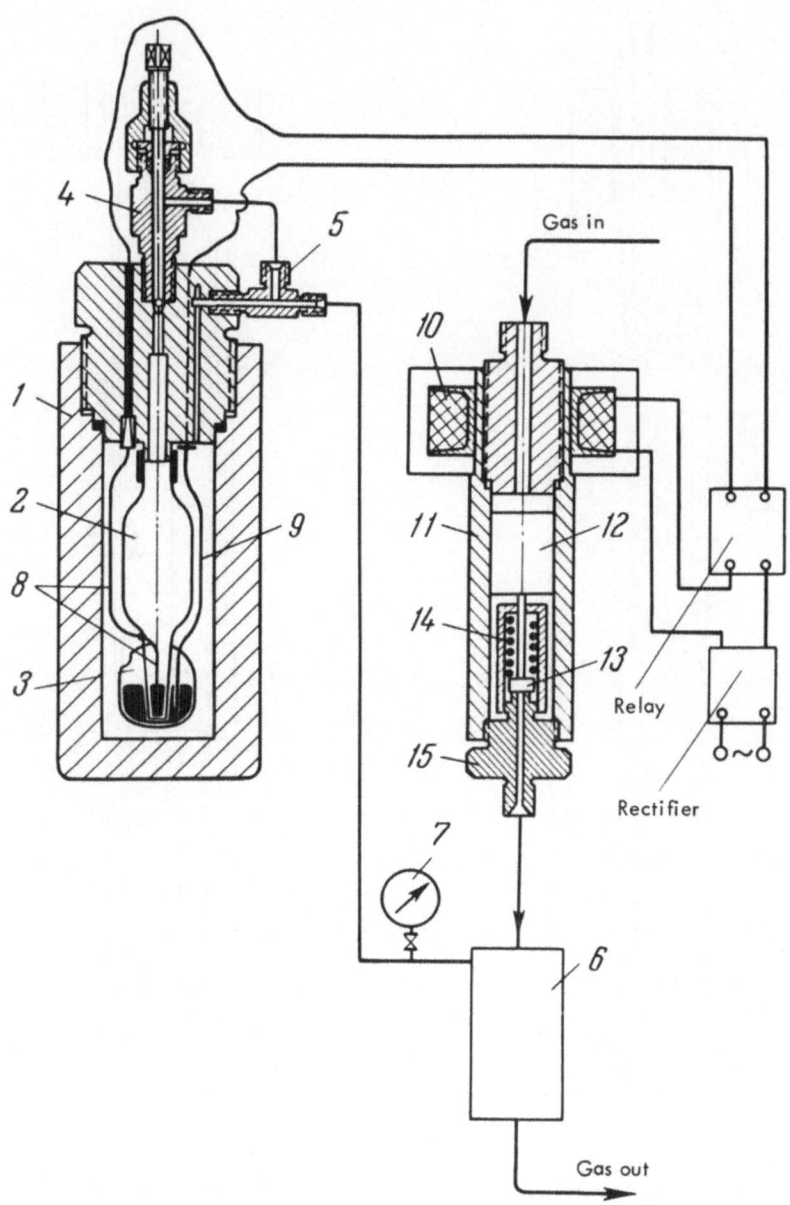

Fig. 120. Sidorov pressure regulator: 1) cylinder; 2, 3) glass vessels; 4) valve; 5) nipple; 6) apparatus; 7) manometer; 8, 9) contacts; 10) solenoid; 11) valve body; 12) armature; 13) valve stem; 14) spring; 15) seat.

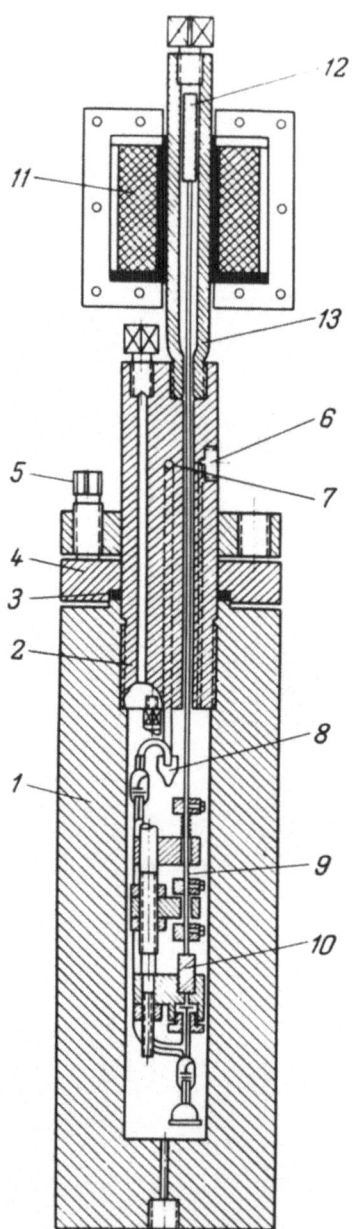

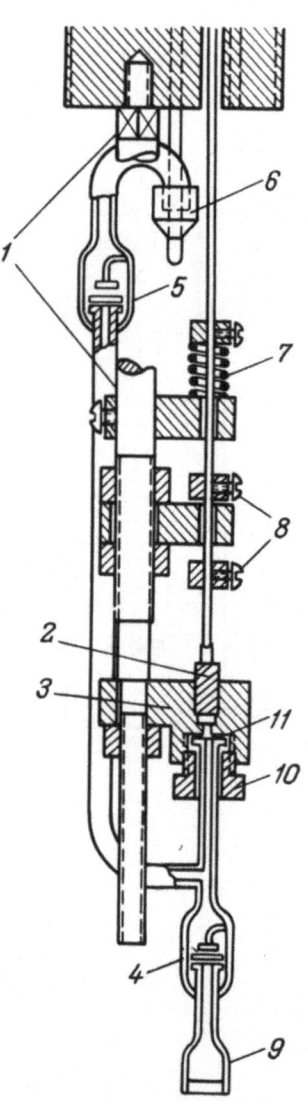

Fig. 121. Bruns-Braude dispenser: 1)
cylinder; 2) head; 3) packing; 4) flange;
5) bolt; 6, 7) openings; 8) funnel; 9) rod;
10) pump; 11) solenoid; 12) core; 13) tube.

Fig. 122. Proportioning pump:
1) rod; 2) pump piston; 3) pump
cylinder; 4, 5) valves; 6) funnel;
7) spring; 8) limiters [stops]; 9)
filter; 10) nut; 11) packing.

filled alternately with compressed gas. As one chamber is filled, the mercury in the other chamber rises and closes a circuit with the platinum contact. This switch operates the valve solenoids, switching the gas input to the chamber filled with mercury and allowing the gas to leave the chamber filled with gas and enter the outlet line. When the second chamber is filled, the operation is reversed.

This device is suitable for measurement of low rates of flow of compressed gas. The literature [7] also contains a description of a high-pressure rotameter based on the movement of a core floating on mercury in the field of an induction coil. As the flow to be measured varies, the core changes position relative to the coil and thereby changes the indication of an instrument connected to the circuit of the coil.

Automatic Pressure Regulator. Steel cylinder 1 (Fig. 120) [8] contains two glass vessels 2 and 3. Vessel 2 is connected to valve 4, vessel 3 to the internal portion of cylinder 1. Vessel 3 is connected through nipple 5 to apparatus 6 in which the pressure is regulated, and to manometer 7. Contacts 8 and 9 enter the glass vessels. Mercury is poured into vessel 3.

Once the required pressure is set in apparatus 6 (as indicated by manometer 7), valve 4 is closed. Then, if the pressure in apparatus 6 is reduced, the mercury level in vessel 3 rises, dropping in the narrow portion of vessel 2. When this occurs, the mercury moves away from contact 8, opening the circuit and closing the contacts of a relay. When the circuit is opened, the relay permits current to flow through the circuit of solenoid 10 of a solenoid-operated valve 11, and armature 12 moves to its upper position. In rising, the armature pulls stem 13 of the valve with it, allowing more compressed gas to flow into apparatus 6. The pressure in the apparatus increases, the level of the mercury in vessel 2 increases, and the relay circuit closes. The solenoid circuit automatically opens, and spring 14 forces the stem back down in support 15, sealing off apparatus 6 from the gas.

The regulator is reliable in operation and permits pressure to be controlled with high accuracy.

Control of Supply [Proportioning] of Liquid to Compressed Gas. An instrument described by B. P. Bruns and G. E. Braude can be used for this purpose [9] (Fig. 121).

Cylinder 1 of stainless steel type 1Kh18N9T is closed by a plug consisting of head 2, copper packing 3 and flange 4.

The cylinder is filled $\frac{3}{4}$ to $\frac{4}{5}$ full with the liquid to be measured out. Gas enters through opening 6 and leaves through aperture 7 (1-1.5 mm in diameter) in a tube terminated with funnel 8. The funnel is placed above the highest level of liquid in the cylinder.

Cylinder 1 contains pump 10, which feeds the liquid to funnel 8, from which it is continuously extracted by the exiting gas. Pump 10 is operated by core 12 connected by rod 9 to the piston of the pump. Core 12 moves when solenoid 11 is actuated. Tube 13 is made of nonmagnetic steel. The pump is fastened onto rod 1 (Fig. 122), screwed into the head of the instrument.

Piston 2 and cylinder 3 of the pump are made of dissimilar metals. When the piston rises, liquid enters through valve 4; when the piston drops, liquid exits through delivery valve 5 to funnel 6 (see position 8, respectively, on Fig. 121). When the solenoid is disconnected, the piston is raised by spring 7. Limit stops 8 determine the length of the piston stroke and consequently the quantity of liquid fed with each actuation of the solenoid. Filter 9 removes foreign matter from the liquid.

The glass portion of the pump (4, 5, 6) is held in the cylinder by nut 10 and packing 11, made of a material which is insoluble in the liquid being pumped.

The pump piston and valves (glass plates 1-1.5 mm thick) should be carefully polished to fit the cylinder and fittings.

MEASUREMENT OF TEMPERATURE
AT HIGH PRESSURES

Methods of measuring temperature in high-pressure apparatus can be divided into two primary classes: 1) measurement of temperature of the thermostated liquid in which the high-pressure apparatus is submerged, or measurement of temperature of the walls of an apparatus heated by a furnace; 2) measurement of temperature within the high-pressure apparatus.

Methods of measuring the temperature of the thermostated liquid are well-known, so they will not be described in detail. It should be noted only that the strength of steels is considerably reduced at external heating temperatures over 500-600°C.

The temperature within an apparatus is measured by thermocouples, resistance thermometers, optical and photoelectrical pyrometers, etc.

Measurement of Temperature with Thermocouples. A thermocouple can be placed in an apparatus through a thermocouple well (pocket). In this case, the maximum temperature which can be measured is determined by the strength of the well material. When a thermocouple is placed directly in the high-pressure cavity, the corroding action of the medium must be considered, as well as the ability of certain gases to be dissolved in metals at high temperatures and pressures (hydrogen dissolves in platinum and palladium, nitrogen in iron, etc.) since this may influence the indications of the thermocouple.

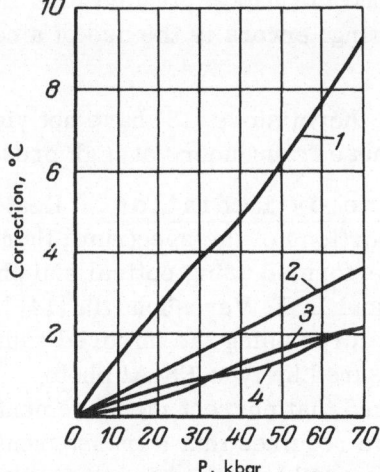

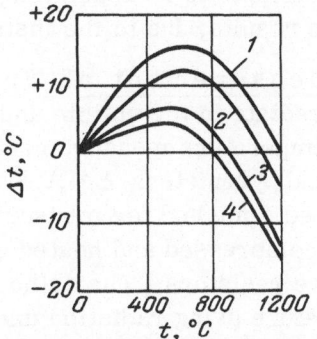

Fig. 123. Variation of correction to indicated thermocouple reading with pressure: 1) platinum—platinum + rhodium
(10%) thermocouple; 2) copper—constantan
thermocouple; 3) nickel—nickel + molybdenum
thermocouple; 4) chromel—alumel thermocouple.

Fig. 124. Dependence of difference in indication of Cr/Al and
Pt—[Pt + 10% Rh] thermocouple
(Δt) on temperature measured by
Pt—[Pt + Rh] thermocouple at various pressures: 1) 42 kbar; 2) 25
kbar; 3) 11 kbar; 4) 8 kbar.

Finally, high pressures cause a change in the thermoelectric emf
of thermocouples. Investigations [10, 11] have been performed to determine the magnitude of corrections which should be made to thermocouple
readings in order to take this [pressure] effect into consideration. Figure
123 shows curves for four different thermocouple junctions, constructed
for a temperature difference of 100°C. If the curve for a Pt-[Pt + Rh]
thermocouple is extrapolated to 100 kbar and 1000°C, the correction under
these conditions [12] reaches 134°C. This extrapolation is not fully justi-
fied, however. We see from Fig. 124 that various thermocouple junctions
give different indications in one and the same medium at different temperatures, the difference in the indications changing with temperature.

In practice, the measurement of temperature with thermocouples requires a solution of the problem of bringing the wire out of the high-pressure region. In some cases, one end of the lead is welded to a steel electrical-lead rod within the apparatus, the other to the outer terminus of this
rod. It is assumed in this case that the electrical lead rod (which may be
in the form of a cone) has uniform temperature over its entire length, and
that the emf arising in the two junctions formed by these connections will
cancel.

In order to eliminate the errors caused by the presence of a temperature gradient at the ends of the electrical lead, differential thermo-

couples and measurement methods described in Chapter III (see Fig. 75) are used. The simplest method of eliminating errors is the use of a continuous conductor (see Chapter VI).

Attempts to measure temperature by thermistors [13] have not yielded positive results, due to the instability of these transducers at high pressure.

Measurement of Temperature by Radiation. Heated media radiate in the visible and infrared portions of the spectrum; therefore, temperature measurements can be performed using optical and photoelectrical pyrometers. Ya. A. Kalashnikov and L. F. Vereshchagin [14] have developed a method for measuring temperatures using the infrared radiation of compressed and heated gas as measured by type FS-A1 photosensitive resistors. The authors established that correct measurement of temperature using radiation under pressure requires that a transparent medium whose density changes as little as possible with change in pressure and temperature be located between the high-pressure window and the measurement point. The authors used a light conductor made of polished quartz for this purpose. Results of investigation showed that use of optical pyrometers under high-pressure conditions is difficult.

Measurement of Temperatures from Thermal Noise. It is known that when a current is passed through an electrical resistance, thermal noise is created in the resistor which is proportional to the temperature. The fluctuation of current in a closed circuit under pressure is related to temperature by the Nyquist formula [15]. The phenomenon of thermal noise has been used for measurement of temperatures at high pressures. Investigations at pressures up to 10,000 atm [16] showed that thermal noise does not depend on the pressure applied to the electrical resistance.

If a dependence of thermal noise emf on temperature is established for a given conductor at atmospheric pressure, the resistor can then be placed in a high-pressure apparatus and the temperature either measured directly or calibrated by a thermocouple under the same conditions.

Measurement of Temperature by Current Loss. All the preceding methods are suitable when the measured temperature does not exceed the melting point of the transducer material. However, cases do occur when a higher temperature must be measured, or at least estimated. In this case, the thermal state in the apparatus is such that the internal portion of the material being investigated and even the heating element may be in the liquid state, retained in their initial form by a colder, strongly compressed surrounding layer. In order to measure the temperature within this sort of apparatus, calibration [17] is performed using a thermocouple, and a graph is constructed with coordinates "temperature vs. current used to reach temperature." Then, the straight line produced is extended into the area of higher temperatures and powers. Thus, temperatures up to 10,000 or even 15,000°C can be estimated.

BIBLIOGRAPHY

1. Zil'berg, G. A., Kamzolkin, V. V., Konstantinov, A. A., and Bashkirov, A. N., Zavodsk. Lab., 16:320 (1950).

2. Nisman, L. N., Vestn. Tekhn. i Ékon. Inform. Nauchn. Issled.-Inst. Teor. i Éksperim. Khim., 2:49 (1961).

3. Barnet, S. S., Jackson, T. W., and Whitsides, R. H., Trans. ASME, Ser. C., 85(2):125 (1963).

4. Kamzolkin, V. V., and Bashkirov, A. N., Zavodsk. Lab., 21:743 (1955).

5. Boulestreau, M., Mesur. et Contr. Ind., 19(202):165 (1954).

6. Aristov, G. E., and Sidorov, I. P., Tr. Gos. Inst. Azo. Prom., No. 1, Goskhimizdat, 1953, p. 258.

7. Vasil'ev, A. V., and Ustinovich, V. V., Nitrogen and Oxygen Industries, Information Bulletin, No. 1, Otdel. Nauk Tekhn. Inst., Gos. Inst.Azo.Prom., 1964.

8. Sidorov, I. P., Tr. Gos. Inst. Azo. Prom., No. 6, Goskhimizdat, 1956, p. 328.

9. Bruns, B. P., and Braude, G. E., Tr. Gos. Inst. Azo. Prom., No. 1, Goskhimizdat, 1953, p. 252.

10. Bridgman, P. W., Proc. Am. Ac. Arts. Sci., 53:269 (1918).

11. Bundy, F., In Collection: Progress in Very High Pressure Research; Svenson, K., High-Pressure Physics, IL, 1963, p. 298.

12. Strong, H. M., High temperature methods at high pressure, In Collection: Modern Very High Pressure Techniques, London, 1962.

13. Grazhdankina, N. P., Domanskaya, P. I.,and Kikoin, A. I., Izmerit. Tekhn., 10:18 (1959).

14. Kalashnikov, Ya. A., and Vereshchagin, L. F., Zh. Tekhn. Fiz., 26:1802 (1956).

15. Physics Dictionary, Vol. III, Gos. Izd. BSÉ, 1963.

16. Garrison, J. B., and Lauson, A. W., Rev. Sci. Instrum., 20:785 (1949).

17. Hall, H. T., Brown, B., Nelson, B., and Compton, L. A., J. Phys. Chem., 62(3):346 (1958).

Chapter VI

Parts for High-Pressure Installations

Performance of experiments under high pressure requires especially careful assembly of apparatus. The quality of this assembly depends completely on the quality of manufacture of parts of the installation. Especially strict requirements are placed on threads and sealing surfaces. Threads should be rounded, i.e., without a sharp crest. In order to reduce the danger of so-called "seizing," threads should not be dimensioned the same as the sealing surface. Usually, the separation distance left should be 3-5 mm.

Proper threading is shown in Fig. 125a, improper threading in Fig. 125b. The threads on the nut and bolt should be cut according to the requirements set by the standards [1]. Threads which are too free are as bad as threads which are too tight. In the first case, the thread comes lose easily; in the second case, it may "seize." A nut and bolt should be made of different materials, especially if the parts are made of type 1Kh18N9T ductile steels.

Sealing surfaces require equally careful manufacture. The slightest scratch, groove, or crack will break the hermetic sealing of the packing. Hence, sealing surfaces are usually polished.

It is extremely important that the plane of the sealing element be strictly perpendicular to the axis of the thread, i.e., that tightening the nut causes the sealing surface to contact the part simultaneously at all points on the circumference. This requirement must be met more strictly, the harder the parts being sealed. The better this requirement is fulfilled, the less force is required for sealing.

The internal surfaces of high-pressure apparatus sealing rings, glands, etc., should also be made extremely carefully. This is especially

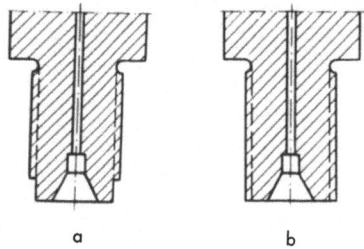

Fig. 125. Cutting of threads on sealing member: a) correct; b) incorrect.

important for the manufacture of seals in such parts as the bores of intensifiers, which are sometimes turned and polished to accuracies of thousandths of a millimeter.

The higher the pressure, the greater the possibility for penetration of the pressure-transmitting medium into microfissures, gaps, and irregularities on surfaces. Finally, stresses are concentrated at fissures, cracks, sharp angles, etc., which may lead to rupture of parts.

Such parts as pistons, sealing rings, pressure boxes, and high-pressure valve pins, made of very hard steels or alloys, are polished after heat treatment.

Special attention should be given to concentricity or eccentricity of mating apertures. Strict concentricity will provide proper operation of valves, agitators, etc. The requirements for concentricity are extremely varied, and depend on the purpose of the parts involved. In some cases, concentricity to 0.01 or even 0.001 mm is required.

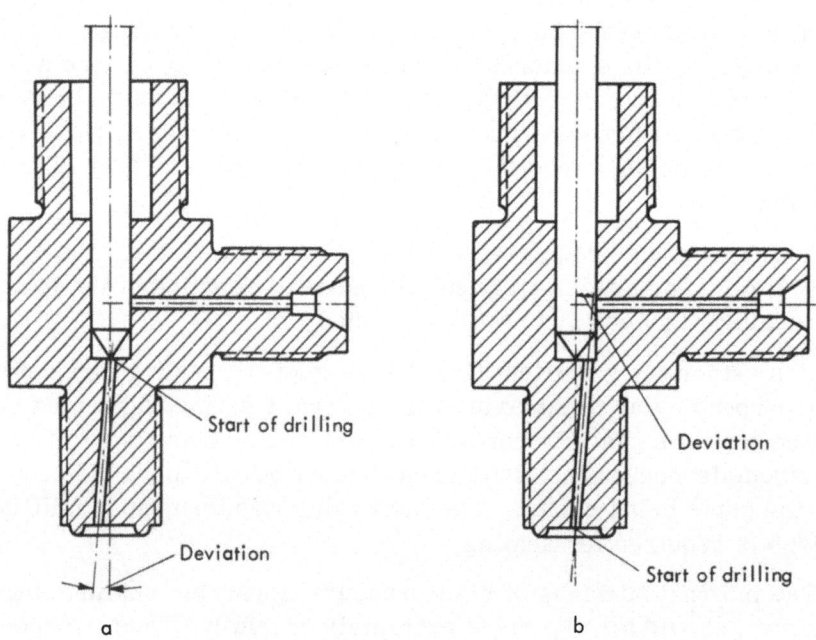

Fig. 126. Drilling of hole in valve: a) correct; b) incorrect.

Cylindricity or conicity of holes or parts is no less important. If the tolerances for conicity are not observed, the operation of piston sealing rings is impeded. Inasmuch as the packing will be compressed when it moves toward the smaller diameter, the sealing will be ineffective in the opposite direction of movement. The same thing is true for a noncylindrical valve shaft which, as it moves in the gland, increases its diameter and destroys the seal.

A certain order must be observed in the handling of parts. For example, it is very important to drill holes for threads and to face the sealing surface without removing the part from the machine tool. This provides the best correspondence between thread and aperture axes, and best maintains perpendicularity of thread axis to sealing surface.

It is very important for proper operation of a valve that the sealing end of the shaft be precisely coaxial with the passage aperture. Otherwise the shaft will be bent and the valve will soon fail. This requirement is not always easy to fulfill, especially in drilling of long holes of small diameter. If a small drill is improperly sharpened, the hole will deviate from the center of the part. The deviation could be several millimeters. Correction of this defect can be performed by placing the part between centers in a lathe and turning the outside diameter, if dimensions permit. It is best to begin the drilling from the end where the hole must be located strictly in the center. Thus, the passage hole of a valve sealed by a stem is always drilled from the end at which the sealing will take place.

Figure 126 shows holes properly and improperly drilled. With proper drilling (Fig. 126a), a slight deviation of the hole does no harm, since the exit opening will still be in the area encompassed by the perimeter of the sealing surface. But, if the operator attempts to make his work easier by beginning the drilling from the outside (Fig. 126b) the part will probably be unreliable, since it will be impossible to close the valve completely.

Before a part is manufactured, the technology of its manufacture should be considered very carefully. It should be recalled that care in manufacture, together with intelligent design, are the primary determining factors in the reliability of operation of apparatus.

VALVES

Control Valves

Valve with Conical Stem. The design of a control valve with conical stem is shown in Fig. 127. The stem, made of hardened tool steel, is hardened and fitted to the sleeve; otherwise it would easily seize in the sleeve and be broken. The cone on its end has an included angle of 7-8°. The gland is sealed with asbestos-graphite packing, leather, polyvinyl chloride, or fluoroplast rings, etc. The remaining parts of a valve

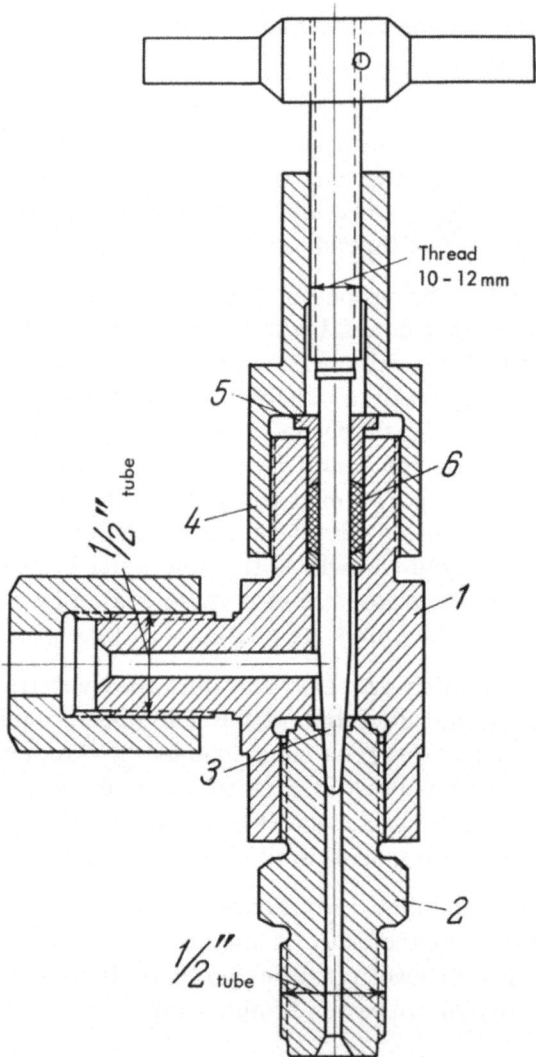

Fig. 127. Control valve with conical stem: 1) body;
2) nipple; 3) stem; 4) pressure-sealing nut; 5) ground
collar; 6) gland.

for pressures of 1500-2000 atm are made of type St. 5 steel. The threads on the stem and the sealing nut should have identical pitch. Then, slight tightening of the gland while the valve is in operation will not change the position of the stem relative to the sleeve of nipple 2.

This type of valve controls gas flow well, but requires care [in use]. The body of the valve, if operating at high pressures, should be made of alloyed, preferably heat-treated, steel.

For smoother control, the valve may be equipped with a long handle or a reducing drive. The use of a reducing drive (regardless of the design of the valve) permits considerably easier and more precise control of gas flow. But, if the experimenter cannot observe the gas flow from the bubbling of gas bubbles through a column of liquid or some other indicator, the use of a reducing drive introduces difficulty, since it is very difficult to determine when a valve is closed by sensing closure through a worm drive. The increased force applied to the valve handle can cause breakage of the stem if care is not exercised.

Smooth control of gas flow can also be achieved using a gas [pressure] reducer described in the literature [2]. The principle of operation of this device is that the pressure to be reduced acts on a piston, moving it from its place. The piston opens a valve and allows a certain quantity of gas to enter the low-pressure line and the so-called working chamber, from which the gas is fed to the apparatus. The pressure in the working chamber raises the piston and closes the valve. This automatically creates an equilibrium, and pressure in the chamber is reduced. This device can be used to reduce pressure from 1000 to 800-20 atm.

Valve with Spherical Stem.*   When working with pressures above 2000-3000 atm, it is most convenient to use a stem with a sphere pressed into its end (a bearing ball).

The ball rotates freely and therefore does not abrade against the seat when the stem is rotated downward, rather pressing straight into the seat, as a result of which the stem does not "seize."

These valves work well, but the ball causes the aperture to be pressed down, and the sealing area increases. Hence, the valve seat must be reworked frequently, as by shaving off a strip 0.1-0.2 mm thick. Any of the glands described above can be used to seal the stem of this valve.

Shut-off Valves

Shut-off Valve with Free Cone.   The impression of the sphere and the depression of the passage opening and resultant increase

---

* The design of the ball stem was developed and tested by I. P. Sidorov.

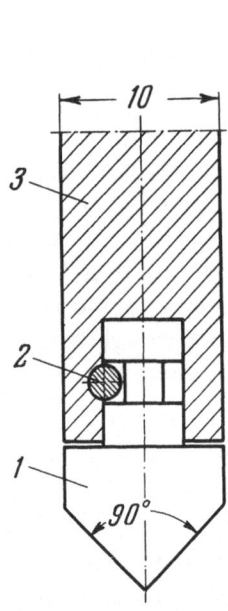

Fig. 128. Freely rotating
cone: 1) body; 2) pin;
3) stem.

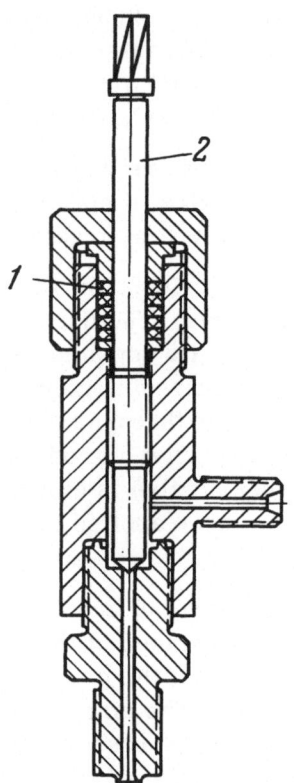

Fig. 129. Valve with removable gland: 1) gland; 2) stem.

of area of tangency of ball with seat can be prevented by using a stem designed by A. M. Rozen. The Rozen stem has a freely rotating cone (Fig.
128) on the end of the stem, fixed with pin 2. When stem 3 is rotated, the
cone is pressed into the hole in the seat, sealing it. In time, area of tangency of cone with seat increases here too; the seat must be reworked so
that sealing is achieved around a thin edge.

When these valves are opened and closed, their stems rotate in a
gland, which is a defect of operation, especially at pressures of over several thousand atmospheres. The operation of the gland can be facilitated
by using valves in which the stem does not rotate, but moves only with reciprocating motion. Valves of this type will be described below.

If the stem must be removed frequently, together with the gland, (for
example, to clean the valve of precipitated solid material), the valve shown
on Fig. 129 can be used. Gland 1 is located above the thread in the stem
2, so that when the stem is removed, the gland is removed as well, and can
be easily replaced.

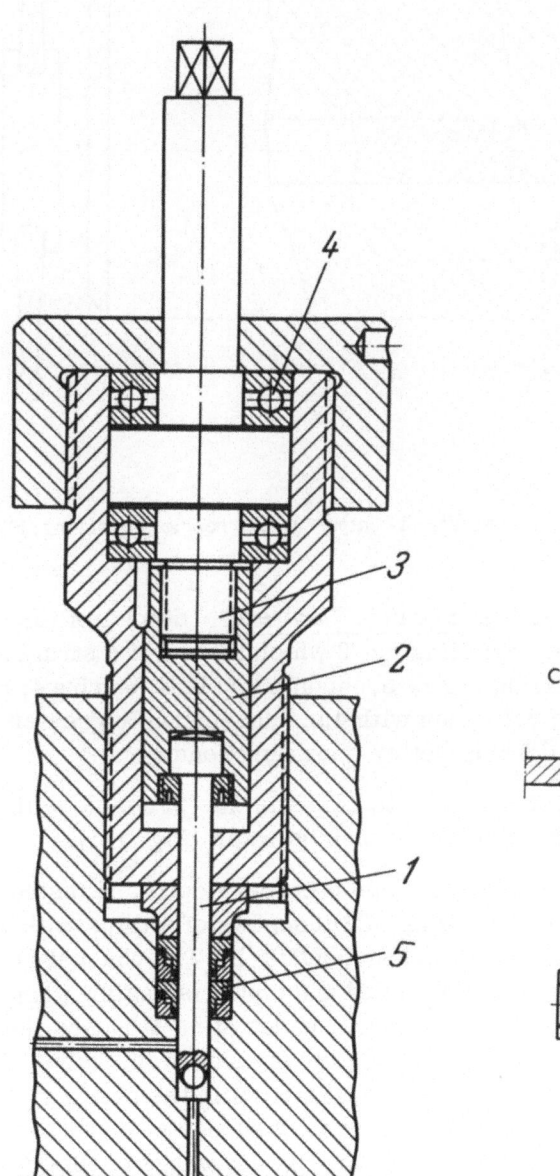

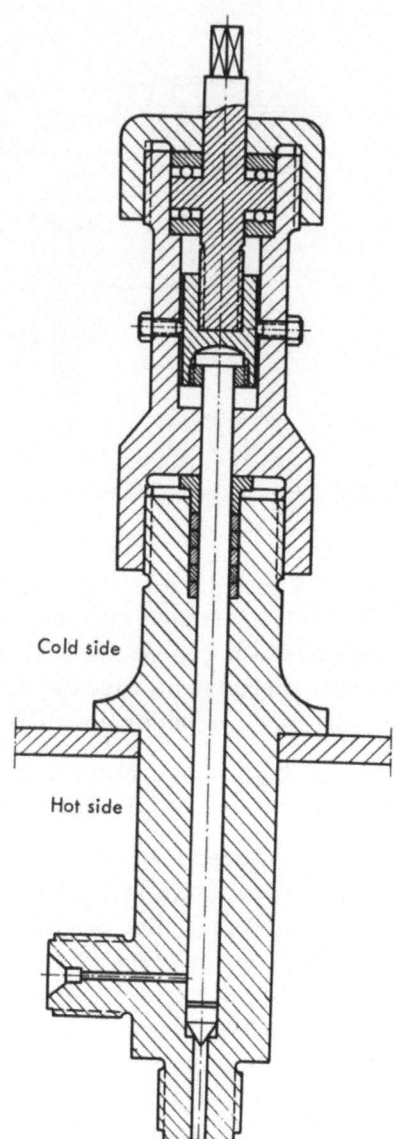

Fig. 130. Press [nonrotating stem] valve:
1) stem; 2) travelling nut; 3) drive screw;
4) ball bearings; 5) gland.

Fig. 131. Valve with cooled gland.

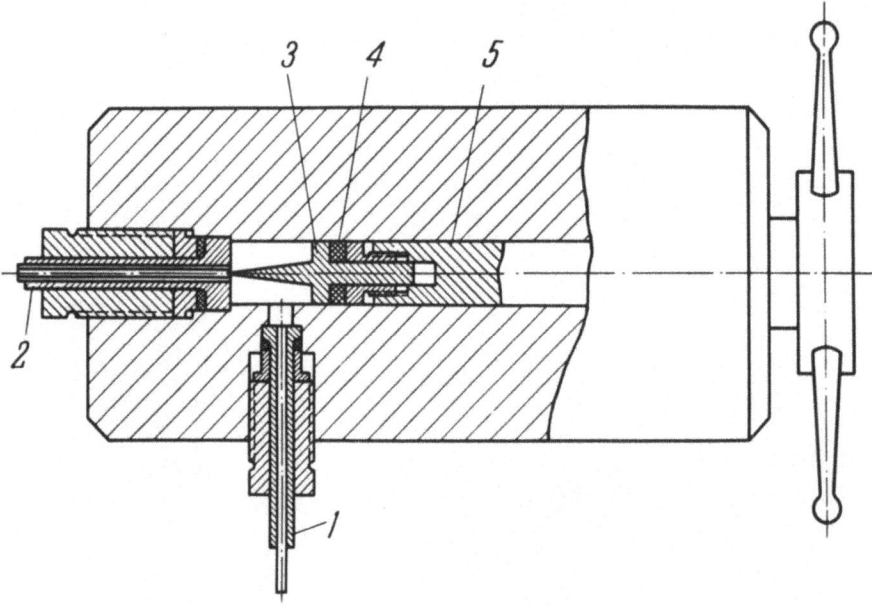

Fig. 132. Vereshchagin press valve: 1) input tube; 2) output tube; 3) needle; 4) packing; 5) stem.

Press [Reciprocating Stem] Valves.   Stem 1 of this valve (Fig. 130) is fastened into the travelling nut 2 which moves in a straight line without rotating when the driving screw 3, mounted on ball bearing 4, is turned. Because of this, the valve can withstand considerably greater loads, and the gland is under considerably better operating conditions.*

At high temperatures, it is useful to combine this principle with the remote glar [3] which can be placed in a refrigerator (Fig. 131).

An interesting press valve design has been developed and tested by L. F. Vereshchagin [4]. The valve (Fig. 132) consists of a thick-walled block, containing inlet conduit 1 and outlet conduit 2. The channel drilled in the body of the block is carefully finished and contains sealing needle 3, resting against rubber packing 4. Turning threaded stem 5 causes the needle to move forward or back and thereby close or open the valve. The needle does not rotate due to the tremendous friction in the packing. In this valve, the ordinary gland is replaced by an unsupported-area seal. The packing is located at the end of the stem. An advantage of this design is the absence of the ordinary gland, which requires heavy tightening and which wears extremely rapidly, especially at high pressures.

---

*A valve of this design can withstand repeated closing at pressures on the order of 8000 atm.

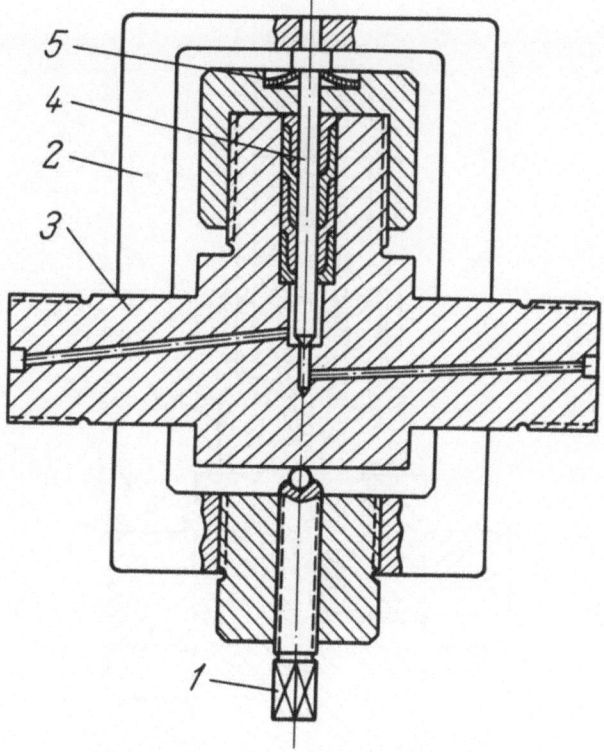

Fig. 133. Ultrahigh-pressure valve: 1) stem; 2) collar; 3) body;
4) pin; 5) spring washer.

Ultrahigh-Pressure Valve [5].   Figure 133 shows a valve
designed by L. F. Vereshchagin and V. E. Ivanov. In this valve also, the
stem has reciprocating movement. The valve is constructed on the elastic
bushing principle (see section on "glands"). When stem 1 is turned, collar
2 moves relative to body 3 and thereby moves pin 4, which closes the passage through the body of the valve. The valve is opened by expansion of
spring 5. This valve, designed for operation with gas at pressures up to
6000 atm, is described in the literature [6].

Hydraulic Valve.   In all the valves described above, gradual
increase of the sealing area occurs, resulting in an increase in the area
of tangency [contiguity] of pin to seat; closing of the valve requires ever increasing force.

This deficiency is eliminated to a considerable extent by the so-called
hydraulic valve [7] (Fig. 134), which is closed by forcing oil into cylinder 1.
When this occurs, piston 2 moves in the direction shown by the arrow, moving pin 3 in gland 4. Spring 5 opens the valve when the pressure in cylinder

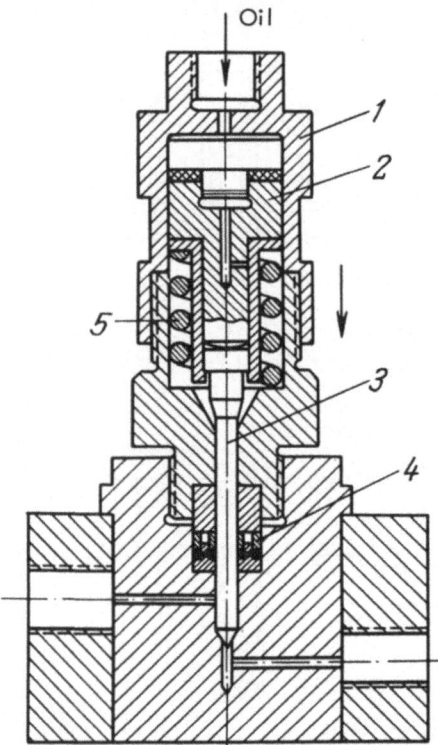

Fig. 134. Hydraulic valve: 1) cylinder; 2) piston;
3) stem; 4) gland; 5) spring.

1 is reduced. In this valve, which is actually a small intensifier, the force required to close off the channel with the pin can be accurately determined and controlled. If the pressure in cylinder 1 exceeds the predetermined pressure, deterioration of the seat takes place considerably more rapidly than with mechanical closing.

Valves with Washable and Self-Sealing Stems. If the cavity of a valve must be blown out or washed with a gas or solvent in order to remove material collected at the seat, a stem with a hole drilled through it may be used [8] (Fig. 135). This type of valve [9] may also be used as a control or drain valve and is convenient in that it is considerably more compact than valves with glands   A drilled hole in the stem [10] (Fig. 136) can be used to improve considerably the conditions of sealing of the conical portion of the stem against the valve seat, since in this case the pressure of the medium will expand the cone tip and seal against the seat.

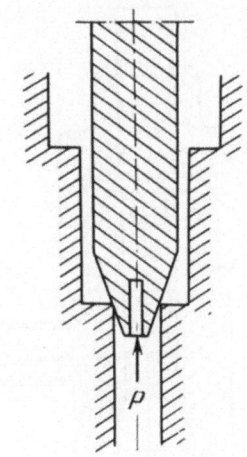

Fig. 136. Self-sealing stem tip.

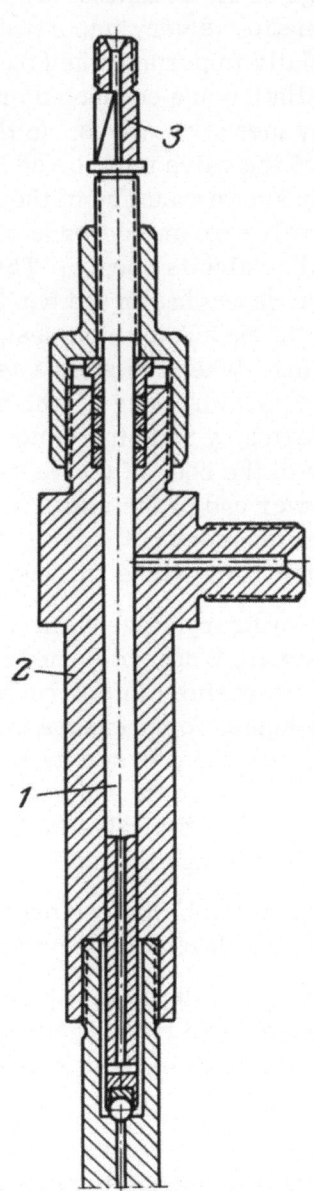

Fig. 135. Valve with washable pin:
1) stem; 2) body; 3) nipple for supply
of washing liquid.

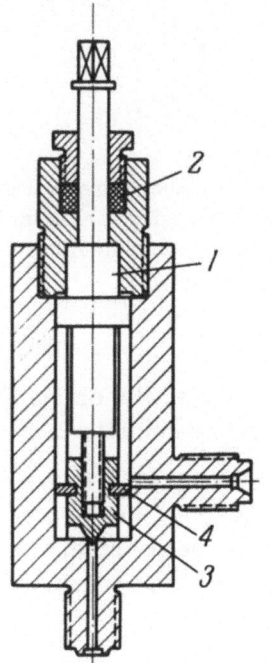

Fig. 137. Constant-volume
valve: 1) stem; 2) gland;
3) sealing cone; 4) guide.

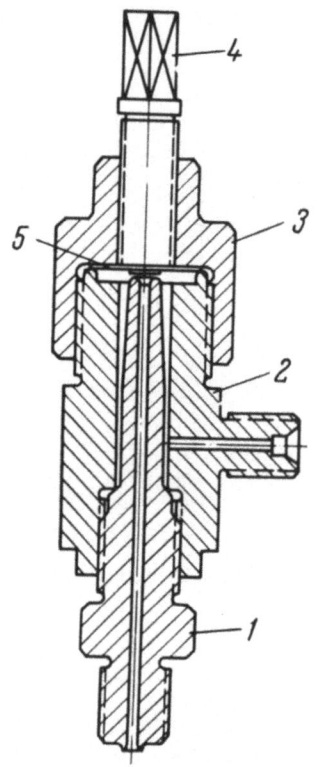

Fig. 138. Glandless membrane [diaphragm] valve: 1) nipple; 2) body; 3) nut; 4) stem; 5) membrane [diaphragm].

Valve with Constant Volume. When making precise investigations with volumetric measurements, knowledge of the volume of all parts in an installation is very important. This is especially important when parts of the installation are connected or disconnected by means of valves. In this, the volume of the valve must be added to the precisely known volumes of the apparatus; the valve volume depends on the amount the valve is opened. The change in volume depending on the position of the stem is excluded in the design shown on Fig. 137. When the stem 1 is rotated in gland 2, sealing cone 3, which is held from rotation by two pins 4 moving in the slots in the body of the valve, moves on the lower end of the stem. Since the primary stem does not leave the valve, the volume of the valve is constant.

An ordinary valve stem with no gland and with a sphere turned on its tip is used successfully in control valves or valves designed for pressure release. This stem has two intersecting holes drilled into it, one on its axis and one through its diameter. When the valve is opened, the liquid exits through the hole.

Of course, the elements of the designs which have been presented can be used in any combination, depending on the conditions of operation.

Membrane [Diaphragm] Valves. Gland valves have certain deficiencies. The stresses created in the gland from tightening and pressure cause considerable friction when the stem is rotated, which hinders control of the valve.

The higher the temperature at which the valve operates, the more difficult it is to select material for packing the gland and to achieve reliable sealing. The use of a remote, cooled gland interferes with the isothermal nature of the system. A glandless, membrane [diaphragm] valve does not have these defects.

Figure 138 shows a valve designed by the author and tested at pressures up to 300 atm. Body 2 of the valve contains nipple 1. Nut 3 seals against membrane 5. Stem 4, when rotated, presses on membrane 5, closing the valve. Pressure cannot burst the membrane, since its entire area is supported by parts 3 and 4.

The thickness of the membrane depends on the pressure being used; in this design, it is 0.1-0.2 mm. A membrane 0.5 mm thick can be used at pressures up to 3000 atm.

The membrane valve allows operation at very high temperatures. It is also a constant-volume valve, since the change in volume as the membrane moves can be ignored.

Due to the small movement, the membrane valve has low throughput capacity. Also, if the pressure in the apparatus is decreased to below atmospheric pressure, air moving through the thread of stem 4 will create an excess pressure which trips the membrane and hinders further

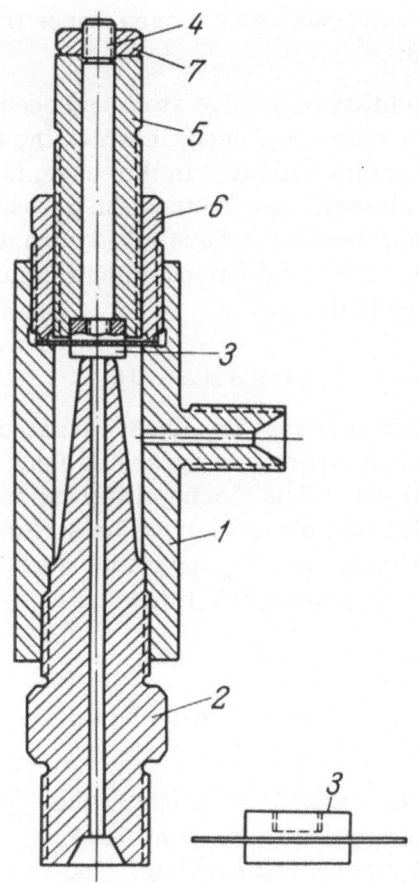

Fig. 139. Membrane valve operating under pressure and under vacuum: 1) body; 2) nipple; 3) membranes; 4, 5) stems; 6, 7) nuts.

evacuation of the apparatus. This deficiency is eliminated in the valve design shown on Fig. 139 [11].

The valve consists of body 1, into which nipple 2 is screwed. Membrane 3 is turned from one piece of metal or welded up by precise welding from two plates with a membrane held between them consisting of stainless steel sheet. Stem 4 is screwed against the membrane. To close the valve, stem 5 is turned into nut 6. This causes the stem to press the membrane against nipple 2. To open the valve, stem 5 is turned back two to three turns and stem 4 is raised by turning nut 7. At the same time, the membrane rises, opening the valve; the valve cannot be closed, since the position of the membrane is firmly fixed.

When an experimenter requires mobility of a valve stem and accurate control of stem travel, glandless bellows valves are used, in which the stem moves due to the flexibility of an elastic metal bellows. In this case, however, the bellows should have sufficient strength to withstand the pressure drop between working pressure and atmospheric pressure. One such design is described in reference [12]. Descriptions of various modifications of this valve can be found in the literature [13].

SEALS FOR LABORATORY APPARATUS

The sealing of high-pressure apparatus is very important to the technology of high-pressure experimentation. Apertures in apparatus may have extremely varied diameters and purposes, thus placing different requirements on the sealing closures used; these closures must be strongest for apparatus requiring frequent assembly, and for apparatus working with great fluctuations of temperature and pressure [14-16].

Seals with Compensated Area*

Closures with Flat Packing. Various designs of plugs are shown on Fig. 140.

Type I. The packing 1 (Fig. 140a) is placed in a recess in the body of the apparatus 2. The packing is made of brass, aluminum, lead, or often (with low pressures, and mainly at low temperatures) rubber. The sealing surfaces of the body and nut 3 carry one or two annular triangular-section channels which fill with the material of the packing. When nut 3 is tightened, there is always danger of destroying the packing. This can be avoided by using closures of other designs.

Type II. Packing 1 (Fig. 140b) is placed in a space formed by ring 4 and flange 5. The sealing is performed by tightening nuts on studs 6.

Another variant of this closure is the so-called bracket cap, described by S. N. Ganz [17] (Fig. 140c). Packing 1 is compressed between ring 4, the body of the apparatus 2 and cover 7. The sealing is produced by tightening bolts 8.

An advantage of this type of closure is the ease of changing packing.

_____

*Editor's comment: Bridgman developed the "unsupported-area" seal (a self-energizing seal) which develops a force of reaction on an area smaller than the area exposed to the pressure medium. In Russian this kind of seal is termed "non-compensated area," and all seals with equal areas (not self-energizing) are termed "compensated-area" types.

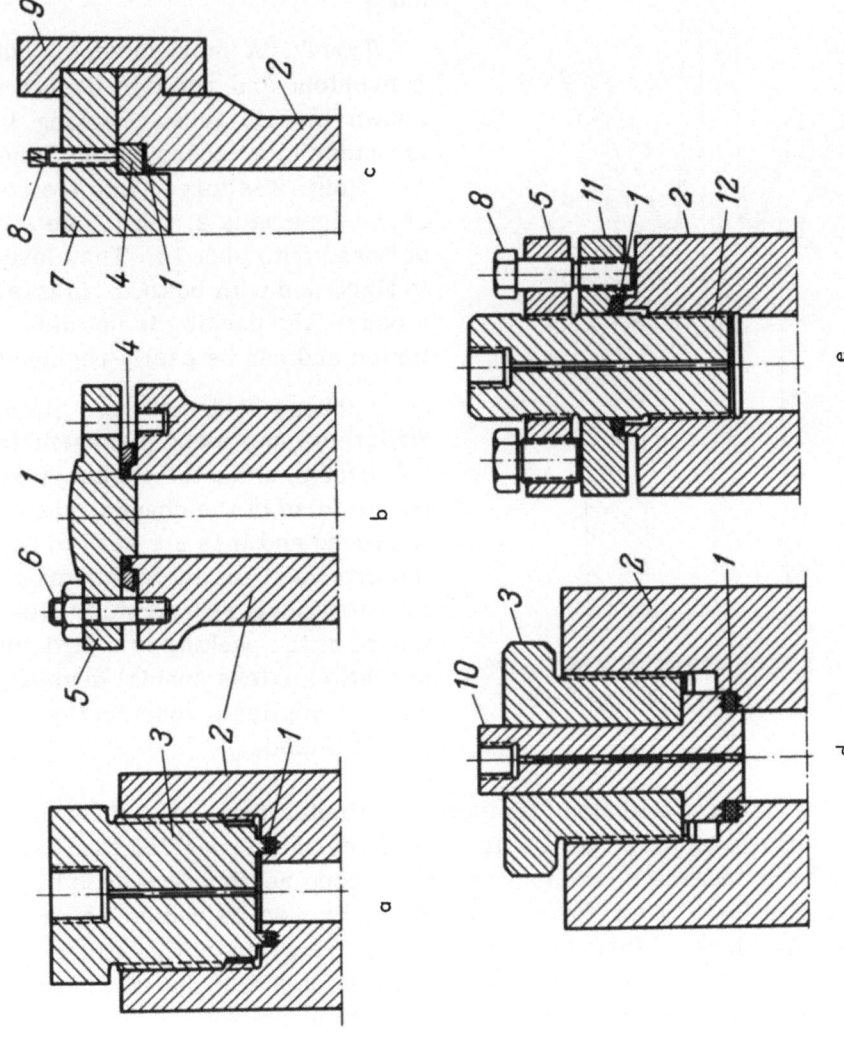

Fig. 140. Closures with flat packing and compensated area (a, b, c, d, e, types I–IV plugs): 1) sealing material (packing); 2) body; 3) nut; 4) rings; 5, 11) flanges; 6) stud; 7) top; 8) bolts; 9) bracket; 10) head; 12) plug.

Type III. **In this design, damage to packing 1 is prevented by the use of a head spindle (Fig. 140d), under which the packing is placed. The head is tightened by nut 3.**

Type IV. **A more complex but convenient and reliable design is shown in Fig. 140e. Packing 1 is placed in a recess and in flange 11. Plug 12 is forced into the body of the apparatus 2, and flange 5 is screwed onto plug 12. The closure is tightened with bolts 8. In this closure, the packing is not distorted and can be easily replaced.**

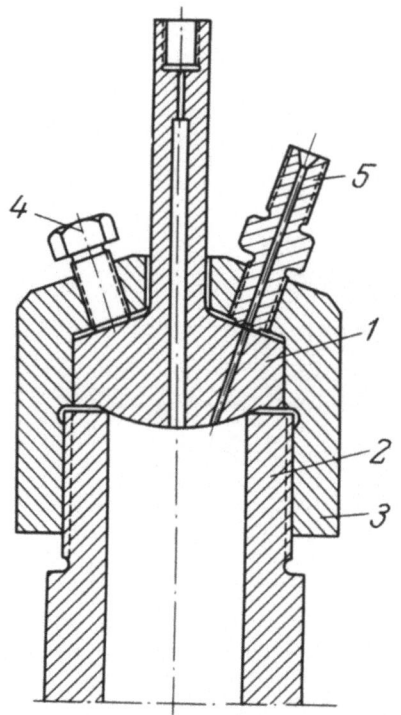

Fig. 141. Closure without packing: 1) plug; 2) vessel; 3) nut; 4) bolt; 5) nipple.

These seals are very simple. Still, they must not be disassembled frequently, since the packing takes on the form of the channel when tightened and it is not easy to fit the grooves back into their proper locations upon reassembly. Removal of the packing is a difficult operation unless special devices, which complicate construction, are provided.

These closures withstand temperature fluctuations well, since the packing generally has a larger coefficient of thermal expansion than the material of the apparatus, and expands more upon heating than does the slot in which it lies. Therefore, when a heated vessel is cooled, the sealing may tend to leak. Closures therefore require periodic tightening of these nuts or bolts and can be recommended only for operation at pressures not over 500-800 atm, and at normal temperatures.

The design of closures of this type consists in determining the dimensions of the packing and of the tightening bolts or threaded nuts.

When a nut is tightened, the following stress is generated in the packing:

$$\sigma = \frac{P_1}{S_1}$$

where P_1 is the tightening force; S_1 is the area of the packing.

In order to provide sealing, the packing should flow and fill the bore. In other words, the tightening force should be great enough that the stress in the packing exceeds the yield point of the packing material.

If pressure p is created in the apparatus, then force P_2 acts on area S_2 of the plug; this force is equal to

$$P_2 = pS_2$$

which must be considered in calculating the tightening force. The total tightening force, without consideration of temperature stresses, the effect of thread expansion, etc., is equal to :*

$$P_t = pS_2 + \sigma S_1$$

Closure Without Packing. The packing is generally the weak point in any closure which must be frequently removed or which operates at varying temperatures. Therefore the use of packing should be avoided when possible. A closure without packing designed by the author [18] is shown in Fig. 141.

Plug 1, which has a spherical surface, is fixed in vessel 2 by nut 3. Sealing is achieved by six bolts 4; 1 or more of the bolts may be replaced by nipples 5. Due to the inclined position of the bolts, the apparatus is easy to service (bolts can be easily removed, valves can be easily connected to nipples 5, etc.). This closure can undergo considerable temperature fluctuations without losing its ability to seal, and will still function properly after numerous assembly-disassembly cycles.

In order to place nipples 5 in position in place of bolts 4, the seal should be assembled with the plug undrilled; the plug should be fixed in relation to nut 3; the nipple should be screwed into the threaded bolt aperture and the location for drilling the nipple aperture should be marked through the nipple.

Unsupported-Area Closures

.*Type I.* The difference in operation of this type of closure [19] (Fig. 142) is that the pressure received by the [projected] area of head 1 is transmitted [through head 1 only] to the area of packing 2, which is less than the

---

* Editor's comment: This formula assumes that the packing does not begin to be unloaded as the cap takes increasing amounts of load (due to pressure increase). Hence this formula yields conservative values of working stress.

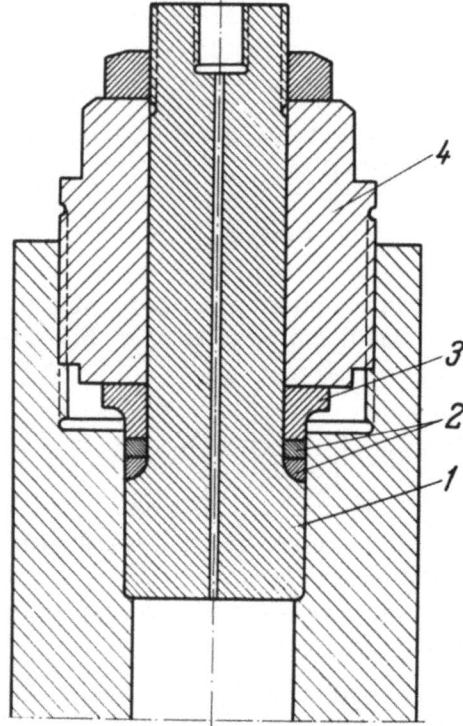

Fig. 142. Unsupported-area closure (type I):
1) head; 2) packing; 3) sleeve; 4) nut.

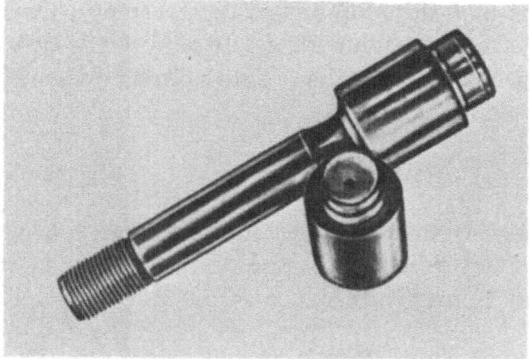

Fig. 143. Beginning of bite (failure) [pinch-off] of
head shank, and severed head.

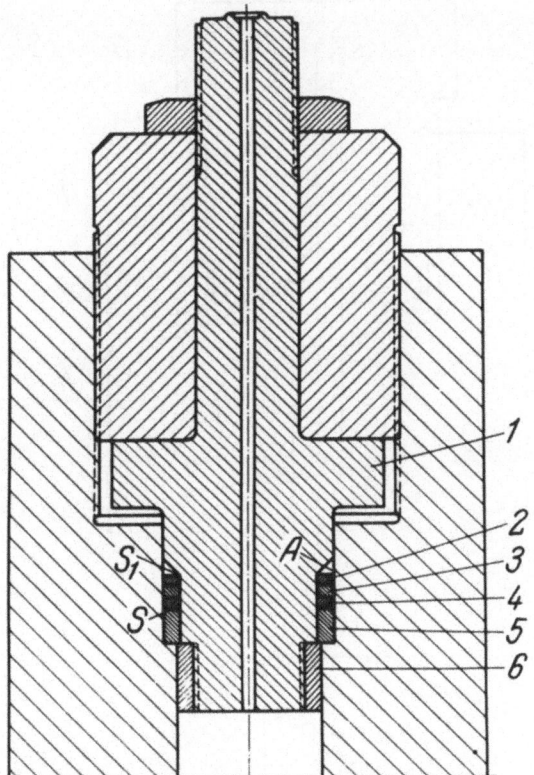

Fig. 144. Closure of unsupported-area type (type II):
1) head; 2, 4) steel packing; 3) copper packing; 5) bearing ring; 6) extraction ring.

area of the head. Therefore, the stresses generated in the packing are greater than the working pressure. A pressure differential is automatically maintained as long as the pressure in the apparatus remains higher than the limit for which the seal is designed.*

These closures are very convenient to use, especially when working with pressures above 500-800 atm. They do require proper design, however. Otherwise, the closures will not work at all or will fail rapidly.

Design of the closure consists of selection of a packing area such that the stress in the packing will exceed the working pressure by 10-15%. If the force acting on the plug is equal to the working pressure P multiplied

_____

*Editor's comment: Some unsupported-area seals do not begin to seal until a threshold pressure is exceeded; this pressure is probably the limit mentioned.

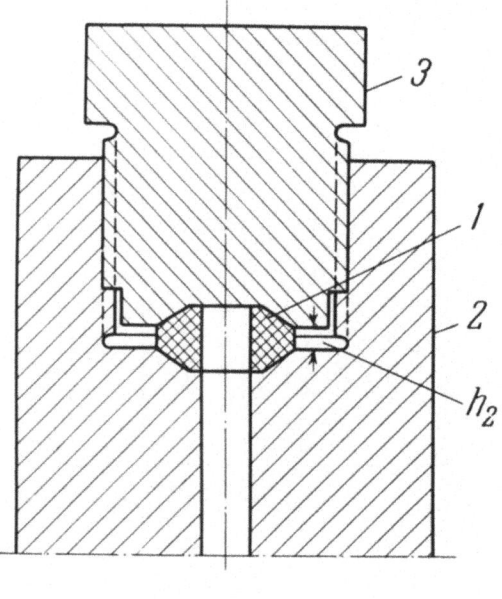

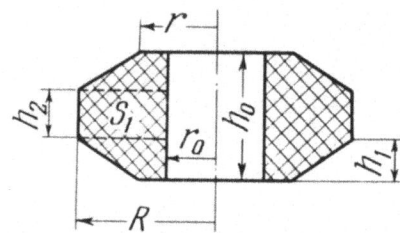

Fig. 145. Lens closure: 1) lens; 2) body; 3) nut.

by the plug area S_2, this force will be resisted [only] by area S_1 of the packing. The stress in the packing is then

$$\sigma = \frac{PS_2}{S_1}$$

Calculations should be performed such that

$$\frac{\sigma}{P} = 1.15 - 1.20$$

The lower limit for σ is the yield point of the packing material. Therefore, if a vessel with large internal diameter must be sealed for operation at low pressure, it is sometimes impossible to retain the relationship:

$$\sigma > \sigma_s$$

where σ_s is the yield point of the packing material.

In such cases, unsupported-area sealing is not applicable.

At high pressures, when σ exceeds the yield point or even the ultimate strength of the head material, the packing will bite into [tend to pinch off] the shank. This occurs frequently, and should be watched carefully, since a part which breaks off will fly out of the closure like a bullet.*

The beginning of this biting (distortion) [pinch-off] is shown on Fig. 143.

The upper limit of applicability of this type of seal is the pressure at which

$$\sigma = \sigma_s'$$

where σ_s' is the yield point of the plug material.

Within these limits, this seal works quite well and withstands temperature fluctuations very well.†

---

* Editor's comment: The shank (see Fig. 164) is subject to lateral principal stresses that are equal and compressive, say C psi. Assuming a diameter ratio of bore to shank of 8:3, one finds that the lateral stresses on the shank are (64/55) P, where P is the pressure being sealed. Thus the lateral stress is about 115% the value of the pressure $(C \approx P)$.

Now, two equal lateral stresses with a zero longitudinal stress is equivalent to a stress state described by a uniaxial tensile stress C submerged in a pressure C. Thus the steel of the shank should have an ultimate tensile strength no lower than 1.15 times the pressure to be sealed, at least when tested in a pressure environment of magnitude 2 C/3 (the average normal stress in the shank).

When "pinching-off" occurs, the unsupported portion of the shank flies off, but it is stopped by the female portion of the closure. Sometimes the female portion contains a through-bore, in which case the moving piece strikes the component of the machine that bears on the female.

The phenomenon can be complicated in practice by the presence of a chamfer ring on the shank, and by frictional forces. The result sometimes is that pinching-off occurs on the down stroke, as pressure is being reduced, and not on the up stroke.

†Editor's comment: The upper limit of pressure that can be sealed is about 600,000 psi, but life is usually short at such high pressure.

All parts of the seal should be carefully fitted to each other. If this is not done, especially when working with pressures on the order of 10,000 atm, the lid packing will flow into the space between the head 1 (see Fig. 142) and the sleeve 3, ruining the sleeve, which has been observed in practice.

Closures of this type in which the packing is sealed by sealing wedges have been described in the literature [17]. This design of closure permits use of the unsupported-area principle for large diameter (industrial) apparatus.

Type II. The principle of operation of these closures (Fig. 144) is that the working pressure is received [directly] by entire area S of the packing, and is transmitted to area S_1, which is limited to the surface area where the packing contacts the conical portion of the plug (cone angle on each side ~25-30°). Due to this, the difference between the pressure in the apparatus and the stress in the packing is considerably greater than in a type I closure. Also, the danger of biting into the head [pinch-off] is not present, since the packing flows freely in space A between the body of the vessel and the plug.

This type of seal can be placed in an apparatus of minimal bore diameter, so that this type of closure is convenient for work at pressures over 10,000 atm. The packing (for high-pressure operation) is made of heat-treated steel.

A type II closure is also used together with an O-ring [20] which is inserted into the side of the apparatus and used for preliminary sealing.

The design of a type II closure is analogous to the design of a type I closure. A deficiency of the closure is that, after space A is filled with the packing material, the closure ceases to work as a self-sealing closure, becoming an ordinary closure.*

Lens Closure. In a lens closure, primary sealing is created when the nut is tightened due to distortion of the contiguous spherical [conical] and plane surfaces (see Fig. 145). Then, the pressure acts on the internal surface of the lens and presses it against the sealing surfaces, as in unsupported-area closures. The lens closure can be disassembled repeatedly. It is convenient to use.

The "noncompensation" of the lens seal changes during the course of operation. The first sealing is achieved around the belt formed as the lens touches the cone of the sleeve. As soon as the pressure in the apparatus begins to increase, the lens is forced into the space between the

---

* Editor's comment: The closure may continue to function as an unsupported-area seal if the clearance between head 1 and body (unnumbered) in Fig. 144 is adequate.

planes, taking on the form of a sleeve. The area of tangency between lens and sleeve increases rapidly at this stage, i.e., the "noncompensation" is decreased. In designing a lens, this fact must be considered.

An extremely important parameter is the dimension of gap h_2. If the gap is too great, the lens will burst radially. Force P acting on the internal surface of the lens is equal to

$$P = 2\pi r_0 h_0 P_0$$

where P_0 is the pressure.

This force is received by area S, which is equal to twice the area of the surface of the truncated cone with height h_1 and bases R and r (Fig. 145). The stress in the lens is

$$\sigma = \frac{P}{S}$$

It is now important to calculate the stress, σ_S, generated in the cross-section of the lens which causes it to burst. The area of the section is equal to

$$S_S = 2\pi (R^2 - r_0^2)$$

The section can receive force P_1 acting on the annular area given by height h_2:

$$P_1 = 2\pi r_0 h_2 P_0$$

From this,

$$\sigma_S = \frac{P_0 r_0 h_2}{R^2 - r^2}$$

All these calculations are more or less approximate. The most efficient lens dimensions must be determined experimentally.

Lens closures are convenient where the diameters and pressures involved are not too great. The material for the lens in these cases can be heat-treated steel with a tensile strength of 10,000-12,000 kg/cm^2.

In those cases when the lens is not operating on the unsupported-area principle, it becomes an ordinary conical closure, which will be described below.

Closure with Wavy Ring. The closure with wavy ring (Fig. 146) is also constructed on the unsupported-area principle. The sealing ring is compressed in the radial direction, not the axial direction.

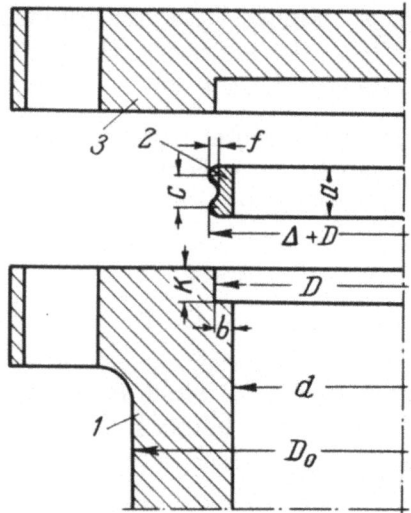

Fig. 146. Closure with wavy ring: 1) body;
2) wavy ring; 3) cover.

The packing itself is in the form of a cylinder with two wave-shaped ridges whose diameter is somewhat greater than the diameter of the seat. The relationships between some of the dimensions represented on Fig. 146 are:

$$a = 3b - 5b$$
$$c = 0.6a$$
$$f = 0.1a - 0.2a$$
$$k = 1.01a/2$$
$$\Delta = 0.0002d - 0.0005d$$

The seal is driven into the seat. It has been empirically established [21] that for diameters between 1.5 inches and 6 inches, the greatest ring thickness at which the ring material will begin to flow should be between $pd/2\sigma_S$ and $pd/5\sigma_S$, where p is the pressure in the apparatus.

According to the data of T. A. Makhnev [22], a closure of this type is convenient in operation, since it requires no preliminary application of axial force. A properly made wavy ring is not deformed, permits multiple assembly and disassembly of the closure and can be used over a wide range of pressures and temperatures. This design is less cumbersome than others. However, the use of a closure of this type requires considerable preparatory work in selection of the material and dimensions of the rings.

Sometimes, the wavy ring is replaced by cylindrical sleeves with sharp turned edges, which protects the sleeve from bending inward (L. F.

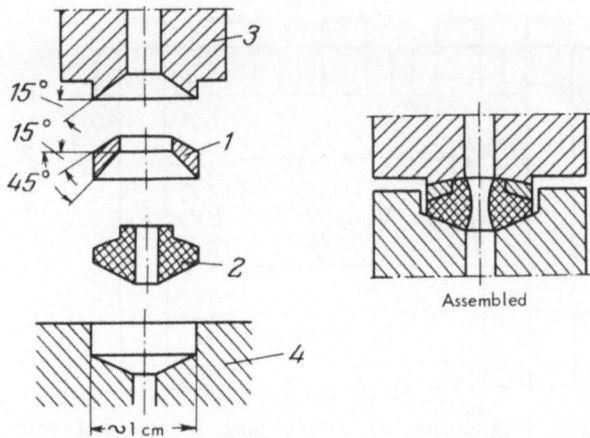

Fig. 147. Closure with rubber packing: 1) metal ring; 2) rubber
packing; 3, 4) parts.

Vereshchagin [23]). This type of sleeve is simpler to manufacture,
and works just as well as a ring.

Closure with Rubber Packing [24]. The closure with
rubber packing consists of a metal ring 1 (Fig. 147) and rubber packing 2,
both compressed by parts 3 and 4. The closure as assembled is shown on
the right side of the figure. According to the data of the author, this closure
will withstand a pressure of 15,000 atm.

Closures for Large-Diameter Apparatus

Various types of seals for large-diameter apparatus have been described in the literature [25]. Figure 148 [26] shows closures suitable for
closing large-diameter apparatus. Figure 148a shows a conical closure for
closing lined apparatus; Fig. 148b shows a closure with so-called Δ-ring.
Here, sealing ring 3 (the packing), which has the form of a right triangle
before the pressure is applied to the apparatus, takes on the form shown in
the figure after the pressure enters the apparatus. Fig. 148c shows a

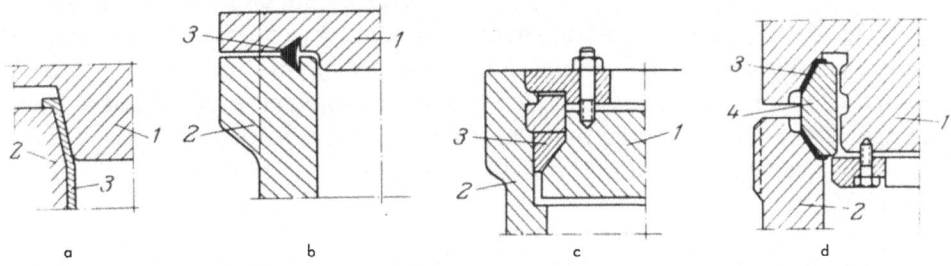

Fig. 148. Seals for large-diameter apparatus: 1) plug; 2) body of apparatus; 3) packing; 4) lens.

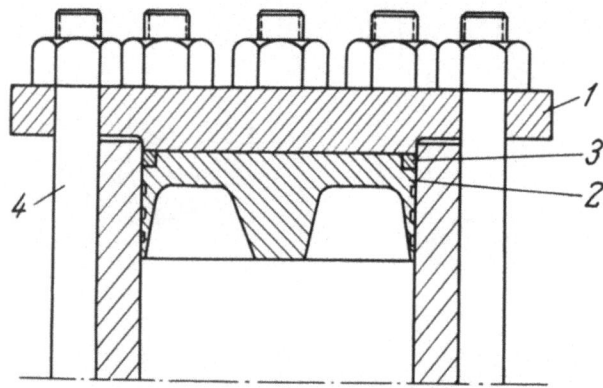

Fig. 149. Plug closure: 1) cover; 2) plug; 3) packing; 4) bolts.

closure of unsupported-area type I designed for large-diameter apparatus; Fig. 148d shows a lens closure.

Figure 149 shows a design of closure for very large diameter apparatus [27]. The closure consists of cover 1, which holds plug 2 of the "expanding section" type. The plug has thin walls with grooves. The initial sealing is created by tightening the nuts. Then, pressure expands the thin walls of the plug, sealing the apparatus. The cover is held down by six tension bolts 4.

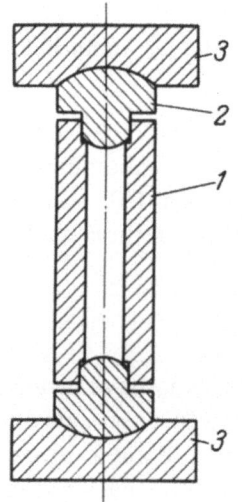

Fig. 150. Hydraulic closure: 1) body; 2) cover; 3) hydaulic-press platens.

Hydraulic Closures

In hydraulic closures [28], sealing is performed not by tightening nuts or bolts, but by a hydraulic press, which creates the force required for counteracting the internal pressure and for distortion of packing.

Hydraulic closures have advantages over ordinary closures. In an apparatus closed by a hydraulic closure, the pressure cannot exceed the predetermined experimental pressure. If the pressure becomes too great, the closure of the apparatus is raised and the pressure is reduced to a value corresponding to the pressure in the cylinders of the hydraulic press. For this reason, this type of device is explosion safe.

Hydraulic closures are more convenient for use at high temperatures, when there is a danger of seizure of the thread of sealing nuts. This type

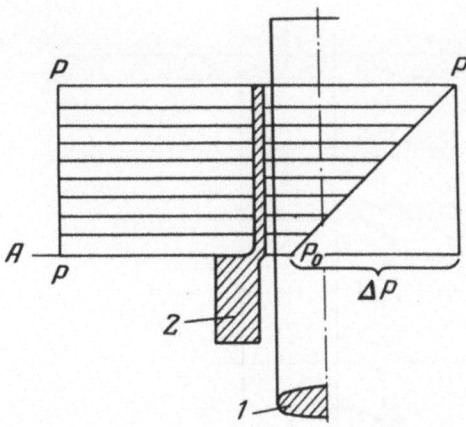

Fig. 151. Diagram of forces acting on elastic cylindrical
sleeve: 1) shaft; 2) sleeve.

of device allows rapid opening and closing of high-pressure apparatus, not
requiring the expenditure of physical force for tightening nuts and bolts.

The apparatus may be sealed by placing it between the platens of a
hydraulic press (Fig. 150) or by installing a special intensifier [29] on the
apparatus. If we know the force required for sealing an apparatus of a
given diameter, we can calculate the dimensions of the intensifier required
for the hydraulic seal.

In principle, a hydraulic seal can be used with any type of high-pressure apparatus.

GLANDS

Glands are devices through which moving parts such as valve stems,
pistons, compressor shafts, pump shafts, agitator rods, etc., are introduced into high-pressure apparatus. These devices prevent leakage, or
reduce leakage to certain values [30]. This factor is highly significant,
since it is at times not important to prevent leakage completely, but only
to limit it.

Glands can be divided into two large groups: 1) glands with natural
sealing; 2) glands with packing.

Glands with Natural Sealing [31]

A gland with natural sealing is a device in which there is always a
gap between the body of the gland and the moving part. In some cases, the
tangent surfaces are carefully machined to create the required minimal gap,

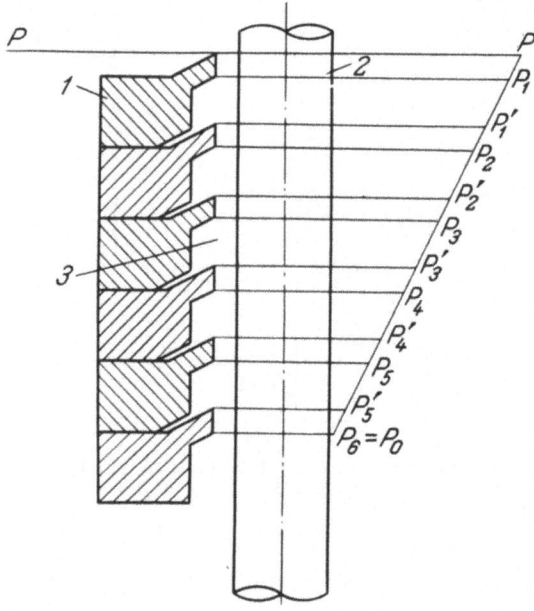

Fig. 152. Labyrinth gland: 1) sleeve; 2) shaft; 3) expansion space.

in other cases the gap is a labyrinth, consisting of alternating ridges and expansion chambers in which the medium is choked and the pressure drops.

Elastic Cylindrical Sleeve.   Mobile shaft 1 (see Fig. 151) is located within an elastic cylindrical sleeve 2, whose sealing surface is located at A. Identical hydrostatic pressure P acts over the entire length of the external surface of the sleeve. But, since the pressure of the medium in the gap between sleeve and shaft drops from P to P_0 (by ΔP) due to viscous flow, the sleeve is partially compressed around the shaft, which in turn helps to reduce the pressure in the gap.

In designing this type of packing, one must strive to increase ΔP to the maximum while maintaining the elastic deformation of the sleeve at a low enough level that the pressure in the gap will not force all of the lubricant out, resulting in metal-to-metal friction.

By determining the elastic radial displacement of the radius of the sleeve, we can show that a sleeve 6 mm in diameter with a wall thickness of 1.5 mm will be fully compressed when $\Delta P = 6000-7000$ atm. Seals of this design can be used with both liquid and gas media. In the latter case, the viscous medium flowing in the gap is the lubricant.

Labyrinth Gland.   This gland [31] (Fig. 152) consists of a series of short sleeves 1, which are not deformed under pressure. Hydraulic re-

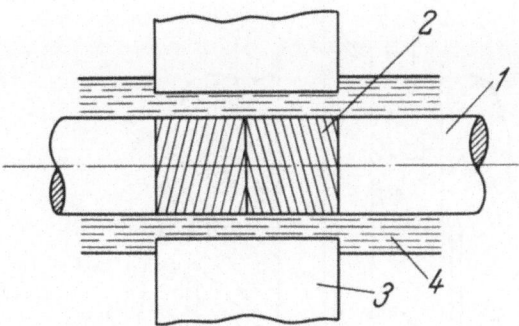

Fig. 153. Hydrodynamic seal: 1) shaft; 2) sleeve; 3) body of gland; 4) pressurized medium.

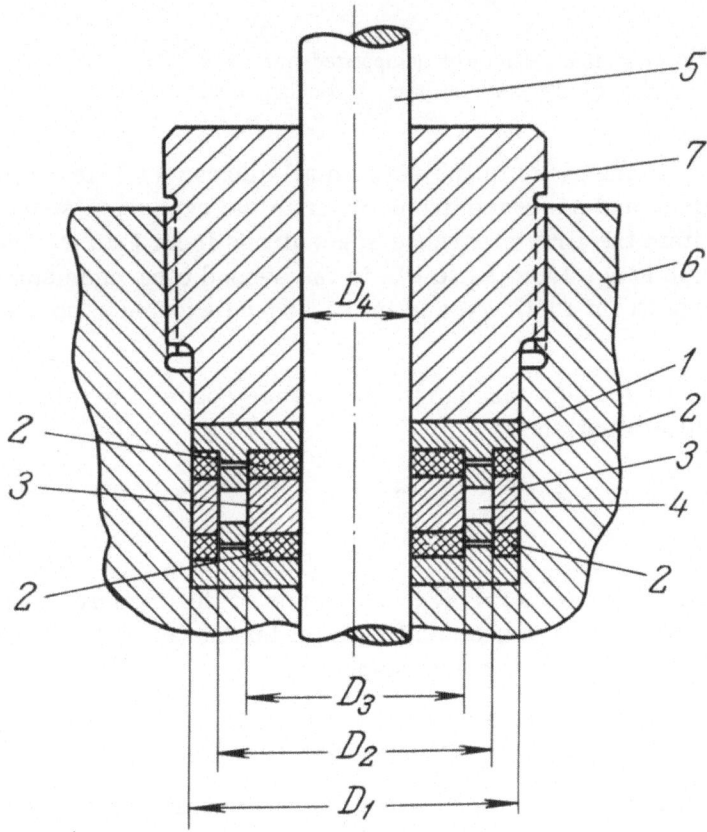

Fig. 154. Self-sealing gland: 1) pressure ring; 2) packing; 3) bearing rings; 4) free space; 5) shaft; 6) body; 7) nut.

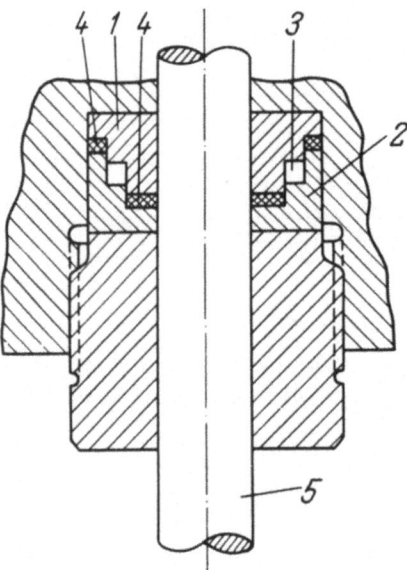

Fig. 155. Gland with unsupported area: 1, 2)
shaped rings; 3) free space; 4) packing; 5) shaft.

sistance is set up in the gland in the very small clearances between the
sleeves and shaft 2, and pressure drops occur in the expansion spaces 3.
The pressure within the first clearance space drops from P to P_1, then in
the first expansion space from P_1 to P'_1; in the second clearance space,
the pressure drops from P'_1 to P_2, and in the second expansion space from
P_2 to P'_2, etc.

The amount of leakage through the clearance space is determined
from the Poiseuille equation

$$v = \frac{\pi c^3 r \Delta P}{6 \eta L}$$

where c is the clearance; r is the average radius of the annular clearances;
ΔP is the pressure drop; η is the coefficient of [absolute] viscosity of the
medium; L is the length of clearance space.

Experiments have shown that if r = 7.5 mm and c = 0.01 mm, the leakage where ΔP = 10,000 atm will not exceed 0.001 cm^3/sec.

Hydrodynamic Sealing [32]. A seal of this type (Fig. 153)
is designed for sealing rapidly rotating shafts. Shaft 1 is attached to sleeve
2 which has a pair of converging threads, a left-hand and a right-hand
thread. When the shaft rotates rapidly, the threads act as a screw pump

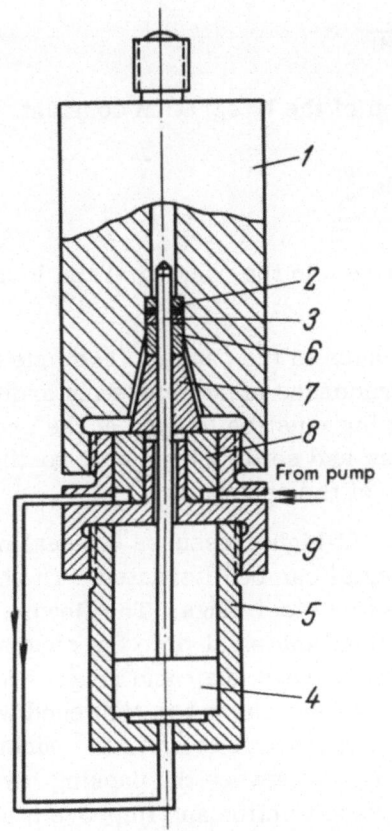

Fig. 156. Self-tightening gland: 1) high-pressure cylinder; 2) gland; 3, 4, 8) pistons; 5, 9) intensifiers; 6) sleeve; 7) cone.

and force fluid 4 into the space between sleeve 2 and the body of the gland 3, creating excess pressure in the gland, sealing it and preventing leakage of the working medium.

Packed Glands

It is usually considered that the sealing action of packed glands is created by the initial tightening of the gland. This is true only at very low pressures. When the pressure drop in the gland increases, forces arise within the gland which cause the actual sealing. This is especially true of glands operating on the unsupported-area principle.

The packing of a gland seals the gland when it [the packing] expands in the radial direction. If we represent the lateral force created by the packing as it is compressed by axial force P as Q [30], in the first approximation we can consider that

$$Q = KP$$

where K is a constant whose value depends on the material and the form of the packing. K changes from 0 for absolutely hard packing to unity for absolutely soft packing.

It can be shown that with force Q in a packing element of length L, the pressure gradient created is

$$\frac{\partial P}{\partial L} = \frac{Q}{A}$$

where A is a coefficient depending on the packing material and properties of the medium being compressed.

In other words,

$$\frac{\partial P}{\partial L} = K \frac{P}{A}$$

After integration and determination of the integration constant, we get

$$L = \frac{A}{K} \ln \frac{P}{P_0}$$

where P_0 is the residual compressive stress in the packing at the location of minimum tightness.

Since the compressive force increases in the packing from one packing ring to the next, rings of various hardnesses must be used in order to equalize the compression: the hardest ring must be located at the ground collar and softer rings next to the body of the apparatus.

At high pressures the sealing material can be alternated with shaped brass or steel rings. The flowing of the gland material [into the clearance space between shaft and rings] which occurs when the nut is tightened will tend to lessen friction in the packing. The rings between the packing layers decrease friction and improve stress distribution in the gland.

Self-Sealing Glands. One such gland [33] is shown in Fig. 154. It consists of pressure rings 1, and packings 2 which prevent leakage of the medium between the packing, shaft, surface of the bore, and surfaces of the two bearing rings 3. The noncompensation [unsupported-area] is caused by the presence of free space 4.

The design of this type of gland consists of determining the diameters of the packing (see Fig. 154):

$$\frac{D_1^2 - D_4^2}{n} = (D_1^2 - D_2^2) + (D_3^2 - D_4^2)$$

where n is the ratio of stress in the packing to the operating pressure.

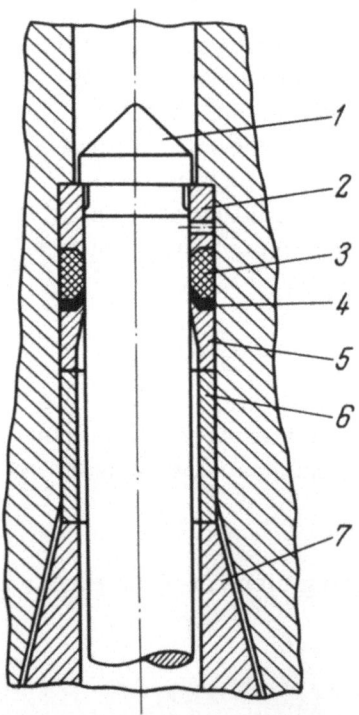

Fig. 157. Sealing of self-tightening piston: 1, 7) cones; 2) bearing ring; 3) fluoroplast ring; 4) bronze ring; 5) compression ring; 6) sleeve.

Also, the area of both [inner and outer] packings should be identical.* Therefore:

$$D_1^2 - D_2^2 = D_3^2 - D_4^2$$

If the values of D_1 and D_4 are fixed (from design considerations), it is not difficult to determine the dimensions of the packings and rings. It should be recalled, however, that at very high values of n, [unequalized] stress in the packing may burst ring 1. In order to improve the conditions of operation of this ring, three or four radial holes 1-1.5 mm in diameter should be drilled around its circumference; these holes will equalize the stresses on both sides of the ring.

The rings for this gland should be made of hardened and tempered steel (such as type 45KhNMF) with a Rockwell hardness of 40-45. The rings should be carefully polished (slip fit) to fit each other, the piston, and the cylinder. It is recommended that the surface of the cylinder be lapped with a special lap. Glands are used containing one or several elements, but it should be kept in mind that an increase in the number of elements also increases friction.

The packing in a self-sealing gland is made of fluoroplast or elastic polyvinyl chloride. The surface of the polyvinyl chloride sheet must be polished on both sides; then the sheet must be glued or fastened to the end of a wooden block and the packing ring is cut from it on a lathe, using a sharp blade. Steel punches of the required diameters can also be made, and the packing can be stamped out. Glands of this design can withstand gas pressures on the order of 10,000 atm with repeated movement of an intensifier piston.

Figure 155 shows another design of gland with unsupported area. Shaped rings 1 and 2 are made and lapped to fit each other, the cylinder, and the piston just as in the gland shown in Fig. 154. The packing is made of polyvinyl chloride or fluoroplast. The unsupported-area in this design is provided by free space 3. This design is similar to the design of the gland described above. Its advantage is that this gland can be placed in a very small volume.

Self-Tightening Gland [7]. In the design shown in Fig. 156, self-sealing of the gland packing is replaced by self-tightening. High pres-

---

* Editor's comment: In American practice, this is not essential.

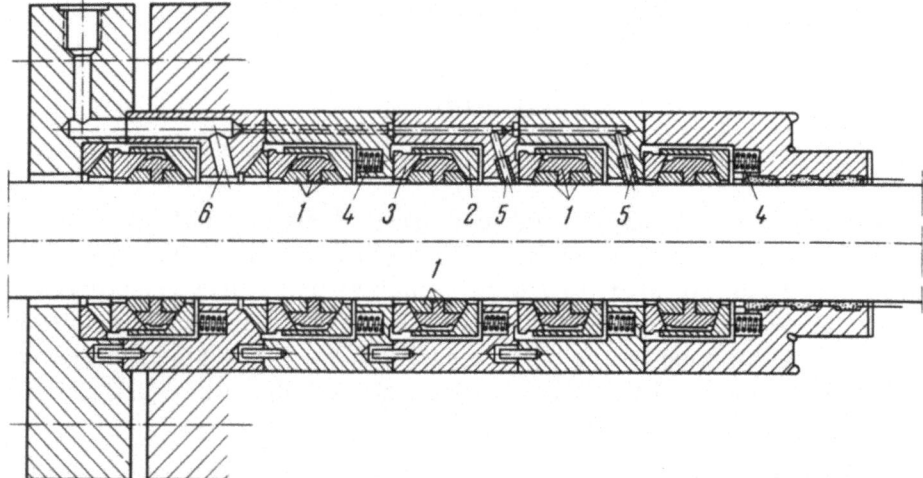

Fig. 158. Self-sealing gland for shafts with reciprocating motion: 1) three-part rings; 2, 3) conical
rings; 4) spring; 5) lubricating holes; 6) gas vent.

sure cylinder 1 contains gland 2, in which piston 3, which is connected to
piston 4 of pressure booster 5, moves. The sealing of the gland is achieved
by sleeve 6, bearing on cone 7. Movement of cone 7 occurs when piston 8
of intensifier 9 moves.

When oil is fed into intensifier cylinders 5 and 9 by the hand press
(pump), pistons 3 and 4 move simultaneously, sealing gland 2. Piston 3 may
also act as an intensifier shaft, valve stem, etc. The gland is designed so
that the stress in the packing is greater than the working pressure.

The design of the gland itself is shown in Fig. 157. The gland consists of bearing ring 2 and compression ring 5, between which is located
the packing (fluoroplast ring 3 and thin bronze ring 4). Rings 2 and 4 are
carefully fitted to the shaft and supplied with circular knife edges about 1
mm high to prevent the fluoroplast from being extruded. A hole is drilled
through ring 2 to protect it from bursting due to internal pressure.

A slipping gland for rotating shafts is described in the literature [34].
A self-sealing high-pressure gland for shafts with back-and-forth movement is shown in Fig. 158 [35]. A gland element consists of three-piece
ring 1 and conical rings 2 and 3. The rings can move in both axial and
radial directions. The working pressure forces the three-part ring against
the piston shaft. The compressive force depends on the pressure, and on
the angle of the cone, which is made less in each succeeding element in the
direction of decreasing pressure. Due to this, the three-part rings are
pressed against the shaft with identical force, even though the pressure

along the shaft decreases. Spring 4 prevents the sealing rings from moving back and forth with the shaft. The gland is lubricated through hole 5. Hole 6 is used for bleeding gas from the first sealing element. Various gland designs have been described in the literature [13, 30].

INTENSIFIER-PISTON SEALS

An essential element of any intensifier that provides continuous failure-free operation are the seals for the small (high-pressure) and large (low-pressure) pistons.

Small-Piston Seals

Ordinary Sealing. The sealing (Fig. 159) consists of alternating brass and leather rings or rings of polyvinyl chloride, fluoroplast, etc. (their form may vary), applied around the narrow portion of the piston, fastened with a nut. Conical rings with cone angles of up to 150° are the most convenient. The wide base of the cone is directed against the direction of piston movement, so that the pressure presses the sealing against the cylinder bore.*

This packing works well at gas pressures up to 500 atm and liquid (oil, kerosene) pressures up to 8000-10,000 atm. This type of seal most usually fails due to rupture of the entire tail portion or stripping of the threads in a nut. Therefore, the length of the nut should be not less than 10 mm, the thread diameter no less than 5-6 mm. The nut should be backed up by a jam nut.

Unsupported-Area Type Seal. The design of this seal was suggested by I. P. Sidorov [36] (Fig. 160). A blind threaded axial hole is made in the body of piston 1. The seal, consisting of shank 2, carrying packing in the form of alternating leather, vinyl chloride, or fluoroplast rings 3 and conical brass rings 4, is placed in this hole. The cone angle is less, the higher the pressure. The maximum angle is 140-150°. The packing is terminated by ring 5 which is threaded, and is compressed by nut 6. Ring 5 is screwed into the body of the piston. Pressure acting on the area of the head of the shank tends to force it into space 7. The pressure acts against the projected area

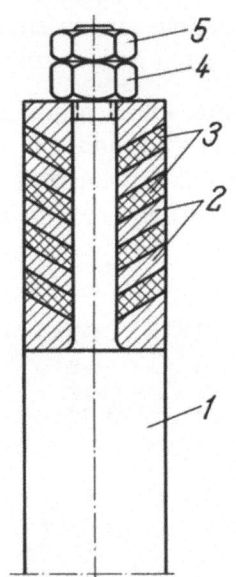

Fig. 159. Ordinary seal: 1) piston; 2) brass rings; 3) leather rings; 4) nut; 5) jam nut.

---

*This type of seal was developed and tested by I. P. Sidorov.

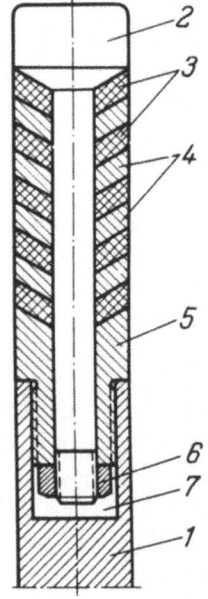

Fig. 160. Sidorov seal of unsupported-area type for pressures up to 10,000 atm: 1) piston; 2) shank (ShKh12 steel); 3) leather rings; 4) brass rings; 5) ShKh12 steel ring; 6) nut; 7) shank receiving space.

of the packing which is smaller than the area of the head of the shank by the area of the shank itself, i.e., in this case a type I unsupported-area seal results.

This type of seal can be successfully used in compression of gases and liquids. For example, it allows the piston to be dropped not only by counter pressure on the small piston, but also by counter pressure on the large piston [since it seals in both directions]. It should be noted that leather packing is not recommended for work at pressure on the order of 10,000 atm, since the fatty substances contained in the leather will be hardened under pressure and the leather will lose its elasticity [flexibility].

A deficiency of this seal is the weakening of the body of the piston by the thread. This hastens breakage of the piston at high pressures. Piston breakage is a frequent occurrence at pressure over 15,000 atm, since at 15,000 atm a piston of 20 mm diameter carries a load of 75,000 kg. Under these conditions, the pressure rings and other parts under high compressive stress should be made of cemented carbides such as pobedit.

Figure 161 shows a photograph of a piston broken at 18,000 atm; the piston was made of ShKh12 steel. In order to prevent breakage of pistons made of this steel, they should be made cylindrical, and not weakened by any threads, indentations, etc.

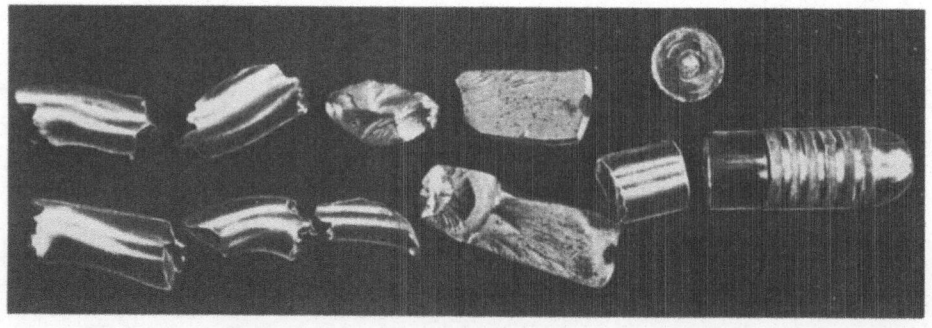

Fig. 161. Piston broken at pressure of 18,000 atm.

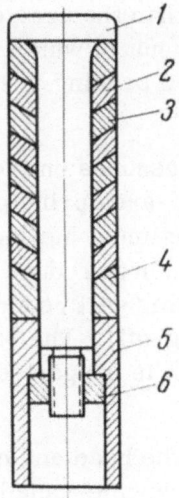

Fig. 162. Tsiklis
seal for pressures
of 10,000 atm:
1) shank; 2) brass
ring; 3) plastic
ring; 4) washer;
5) shaped washer;
6) nut.

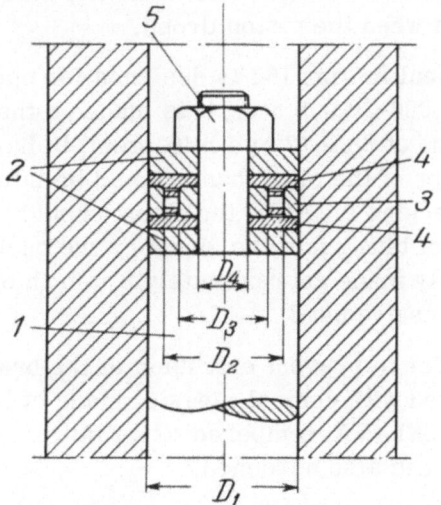

Fig. 163. Unsupported-area gland: 1) piston;
2, 3) shaped rings; 4) sealing packing; 5) nut.

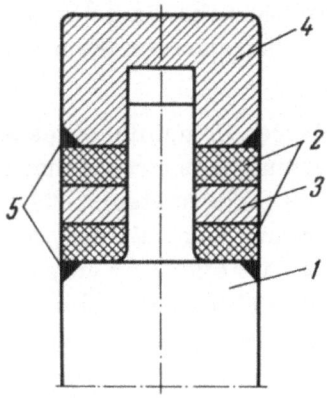

Fig. 164. Unsupported-area seal for
pressures over 10,000 atm: 1) plug;
2) brass ring; 3) sealing ring; 4) cap;
5) steel rings.

Therefore, the seal design [37] consisting of shaft 1 (Fig. 162), made of ShKh12 hardened steel, brass rings 2, and plastic rings 3 mounted on the shaft and compressed by ring 4, also made of ShKh12 steel, is more suitable. Shaped ring 5 (also made of ShKh12 steel) and nut 6 are used to compress the entire structure. This seal is forced into the cylinder bore, nut 6 is additionally tightened by a wrench, and the piston is then placed in the bore.

When reverse travel of the piston is produced by counter pressure on the large piston, the use of packing resting freely against the piston is impossible. In this case, a gland constructed on the unsupported-area principle is recommended (Fig. 163). Piston 1 carries

shaped steel rings 2 and 3, compressing packing 4. The unsupported-area
results from the fact that the area of the packing is less than the area of
the sealing rings. The entire structure is compressed by nut 5, which produces the initial sealing. This same nut carries the entire packing system
down with it when the piston drops.

The seal in Fig. 164 is designed for operation at pressures on the
order of 20,000-30,000 atm; it is made on the unsupported-area principle.
Plug 1, made of ShKh12 or KhVG steel, is hardened to maximum hardness.
The diameter of the plug shank should be minimum, in order to reduce the
stress in packing 2. Also, the shank should be short, as this will reduce
the danger of biting into the shank by the packing [pinching-off]. The packing is usually made approximately 3 mm thick. The shank is supported by
the more massive head.

In order to prevent extrusion of the brass packing, the head and cap
4 are covered with conical rings 5, made of heat treated chrome-vanadium
steel type 18KhNVA, tempered to its initial hardness. Part 3 is made of
KhVG steel and also hardened.*

The higher the pressure, the more careful should be the finishing of
the intensifier bore when these seals are used. However, a gap should exist between the piston and the wall of the intensifier bore or the wall of the
bore may be damaged if a metal particle comes between the piston and the
wall. Also, if the piston is insufficiently hardened, a large load will cause
it to expand and its diameter will increase. If the clearance is very small,
the piston may become wedged in the intensifier bore. Therefore the diameter of the piston, the hardened pressure rings, washers, etc., should
be 0.1-0.2 mm (see calculations in Chapter II) less than the diameter of
the bore. If brass and copper rings are used, the clearance for them may
be less (on the order of 0.01 mm).

In packing that operates on the unsupported-area principle, stresses
arise which are considerably greater than the working pressure so that it
is easy to exceed the yield strength of even the best modern steels. Therefore the bore of an intensifier, after careful machining and finishing, may
be found to be "expanded" after its first usage, i.e., its diameter may be
increased. This disrupts the operation of the seal, and the painstaking work
of lapping-in the bore and manufacture of a piston must be begun again. In
order to prevent "expansion" of the bore, autofrettage is used. However,
it is impossible to completely avoid "expansion" if the seals described
above are used.

------

*When gases are compressed, it is recommended that a liquid be poured
  over packing of this type (if the conditions of investigation permit) [38]; the
  liquid, filling the gaps, will reduce the gas leakage and also reduce friction
  between the packing and the intensifier wall.

When operating with a pressure on the order of 30,000 atm, after just a few experiments (if the conical vessel has not been burst by this time), the bore of the vessel must be made uniform, which is done by processing it with special reamers with gradually increasing diameters.

We should note briefly the possibility of bursting of apparatus working at pressures over 20,000 atm. Both the piston and the conical vessel with its supports are subject to bursting.

Rupture of a piston begins with the appearance of small fissures around its edges. If a piston with these fissures is not removed from the installation, it may burst into small parts and, since this disrupts the equilibrium in the apparatus and the pressure in the intensifier cannot drop, strong compression of the conical vessel will result, possibly making it unsuitable for further use.

The life of a piston can be increased by applying a reinforcing ring of hardened alloy steel around its end. The other end, which is in the conical vessel is in better operating condition since it is supported by the walls of the vessel itself.

When cemented carbides are used for manufacture of pistons, it is important to prevent local overheating, which will weaken the material. It is therefore necessary to use diamond cutting tools.*

In order to avoid expansion of the bore and to make the quality of operation independent of the quality of machining of the bore and of the uniformity of the bore diameter, the packing can be moved from the piston to the cylinder, and one of the above-described self-sealing glands can be used. These glands improve the operating conditions of the piston and its resistance to longitudinal bending as well (see Fig. 154).

Large-Piston Seals

The pressure in the large cylinder of the intensifier is considerably lower and therefore considerably less stringent requirements are placed on the sealing around the large piston. Under ordinary conditions, the sealing can be achieved using rawhide leather sleeves applied to the piston and compressed by pressure rings. In order to improve the sealing, the rawhide sleeves can be placed with their larger ends directed against the pressure. The design of a piston is presented in Fig. 50. Sleeves 3 are placed around piston 2 and compressed by rings 4. Washer 5 and nut 6 compress the entire assembly. The low-pressure piston has a recess into which the high-pressure piston is placed. If the pistons are to be firmly attached, the upper portion of the recess carries a thread to hold attachment nut 7.

---

* This problem is analyzed in greater detail in the literature [39].

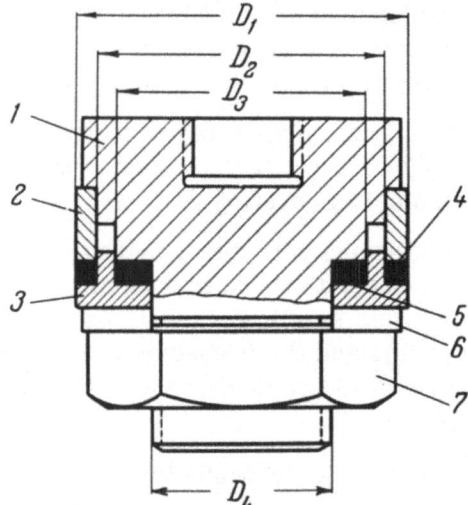

Fig. 165. Large-cylinder piston of intensifier:
1) piston body; 2) bronze ring; 3) shaped rings;
4, 5) packing; 6) rings; 7) nut.

The end of the upward stroke of the piston is determined by the sharp
increase in pressure beneath the large piston as its top touches the bottom
of the high-pressure cylinder. If nut 7 extends above the surface of the
piston, the tremendous force applied to the large piston may distort the
threads.

In order to avoid skewing of the pistons, the head of the high-pressure
piston is made spherical. A flat washer of a hard material, capable of
withstanding the pressure developed at the point where the sphere touches
the plane must be placed beneath the sphere. The lower portion of the low-
pressure piston should have a cross-shaped cut to permit passage of oil
when the piston is located at the bottom of the cylinder. Otherwise, the oil
channel will be closed and very high pressure will be required to break the
piston away from the bottom of the cylinder.

A piston whose reverse travel is achieved by forcing oil into the space
above the low-pressure piston (while the oil in the space beneath the low-
pressure piston is removed) has four sealing rings, two each turned upward and downward. The upper cavity of the cylinder is sealed using either
a gland around the high-pressure piston or packing between the intensifier
cylinders. In the first case, the high-pressure piston must be made in two
parts.

Sometimes, it is necessary to provide complete hermetic sealing in
the low-pressure piston seal as well (for example, when compressibility is

to be determined by the Krichevskii-Tsiklis method [37]).* In these cases, unsupported-area seals are used. Figure 165 shows the large piston of an intensifier with this type of seal.

A bronze or cast-iron ring 2 is shrunk onto piston 1, made of St. 5 steel; the ring is carefully lapped to the cylinder bore. Shaped ring 3, made of bronze or cast iron, compresses double plastic packing 4 and 5. The initial compression of the packing is obtained by nut 7, pressing on ring 6. The dimensions of the packing are determined as was described above, with the values of D_1 and D_4 fixed.

The ratio of cross-sectional areas of the piston in the intensifier may vary over very wide ranges. We have used intensifiers with area ratios from 17 to 35. The load on a large piston 100 mm in diameter with a pressure of 500 atm reaches 40,000 kg. This is why gear-driven jacks are rarely used to create high pressures.

CHECK VALVES

Check valves are one of the most important parts of a high-pressure apparatus. They are the determining factor in the operation of a multi-stroke intensifier. When the pressure is being created by a liquid, a ball valve (Fig. 166) is sufficiently reliable. This valve consists of two balls 1 and 2, 3 and 6.3 mm in diameter, located in two bores. This makes the operation of the valve more reliable. The upper ball is pressed into its seat by plate 3 and spring 4. The travel of the ball is determined by limiter 5, on which the spring rests. The limiter is a threaded tube with a center aperture for the shaft from the plate and with holes for liquid flow.

Valves of this design operate well as long as the viscosity of the liquid is not too great. For example, for mixtures of oil and kerosene these valves can operate up to 10,000 atm at room temperature. With higher pressure or when working with liquids whose viscosity increases sharply with pressure, the ball will drop so slowly that the valve will not be able to close the bore.

Another deficiency of this type of valve is the possibility of the small ball entering the bore of the large ball, which will cause the

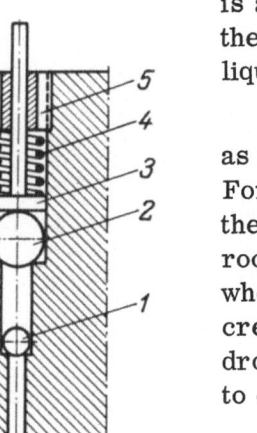

Fig. 166. Ball valve: 1, 2) balls; 3) plate; 4) spring; 5) limiter.

————

*See also Chapter X.

valve to fail. This can be avoided by making the height of the bore of the small ball sufficiently great or by limiting the rise of the large ball.

From time to time, ball valves require truing of their bores. This can be easily done by removing a layer of metal 0.1-0.2 mm thick from the internal surface of the bore with a flat drill.

An advantage of this valve is its design simplicity, which makes it possible to place the delivery valve in the body of an apparatus.

L. F. Vereshchagin notes that ball valves with 12-14 mm diameter balls work well in a new bore the first time only. After that, wear (distortion) begins on the ball at the edge of the bore and the valve will stop working if the ball rotates. Valves fail particularly quickly at pressures of 3000-4000 atm. However, this relates chiefly to the operation of compressors in which valves are opened and closed at high frequency. In intensifiers, the valve operates under considerably easier conditions and will wear less. We have observed this in practice. Where gases are being compressed, carefully lapped plate valves are to be preferred.

JOINTS

Joints for small diameter tubing, valves, manometers, etc., operating under high pressure, can be made in various ways. Some elements of such joints are described below.

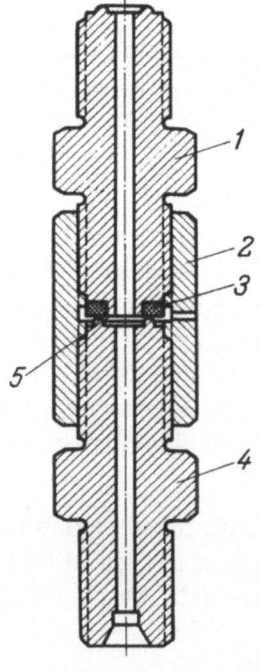

Fig. 167. Joint with packing: 1, 4) nipples; 2) nut; 3) packing; 5) nipple tail.

Packing

At low pressures, two nipples with $\frac{1}{2}$-inch pipe threads can be joined by a packing (made of copper, fibra, aluminum, etc.) using a tightening nut with one-sided or right-left thread (Fig. 167).

In a similar manner, any valve, manometer, connecting pipe, etc., can be screwed down into a thread in the body of an apparatus onto packing. Of course, this joint will have all the deficiencies inherent in the packing used.

If the nipple does not have a special tail ring, high compressive stresses may cause plastic flow of the packing into the hole (the hole may become filled with the packing material) and the opening between the parts being joined may be closed.

Cones

A considerably more convenient type of joint is the cone pressed by a special nut against a conical seat. This type of joint operates equally well at pressures from 20 to 20,000 atm.

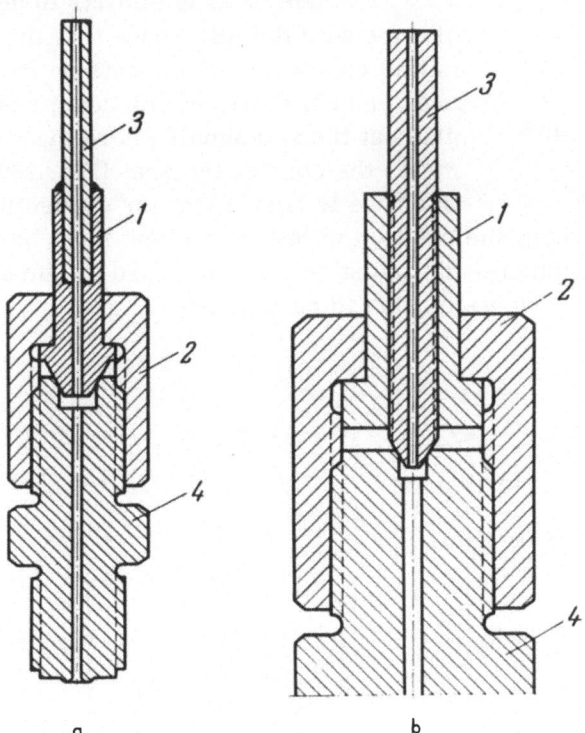

a   b

Fig. 168. Conical joint: a) for pressures up to 2000 atm;
b) for pressures over 2000 atm; 1) nipple; 2) compression
nut; 3) high-pressure capillary; 4) seat.

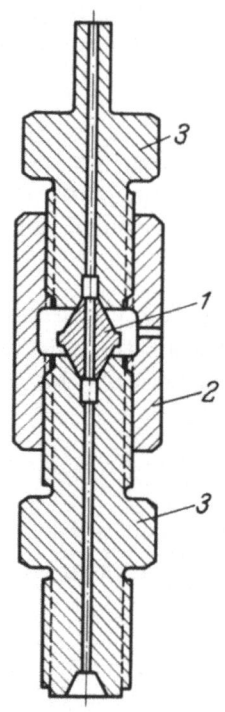

Fig. 169. Lens seal: 1) lens; 2) nut; 3) nipple.

Cones working at pressures up to 2000 atm are made of St. 5 steel. Cone 1 (Fig. 168a) is made with an included angle of 60°, compressing nut 2 is placed over it, and it is welded to high-pressure capillary 3. Then, the cone is placed in seat 4 and tightened with the nut.

For operation at pressures over 2000 atm, the steel capillary (Fig. 168b) is screwed into an alloy-steel nipple which is also shrunk onto the capillary. The end of the capillary is then turned on a lathe to 60°. The sealing is done on the capillary itself, for which the seat is properly prepared. The angles of seat and cone differ by 1-2°. Depending on whether the joint is to be sealed at the large or small diameter, the greater angle is made in the cone or in the seat.

Lenses

The lens seal is similar in design to a lens closure with the difference that the sealing lens in this case does not operate on the unsupported-area principle. As with a cone, sealing takes place at the two annuli where the cone of the lens meets the cone of the seat (Fig. 169). Sealing of this type is convenient for connection of manometers to valves, where the position of the instrument must be fixed (for example, if the manometer dial must be turned toward the investigator). A tightening nut with right-left threading is used.

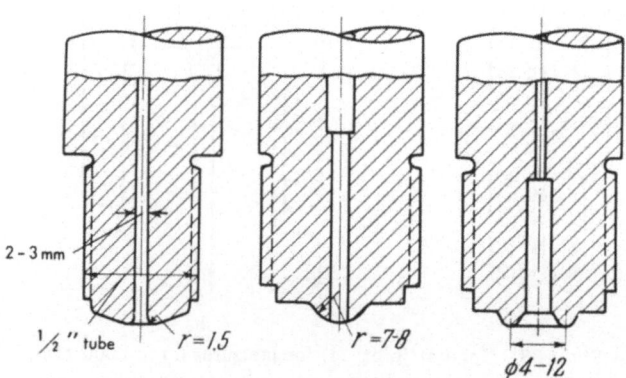

Fig. 170. Sealing by shaped ridges (lips).

The lens joint can be manufactured using the rules presented above for lens seals.

Lips

The most convenient and widely used design for attaching nipples, valves, and manometers are the so-called "lips" (Fig. 170), i.e., shaped ridges at the ends of the parts to be connected. Ordinarily, these lips are made on parts which are adequately hard and carefully finished. They must be made at the same time as the thread without removing the part from the machine tool. Otherwise, the surface of the ridge will not be perpendicular to the threaded side, and tight sealing will be impossible.

Lips should be made with small diameter in order to reduce the load on the thread. The seat material should be 5 to 8 Rockwell-hardness units softer than the lip material.

This type of sealing can withstand considerable load, is simple in manufacture, and is convenient to use. The slight distortion of the lips after long use can be easily corrected (observing all precautions concerning matching with the thread).

Metal — Glass Joints

The most widely used joint for metal-to-glass connection is the asbestos gland (Fig. 171). It is very simple, but can withstand only slight pressure drops (on the order of a few atm).

In the joint designed by I. P. Sidorov [40] (Fig. 171b), a glass cylinder (12 mm diameter, 8 mm high) is fused onto the end of glass capillary 1. Leather ring 2 is applied to the upper base of the cylinder, and nut 3 is applied over the leather ring. A conical ridge is made at the base of the cylinder, then carefully ground and pressed against lead packing 4 (1.5 mm thick). This joint can withstand pressure drops up to 100 atm.

The designs [21] shown in Fig. 171c and d are clear without explanation.

A Teflon plug with an internal conical aperture can be pressed into a metal tube, and a glass tube placed in the Teflon plug. This joint is suitable at temperatures up to 200°C for work in a vacuum of 10^{-5} mm Hg and at pressures up to 30 atm.

For work at higher pressures, the joint designed by Kh. M. Khalilov [41] (Fig. 172) can be used. The end of glass capillary 1 is melted into the form of a sphere This is done by pressing it into the hemispherical recess of heated nipple 3, rotating it while hot. The joint is assembled, as is shown on the figure, using a lead (or fibra) packing 5 and clingerite [an

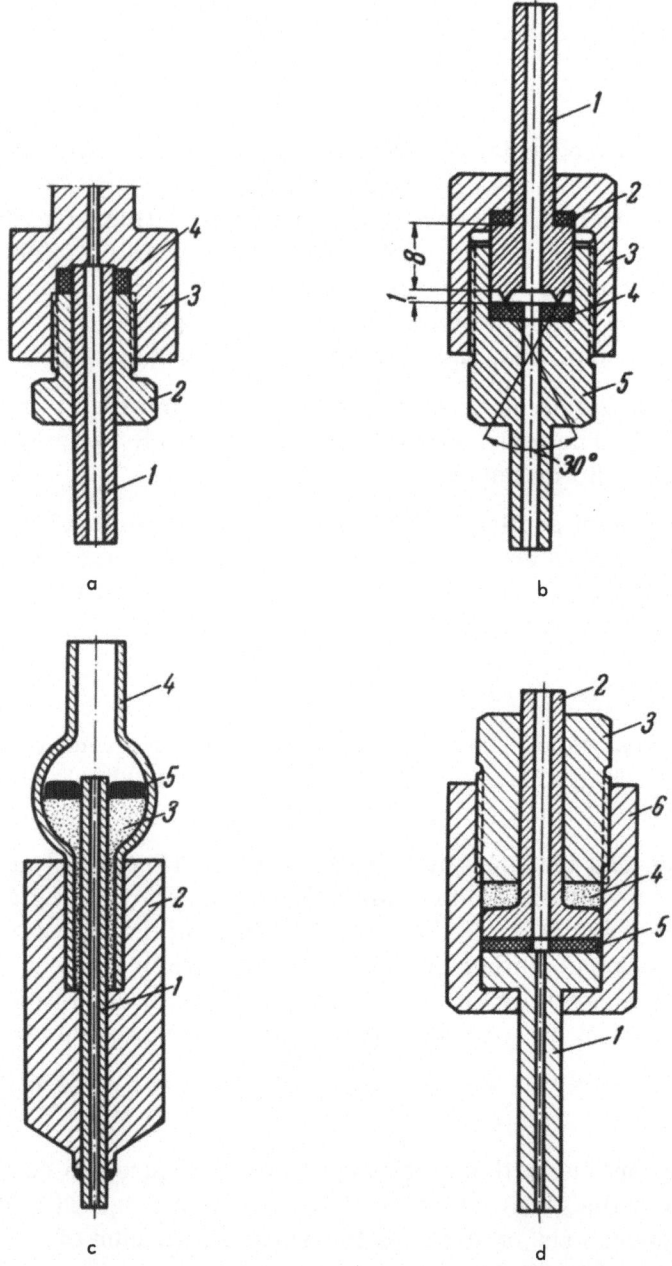

Fig. 171. Joining metal to glass: a) asbestos gland: 1) glass capillary; 2,3) nuts; 4) asbestos fiber; b) Sidorov joint: 1) glass capillary; 2) leather ring; 3, 5) nuts; 4) packing; c) shellac seal with mercury cap: 1) metal capillary; 2) metal well; 3) shellac with resins; 4) glass; 5) mercury cap; d) soft seal: 1) metal capillary; 2) glass capillary; 3,6) nuts; 4) asbestos; 5) soft packing.

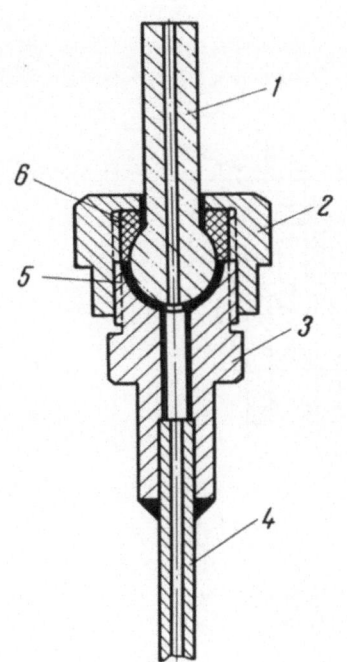

Fig. 172. Khalilov metal-glass joint:
1) capillary; 2) cover nut; 3) nipple;
4) steel tube; 5) lead packing; 6)
clingerite packing.

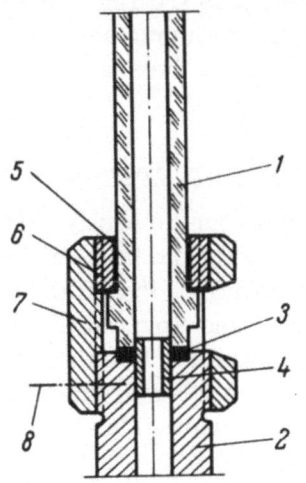

Fig. 173. Metal-glass joint: 1)
glass capillary; 2) nipple; 3) packing; 4) metal capillary; 5) protective packing; 6) split nut; 7)
connecting nut; 8) stopper [lock].

asbestos-rubber cement] packing 6.
Due to the high strength of the glass
ball, the joint will withstand considerable tension as well as high pressure
differences.

A very convenient design [42]
(Fig. 173) has been tested by I. R.
Krichevskii and G. D. Efremova.
Shoulder ridges [shoulders] are fused
onto the end of glass capillary 1, and
the ends are polished, making sure that
the surface of the end is perpendicular
to the axis of the capillary. Fluoroplast packing 3 is placed in the recess
and the end of nipple 2 and capillary 4
are inserted, protecting against flow
of the packing. Protective lead or vinyl
chloride packing 5 and split nut 6 are
placed over the glass capillary. The
split nut is necessary in case both ends
of the capillary are being sealed, i.e.,
when the capillary is made with two
shoulder rings. Then connecting nut 7,
which has an aperture for observation
of the position of the packing, is tightened. Lock 8 is required to prevent
loosening. This joint permits generation of pressures of 150 atm at a temperature of 100°C in a $\frac{4}{12}$ mm diameter
capillary.

The joint shown in Fig. 174 [43]
can operate at temperatures of 200-
300°C and pressures up to 100 atm.
Glass tube 1 is compressed between two
(the figure shows only one) flanges 2.
Shaped copper washer 3 is used for sealing. The flange is pressed to the body
of the apparatus 4 with nut 5. A window
is made in the body of the apparatus for
observation. Bolts 6 are used to locate
the flange.

Figure 175 shows a design [44] of
a collar for joining glass and metal op-

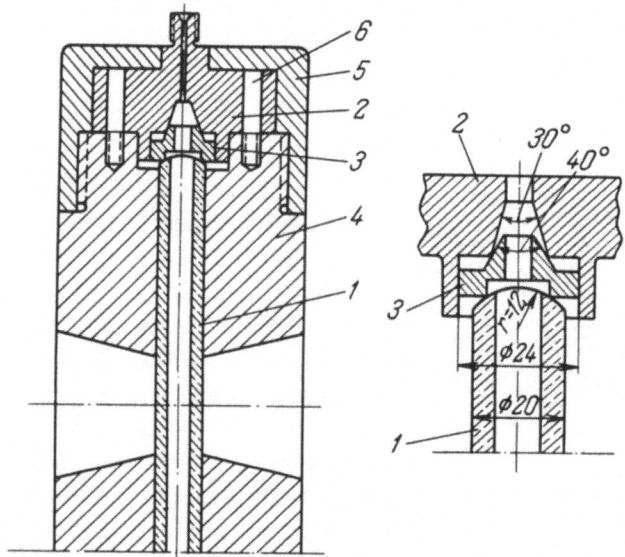

Fig. 174. Joining glass tube to metal parts: 1) glass tube; 2) flange; 3) washer; 4) body; 5) nut; 6) locators.

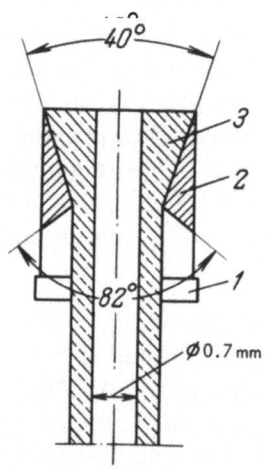

Fig. 175. Collar for joining metal to glass: 1) collar; 2) fluoroplast packing; 3) quartz cone.

erating at pressures up to 300 atm. Collar 1, made of a titanium alloy (6% Al and 4% V) compresses fluoroplast packing 2, designed to seal quartz cone 3. After the quartz tube is manufactured, it is treated in a 10% aqueous solution of hydrofluoric acid for 10 minutes at 25°C to remove the sharp angles and surface fissures.

Metal − Glass Solder Joints

A thin brass tube can be easily soldered to glass. The other end of the tube can be welded to a metal capillary. Pyrex glass can be soldered with Invar and Kovar, and the solder joints will be capable of resisting high pressures. With an internal tube diameter of 9 mm, a pyrex−Invar solder joint will withstand 60 atm [45].

A. I. Shatenshtein [46] recommends joining metal to glass using a special alloy consisting of 26.3% lead, 10% cadmium, 50% bismuth, and 13.7% tin, or another alloy consisting of 56% bismuth and 44% lead. These alloys are well suited for joining metal to glass manufactured by the "Druzhnaya gorka" plant, although they do not work as well with Jena glass, and are completely unsuitable for Pyrex glass.

ELECTRIC-INSULATION TECHNIQUES

Low-Pressure Electrical Leads

It is comparatively easy to seal an insulated electrical lead for pressures up to 1000 atm using various insulating materials. Sealing wax or another cement, melted onto the rod to be insulated, can be used for this purpose. The plug produced in this manner, with the electrode passing through it, is pressed into the seat using a nut.

Good results can be obtained using fibra as an insulator (Fig. 176). This electrical lead design is constructed on principles of ordinary seals. The insulated lead is compressed by the sealing material, and friction prevents the lead from being expelled. If the conductor diameter is greater than 0.5 mm (such as when high current must be passed) and the pressure in the apparatus is over 1000 atm, this method is unsuitable.

Unsupported-Area Electrical Leads

A more reliable seal for an electrical lead is the inverted cone (Fig. 177). Metal cone 1 with conductor 2, surrounded by insulating material 3, is placed in a carefully fitted seat in body 4. The pressure acts on the large base of the cone and forces it into the seat.

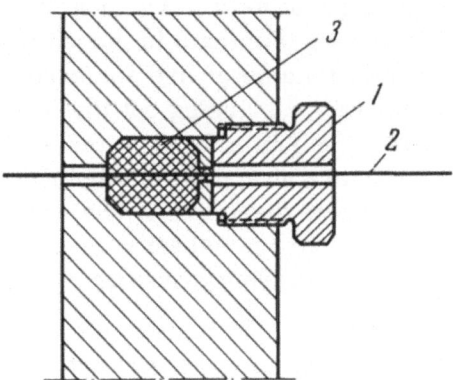

Fig. 176. Electrical lead with fibra insulation:
1) nut; 2) conductor; 3) fibra insulation.

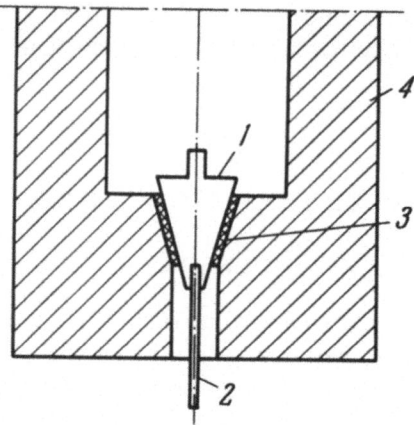

Fig. 177. Inverted cone: 1) cone; 2) conductor;
3) insulating material; 4) body.

The insulator used may be any one of numerous materials such as
ivory, plastic, pyrophyllite, mica, etc. The selection of the material to be
used depends on the pressure and temperature in the apparatus. Ivory,
for example, can be used up to 3000 atm only. The author has used cones
with sleeves made of methyl methacrylate. This type of cone operates
without failure up to pressures of 18,000 atm.

A simple and reliable method allowing the use of an electrical lead
at temperatures up to 800-900°C is winding of the cone with mica. This
design, developed and tested in 1937 by I. P. Sidorov at pressures up to
5000 atm, has been successfully used by us for many years at pressures of
18,000-20,000 atm.

The joint consists of cone 1 (Fig. 178), made of ShKh12 hardened steel,
well polished, with a solid angle of 8-10°. The cone is carefully lapped into
the seat in body 2 and a covering of mica 3 is placed over the cone. Conductors 4 and 5 of the required diameter are soldered or welded to the two
ends of the cone. This method for bringing out the lead is even suitable for
temperature measurement. The emf arising at the solder joints will mutually cancel since the joints are at approximately equal temperatures and
the emfs are directed in opposite directions. The end of the cone which
protrudes from the seat is wrapped in asbestos or glass fiber 6 or thin
wire, in order to protect the mica from cracking.

In order to ensure equality of the angles, a special tool bit in the form
of a cone should be made and used to work the seat. The surface produced
by working with a good bit will be rather clean. Then, the cone is lapped
into the seat. After preparation of seat and cone, a pattern for the mica is
prepared. The insulation layer should wrap around the cone twice and meet

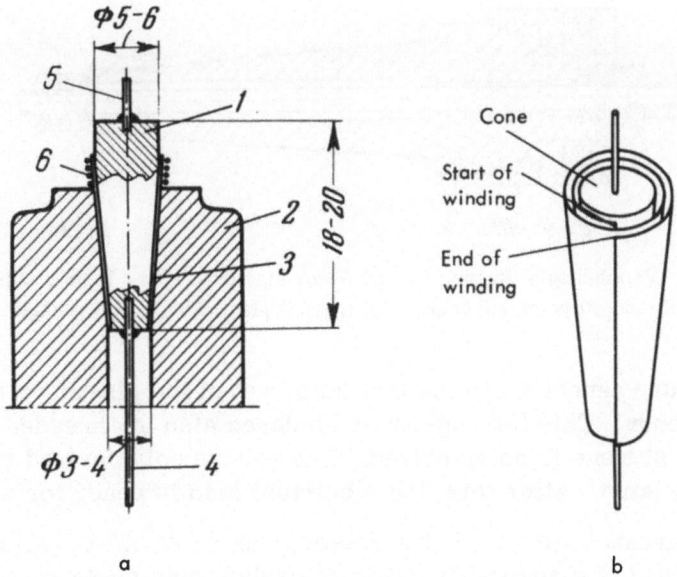

Fig. 178. Mica-insulated electrical lead: a) plan of electrical lead:
1) cone; 2) body; 3) mica; 4, 5) leads; 6) glass fiber; b) winding of cone.

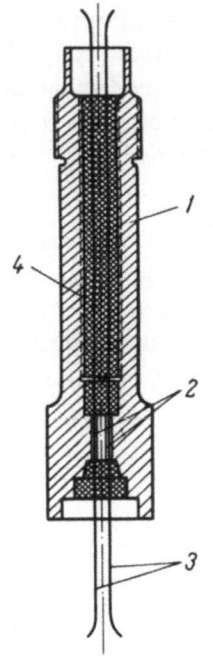

Fig. 179. Kalashnikov
electrical lead: 1) body;
2) holes; 3) conductors;
4) thread.

at the joint (Fig. 178b). The mica should not be broken as it is wrapped. The wrapped cone is placed in the seat and driven into place by light blows with a hammer on a tube placed against the large end of the cone. The exit aperture (for the conductor) from the narrow end of the female cone should have a diameter somewhat greater than the small base of the male cone.

A design for bringing several continuous conductors out of an apparatus was suggested by Ya. A. Kalashnikov [47] (Fig. 179). Body 1, made of 40Kh steel (hardness $R_c = 45-48$) contains two holes 2, whose length depends on the diameter of the wire to be brought out of the apparatus and the pressure to be contained. In particular, for a pressure of 5000 atm and a conductor diameter of 0.5 mm, the channel length should be at least 18 mm.

After passing wires 3 through the holes, body 1 is filled with a styrene monomer, then it is polymerized in a special bomb for two days at 1500 atm and 80°C.

As a result of this, the bond between the leads, polystyrene, and body wall becomes so strong

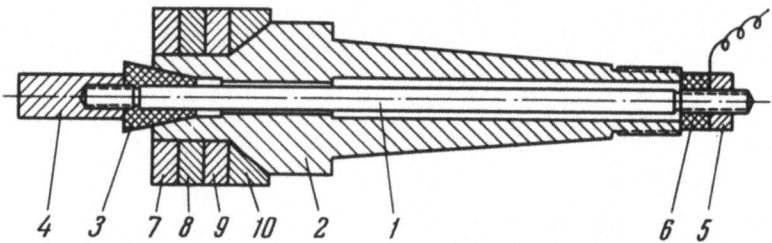

Fig. 180. Vereshchagin, Kuznetsov, and Alaev electrical lead: 1) steel rod;
2) body of lead; 3) porcelain cone; 4, 5) nuts; 6) plate; 7, 8, 9, 10) packing.

that the pressure cannot force the conductor out of the plastic or the plastic out of the body. This forcing-out is hindered also by threaded diameter
4 in which the styrene is polymerized. The excess polymerized styrene
is cut out on a lathe. After this, the electrical lead is ready for use.

L. F. Vereshchagin, L. F. Kuznetsov, and T. A. Alaev [48] have used
an electrical lead of a somewhat different design, also made in the form of
an inverted cone.

The electrical lead consists of steel rod 1 (Fig. 180), passed through
porcelain cone 3 placed in the lead body 2. The steel rod has threads on
both ends fitted with nuts 4 and 5 which are designed for fastening the structure into the body of the lead. Cone 3 is carefully lapped to the cavity in
the lead using boron carbide. Plate 6 is used to insulate the rod from the
body of the lead. The lead is sealed to the high-pressure apparatus using
packing 7, 8, 9, and 10, which press part 2 around the cone and improve
sealing.

The design of an electrical lead with [unsupported] noncompensated
area has been suggested by I. I. Oleinik [49] (Fig. 181). Plug 1 with electrical lead 3 carries vinyl chloride tube 2. The electrical lead rests against
packing 4. The degree of noncompensation is fixed by the ratio of diameters of the wide portion of the electrial lead and the aperture in which it
is placed.

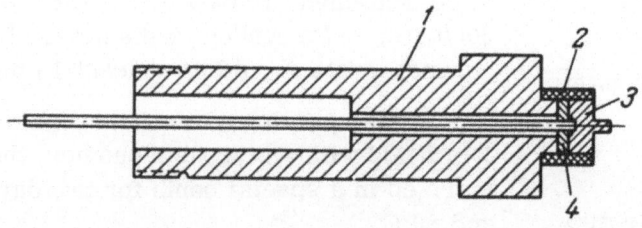

Fig. 181. Oleinik electrical lead: 1) plug; 2) vinyl chloride tube;
3) electrical lead; 4) insulating packing.

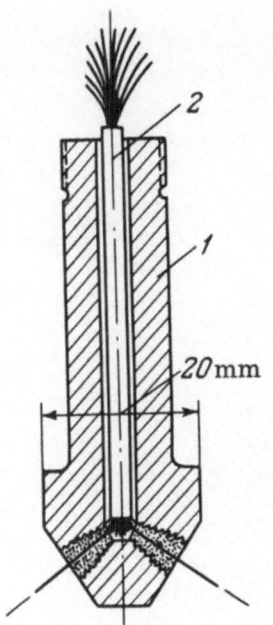

Fig. 182. Conical electrical lead for carrying solid conductors: 1) body; 2) cable.

When it is necessary to bring several insulated continuous conductors out of an apparatus, electrical leads such as those shown in Figs. 182 and 183 [50, 51] can be used. Cable 2, made of enamel-covered piano wire pressed into insulation (magnesia or other heat-stable insulation) is placed in body 1 (Fig. 182). The insulation is contained in a steel shell. The end of the cable is divided, and the individual wires are passed through conical apertures (in this case there are 12 of them), drilled around the outside of the body of electrical lead 1. The apertures have threads which carry packing made of a mixture of epoxy resin and Portland cement.

A very simple design for an electrical lead is shown on Fig. 184 [52]. The conical aperture in plug 1 carries steel cone 2, surrounded by insulated electrical leads 3. The space between the plug and the cone is filled with epoxy resin.

V. A. Borzunov and V. P. Semin [7] describe the design of the electrical lead shown on Fig. 185. The body of the electrical lead 1 contains

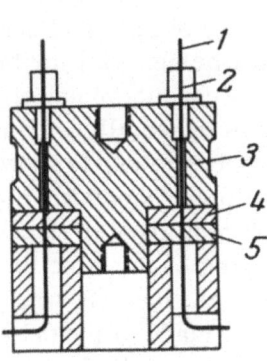

Fig. 183. Electrical lead for connection of continuous conductors: 1) conductor; 2) insulation; 3) body; 4, 5) packing.

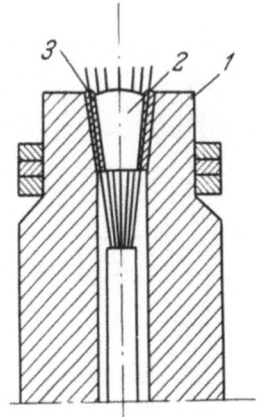

Fig. 184. Electrical lead containing epoxy resin: 1) plug; 2) cone; 3) electrical conductor.

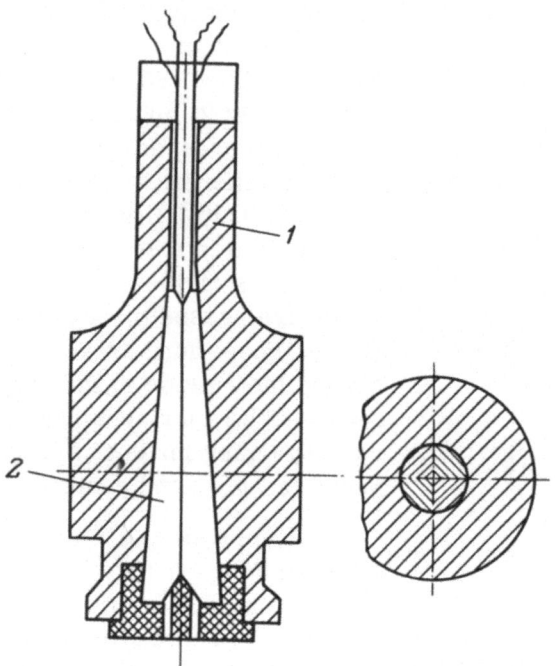

Fig. 185. Borzunov-Semin electrical lead: 1) body; 2) cone.

metal cone 2 consisting of four parts, insulated from each other and from the body of the lead by mica. This design makes it possible to bring four conductors out through one cone.

A very simple and convenient electrical-lead design [53] for bringing a continuous conductor out of an apparatus (this is especially important for use with thermocouples) is shown in Fig. 186. Conical nipple 1 contains a drilled hole 0.01-0.02 mm greater in diameter than the diameter of the wire to be brought out. Conical sleeve 2, wall thickness 0.3 mm, is made of pyrophyllite. The conical cap is applied over nipple 1 and, by striking this cap with a hammer through a mount which protects the thermocouple lead, the wire is pressed into the nipple and the cone is pressed

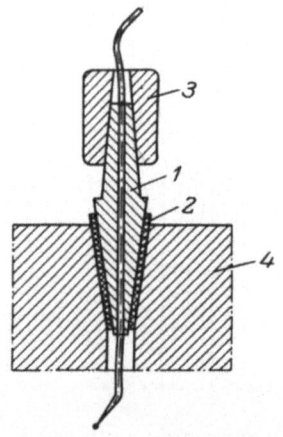

Fig. 186. Electrical lead using pyrophyllite cone: 1) nipple cone; 2) conical sleeve made of pyrophyllite; 3) conical cover; 4) body.

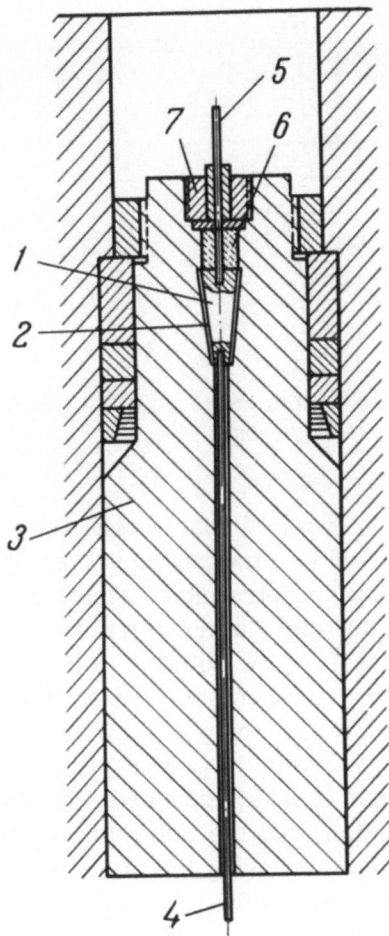

Fig. 187. Bridgman electrical lead: 1)
cone; 2) shell; 3) body; 4) conductor; 5)
lead; 6) insulation; 7) nut.

into the pyrophyllite and the body
of the apparatus.

I. A. Ostrovskii [54] recommends that the cone be wrapped
in silk insulation saturated with
type BF-6 glue.

The electrical lead shown in
Fig. 187 [55] is designed for operation at pressures up to 30,000 atm.
For simplicity, one lead is shown
in the illustration. Cone 1, insulated
with shell 2 made of pipestone, is
placed in body 3 of the electrical
lead, which consists of a cap plug
using the unsupported-area principle (type II). The conductor 4 is
insulated from the body 3. The lead
5 is insulated from the body by insulation 6, held down by nut 7.

This type of electrical lead can
withstand 100 applications of pressure up to 30,000 atm without flowing or rupture of the insulation. It
is interesting to note that this electrical lead does not differ in principle from that which we have described. Bridgman uses this lead,
rejecting more complex designs described in the first edition of this
book.

VIEWING WINDOWS*

Windows for visual observation at high pressures are made of
glass, transparent quartz, sapphire,
diamond, and even cooking salt (see
Chapter XII).

Glass windows are extremely
sensitive to the action of water. It

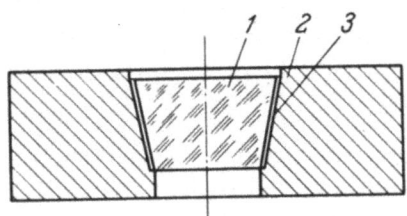

Fig. 188. Viewing window: 1) cone;
2) seat; 3) packing.

*We will not describe ordinary water
  gauge glasses and levels.

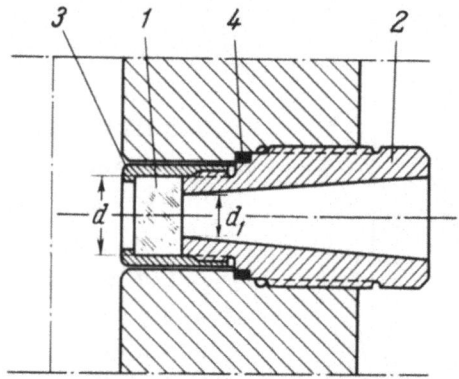

Fig. 189. Viewing window with unsupporting area:
1) glass cylinder; 2) nipple; 3) cap; 4) packing.

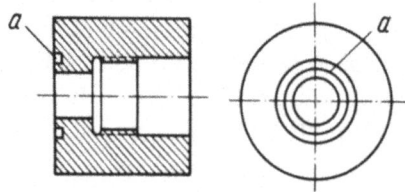

Fig. 190. Female piece for lapping nipple
(a = groove).

is known [55] that water is dissolved into the surface layer of glass, penetrating to depths of up to 200 molecular layers. At high pressure, the solution is increased and water penetrates even more deeply into the glass. When the pressure is dropped rapidly, the glass breaks, apparently due to the fact that the water is not able to escape from the glass and expands within it, breaking the glass [56]. Other liquids, such as oil and alcohol have no such effect on glass.

The simplest viewing window design [57, 58] (Fig. 188) consists of a cone 1 made of glass or transparent quartz, lapped to metal seat 2. Sometimes, packing 3 is placed between the glass and metal. The large end of the cone is placed toward the inside of the apparatus, so that the pressure forces the cone into the seat.

The reliability of this type of construction depends on the care with which the lapping is done and the quality of the glass or quartz cone. Even very well-made viewing windows of this type are burst due to chipping of the edges of the cone. These windows are usually used at pressures on the order of 1000–1500 atm, although in individual cases quartz cones have withstood pressures up to 10,000 atm [59].

A considerably more reliable viewing window can be constructed on the unsupported-area principle [60]. Glass cylinder 1 (Fig. 189) with optically parallel ends, is pressed against the polished surface of nipple 2. Cap 3 retains the glass. The nipple is screwed into the body of the appara-

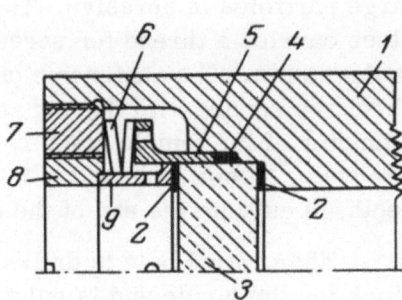

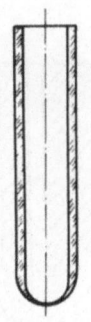

Fig. 191. Mounting of sapphire window: 1) body; 2, 4) packing; 3) window; 5, 9) sleeves; 6) spring; 7, 8) nuts.

Fig. 192. End view- ing window.

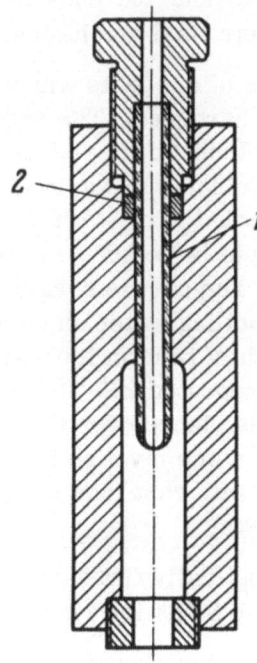

Fig. 193. Installation of end glass in apparatus: 1) capillary; 2) packing.

tus and packed* using packing 4. The pressure created in the apparatus acts on the area

$$F = 0.785d^2$$

The force of this pressure is resisted by a ring whose area is equal to

$$F_1 = 0.785\,(d^2 - d_1^2)$$

Since $F > F_1$, the stress on the surface of the ring is always greater than the pressure in the apparatus.

The manufacture of these nipples has been described in detail by P. E. Bol'shakov [61]. Nipple 2 is made of hard steel, well hardened and polished (for example, type ShKh12). Stock for the nipples are heat-treated three to four hours at 450°C and cooled to 100°C over a period of 12 hrs. The same type of steel is used as stock for a female piece.

The female piece (Fig. 190) is designed for protection of the polished surface from "clogging" and penetration of

---

* The sealing of the nipple into the body of the apparatus can be achieved by any of the methods described above.

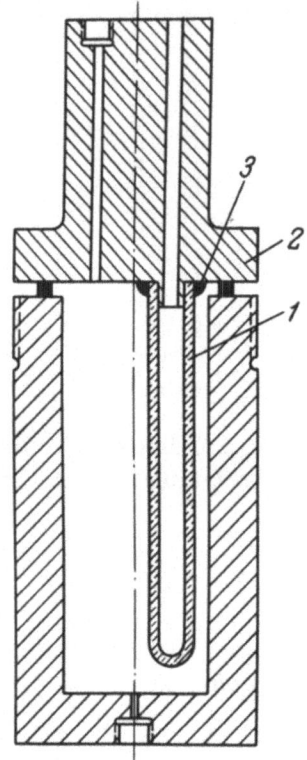

Fig. 194. Autoclave with internal window: 1) glass tube; 2) plug; 3) O-ring (packing).

large particles of abrasive. The female piece carries a thread for screwing on to the nipple. The center line of the thread should be square with the end surface. The diameter of the female piece is 50-60 mm. Groove *a*, 10 mm wide and 1-1.5 mm in depth, is cut into the end of the bore.

The central hole is drilled into the stock for the nipple and is enlarged with a reamer. The part is placed between centers of a lathe and turned, after which, while mounted in the chuck, the threads and final turning are performed in one pass. Then, the female piece is mounted on the large thread and the ends of both parts are given their final machining.

The surfaces of the parts which are to be polished are hardened. The end of the nipple is heated until dark red, then tempered for 2-3 minutes in warm (30-35°C) water and cooled in oil. After the hardening, the nipple is "aged" to eliminate brittleness. The part is heated for 30-60 seconds in boiling water and placed for the same length of time in ice water. This operation is repeated 250-300 times. The female piece is placed over the dried part and set at the same level as the part with Johanson gauge blocks. Grinding is performed with cast iron and glass blocks using M-230 and M-240 powder plus GOI paste.*

After polishing, the viewing window with its optically flat surface held against the dry surface of the metal should show 2 to 3 Newton rings.

The nipple and glass are carefully washed with pure benzine and heated in a drying oven to 80°C. After this, 4-5 drops of Canadian balsam are placed on the polished surface of the nipple and, pressing the glass to the surface, it is rotated 2 to 3 times around its axis. Then, the cap is applied and tightened, after placing a packing of lead foil on the bottom. The nipple is set aside in this form, after which it can be screwed into the apparatus.

---

*75.8% Chromium oxide, 10% stearin, 10% decomposed fat, 2% oleic acid, 0.2% sodium bicarbonate, and 2% kerosene.

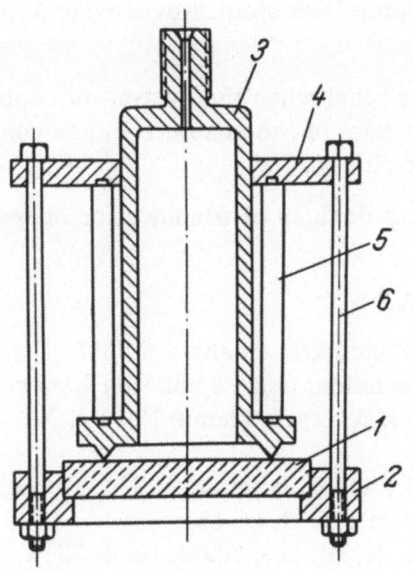

Fig. 195   Window for operation at low temperatures and high pressures: 1) window; 2) flange; 3) chamber; 4) ring; 5) Kovar rods; 6) bolts.

Figure 191 shows a method for mounting a sapphire window [62] designed for operation at temperatures from 0 to 250°C and pressures from 10^{-3} mm Hg to 70 atm. Window 3 is placed in body 1 of the apparatus (made of titanium) against packing 2; the window assembly is sealed by packing 4. The packing is compressed by sleeve 5, which is pressed by spring 6. The compression of the spring is controlled by nut 7. Nut 8 tightens sleeve 9 which retains the sapphire window.

Sometimes, a viewing window must be placed on the end of an apparatus. For this, it is convenient to use the design in which the window consists of a capillary [63]. An annealed capillary made of Pyrex glass with an internal diameter of 12 mm and an external diameter of 13 mm is covered with a hemispherical tip, the glass thickness of which is less than the thickness of the capillary glass (Fig. 192). The capillary is placed in the high-pressure apparatus as shown in Fig. 193. The pressure acts on the external surface of the capillary, so that the capillary can withstand pressures on the order of 700 atm.

Figure 194 shows the design of an autoclave with an internal window consisting of glass tube 1 and a soldered tip [64]. The tube is held against the rib in plug 2. It contains O-ring 3. When pressure is created in the apparatus, the O-ring seals the joint between tube and plug. This type of tube can be used for observation of the level of a liquid at pressures up to 2000 atm and high temperatures.

Windows of this design are not suitable for precise optical measurements. Other windows described above must be used for this type of operation.

A window such as that shown in Fig. 195 [65] has been suggested for operation with pressures up to 100 atm at low temperatures (down to −200°C). Window 1, consisting of methyl methacrylate, is installed in flange 2 and compressed by the sealing knife edges of chamber 3. Compression is achieved by ring 4, made of stainless steel, Kovar rods 5, and bolts 6 made of stainless steel.

The lengths of bolts 6 are so selected that when the system is cooled their absolute contraction is equal to the sum of the absolute linear contractions of the methyl methacrylate and Kovar.

Chapter XII contains descriptions of designs of windows for operation at very high pressures.

BIBLIOGRAPHY

1. Ogloblin, A. N., Lathe Operator's Handbook, Mashgiz (1960).
2. Gaida, A. V., Leading Scientific-Technical and Production Experience from Affiliates of the Nauk Tekhn. Inst., Theme No. 24, No. É-57-60/28, 1957.
3. Sidorov, I. P., Kazarnovskii, Ya. S., and Golydman, A. M., Tr. Gos. Inst. Azo. Prom., No. 1, Goskhimizdat, 1953, p. 48.
4. Vereshchagin, L. F., and Zelinskii, N. D., Izv. Akad. Nauk SSSR, Otdel. Khim. Nauk, 6:443 (1943).
5. Vereshchagin, L. F., and Ivanov, V. E., Pribory i Tekhn. Éksperim. 6:114 (1958).
6. Stepanov, V. A., Pribory i Tekhn. Éksperim., 6:135 (1960).
7. Borzunov, V. A., and Semin, V. P., Tr. Inst. Kom. Standartov Mer i Izmerit. Prib., 46(106):107 (1960).
8. Tsiklis, D. S., Zh. Khim. Prom., 7:404 (1958).
9. Rusching, S. S., and Gebhart, I. E., Rev. Sci. Instrum., 33(8):871 (1962).
10. Daniels, W. B., Rev. Sci. Instrum., 33(1):131 (1962).
11. Tsiklis, D. C., Zh. Fiz. Khim., 35:669 (1961).
12. Astrov, D. N., and Boronely, A. V., Pribory i Tekhn. Éksperim., 3:149 (1960).
13. Beard, C. S., Instrum. Automat., 29(3):490 (1956); Comings, E. W., High Pressure Technology, New York, 1956; Pearson, G. H., Chem. Proc. Eng. & At. World, 44(12):698, 706, 716 (1963).
14. Bomshtein, E. I., Khim. Mash., 8:10 (1939).
15. Maccary, R. R., and Fey, R. F., Chem. Met. Eng., 56:124 (1949); Granvill, I. I., High Pressure Vessels in the Chemical Industry, Proc. Inst. Mech. Engrs., 162(2):199 (1950).
16. Zakharchenko, E. S., Khim. Mash., 4:2 (1935); 5:2 (1935); 9:14 (1939).
17. Ganz, S. N., Zh. Khim. Prom., 14 (1953); Ind. Eng. Chem., 48(5):826 (1956); Jonson, D. P., ISA J., 3:241 (1956).
18. Tsiklis, D. S., Mushkina, E. V., and Shenderei, L. I., Inzhenero. Fiz. Zh., 1(8):3 (1958).
19. Kuzlovskii, D. E., The History of Artillery Material, Artillery Academy of Red Army, Moscow, 1946, p. 166; Bridgman, P. W., High Pressure Physics, Otdel. Nauk Tekhn. Inst., 1935.

20. Babb, S. E., Rev. Sci. Instrum., 31(2):219 (1960).

21. Newitt, D. M., The Design of High Pressure Plant, London, 1940.

22. Makhnev, T. A., Khim. Mash., 9:48 (1939).

23. Vereshchagin, L. F., Zh. Tekhn. Fiz., 16:699 (1946).

24. Anson, D., J. Sci. Instrum., 32:446 (1956).

25. Vollbrecht, H., Dechema Monograph., 34:205 (1959); Ushakov, I. P., Tr. Gos. Inst. Azo. Prom., No. 12, Goskhimizdat, 1961, p. 292.

26. Boon, E. F., and Lok, H. H., Verlag Deut. Ingenieur Z., 100:1613 (1958).

27. Semerchan, A. A., Shishkov, N. Z., and Isaikov, V. K., Pribory i Tekhn. Éksperim., 4:152 (1963).

28. Eitel, W., Die experimentellen Hilfsmittel zur Mineralsynthese unter hohen Drucken und hohen Temperaturen, Berlin, 1925.

29. Plachenov, T. G., Zh. Prik. Khim., 28(3):245 (1955).

30. Thomson, T., Combustion, 29(11):38 (1958).

31. Ivanov, E., Pressure Working of Metals, Izd. Akad. Nauk, USSR, 1958.

32. Boon, F. F., and Tal, S. E., Chem. Ing. Techn., 31(3):202 (1959).

33. Griggs, D. T., J. Geol., 44:544 (1936).

34. Brennstoffchemie, 37(1/2):8 (1956).

35. Vollbrecht, H., Chem. Z., 79(5):139 (1955); 79(6):171 (1955).

36. Krichevskii, I. R., and Khazanova, N. E., Zh. Fiz. Khim., 21:719 (1947).

37. Krichevskii, I. R., and Tsiklis, D. S., Dokl. Akad. Nauk SSSR, 78:1169 (1951).

38. Boksha, S. S., and Shakhovskoi, G. P., Pribory i Tekhn. Éksperim., 3:86 (1958).

39. Bridgman, P. W., Proc. Am. Ac. Sci., 72:45 (1937); 74:21 (1940).

40. Kazarnovskii, Ya. S., and Sidorov, I. P., Zh. Fiz. Khim., 21:1365 (1947).

41. Khalilov, Kh. M., Izv. Akad. Nauk Azerb.SSR, 3:3 (1950).

42. Davis, P. C., Gore, T. L., and Kurata, E., Ind. Eng. Chem., 43:1826 (1951); Tait, T., and Michels, Op. Cit., 45:233 (1953); Krichevskii, P. R., and Efremova, G. D., Zh. Fiz. Khim., 30:1877 (1956).

43. Grishchenko, A. P., Pribory i Tekhn. Éksperim., 6:116 (1957).

44. Gill, J. S., and Marschall, W. L., Rev. Sci. Instrum., 34(4):442 (1963); Shorr, L. M., Rogosinski, M., and Hashman, U., Chem. & Industr., 2:52 (1964).

45. Scholl, W., and Devis, R., Ind. Eng. Chem., 12:1299 (1934).

46. Shatenshtein, A. I., 15:246 (1941).

47. Kalashnikov, Ya. A., Zh. Khim. Prom., 8:491 (1954).

48. Vereshchagin, L. F., Kuznetsov, L. F., and Alaeva, T. A., Zh. Eksperim. i Teor. Fiz., 30:661 (1956).

49. Oleinik, I. I., Pribory i Tekhn. Éksperim., 6:119 (1957).

50. Ruoff, A. L., Rev. Sci. Instrum., 32(6):639 (1961).

51. Scott, G. L., and Babb, S. E., Rev. Sci. Instrum., 32(7):868 (1961).
52. Blosser, L., Young, H. S., and Cornish, R. H., Rev. Sci. Instrum., 33(9):1007 (1962).
53. Stepanov, V. A., Pribory i Tekhn. Éksperim., 3:179 (1961).
54. Ostrovskii, I. A., Experimental Investigations in the Area of Deep Processes, Izd. Akad. Nauk SSSR, 1962.
55. Bridgman, P.W., Investigations of Large Plastic Deformation and Fracture, Izdatinlit, 1955.
56. Poulter, T. C., and Wilson, R. O., Phys. Rev., 40:877 (1932).
57. Amagat, Annal. Chim. Phys., 29(6):30 (1893).
58. Shakhovskoi, G. P., and Tikhomirov, N. A., Pribory i Tekhn. Éksperim., 6:191 (1963).
59. Ryabinin, Yu. N., Zh. Éksperim. i Teor. Fiz., 24:107 (1953).
60. Poulter, T. C., Phys. Rev., 40:860 (1932).
61. Bol'shakov, P. E., Tr. Gos. Inst. Azo. Prom., No. 1, Goskhimizdat, 1953, p. 266.
62. Waggener, W. C., Rev. Sci. Instrum., 30(9):788 (1959).
63. Seren, L., Rev. Sci. Instrum., 19(2):123 (1948).
64. Rose, W., Rev. Sci. Instrum., 29(9):797 (1958).
65. Baldin, S. A., and Gavrilovskii, Pribory i Tekhn. Éksperim., 1:144 (1960).

Chapter VII

Mixing and Circulation Under Pressure

MIXING (AGITATION)

Mechanical Agitators

Agitators with Motors Located Outside Apparatus.
One of the oldest and least-perfected methods of mixing, used less and less, is the introduction of a stirrer into high-pressure apparatus with its shaft passing through a gland and being rotated by a motor. Such stirrers have a number of deficiencies. The presence of a gland limits the pressure difference which can be used, and creates a danger of leakage and of contamination of the contents of the apparatus by the lubricant of the gland and the packing material. The tighter the gland is held, the greater is the friction on the stirrer shaft. This creates the danger of overheating of the shaft, and considerably increases the motor power required to rotate the stirrer. Thus, at a pressure of 200-400 atm, rotation of a laboratory stirrer at 120 rpm requires a 0.25 kw motor.

Thanks to improvements in the design of glands and the use of special packing, friction can be reduced without breaking the hermetic seal. For example, Teflon has a very low coefficient of friction. However, its use allows friction to be reduced only at low surface rubbing speeds [1]. At high speeds, the nature of the surface changes irreversibly, so that even a subsequent reduction in speed does not decrease the coefficient of friction to its previous value. Because of this, other methods must be used for mixing at very high pressures (on the order of several thousand atm).

Stirrers with Motors Located Inside Apparatus.
When the pressure is not too great, and the internal dimensions of the apparatus are large enough, the stirrer and motor can be placed inside the apparatus [2], introducing the motor power current through electrical leads.

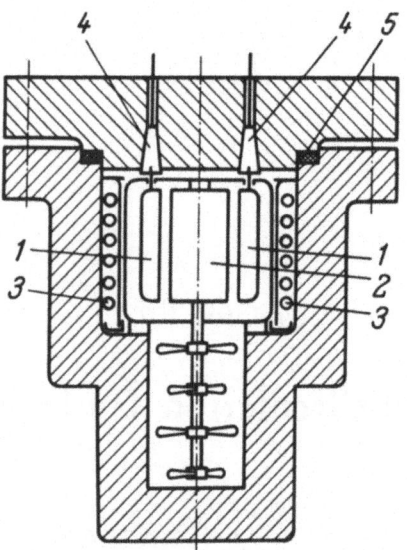

Fig. 196. Location of stirrers with motor within
apparatus: 1) motor winding; 2) rotor; 3) cooling coil; 4) electrical leads; 5) packing.

This is, of course, possible only when the motor winding can withstand the
action of the medium inside the apparatus.

A deficiency of this design is the possibility of contamination of the
medium being mixed by the lubricant of the motor, as well as the necessity
of cooling the motor, which can easily overheat.

Figure 196 shows the location of stirrers rotated at 1500 rpm in a
high-pressure apparatus with a capacity of 750 cm^3 and a diameter of 75 mm.

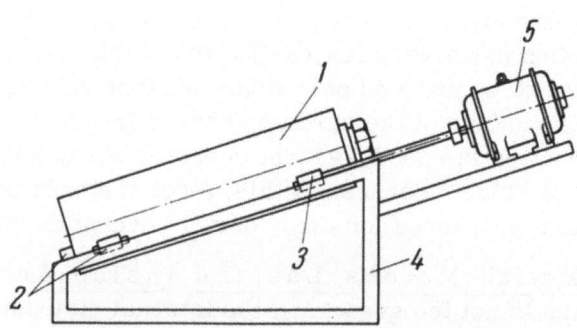

Fig. 197. Rotating autoclave: 1) autoclave; 2) supporting
rollers; 3) friction rollers; 4) stand; 5) motor.

Rotating and Rocking Apparatus

Rotating Apparatus. If the apparatus is not too cumbersome and if the general location of the apparatus does not present a problem, mixing may be performed by rotating the high-pressure vessel around an inclined axis or by rocking it about its vertical axis. This first method is especially effective when working with a two-phase (liquid-gas) system. The apparatus (Fig. 197) is placed on an inclined support and rests on rollers, one of which is connected with the shaft of a slowly rotating reducing gear. With a high gear ratio, the motor power required for rotation of a cylinder approximately 100 kg in weight does not exceed 50 watts.

The rocking autoclave (Fig. 198) can be connected to an analytical installation by a flexible capillary wound in the form of a clock spring. The capillary is connected to the apparatus at the fulcrum. As the apparatus rocks, the capillary winds and unwinds. A steel ball is placed in the apparatus, creating intense stirring action as it rolls back and forth.

In some cases, the high-pressure vessel may be suspended on a system of springs and tension members and rocked by a motor [3]. The rocking frequency of the entire system is selected in accordance with the weight of the vessel. The vibration of the contents of the vessel provides good mixing. Work [3] presents the details of the structure and method of calculating springs (operating and supporting) and the unbalanced load.

A similar method is described in [4]. The high-pressure vessel and apparatus creating the pressure are connected by a capillary with two loops; the vessel is shaken with a frequency of 2000 vibrations per minute.

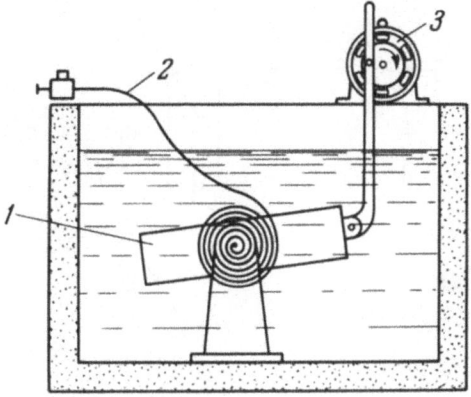

Fig. 198. Rocking autoclave: 1) autoclave;
2) capillary; 3) motor.

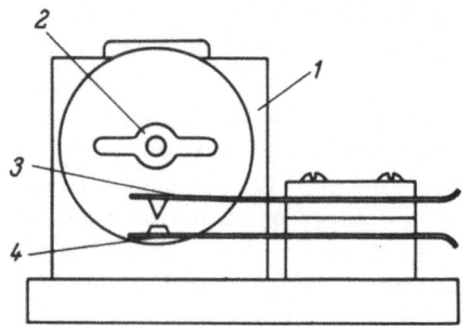

Fig. 199. Current interrupter: 1) motor; 2) lever
wheel; 3, 4) contacts.

Electromagnetic Stirrers

Electromagnetic stirrers are widely used for agitation of the contents of vessels operating at pressures of a few thousand atmospheres. Before beginning our analysis of various designs for such stirrers, we should note a few general problems.

Power Supply of Coil. If there is available in the laboratory a DC power supply of the required voltage and power, it should certainly be used. If a coil is fed by DC power, it will not be overheated by parasitic currents, which eliminates the necessity of making the core of transformer iron, which is not always desirable due to design considerations. A second advantage of DC power is a result of the specific nature of high-pressure apparatus. We know that the conditions of operation of an electromagnet are considerably improved if the external magnetic flux passes through an external framework containing a solenoid. Since the body of

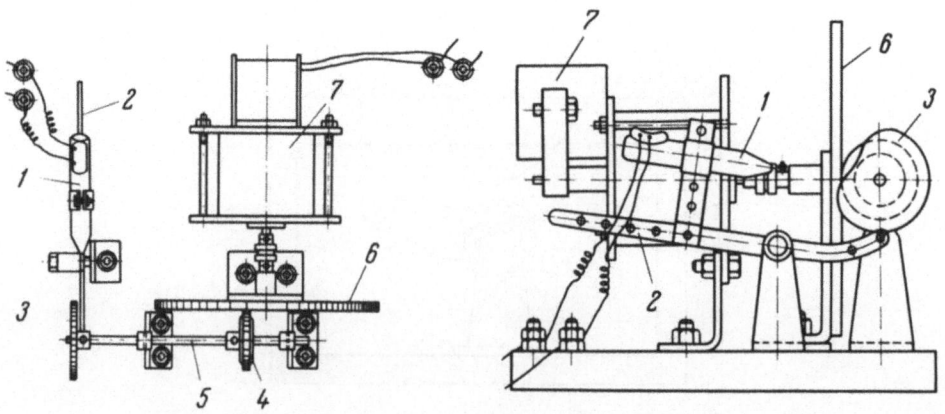

Fig. 200. Mercury interrupter: 1) relay; 2) lever; 3) cam; 4) wheel; 5) axle; 6) disk; 7) motor.

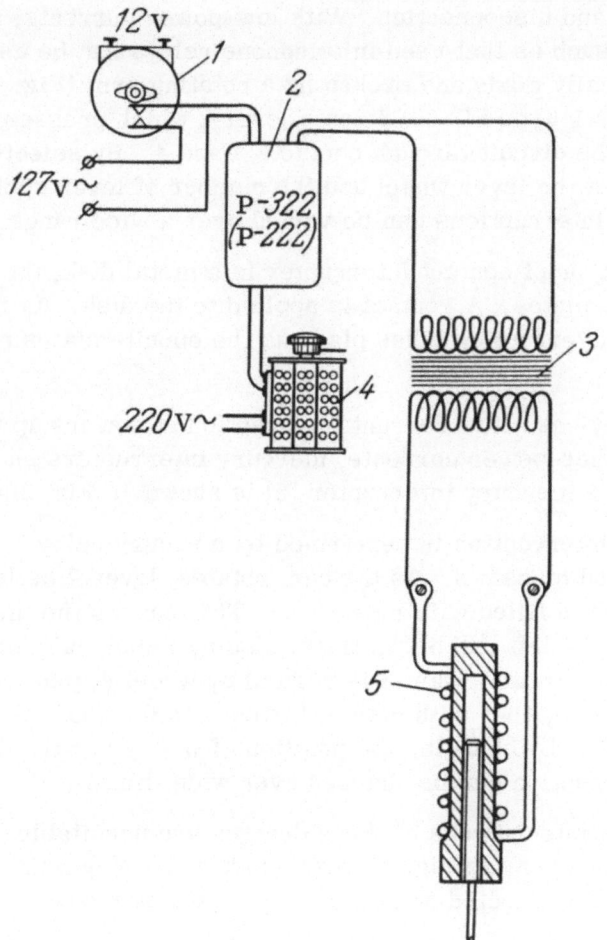

Fig. 201. Power supply of powerful solenoid: 1) interrupter; 2)
magnetic starter; 3) transformer; 4) autotransformer; 5) solenoid.

the high-pressure apparatus represents an excellent path for magnetic flux
(especially for stirrers with internal solenoids), obviously direct current
is most suitable. Finally, when working with DC the danger of coil shorting is not as great.

The use of alternating current also has its advantages. When working with alternating current, there is no requirement for special power
sources. The possibility of transforming AC permits the most suitable
selection of voltages required for operation of the agitating units (with
fixed coil resistance).

Current Interrupters.   Solenoids require a special structure for connection and disconnection. With low-power currents, a tungsten or silver contact such as that used in telephone relays can be used; the contact is periodically made and broken by a rotating arm (Fig. 199). Synchronous motor 1 (type SD-60) rotates lever 2 which presses on the upper plate to close the circuit through contacts 3 and 4. By selecting the rotational speed of the lever wheel and the number of lever spokes, the number of current interruptions can be varied over a wide range.

Another type of contact interrupter is a metal disk, the end of which carries ebonite plates. A contact is applied to the disk. As the disk rotates, current interruption takes place as the ebonite plates pass under the contact.

These devices allow current interruption at powers up to about 100 watts. For higher-power currents, mercury interrupters should be used. One design for a mercury interrupter [5] is shown in Fig. 200.

Current interruption is performed by mercury relay 1, fastened to lever 2, actuated by cam 3. As the cam rotates, lever 2 deflects the flask of relay 1 which is filled with mercury.* This causes the mercury to move either to the left end of the flask, closing the circuit, or to the right end, opening the circuit. Cam 3 is rotated by wheel 4, placed on axle 5. Wheel 4 rotates together with disc 6, fastened to the shaft of synchronous motor 7 (SD-60). By changing the position of wheel 4 on the disc, the frequency of interruption can be changed over wide limits.

However, interrupters of these designs are unsuitable for connection to solenoids carrying high powers. Industrial magnetic starters (types P-222 or P-322) connected as shown in Fig. 201 are used for this purpose.

An ordinary interrupter 1 closes the circuit of the coil of magnetic starter 2 and thereby turns on the current. When this occurs, the mechanism of the starter operates, closing the primary circuit of powerful transformer 3. By changing the voltage in this circuit using autotransformer 4, the current passing through the secondary winding of transformer 3 can be varied. This secondary is connected to solenoid 5, mounted on the (nonmagnetic steel) head of the high-pressure apparatus. The number of turns in the secondary is selected so that a high-power, low-voltage current passes through the winding.

In order to reduce the heating in the solenoid, the time during which current passes through it must be reduced to $^1/_{10}$ the duration of the entire close-open cycle.

---

*Editor's comment: Suitable mercury relays are available commercially in the USA.

Internal Coil [6]. When agitation must be performed in apparatus working at pressures over the yield point of the nonmagnetic steel of which the apparatus head is made, the coil is placed within the apparatus. In this case, the dimensions of the coil must be very small; also, reliability of the insulation of the wire is very important (the insulation should withstand not only high pressure, but also the action of the compressed medium).

When gases are compressed, ordinary insulation cannot be used at all. Also, any fibrous insulation will absorb the moisture contained in compressed gases, leading to short-circuiting of the turns. The use of lacquer insulation is impossible. The medium under pressure will penetrate beneath the lacquer coating, which will then separate from the conductor when the pressure is reduced.

In general, solution of the problem of introducing solenoids (and motors) within high-pressure apparatus amounts to a search for reliable conductor insulation. A satisfactory insulation is a moisture-proof, heat-stable plastic or chemical coating like the oxide films on aluminum. These films have good insulating properties, but are brittle and will not withstand bending of the wire.

Solenoid placement within apparatus can be most expediently made using the least possible number of layers of winding, consisting of a small number of turns of thick wire, not contacting each other. The small number of turns can be compensated for by using high current values. Therefore, the total number of ampere-turns, which determines the pulling force of the solenoid under otherwise equal conditions, remains unchanged. The insulation is made of porcelain or glass beads with minimal wall thickness. These beads, freely strung, prevent shorting of the turns but do not protect the conductor from corrosion, since they do not insulate the conductor from the medium.

Design of Coils. When using an agitator with an external coil, the experimenter is not limited in dimension, so that the design of this type of solenoid consists of determination of the quantity of conductor required and the current parameters. When the solenoid is placed within the apparatus, however, design becomes more significant. Usually, design consists in using the dimensions of the agitator to find its mass and calculate the force required, fixing the number of ampere-turns required for the coil. We present below an empirical formula* for calculation of the tractive force P(kg) of a solenoid with open magnetic circuit [7]:

---

*This formula is suitable only for unsaturated cores, where the tractive force is actually proportional to the square of the current. When saturation is achieved, the quadratic dependence becomes linear, which must be taken into consideration.

$$P = 18.4 \frac{l^2}{h} [x' - 0.89 (x')^3] (L_e - L_0)$$

$$h = 0.7l + 0.3l_e$$

$$x' = \frac{x}{h}$$

where I is the current, a; l is the length of the coil, cm; l_e is the length of the core, cm; x is the length of the core inserted into the coil, cm:

$$L_e = \frac{\pi^2 w^2 d_e}{0.44} k \cdot 10^{-9} H$$

$$L_0 = \frac{\pi^2 w^2 d}{0.44 + \frac{l}{d}} \cdot 10^{-9} H$$

$$k = 1 + 0.1 \frac{l}{d_e} + \frac{0.6 (l_e - l)}{d_e}$$

where w is the number of turns in the coil; d_e is the diameter of the core, cm; d is the average diameter of the coil, cm; H is the magnetic field intensity, Oe.

If $l_e = l$, then h = l, and x' = x/l. The maximum tractive force will be available where x/l = 0.61, i.e., when the core is $^2/_3$ inside the coil.

Then,

$$P_{max} = \frac{7.55 l^2 (L_e - L_0)}{l}$$

In the first approximation, the difference $L_e - L_0 = 0.9 \, L_e$.

Then,

$$P_{max} = \frac{7.55 l^2 \cdot 0.9}{l} \cdot \frac{\pi^2 w^2 \cdot 10^{-9} d_e}{0.44} \left(1 + 0.1 \cdot \frac{l}{d_e} \right) H$$

$$Iw = \sqrt{\frac{0.44 \cdot 10^9 P_{max} l}{7.55 \cdot 0.9 \pi^2 d_e \left(1 + 0.1 \cdot \frac{l}{d_e} \right) H}}$$

If the tractive force P_{max}, the diameter of the core d_e, and its length l are fixed, the number of ampere-turns Iw can be calculated.   Then the diameter of the wire and the current density are fixed, and the number of turns, total wire length, and coil resistance are calculated. Then, L_e and L_0 are calculated and their difference is compared with that assumed at the beginning of calculation. It is thus possible to determine the primary dimensions of the coil. If the coil resistance is known, the voltage required to feed the coil with the assigned current can be determined. Data on the resistance of copper wires are presented in Appendix VI.

The results of calculation using the formulas presented above correspond well with experiment, but this method requires considerable expenditure of time. The maximum tractive force can be determined approximately using the Roters formula [8] for a solenoid with a firmly contacting framework:

$$P = 2.73 \cdot 10^{-4} H B_p r_1^2$$

where P is the tractive force, kg; H is the magnetic field intensity, Oe; B_p is the maximum induction in the core, the value of which can be taken equal to 10 (in the CGS system); r_1 is the radius of the core, cm.

Since $H = i(r_2 - r_1)$, then

$$P = 2.73 \cdot 10^{-4} i (r_2 - r_1) B_p r_1^2$$

where i is the current density, a/cm^2; r_2 is the external radius of the coil, cm.

With this calculation, however, somewhat elevated values are produced.

Results which agree well with experimental data are given by the following formula, presented by O. G Katsnel'son [9];

$$P = \frac{sr}{l} \mu_0 \mu_g \frac{(i w_0)^2}{4} [2f(y_1) + f(m - y_1) - f(m + y_1)] [\varphi(m + y_1) + \\ + \varphi(m - y_1) - 2\varphi y_1]$$

where P is the tractive force (if the core length is equal to the coil length, n), s is the cross section, m^2; r is the average coil radius, m; l is the core length, m; μ_0 is the initial magnetic permeability (where $H \rightarrow 0$); μ_g is the magnetic permeability of the core; i is the current, a; w_0 is the number of turns per unit length of the coil; $y_1 = x_1/r$ (x_1 being the amount of the core located inside the coil); $m = l/r$ is the ratio of length of coil to average coil radius.

$$f(y_1) = \frac{y_1}{\sqrt{1 + y_1^2}}, \quad f(m - y_1) = \frac{m - y_1}{\sqrt{1 + (m - y_1)^2}}$$

$$\varphi(m - y_1) = \sqrt{1 + (m - y_1)^2}$$

These functions can be calculated from tables presented in work [10].

$$\mu_g = \frac{\lambda^2}{2.72\log\lambda - 0.69}$$

where $\lambda = l_c/d_c$ is the ratio of coil length to coil diameter.

At the present time, external solenoids can be used in apparatus working under pressures up to 10,000 atm [11]. For this, the head of the apparatus should be made of nonmagnetic steel type ÉI437B or ÉI698, which have sufficient strength to withstand the required pressure.

External solenoids should be manufactured using wires of the largest possible cross section, carrying the highest possible current; it is then sufficient to wind a few dozen turns around the coil, thereby reducing its dimensions.

For example, a solenoid made of copper tubing $^5/_3$ mm in diameter (after Katsnel'son) can be used. The secondary winding of the powerful transformer (welding transformer type) can be made of the same tubing. Then, feeding a current of less than 10 a to the primary, currents of thousands of amperes at a few volts can be taken from the secondary and fed to the solenoid. This type of solenoid need not be carefully insulated, and can be used to produce thousands of ampere turns with a single layer winding. Cooling water can be passed through the copper tubing of the transformer secondary. Figure 201 shows the connection of a solenoid of this type.

Drop Stirrer. Coil 1 (Fig. 202) is applied to head 2 of an apparatus, made of nonmagnetic steel (1Kh18N9T or ÉI437B). Core 3 of the coil is connected to shaft 4 of the stirrer, located in apparatus 5. When current is passed through the coil, the core returns symmetrically relative to the coil, raising the stirrer. When the current is turned off, the stirrer drops by its own weight to the bottom of the apparatus. Thus, agitation is achieved by turning the current on and off.

This type of agitation requires a considerable amount of time, but gives good results, especially if the amplitude of the stirrer oscillations is rather great. The intensity of the mixing action depends also on the frequency with which the stirrer is dropped and the form of its blades.

The optimal blade form for drop stirrers is determined to great extent by the nature of the medium to be mixed. For mixing of two gases, round, star-shaped, or spoon-shaped blades should be placed at certain intervals along the shaft. If a liquid is being mixed with a gas, an Archimedes screw is considerably more effective. In this case, a blade will break the surface of the liquid, no matter what the liquid level. Individual blades located at a certain distance from each other may agitate the gas and the liquid separately without mixing them.

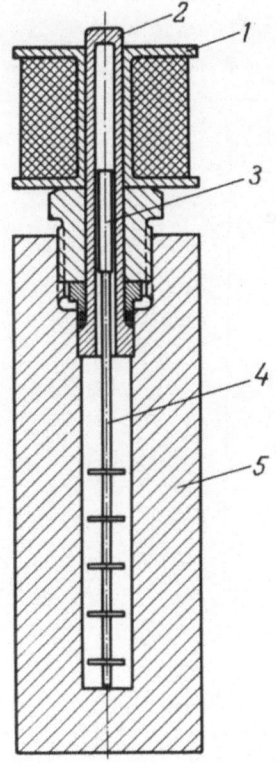

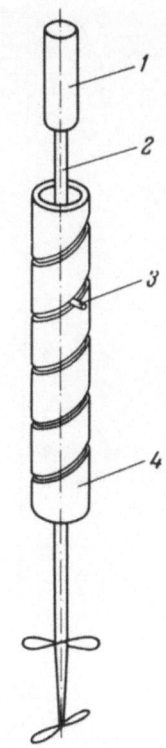

Fig. 202. Drop stirrer: 1) coil;
2) head; 3) core; 4) shaft; 5)
body of apparatus.

Fig. 203. Stirrer with reversing
rotation: 1) core; 2) axle;
3) finger; 4) tube.

Finally, a system of a solid body and a liquid may be mixed using no blades at all, by fastening a small container filled with the solid material onto the shaft; the solid material will be dissolved in the liquid as it moves through holes in the container, and the solution will be mixed.

Rotating Electromagnetic Stirrers. The reciprocating motion of a stirrer can be combined with rotation. For this (Fig. 203) core 1 is joined to a stirrer shaft 2 with finger 3 which moves in a spiral cut in tube 4, fastened firmly into the apparatus. As the core is drawn upward, the motion of the axle will include rotation, due to finger 3.

Figure 204 shows an agitator with an external rotor [12]. Axle 1, which enters the apparatus through capillary 2, holds magnet 3, enclosed in a body of nonmagnetic steel and operating under the working pressure. Electromagnet 4, rotated by pulley 5, is under atmospheric pressure. The rotating magnetic field causes magnet 3 and agitator 6 to rotate. The coils

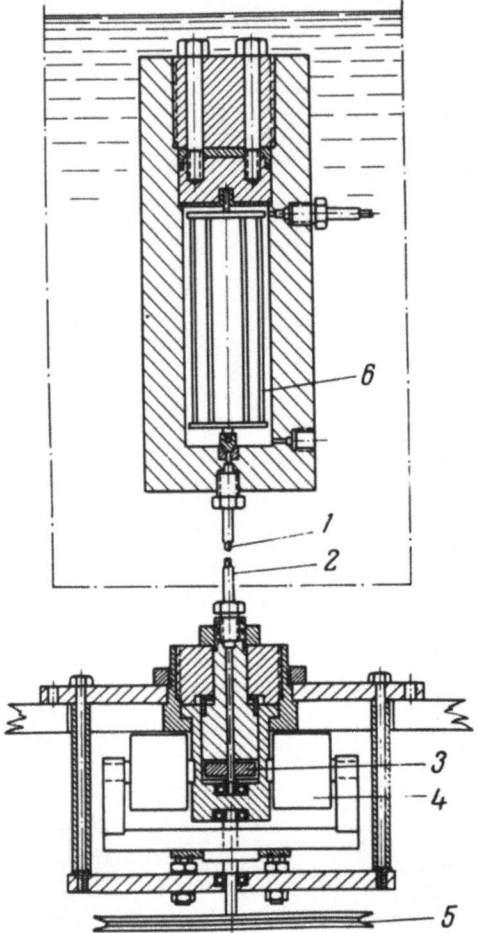

Fig. 204. Stirrer with external rotor: 1) axle; 2) capillary;
3) magnet; 4) electromagnet; 5) pulley; 6) stirrer.

of the electromagnet are supplied with 0.5 a DC, and rotate at 100 rpm. The current is fed in through carbon brushes sliding on bronze contacts. The agitator itself 6 is a metal basket placed directly in the apparatus. The location of the remaining parts is obvious from the figure.

This type of stirrer is distinguished by the high intensity of agitation it produces and combines all the advantages of mechanical and electromagnetic agitation.

The most effective stirring is produced using a device [13] in which only the electric motor rotor is under pressure.* The stator is placed around

---

*This principle is widely used for creation of glandless pumps for transferring corrosive liquids.

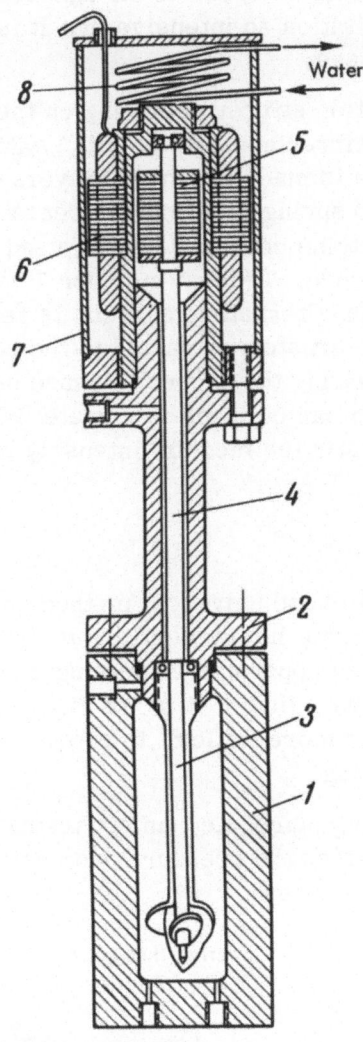

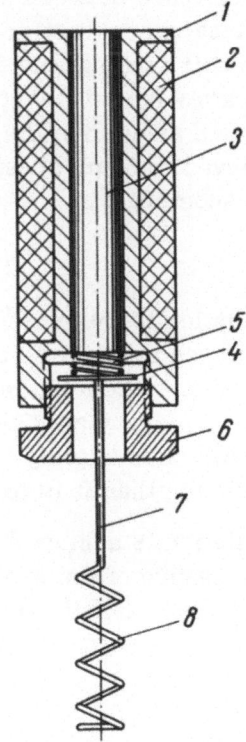

Fig. 205. Agitator with isolated stator:
1) high-pressure vessel; 2) cover; 3)
agitator; 4) axle; 5) rotor; 6) stator; 7)
cover; 8) coil.

Fig. 206. Vibration stirrer:
1) coil; 2) winding; 3) core;
4) armature; 5) spring; 6)
screw; 7) tension members;
8) stirrer.

a nonmagnetic steel tube. This type of agitator is shown in Fig. 205. The
high-pressure vessel 1 is closed with cover 2 which carries agitator 3 on
a bearing. Axle 4 of the agitator is connected with rotor 5, which rotates
on a second bearing in the upper portion of the high-pressure vessel. Stator
6 with its winding placed in oil-filled cover 7 is placed on the upper portion
of the vessel. Cooling water circulates through coil 8.

It is thus possible to achieve high rotational speed of the agitator and, using the proper blade form, to produce agitation so intensive that it is possible to effectively mix liquid and solid phases.

Figure 206 shows a design of a vibration stirrer operating on the electromagnetic vibrator principle [14]. The stirrer consists of coil 1 with winding 2, which contains core 3. The core is made of several layers of transformer iron. Armature 4, soldered to spring 5, is located beneath the core. The lower portion of the coil contains control screw 6, used to change the distance between armature and core. Tension member 7 connects the armature with stirrer 8. When an alternating current is fed through the coil winding, the armature is alternately attracted to the core and pushed away from it by spring 5. By setting the proper distance between core and armature, the entire system can be made to vibrate. When this is achieved, the spiral (stirrer) 8 will stir the medium intensely in which it is submerged.

CIRCULATION

Circulation is the most highly perfected and universal method of mixing. If a gas can be made to circulate through a liquid, equilibrium can be produced in this system rather rapidly. If an apparatus containing a compressed gas or solid material is showered with liquid, it is not difficult to produce equilibrium rapidly. However, it is more difficult to provide reliable circulation than it is to provide stirring.

Circulation is achieved by a circulating pump, i.e., an apparatus working at the pressure of the system which creates the pressure drop necessary to overcome the hydraulic resistance of the system.

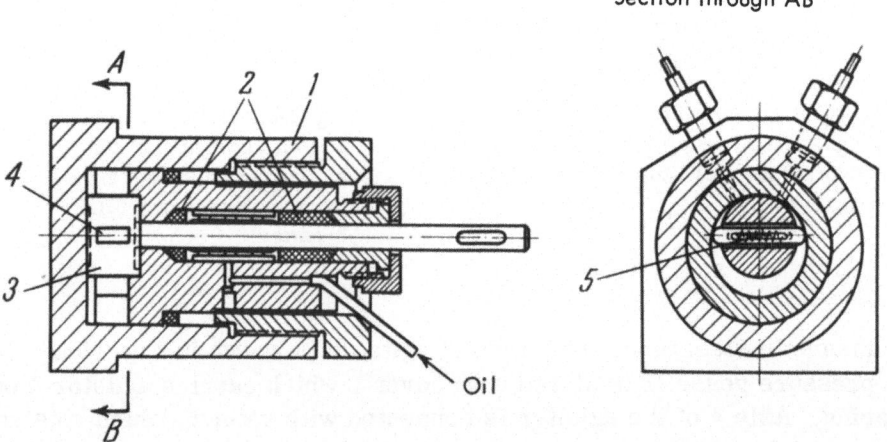

Fig. 207. Gonikberg, Fastovskii, and Gurvich circulation pump: 1) cover; 2) glands; 3) piston; 4) blades; 5) spring.

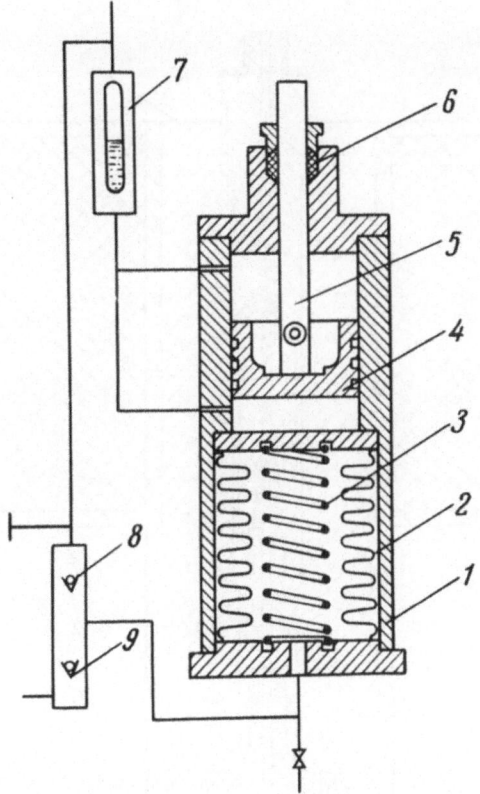

Fig. 208. Bellows pump: 1) cylinder; 2) bellows;
3) spring; 4) piston; 5) shaft; 6) gland; 7) reservoir;
8, 9) valves

Gonikberg, Fastovskii, and Gurvich Pump. Figure
207 shows the design of a pump suggested by M. G. Gonikberg, V. G. Fastovskii, and I. G. Gurvich [15]. Piston 3, carrying blades 4 [loaded and] moved
by springs 5 is fastened to a shaft which rotates in gland 2 within thin-walled
cover 1. Gland 2 consists of two parts, the gap between them being filled
with oil under a pressure greater than the pressure in the pump chamber.
The pump shaft rotates at up to 100 rpm. The discharge rate of the pump
at 100 rpm is two liters of gas per minute at a pressure of 200 atm.

A circulation pump described by G. E. Aristov and I. P. Sidorov [16]
operates on the same principle.

Bellows Circulation Pump. The design of this pump is
shown in Fig. 208 [17]. Bellows 2, extended by spring 3, is located in high-
pressure cylinder 1. The cylinder contains piston 4, moved by shaft 5,

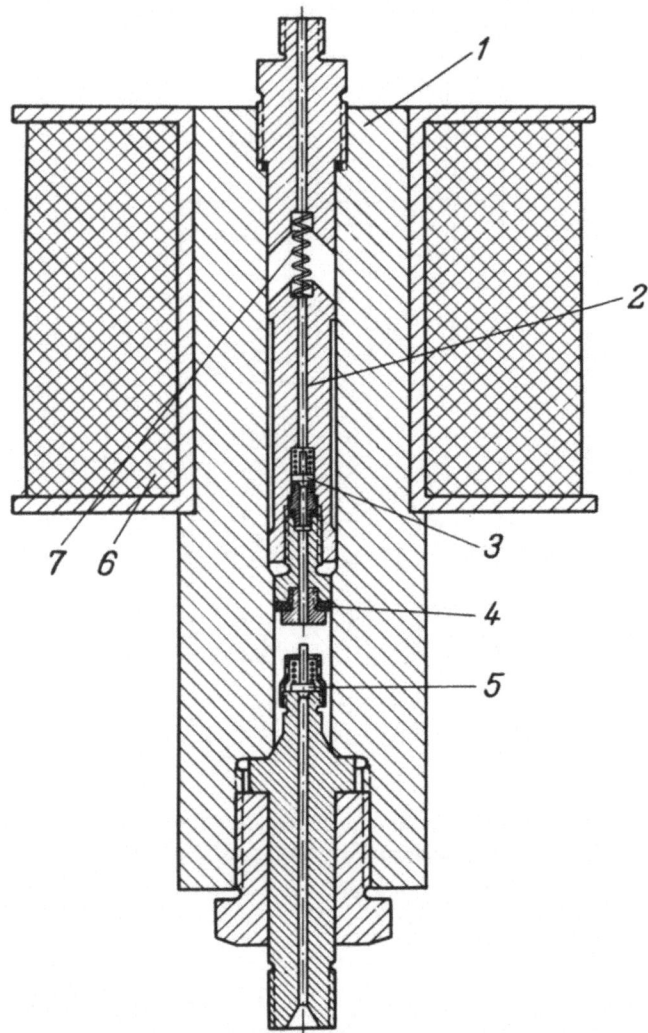

Fig. 209. Electromagnetic circulation pump: 1) body; 2) piston;
3) delivery valve; 4) packing; 5) inlet valve; 6) coil; 7) spring.

which passes through gland 6. The entire space within the cylinder is filled with oil, supplied from reservoir 7.

The gas pumped is supplied at its initial pressure through input valve 8 and exits under its final pressure, that created by the pump, at delivery valve 9. The valve box, separate from the body of the pump, is connected with the gas space of reservoir 7.

Electromagnetic Circulating Pump. The pump (Fig. 209) consists of body 1, made of nonmagnetic steel (ÉI183), containing iron

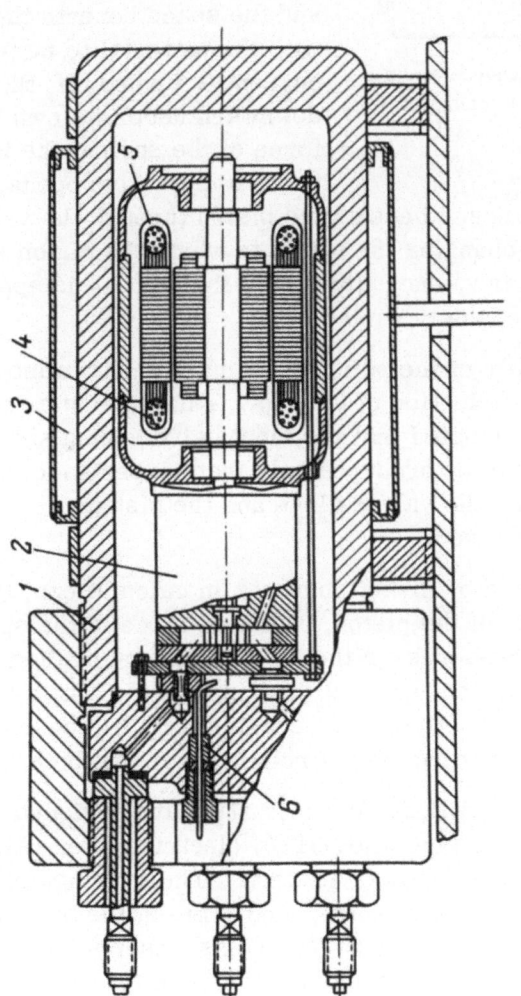

Fig 210   High-pressure circulation pump: 1) pump body; 2) gear pump with reducing drive; 3) cooler;
4) electric motor winding; 5) MBK compound; 6) sealing of electrial lead.

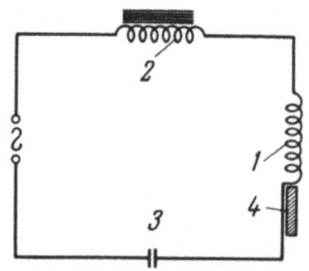

Fig. 211. Ferroresonance circuit:
1) thrust solenoid; 2) choke; 3) condensor; 4) core.

piston 2, supplied with delivery valve 3 and packing 4. The lower part of the pump contains the input valve 5. When coil 6, applied to the external portion of the body, is activated, the piston moves into the coil and, compressing spring 7, occupies its upper position. This opens the input valve, and the space beneath the piston is filled with the material to be pumped. When the current is turned off, the piston moves downward under its own weight and the force of the spring, the inlet valve closes, the delivery valve opens, and the material is pumped out the space beneath the piston through the valve into the body of the piston. By changing the height to which the piston is raised, its diameter, and the frequency with which the current is applied, the delivery of the pump can be widely varied.

The reliability of the pump is primarily determined by reliability of valves and proper selection of springs. This particular design uses plate valves. It should be noted that the input valve spring should be weak. Precise data on this can be produced only from experience, by selecting the spring force so that the valves close and the pistons drop into the body of the cylinder as required.

The coil is fed by direct current. In calculations, the lifting force of the coil, the weight of the piston, resisting force of the spring, and pressure drop (in atmospheres) in the installation, multiplied by the cross-sectional area of the piston, must all be considered. The presence of a framework closing the magnetic circuit considerably increases the tractive force of the pump, making it capable of reliably pumping both liquid and gas.

A. M. Sirota and B. K. Mal'tsev [18] have designed a circulation pump (Fig. 210) driven by a type AOL-012/4 electric motor (All Union State Standard GOST 186-52), power 80 watts. The windings of the motor are sealed with thermo-reactive compound MBK-2, the ball bearings are replaced by friction bearings with fluoroplast inserts, and the ventilating vanes are eliminated. The pump and motor have operated in water for 300 h at a total pressure of 500 atm.

We will not discuss pumps of other designs which do not differ essentially from those described (in positioning of valves, piston, etc.) For example, valves are frequently used (by I. P. Sidorov) with forced closure or opening, which can be achieved by wrapping a coil around the valve to apply force at the required moment. Good results have been produced by synchronizing the actuation of the coil with the stroke of the piston.

Finally, we might note the design described in Chapter IX, amounting to a compromise between a pump and an agitator. It is used in the installation of Kricheskii and Efremova [19]. It is called the agitator-pump, and works on the dropping-piston principle.

All electromagnetic devices require periodic operation of the solenoids. O. G. Katsnel'son, B. P. Bruns, and D. Yu. Gamburg [20] suggested a pump design in which a so-called ferroresonance circuit is used, allowing complete avoidance of current interruption. In this circuit (Fig. 211) the phenomenon of sharp current change in a ferroresonance circuit upon change in voltage is used. Thrust solenoid 1 is connected in series with choke 2 and condenser 3. In the nonworking condition, core 4 is almost totally extended from the coil, whose self-induction is slight, as is its reactance. A voltage equal to the resonance voltage is fed to the ferroresonance circuit connected in series with the coil. When this occurs, the current increases sharply, and the core is pulled up into the solenoid. The reactance of the solenoid immediately increases sharply, and the voltage in the ferroresonance circuit is reduced to the so-called dissonance voltage or even lower. Then, the current sharply decreases, and the coil falls back out of the solenoid. The entire cycle is then repeated.

This type of circuit makes it possible to control the frequency and amplitude of oscillations by changing the supply voltage.

BIBLIOGRAPHY

1. Flom, D. G., and Porill, N. T., J. Appl. Phys., 26(9):1088 (1955).
2. Kiebler, M. W., Ind. Eng. Chem., 37(6):540 (1945).
3. Björkman, A., Ind. Eng. Chem., 44:2359 (1952).
4. Payne, I. W., Sreed, C. W., and Kent, E. R., Ind. Eng. Chem., 50(1):47 (1958).
5. Bruns, B. P., and Braude, G. E., Tr. Gos. Inst. Azo. Prom., No. 1, Goskhimizdat, 1953, p. 252.
6. Tsiklis, D. S., Dokl. Akad. Nauk SSSR, 86:993 (1952).
7. Küchler, Jahrbuch der AEG Forschung, June, 1939.
8. Roters, Electromagnetic Mechanisms, GÉI, 1949.
9. Katsnel'son, O. G., Chemistry and Technology of Nitrogen Fertilizers, Works of the State Institute of the Nitrogen Industry, Tr. GIAP, No. 12, Goskhimizdat, 1961, p. 323.
10. Semendyaev, K. A., et al., Five Place Mathematical Tables, Izd., second edition, 1959.
11. Tsiklis, D. S., and Maslennikova, V. Ya., Dokl. Akad. Nauk SSSR, 157:427 (1964).
12. Sage, B., and Lacey, W., Trans. Am. Inst. Min. Met. Eng., Petr. Div., 136 (1940).

13. Bishnevskii, N. E., Zh. Khim. Prom., 2(102):38 (1956).
14. Tsiklis, D. S., Zavodsk. Lab., 13:242 (1947).
15. Gonikberg, M. G., Fastovskii, V. G., and Gurvich, I. G., Zh. Fiz. Khim., 13:1669 (1939).
16. Aristov, G. E., and Sidorov, I. P., Tr. Gos. Inst. Azo. Prom., No. 1, Goskhimizdat, 1953, p. 258.
17. Beck, R., Ranbo, M., Sensil, E., and Higrave, P., Ind. Eng. Chem., 42:144 (1950).
18. Sirota, A. M., and Mal'tsev, B. K., Inzh. Fiz. Zh., 2(1):93 (1959).
19. Kricheskii, I. R., and Efremova, G. D., Zh. Fiz. Khim., 22:116 (1949).
20. Katsnel'son, O. G., Bruns, B. P., and Gamburg, D. Yu., Zavodsk. Lab., 12:379 (1946).

Chapter VIII

General Laboratory High-Pressure Equipment

The equipment of a high-pressure laboratory has specific features differing from the equipment of ordinary physical and chemical laboratories. The primary difference is the combination of cumbersome metal machinery and apparatus, glass analytic apparatus, and precise measuring instruments. This combination makes it especially necessary to think through the problem of efficiently locating instruments and apparatus available for use.

Obviously, general instructions for the location of apparatus and instruments cannot be given for all cases. However, some conditions must be observed.

BUILDINGS

A modern, more or less independent high-pressure laboratory should be located in a building whose design is sufficient for placement of heavy equipment (such as compressors, pumps, hydraulic presses, etc.). This building should be capable of withstanding not only the static, but also the dynamic loads (vibration caused by the operation of compressors, shock in the operation of adiabatic installations, etc.).

It is best to place the laboratory in a one-story building with reinforced foundation and light roof, which will not create heavy fragments capable of doing additional damage in case of an explosion. If this is impossible, the laboratory should be placed on the top floor.

It is necessary, of course, that the floor below the laboratory be strong, capable of withstanding the load from the weight of the equipment and vibration created by the operation of compressors. Also, the room should be suitable for installation of sensitive instruments, so that the op-

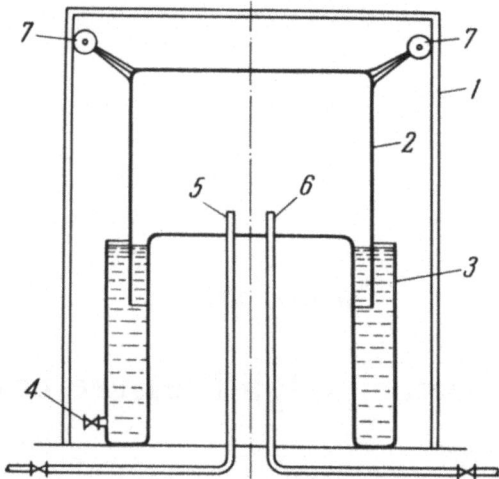

Fig. 212. Wet gas holder: 1)frame; 2) bell; 3) cylinder;
4) outlet valve; 5) filler line; 6) line to compressor; 7)
rollers.

eration will not be disrupted by the operation of compressors, pumps, etc.
Therefore, a building with a reenforced concrete shell is most desirable.

EQUIPMENT

Gas Holders

Gas holders for supplying compressors (there should be at least two)
should be installed outside the laboratory. For this, a separately heated
room should be built, containing the so-called wet gas holders, i.e., bells
floating in reservoirs filled with water (Fig. 212). If the wet gas holders
are installed outside, they should be supplied with steam conduits for heating the water in the winter. A layer of oil is usually placed over the water
in gas holders, in order to reduce solution of the gas by the water. In-
cidentally, this does not eliminate the necessity of removing water droplets
and vapor from the gas before feeding it to the compressor. Wet gas
holders are equipped with protective relays to disconnect the compressor
motor when the gas-holder bell drops to its bottom position.

Dry gas holders or pressure vessels can be used in place of wet gas
holders. Dry gas holders are less convenient than wet gas holders, however, since the pressure vessels require extremely accurate control of
gas supply to the compressor.

Compressors

The compressor is installed on a special foundation. If the compressor has a belt drive, a guard constructed according to corresponding safety rules must be included.

The compressor should be turned on through a start—stop switch. The conductor, like all the other electrical equipment of the laboratory, is selected depending on the class of the room in which the laboratory is located (according to the rules for installation of electrical equipment).

Gas Purification

After compression, a gas contains considerable quantities of water and oil vapor and droplets; in order to remove them, the gas is passed through a series of high-pressure columns filled with activated carbon, cotton wool, silica gel coated with calcium chloride, melted alkali, etc., after going through the oil separator. After this purification, the gas is practically free of oil and water droplets. Complete purification of a compressed gas from all dissolved oil and moisture is difficult.

Installation of Apparatus

Apparatus operating under pressures over 1000 atm must generally be installed behind special protective walls. A typical protection is the chamber described below. Vessels with large capacity should also be screened, even if the working pressure is less than 1000 atm.

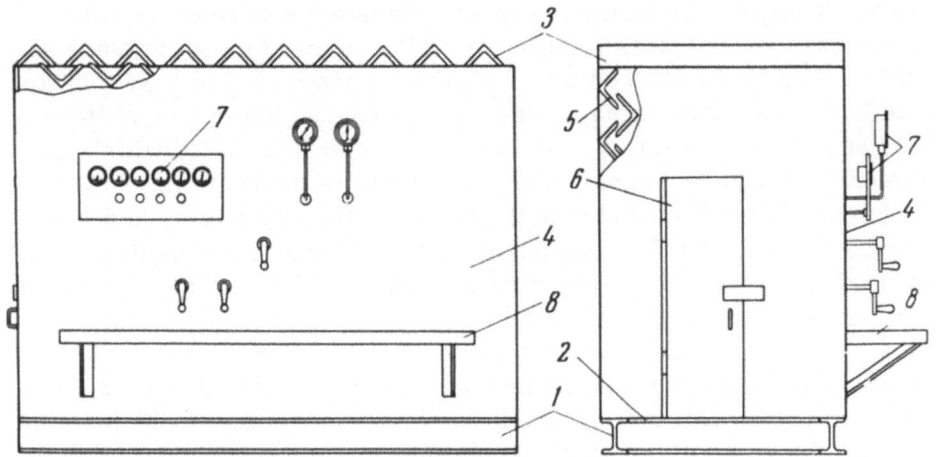

Fig. 213. Schematic drawing of shielded chamber: 1) H-beam; 2) floor (sheet steel); 3) roof (angle steel); 4) front wall; 5) rear wall (angle steel); 6) door; 7) instruments; 8) bench.

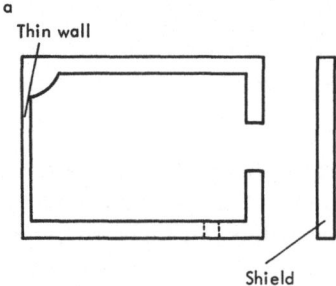

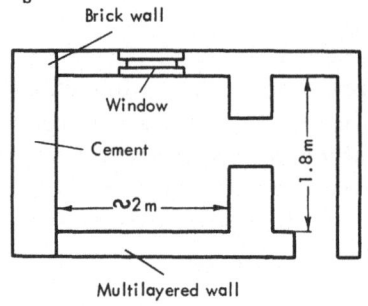

Fig. 214. Plan of chamber for operation with high-pressure apparatus.

All apparatus and vessels working under compressed-gas pressure should be specially observed. According to the existing rules [2], "vessels used for scientific experimental work with capacities of 25 liters and less, regardless of the working pressure and temperature" need not be inspected by the state. Therefore, all work on periodic testing of the apparatus should be performed by the administration of the enterprise where the vessel is installed.

It should not be thought that the primary danger is from bursting of the apparatus and resulting fragmentation. Individual parts such as nipples, valves, etc., fly off much more frequently, either due to weakening of threads or biting into shanks [pinching-off] by self-sealing closures, etc. For protection from these flying parts which may be "fired" even by vessels working under liquid pressure, apparatus should be installed in special metal-shielded chambers, as noted above.

Chambers. There are many designs for protective shielding of chambers. They may be stationary or mobile (steel spheres), made of steel or bronze sheets, whose thickness is determined by the pressure being used. It can be considered that a chamber constructed of 8-12 mm steel is reliable for protection from rather high pressures (up to tens of thousands of atm).* The chamber may be welded up or assembled with bolts. The dimensions of the chamber depend on the dimensions of the installation and methods used to assemble it. However, the assembly of high-pressure apparatus, tightening of sealing nuts, etc., requires the application of considerable force, which in turn requires the use of wrenches with handles up to 1-2 m long. If the installation is to be permanently placed in a shielded chamber, sufficient space must be provided for work with these wrenches. The chamber is assembled on a frame of channel iron or T-section beams. The dimensions are determined by the size of the installation and the thickness of the steel. Steel sheet 10 mm thick is fastened to a base of dual T-beams (No. 8-12).

―――――――
* Unless work is performed with large quantities of hot gases.

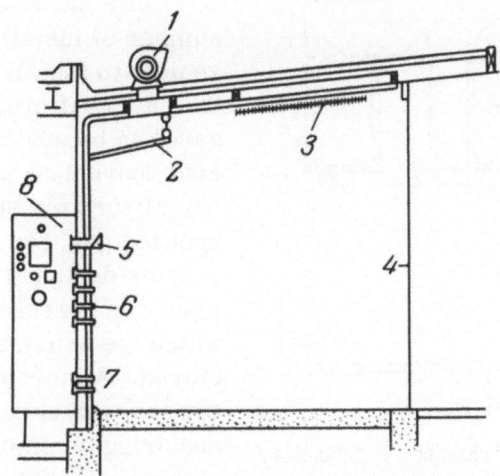

Fig. 215. Chamber for high-pressure apparatus: 1) exhaust fan; 2) window; 3) steam heater; 5) break-away wall; 5) prism; 6) apertures for valves; 7) apertures for tubing; 8) instrument panel.

The front wall of the chamber should be free of all structures, and is assigned for through-passage of valve handles, etc. The front wall may also have removable plates, which can be conveniently placed and drilled after the installation is assembled. This type of sheet can be replaced when another installation is used within the shielded chamber.

The rear wall and cover of the chamber are made of angle iron, located so that the gas formed by the explosion of an apparatus within flies freely from the chamber (see Fig. 213). The floor of the chamber is made of sheet steel. The doors are made in side walls. The door should open into the chamber.

A shielded chamber may be built of concrete [3], brick or other similar material. Figure 214a shows a chamber, one wall of which is thinner than the others, so that it breaks away in case of an explosion. Figure 214b shows a chamber whose internal wall is made of wood and boiler plate, with an interlayer of sand (2.5 cm wood, 15 cm sand, 0.6 cm boiler plate). Often, a shield is placed in front of the entrance to the chamber.

Figure 215 shows the design of a chamber [4] whose rear wall 4 acts as a protective membrane, designed to allow the products of explosion to escape. Several such chambers can be placed in a row (Fig. 216). Various designs for buildings and chambers have been described in the literature [5, 7].

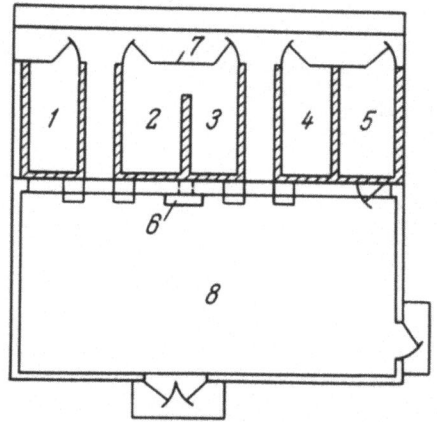

Fig. 216. Location of chambers in laboratory room: 1-5) chambers; 6) instrument panels; 7) break-away walls; 8) laboratory room.

If a chamber contains a large number of installations, it is convenient to install a manifold after the gas purifiers, allowing various gases to be available on tap. It is also convenient to have special accumulators for storage of compressed gas, i.e., large-volume vessels designed for the maximum compressor pressure at which the compressed gas is stored. An investigator can use gas from an accumulator, without requiring a compressor.

A mercury adjustor or an adjustor with a metal bellows can be connected to the distributing block. It is also convenient to have a pressurized-oil manifold, to supply oil under pressure to the chambers. Electric power and water should be supplied to each chamber.

Remote Control. It is absolutely forbidden to enter a chamber while the apparatus is in operation. It is therefore necessary to be able to open and close all valves and observe pressure and temperature without entering the chamber.

When pressure is measured by manganin manometer and temperature by thermocouple, this is not difficult. If Bourdon gauges are used, pressure can be observed by observation through a mirror.

Long-handled shafts are used for turning valves where the valves are installed perpendicular to the front wall. If some valves cannot be located in this common plane (which often occurs), gear drives must be used to rotate the valves. A reducing gear with a worm-to-wheel ratio of 80 : 1 or 100 : 1 is very useful when gas supply must be precisely regulated or when gas pressures are being reduced from a very high pressure. Also, obviously, when high pressures are being used and a great deal of force is required to rotate the stem, a reducing gear greatly facilitates the work. However, a reducing gear must be used carefully in order to avoid damaging valve parts.

Installation of Manometers. Bourdon gauges installed outside the chamber should be elevated to a height no less than 50 cm higher than a man's head. If this is impossible (as when manometers are installed on compressor stages) the dials should be covered with a loose but strong

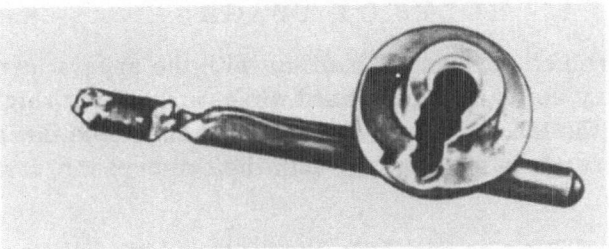

Fig. 217. Sleeve and valve stem burned in oxygen.

metal screen or transparent plates of methyl methacrylate. Manometers burst fairly regularly, and safety measures must be taken to protect personnel from metal and glass fragments.

Safety Valves. In correspondence with the requirements of the safety regulations for high-pressure apparatus, safety valves or bursting membranes [rupture discs] must be installed. In laboratory practice, these devices are usually used only in comparatively large installations [8], working under pressures of hundreds of atmospheres.

Usually, laboratory high-pressure vessels are made with considerable strength reserve, so that slight excesses in pressure will not burst the apparatus. When working with gas under pressures of 10,000-15,000 atm, the apparatus does not have this reserve of strength. A sudden sharp increase in the pressure in such an apparatus is a rather rare phenomenon (excluding, of course, the possibility of explosion). If an explosion occurs within an apparatus, neither a safety valve or membrane can operate fast enough to reduce the pressure. The design of valves and bursting membranes has been described in a number of works [9-13], although this problem cannot be considered finally solved.

The reason for an explosion of a high-pressure apparatus may be insufficiently careful removal of traces of organic materials from apparatus parts when working with compressed oxygen. Flaming of traces of organic material can cause burning of massive steel parts (Fig. 217). In these cases, only the chamber provides reliable protection.

Explosions in high-pressure apparatus may also occur due to a change in the detonation limits of compressed mixtures with pressure. It should be noted that when working with such mixtures (for example, with hydrocarbons mixed with air or oxygen), great care must be exercised. Thus, rapid exhaust of a gas jet from a high-pressure vessel causes formation of an electric charge in the jet (triboelectricity). If the high-pressure apparatus is not grounded [14], the charge will accumulate until a spark jumps. This may cause the vessel to explode.

RULES OF USAGE

Only well-trained personnel, familiar with the apparatus and acquainted with safety measures to be used when working with high pressure, can be allowed in the laboratory. The personnel should be thoroughly familiar with the instructions for working with the compressor, manifolds and every installation.

All equipment must be carefully checked, and all manometers must be periodically replaced and checked by the State Committee of Standards, Measures, and Measuring Instruments of the State Planning Institute, USSR; also, hydraulic testing of equipment must be periodically performed. It is especially important to test apparatus which are not accounted for in the State Technical Accounting System.

All rules published by the Regulations Publishing Houses must be strictly adhered to [1, 2, 15].

Work at high pressure requires special care and strict observation of all safety rules. It is most dangerous to ignore safety measures after a long period of work without accidents.

BIBLIOGRAPHY

1. Rules for Installation of Electrical Apparatus. (PUÉ), Izd. "Energiya," 1964.
2. Rules for Installation and Safe Operation of Vessels Operating under Pressure. Gosgortekhizdat, 1963.
3. Porter, R. L., Lobo, P. A., and Sliepcevich, C. M., Ind. Eng. Chem., 48(5):841 (1956).
4. Graig, L. E., and Dew, I. E., Ind. Eng. Chem., 51(10):1249 (1959).
5. Knop, Chem. Ing. Techn., 30(10):664 (1958).
6. Bowen, I. S., and Jenkins, R. L., Ind. Eng. Chem., 49(12):2019 (1959).
7. Sliepcevich, C. M., and Porter, R. L., Ind. Eng. Chem., 54(7):44 (1962).
8. Weber, G. G., Chem. Eng., 62:170 (1955).
9. Kondrat'eva, T. F., Khim. Mash., 4:29 (1959).
10. Lowenstein, I. G., Chem. Eng., 65(1):157 (1958).
11. Stephens, H. R., and Walker, K. E., Ind. Eng. Chem., 49(12):2022 (1957).
12. Munday, G., and Newitt, D. M., The Physics and Chemistry of High Pressure (papers read of the Symposium 1962), London, 1963.
13. Jennett, E., Chem. Eng., 70(17):151 (1963).
14. Rules for Protection from Static Electricity in Chemical Industry Production, Goskhimizdat, 1963.
15. Solov'ev, P. V., Ermilov, P. I., and Strel'chuk, N. A., Principles of Safety Techniques and Fire Safety Techniques in the Chemical Industry, Goskhimizdat, 1960.

Chapter IX

Methods of Investigating Phase Equilibria
at High Pressures

Methods of investigation of phase equilibria can be divided into two large groups: analytic and synthetic. Analytic methods of investigation consist of determination of the composition of the phases present by analysis of samples selected from each phase. With this method, the investigator should solve two primary problems: 1) how to create the conditions providing for establishment of equilibrium between the phases, and 2) how to analyze the phases existing in equilibrium without disrupting the equilibrium conditions.

The synthetic methods are based on a principle suggested by V. F. Alekseev [1]. According to this principle, a mixture is made up of two components (liquid and gas, or solid and liquid), the quantities of which are precisely known. The mixture is placed in a closed volume and, measuring the temperature, the moment when the system becomes a single-phase system is noted. The qualitative composition of this phase and its volume, as well as the temperature and pressure are known; consequently, the experimenter has exhaustive knowledge concerning the phase and volumetric behavior of the system without the necessity of analyzing the phase itself. The experimental and theoretical investigations performed in recent years have shown that an understanding of the thermodynamic phase equilibria requires, in addition to data on the solubility, information on the volumetric behavior of solutions under pressure.

In connection with this, many modern installations for investigation of phase equilibria by analytic methods are capable of measuring phase volumes. The use of synthetic methods makes it possible to produce such data without resorting to special devices.

Analytic methods of investigation, in turn, may be divided into dynamic, static, and circulation methods.

Dynamic methods are characterized by the passage of a gas through a layer of liquid. In this, the gas, displacing the liquid, is dissolved in the liquid and is simultaneously saturated with the liquid. These methods reduce the time required for establishment of equilibrium and, it seems at first glance, eliminate the necessity of introducing special devices into the apparatus for mixing of the gas and liquid phases.*

The more rapidly the gas is passed through the liquid, the more rapidly the liquid is saturated with the gas, but the more slowly the liquid is dissolved in the gas and the less likely it is that the gas phase will reach saturation. On the other hand, if the gas is passed through the liquid slowly, conditions for saturation of the gas phase are improved, but conditions for saturation of the liquid phase are worsened. The condition of equilibrium between phases assumes both saturation of the gas with liquid and saturation of the liquid with gas, regardless of which phase is being investigated. If only one phase is saturated, while the coexisting phase is unsaturated, errors in determination of solubility are unavoidable. These errors are particularly great when the gas is easily soluble in the liquid or the liquid is easily soluble in the gas.

The volatility of each of the two components in the different phases are equal at equilibrium. If, for example, the gas phase is not saturated with liquid, the volatility of the gas will be greater and the solubility of the gas in the liquid will also be greater than at equilibrium. On the other hand, if saturation of the liquid phase is not achieved during solution of an easily soluble gas, an analogous error is produced in the determination of the solubility of the liquid in the compressed gas.

Consequently, contradictions are inherent in the dynamic methods, which can lead to errors. These errors are insignificant and can be ignored if the gas is poorly soluble in the liquid and the liquid is poorly soluble in the gas. In other cases, the dynamic method must be combined with mechanical agitation.

The deficiencies of the dynamic method also include the necessity of having a large quantity of gas for passage through the liquid.

During dynamic-method investigations with a solvent which has high vapor pressure, a considerable quantity of the liquid will be carried away by the gas, which makes it necessary to add additional liquid to the appara-

---

*There is no doubt that mixing is still necessary. The improper data produced by the authors of works [2, 3], who assumed that diffusion of a dissolved gas can replace mixing, having demonstrated that diffusion alone is not sufficient to achieve equilibrium, even over durations of many days.

tus and to purify the gas of this liquid before compression if it is returned to the cycle.

The advantage of the dynamic methods is that the liquid is in equilibrium with a gas mixture which always has constant composition, whereas in static methods the composition of the gas mixture changes as the components of the mixture dissolve unequally in the liquid. Of course, in the study of the solubility of pure gases, this advantage disappears.

The static methods consist in placing a gas and liquid in contact in a closed volume at defined temperature and pressure and intensively mixing the two phases. These methods are more complex, since special apparatus for agitation under pressure are required. But, they do have advantages: a small quantity of gas is required, and reliable results are given if the agitation is thorough. The static methods differ in their methods of sample selection and mixing.

Circulation methods are analytic methods somewhere between the dynamic and static methods. They are distinguished in that the gas bubbling through the liquid is not valved, but is recycled for bubbling with special circulation pumps. These methods combine the advantages of static and dynamic methods, although they also have some of their deficiencies. The liquid phase may also be circulated.

LIQUID-GAS SYSTEMS

Analytic Methods. Dynamic methods [4, 5] for determination of the solubility of gases in liquids have been rather frequently used. Let us analyze the Krichevskii, Zhavoronkov, and Tsiklis apparatus, which these authors used to investigate the solubility of the oxides of carbon, hydrogen, and their mixtures in methanol [6].

The installation (Fig. 218) consists of a high-pressure vessel 1 submerged in thermostat 2. The vessel has valves for inlet 3 and discharge 4 of gas, and for taking liquid samples 5. Valve 5 is connected to the analytic device.

Gas, compressed to the required pressure and carefully purified, is introduced into the high-pressure vessel through a tube which extends to the bottom and is twisted at its end. Glass coil 6 is applied to the twisted end of the tube; gas bubbles exiting from the tube enter this coil. As they move through the coil, the gas bubbles entrain the liquid, pumping it from the bottom of the vessel to the surface of the liquid. The gas leaving the coil bubbles through the remaining liquid column and leaves the vessel through the outlet valve, which throttles it down to atmospheric pressure. This gas is purified from solvent vapors, and can be used again for saturation of the liquid.

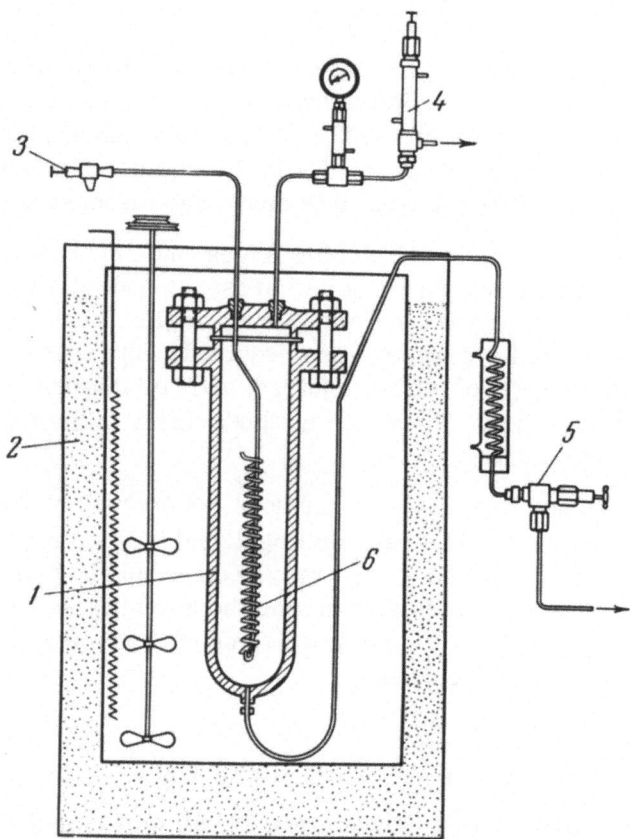

Fig. 218. Plan of Krichevskii, Zhavoronkov, and Tsiklis installations
for determination of solubility of gases in liquids under pressure: 1)
high-pressure vessel; 2) thermostat; 3, 4, 5) valves; 6) glass coil.

When working with gas mixtures, the composition of the gas leaving
the vessel differs from the composition of the initial mixture, due to the
non-identical solubility of the components. Therefore, in addition to purification of the mixture of solvent vapors, the corresponding components
must be added in order to restore the initial composition. The establishment of equilibrium in the system can be judged from the results of analysis of the liquid phase. Usually, the equilibrium mixture at the required
pressure is produced by increasing or reducing the pressure. In the latter
case, a pressure somewhat greater than the predetermined pressure is
created; then, reducing this pressure, gas dissolved in the liquid is removed from it. The correspondence of data produced by this method with
data of direct saturation allows conclusions to be made on the establishment of equilibrium.

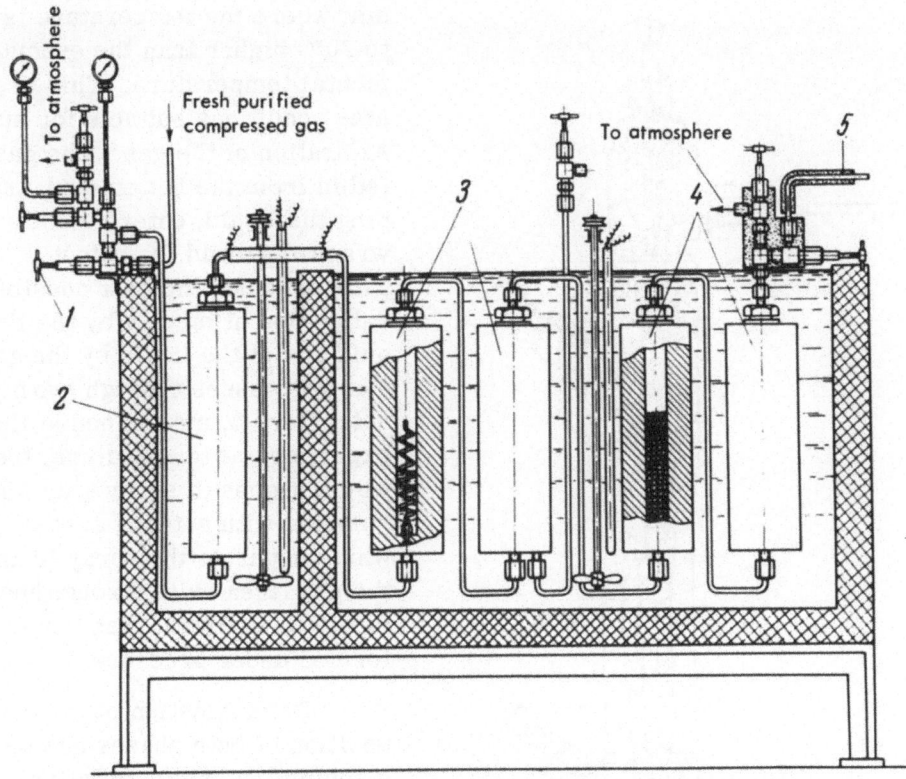

Fig. 219. Krichevskii-Gamburg installation for determination of solubility of liquids in compressed gases: 1) valve; 2, 3) saturators; 4) spray traps; 5) output tube.

Dynamic methods have become especially widespread in determination of the solubility of materials in compressed gases [7-14]. In this case, static methods are less reliable, since the small quantity of saturated gas located in the apparatus over the liquid hinders the removal of samples for analysis. With slight solubility of the liquid (or solid material), precise results can be produced by taking a considerable quantity of the gas phase for analysis. This results in a pressure drop in the system and leads to errors. Dynamic methods allow any quantity of gas to be taken for analysis.

One of the first dynamic installations was the Krichevskii-Khazanova apparatus [11], used in a study of solubility of ammonia in compressed nitrogen and hydrogen. With the exception of the analytic portion, the installation differs little from that presented in Fig. 218.

A more efficient design was developed by I. R. Krichevskii and D. Yu. Gamburg [15] (Fig. 219). The purified gas enters through valve 1 to saturator 2 (high-pressure vessel filled with liquid being investigated), lo-

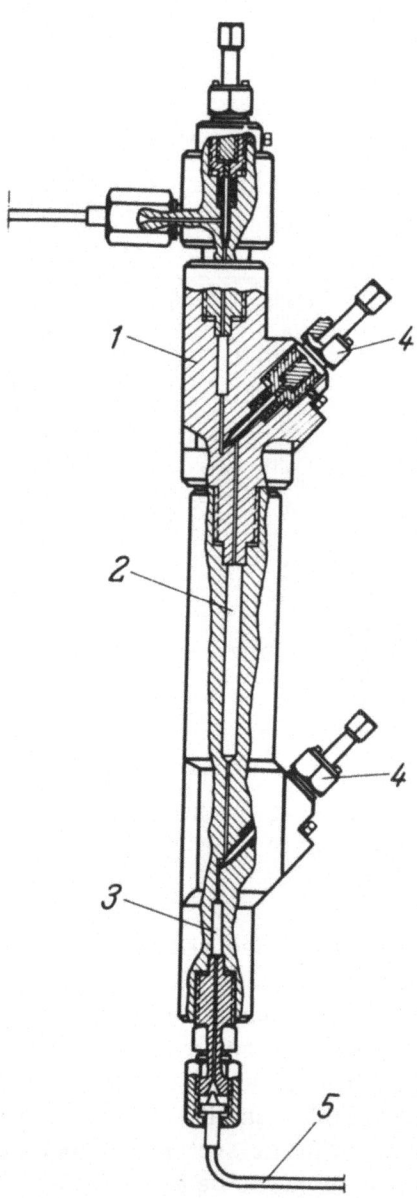

Fig. 220   Ipat'ev high-pressure installation
for studying solubility of gases in liquids and
liquids in gases under pressure: 1) upper portion of the vessel; 2) middle portion of vessel;
3) lower portion of vessel; 4) sealing valves;
5) gas inlet.

cated in the section of the thermostat where the temperature is 10
to 20°C higher than the experimental temperature. This creates conditions suitable for super-
saturation of the gas. The gas is
fed in from the bottom and, trap-
ping the liquid, enters a coil
wrapped around the output 2.
This provides the best possible
saturation of the gas by the liquid and of the liquid by the gas.
The gas passes through two more
saturators 3, maintained at the
experimental temperature, from
which it goes to spray trap 4 in
columns with settling areas
which separate the spray from
the gas, then exits through heated
tube 5 to analysis, also performed under pressure.

Determination of the com-
position of both phases can be
combined in one installation.
This type of installation includes
the installation devised by V. V.
Ipat'ev et al. [16] (Fig. 220),
which was used to study the solubility of hydrogen in benzene at
pressures up to 3000 atm. The
vessel for establishment of
equilibrium consists of three
parts 1, 2, and 3, separated by
valves 4. The gas, passing from
the input 5 through the liquid
filling the lower portion of the
vessel 3 and the middle portion
2, saturates and dissolves the
liquid. After equilibrium is
established, a gas sample is removed, then the liquid phase is
extracted. In order to determine the density of the liquid

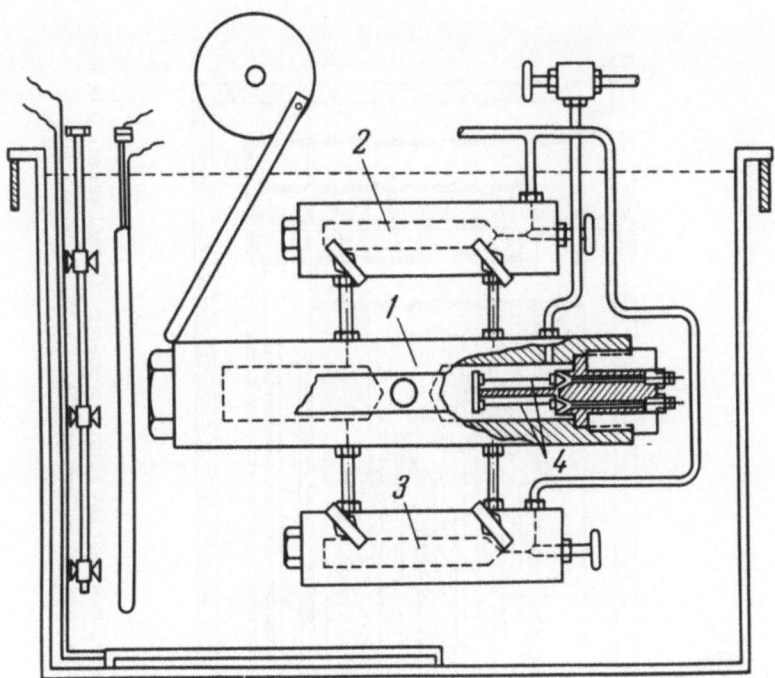

Fig. 221. Rocking autoclave for determination of solubility of gases in liquids
and liquids in gases: 1) main column; 2) gas piezometer; 3) liquid piezometer; 4) platinum wires.

in gas, valve 4 is closed and the samples are separated from each
other.

In studying the solubility of easily soluble gases, it is expedient, as
we noted above, to combine the dynamic method with mechanical agitation.
The most convenient type of agitation in this case is electromagnetic stirring. This type of agitation was used in studying the solubility of carbon dioxide in benzene[17]. The carbon dioxide is passed through several benzene
saturators, after which the gas saturated with the benzene enters a vessel
for establishment of equilibrium, supplied with electromagnetic stirring of
an ordinary type. This method allows complete and rapid attainment of
equilibrium.

Dynamic methods used at the present time for establishment of
equilibrium in a liquid medium differ from each other only in details.

Static methods differ mainly in the method of mixing the gas with the
liquid (mixing with mechanical stirrer, mixing by rocking autoclave, mixing by electromagnetic stirrer, etc.).

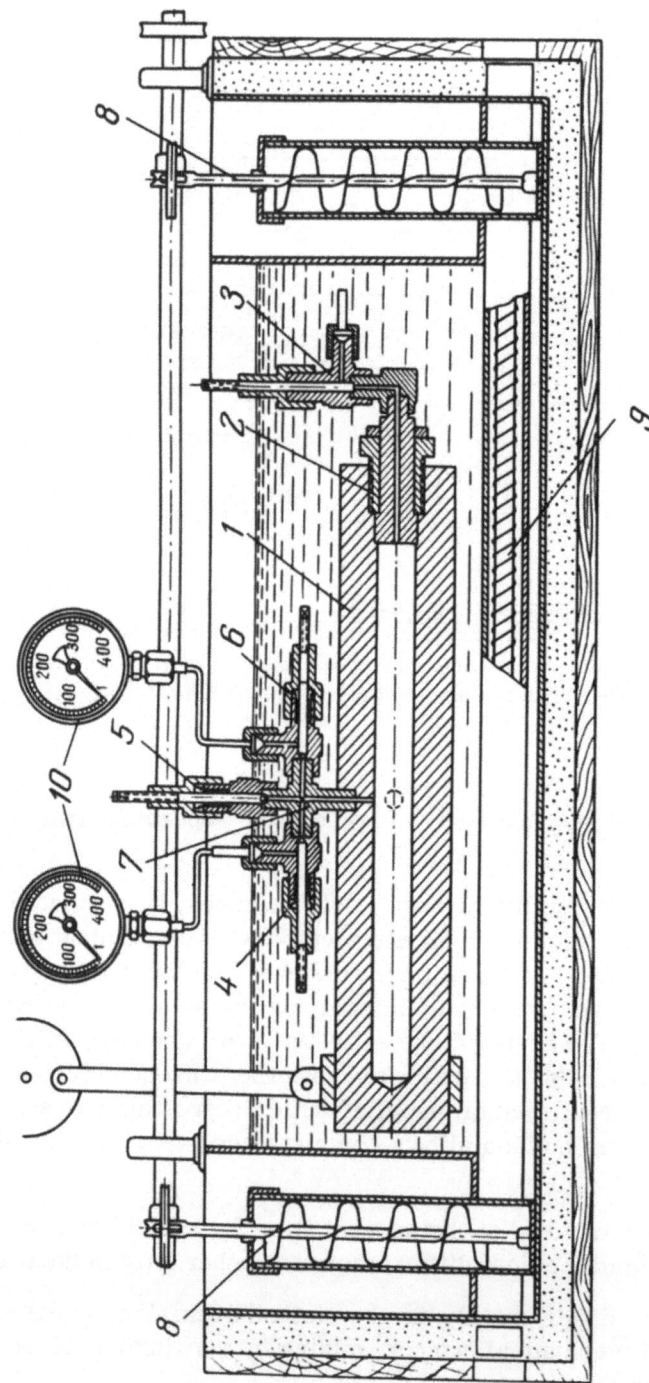

Fig. 222. Bol'shakov-Linshits installation for setting phase equilibria between liquid and gas at high-pressures: 1) steel vessel; 2) head; 3, 4, 5, 6) valves; 7) cross; 8) screw stirrer; 9) heater; 10) pressure gauges.

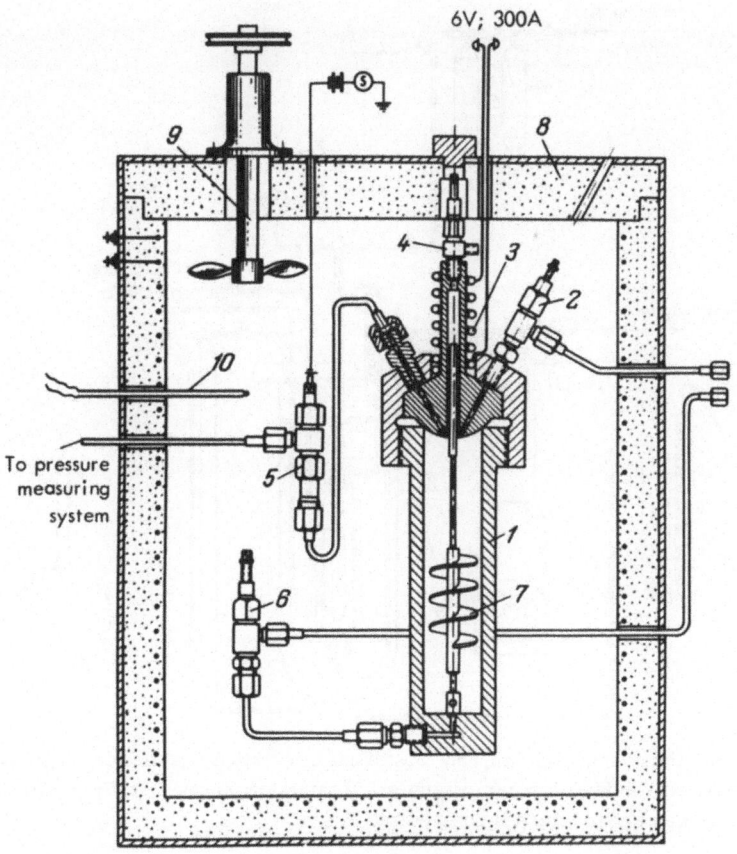

Fig. 223. Tsiklis installation for determination of liquid-gas equilibria: 1) high-pressure vessel; 2, 6) membrane valves for removal of gas and liquid phases; 3) solenoid; 4) loading valve; 5) membrane null instrument; 7, 9) stirrers; 8) thermostat; 10) thermocouple.

Mechanical stirrers have a number of deficiencies: leakage through glands is possible; it is difficult to achieve high revolution rates, etc. The authors of works [4, 18], working with such apparatus, produced clearly incorrect data, due to poor mixing.

Some investigators have used rocking autoclaves [19, 20]. We will describe one such installation [20], which is interesting in that both phases can be investigated. The autoclave (Fig. 221) consists of main column 1, placed horizontally, connected to two vessels — piezometers 2 and 3. The column consists of two vessels separated by a massive barrier. During rocking, the liquid pours from one portion of the column through the lower piezometer to the other half, while the gas passes through the upper piezometer from one half of the vessel to the other. The column is rocked in

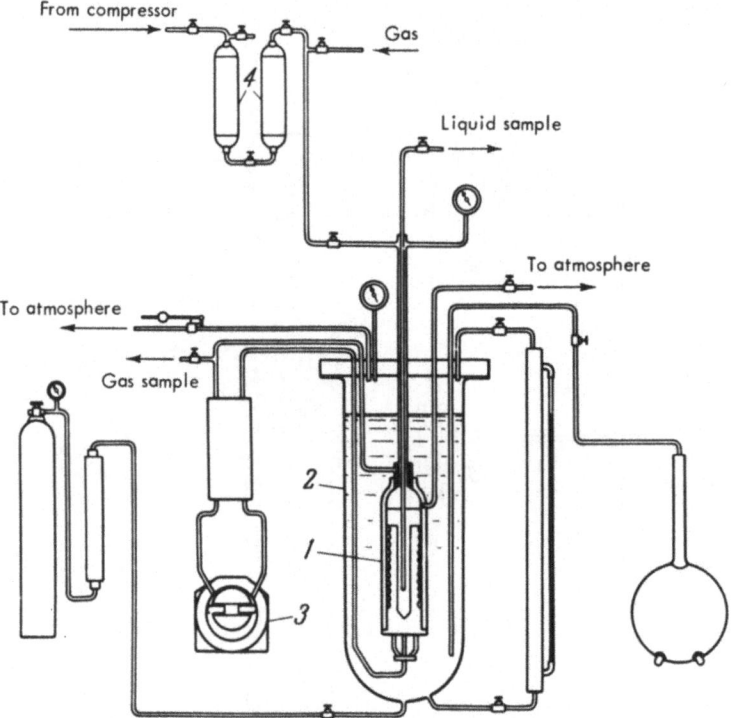

Fig. 224. Gonikberg, Gurvich, and Fastovskii circulation installation for studying phase equilibria at low temperatures: 1) saturator; 2) cryostat; 3) circulation
pump; 4) water-glycerine gas compressor.

such a way that it remains at its furthest tilted positions for 0.5 minute
in order to allow the liquid to flow from one half of the column into the other.

If the primary parameters (pressure, temperature) in the process being investigated are near the critical parameters (when the densities of the
corresponding phases become almost identical) the lower piezometer includes a piston which, moving during rocking, forces the liquid back and
forth through the system. In order to measure the level in the column,
the plug closure of the column contains an indicator operating on the principle of changing resistance of platinum wire due to the different heat conductivity of the liquid and gas phases. The indicators are two platinum
wires, acting as arms of a bridge placed within the column and connected
with electrical leads. The wire is heated by a current of 0.06 a. In this state,
the bridge is balanced using a sensitive null instrument. As soon as one of
the wires touches the liquid, it is cooled, its resistance changes, and the
bridge is unbalanced.

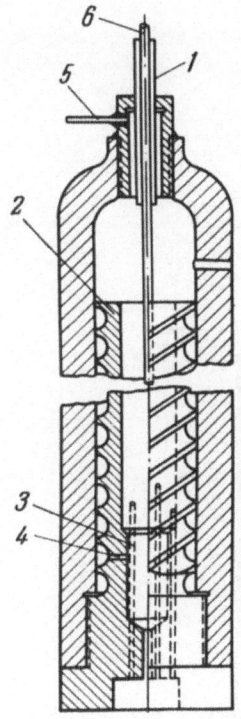

Fig. 225. Saturator: 1) high-pressure tube; 2) hollow rod; 3) tube from circulation pump; 4) aperture; 5, 6) sample removal tubes.

This installation can be used to analyze both liquid and gas phases, and also to produce data on the solubility of the liquid in the compressed gas. The piezometers also allow determination of the density of the corresponding phases, which is very important for subsequent calculations.

A simple and convenient installation (Fig. 222) has been used by Bol'shakov and Linshits [21]. This installation consists of a steel vessel 1 closed with head 2. Valve 3 for removal of the liquid phase is screwed into the head. At the other end of the vessel there is a collar connected to an eccentric driven by a reducing gear. Cross 7 carries valve 5 for removing samples of the gas phase and two pressure gauges 10 for approximate and precise measurement of the pressure in the vessel. Valves 4 and 6 can be used to disconnect the pressure gauges and exhaust the gas from them in order to protect the curved hollow springs of the gauges from deformation. Vessel 1 is submerged in the liquid filling the thermostat. The thermostat carries stirrer 8 and heater 9. As the eccentric drive turns, vessel 1 rocks in the thermostat (about the transverse horizontal axis), causing mixing of the phases, which may be intensified by placing a ball within the vessel.

An installation with electromagnetic stirring designed by the author [22] is shown in Fig. 223. High-pressure vessel 1, closed with a packingless spherical cap (see also Fig. 141) is supplied with two membrane valves 2 and 6 for removal of gas and liquid phase samples. Phase mixing is achieved by stirrer 7, whose shaft is drawn up into the head of the cap as current is passed through solenoid 3. Valve 4 is used to fill the vessel, the pressure in which is measured by a manometer using the membrane null instrument 5. The entire installation is placed in air thermostat 8.

The liquid phase is poured into the vessel through a conduit containing valve 4, then the vessel is closed, the liquid is frozen, the air is pumped out of the vessel, and the gas being investigated is admitted through valve 4. Then, the vessel is placed in the air thermostat, the required temperature is established, the phases are mixed, the pressure is measured using null instrument 5 (as described in Chapter IV) and the liquid and gas phases are sampled for analysis.

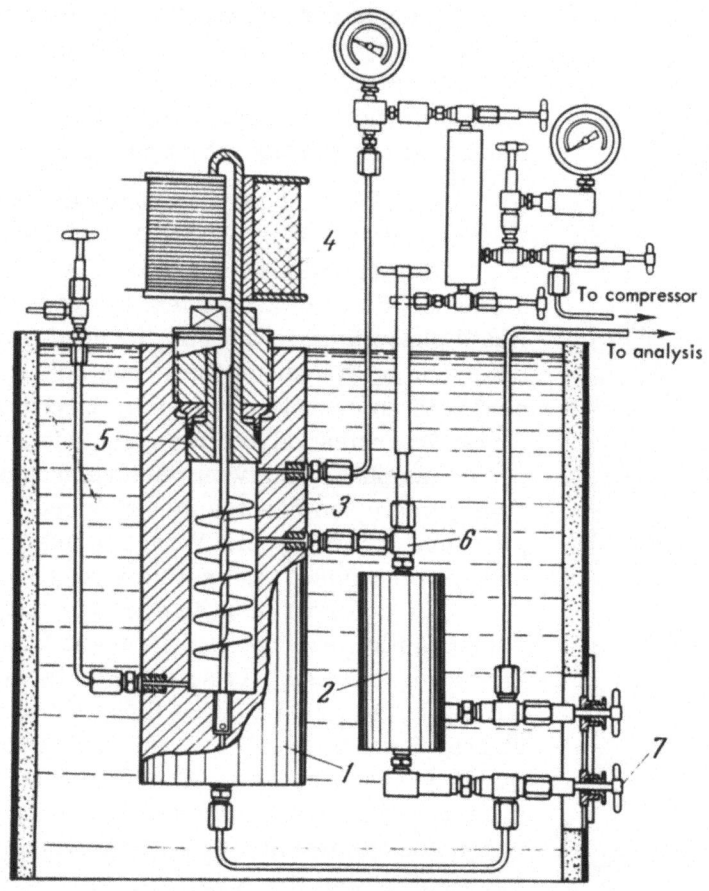

Fig 226. Krishevskii-Efremova circulation installation with piezometer
for removal of sample: 1) steel vessel; 2) piezometer; 3) stirrer; 4)
solenoid; 5) plug closure; 6, 7) valves.

Circulation methods, as was noted above, are convenient in those cases
when the investigator has a small quantity of gas. The liquid phase may
also be circulated. A Gonikberg, Fastovskii, and Gurvich installation [23]
for determination of the solubility of hydrogen in liquid nitrogen will be described below (Fig. 224). The vessel 1 (saturator) in which equilibrium is
established is placed in cryostat 2.

Circulation is achieved by pump 3, the design of which was described
above (see p. 286).

When gas must be compressed to a pressure of over 150 atm, a pressure above which an ordinary gas-pressure cylinder can no longer be used,
water-glycerine compressor 4 is used. Analysis of samples collected over

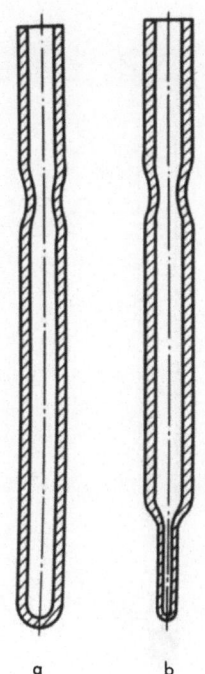

a   b

Fig. 227. Ampules
with constriction.

glycerine is performed using a molecular gas balance. The liquid in the cryostat is agitated by compressed gas.

The saturator (Fig. 225) is a bronze cylinder, connected to high-pressure line 1. Hollow rod 2 is screwed into the cylinder. The outside of this rod is triple threaded; each turn of the thread carries tube 3 from the circulation pump, supplying the gas phase to the vessel. Each turn is therefore a coil, continuously receiving liquid from the internal portion of the vessel through aperture 4 plus the gas from the supply line. The gas bubbles entrain the liquid, carry it upward, and thereby assure intense agitation. Tubes 5 and 6 are used for removal of liquid and gas phase samples and for feeding the gas to the circulation pump. Thus, circulation of both liquid and gas phases is attained.

Circulation of the liquid is provided in the Krichevskii – Efremova installation [24]. Steel vessel 1 (Fig. 226) of about two liters capacity is connected with 25 ml piezometer 2. The vessel is filled with liquid until its level is above the top of the piezometer. Electromagnetic stirrer 3 moves within the vessel, moved by solenoid 4, fitted over plug 5, made of stainless, nonmagnetic steel. The stirrer has two purposes: it agitates the contents of the vessel and pumps the liquid from the vessel to the piezometer. The liquid is pumped by the container at the end of the stirrer, which acts as a piston. The stroke of the stirrer is 5 cm; in one stroke, it can pump approximately 10 cm^3 of solution.

The lines connecting the piezometer with the vessel includes valves 6 and 7, which are used to disconnect the piezometer from the vessel after equilibrium is established (the purpose of this operation will be explained in describing the analytic portion of the installation).

The sequence of operations using this installation is as follows: first, the liquid is poured into the installation and the stirring action is started; then, the vessel is filled with the gas to be investigated at the required pressure, after which the liquid is circulated until equilibrium is established.

Synthetic Methods – Method of Sealed Ampules.

As was noted above, when synthetic methods are used, the investigator must be able to note the moment when the two-phase system becomes a one-phase system. Naturally, the best method is visual observation. How-

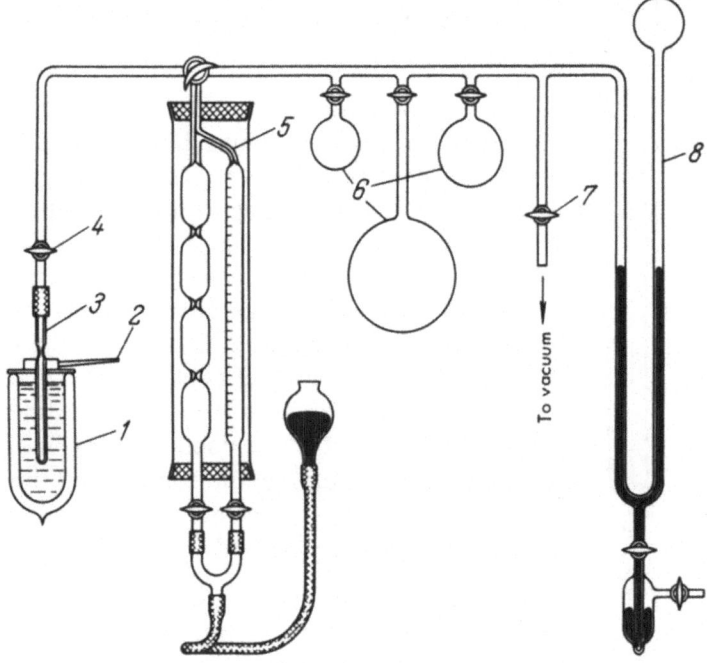

Fig. 228. Metering device: 1) Dewar flask; 2) retainer; 3) ampule;
4, 7) stopcock; 5) burette; 6) flask; 8) manometer.

ever, this limits the investigator as to his selection of media, pressures, and temperatures which can be contained by a transparent material.

One very convenient method for investigating phase equilibrium (and volume relations) at pressures on the order of 100-200 atm and temperatures up to 300°C is the method of sealed glass ampules [25]. An ampule is a thick-walled glass capillary sealed at one end. A known quantity of components is placed in the ampule, the other end of the capillary is sealed, the ampule is placed in a thermostat and the temperature at which one of the phases disappears is noted.

The success of application of this method depends on the accuracy with which the metering of components is achieved; therefore, let us discuss this operation in more detail.

The required quantity of liquid (or solid) is placed in a weighed ampule with a constriction (Fig. 227), after which the ampule is weighed again. The internal diameter and wall thickness of the capillary used to make the ampule are selected in view of the pressure which may develop in the ampule after it is heated to the experimental temperature. Then, the ampule is connected to the metering device (Fig. 228). This installation consists of burette 5, several calibrated flasks 6 and manometer 8. The gas to be investigated is placed in the calibrated flask. The gas can be metered directly from the flask or from the burette. In either case, ampule 3 is connected to vacuum stopcock 4, the contents of the ampule are frozen with liquid nitrogen, and air is evacuated from the system. In order to feed gas from a flask into the ampule, the tube leading to stopcock 4 is filled with the gas. Pressure sufficient to produce the mixture desired when the gas is transferred from the tube to the ampule is then created in the system. This pressure can be calculated if the capacity of the tubes and the variation in tube capacity depending on the change in position of mercury in the manometer column connected with the tubes are known. When the metering of the components is completed and the gas has been introduced to the ampule, the ampule is sealed off at the constriction with a thin flame from a glass-blowing jet. Then, retainer 2 raises the ampule slightly out of the liquid nitrogen, as heat continues to be applied to its upper end. This causes a portion of the gas condensed in the ampule to be evaporated and blows a sphere at the end of the ampule, as a result of which the ampule can withstand the maximum possible pressure for the given type of glass.

Naturally, the sphere can be blown on the end of the ampule only if the components of the mixture being investigated are not decomposed upon touching the hot glass. The procedure for sealing off the ampule and forming the sphere requires considerable dexterity from the experimenter, which can be attained only by practice.

After the ampule is sealed, it is left in the liquid nitrogen until the hot end cools. Then, the ampule is placed in a closed metal container, where its temperature is reduced to room temperature. Before beginning work with ampules, the experimenter should don a glass safety mask. If the ampule is not broken after heating, it is weighed together with the part separated at stopcock 4, in order to determine the amount of gas which was placed in the ampule. If very small quantities of material are sealed off, the quantity of air located in the ampule before it was evacuated must be taken into consideration.

In selecting quantities of materials to be sealed into ampules, an attempt is made to produce a series of ampules with mixtures of identical composition but different densities. It should be kept in mind that the pressure in the ampule is limited by the strength of the material of which the ampule is made.

The filled ampule is placed in a thermostat or furnace with viewing windows and the temperature at which the system becomes a single phase is determined. In order to speed up the attainment of equilibrium, an agitator may be placed in the ampule before it is sealed (a glass agitator if the ampule is rotated while in the thermostat, a metal agitator covered with

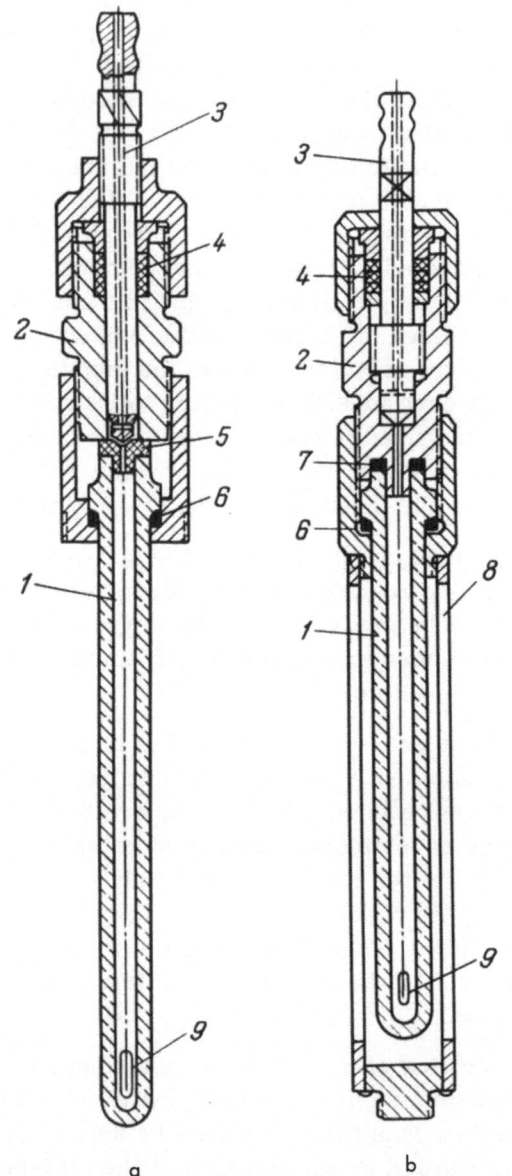

Fig. 229. Ampules with valves: a) ampule designed
by Lebedeva and Khodeeva; b) ampule designed by
Tsiklis; 1) glass ampules; 2) valve; 3) pin; 4) gland;
5) fluoroplast disc; 6, 7) fluoroplast packing; 8) cover;
9) agitator.

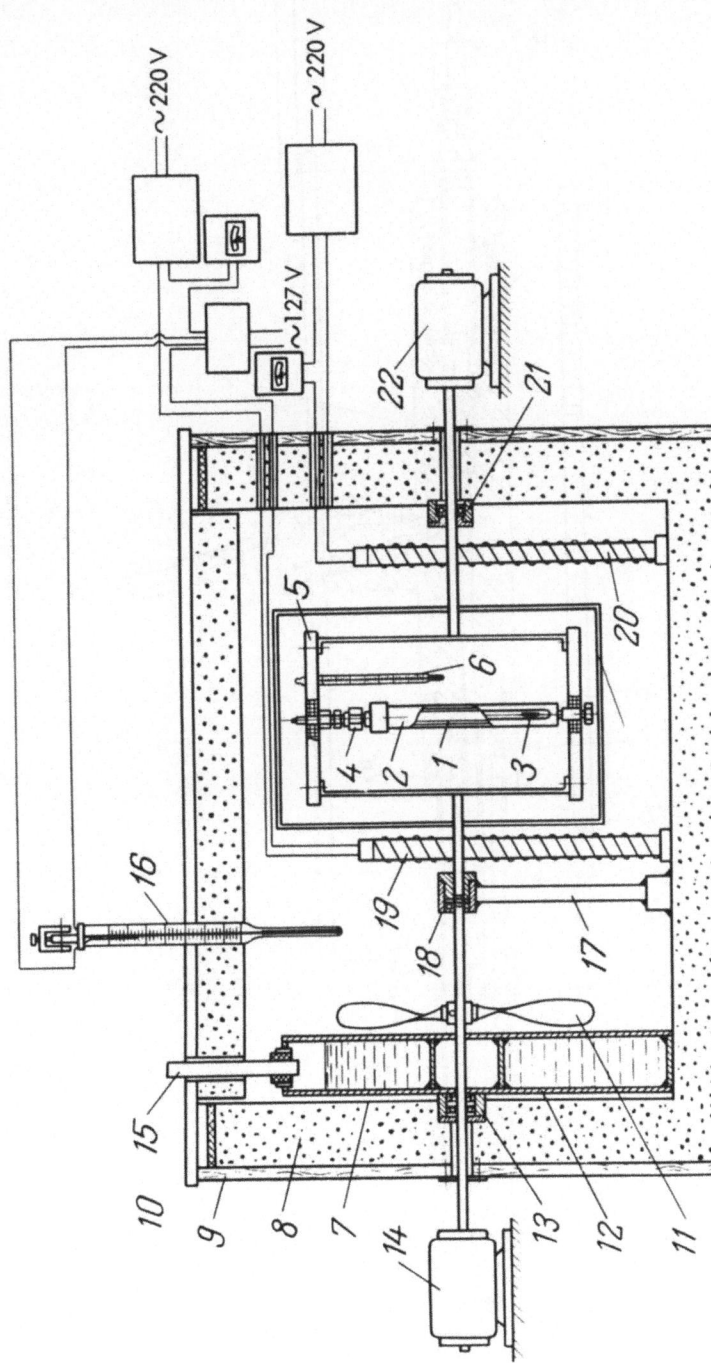

Fig. 230. Thermostat for handling of ampules: 1) ampule; 2) cover; 3) mixer; 4) valve; 5) frame; 6) thermometer; 7) thermostat cover; 8) insulation; 9) frame; 10) cover; 11) fan; 12) tank for cooling fluids; 13) shaft gland; 14, 22) motors; 15) tube; 16) contact thermometer; 17) brace; 18, 21) bearings; 19, 20) heaters.

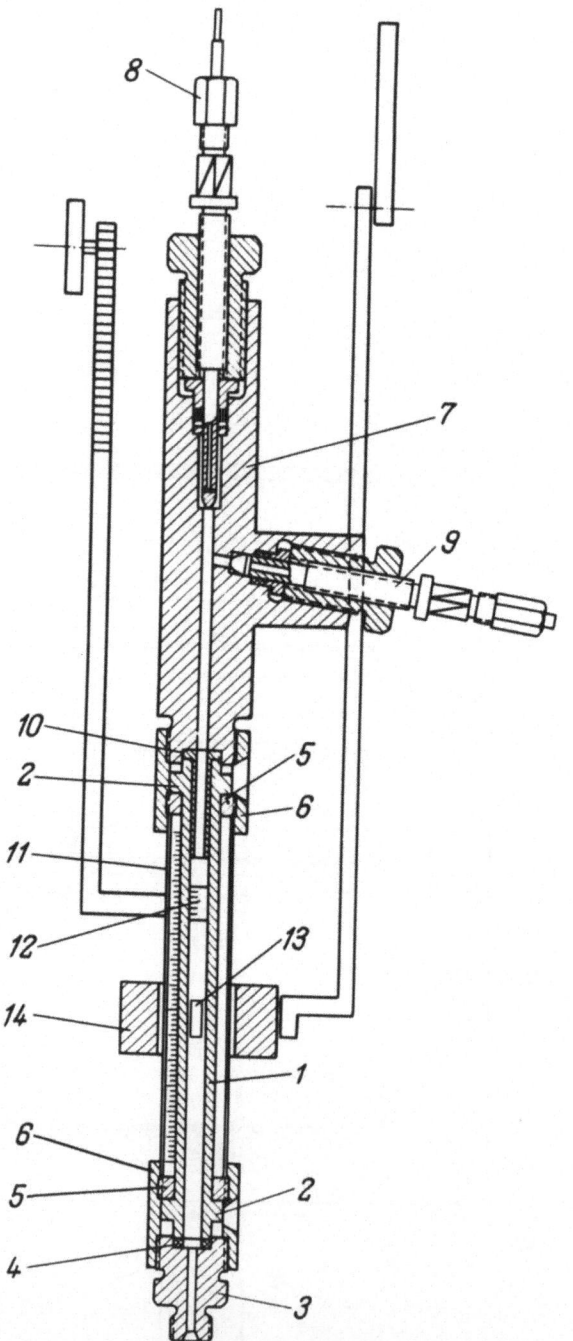

Fig. 231. Tube for investigation of phase transformations by
visual observation: 1) glass tube; 2) flanges; 3) transfer nut;
4) packing; 5) split nut; 6) nut; 7) head; 8, 9) valves; 10)
tube (insert); 11) sheath; 12) nonius; 13) bar; 14) magnet.

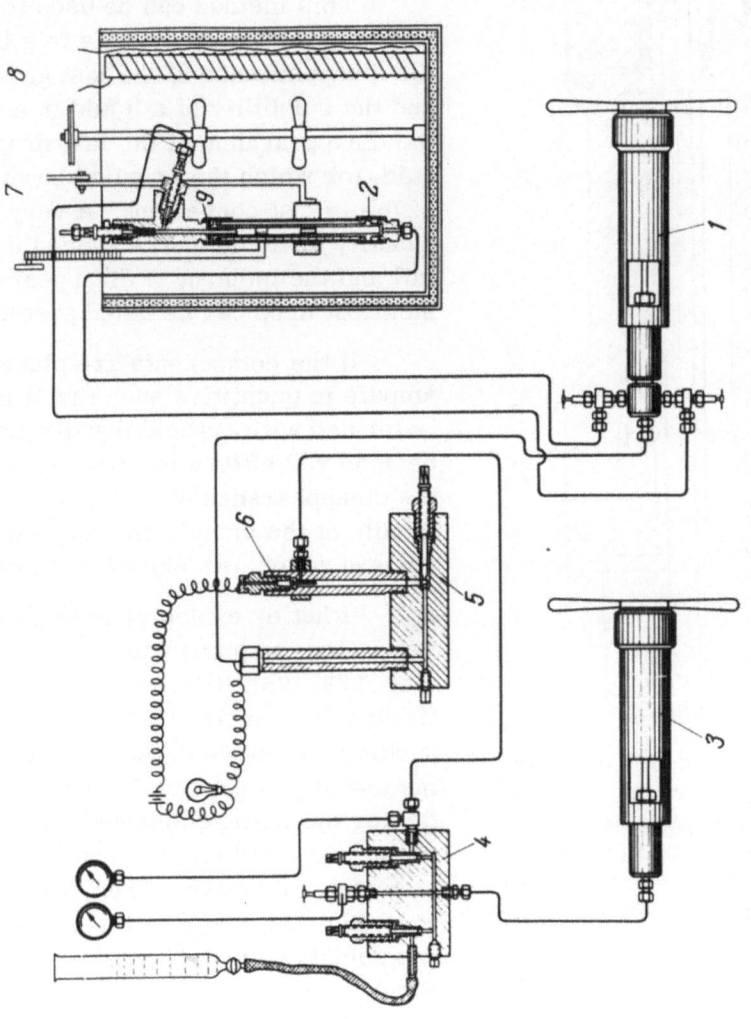

Fig. 232. Plan of installation for investigation of phase equilibria by visual observation: 1) mercury press; 2) tube; 3) oil press; 4) manifold; 5) mercury equilibrator; 6) contact; 7, 8) valves; 9) insert.

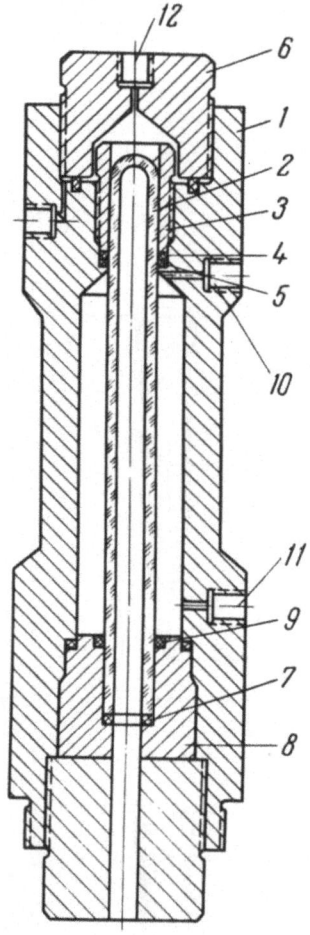

Fig. 233. Apparatus with visual observation for investigation of solubility of gas in liquid: 1) steel vessel; 2) glass tube; 3, 6) nuts; 4, 5, 9) packing; 7) copper packing; 8) ring; 10, 11, 12) connecting pipe taps.

glass if the agitation is to be performed using a permanent magnet). After the experiment is performed, the ampule is frozen, precisely cut, and the capacity of both resulting portions is determined by mercury calibration.

This method can be used to determine the solubility of a gas in a liquid by the disappearance of the last gas bubble, and the solubility of a liquid in a gas by the disappearance of the last drop of liquid, for which the ampule shown in Fig. 227b is most convenient. A very small quantity of liquid can be placed in the small tail and the moment of disappearance of the smallest drop can be noted precisely.

If the components are placed in the ampule in quantities such that a mixture is formed with critical density (this can be achieved after a few trial experiments), the disappearance of the meniscus in the middle of the ampule and the critical phenomena which accompany this are observed.

If hot or explosive mixtures are investigated, ampules with a valve are used (Fig. 229) [26]. Glass ampule 1 with a flange* is connected through fluoroplast packing 5 to valve 2, pin 3 of which closes the opening in packing 5. The ampule containing the liquid component is connected to the valve and pin 3 is connected to the metering apparatus. Then, the second component is placed in the ampule as was described above and the valve pin is closed. Figure 229b shows another type of ampule which differs in that the valve pin closes off a hole in metal; this allows operation without frequent replacement of packing 5.

After the ampule is heated, it is placed in the thermostat (Fig. 230), rotated in a frame for agitation of the phases or placed within a circular

_____

*The design of the flange seal is described on page 189.

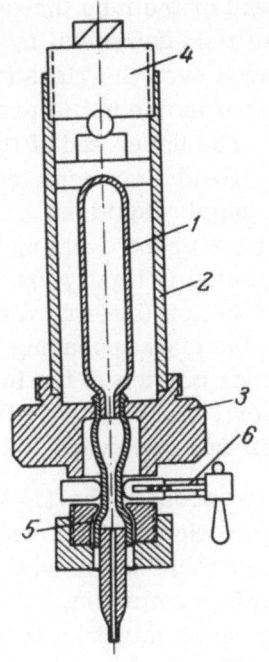

Fig. 234. Device for investigation of phase equilibria under pressure: 1) quartz burette; 2) support; 3) valve body; 4) nut; 5) rubber tube; 6) pressure clamp.

magnet which is used to cause a glass-covered steel agitating ball to move within the ampule.

Placing the ampule with the valve downward, the disappearance of the last gas bubble is observed.

If sealed ampules are used to study phase equilibra in liquid-liquid systems, the following difficulties may be encountered: at the moment when one of the liquid phases disappears, a gas phase remains in the ampule, which may contain a considerable quantity of vapors of the liquid. In this case, neither the volume nor the composition of the liquid phase is known. In order to avoid this difficulty, the following is done: first, observation is performed in the usual manner, to determine the height of the remaining vapor phase. After this, the ampule is cooled and shortened by sealing off the tip in order that upon heating the gas phase will be tiny enough to be ignored.

Using another method, a calibrated tube sealed at one end and containing a known quantity of components is connected at the other end through a mercury seal to the device creating the pressure Using this method, it is possible, by changing the temperature of the system, to measure the pressure at which one of the phases disappears, i.e., to determine the equilibrium pressure corresponding to the given composition of the system.

The necessity of observing the disappearance of the last bubble of gas has forced investigators to seal the upper end of the tube, which has resulted in a number of methodological difficulties in introducing materials to the tube and in connecting the tube to the installation [27].

Installation of Krichevskii and Efremova. These authors [28] used a glass tube open at both ends for investigating phase equilibria in liquid-gas systems. The system being investigated was included between two mercury plugs — an upper and a lower plug.

Flanges 2 are formed at the ends of thick-walled glass tube 1 (Fig. 231).* A steel capillary is fastened to the lower end of the tube through transfer nut 3. A seal between the polished ends of tube and metal is achieved using Teflon packing 4. Split nut 5 is placed over the glass tube; nut 6 is screwed down around it, and nut 3 is screwed inside nut 6, pressing packing 4 to the tube section (see Chapter VI). The upper end of tube 1 is connected to head 7, made of 1Kh18N9T steel. Head 7 contains two valves with drilled pins. Upper valve 8 is used to supply liquid and gas to the apparatus. Side valve 9 is connected to a mercury press creating the pressure, and is also the gas-collecting valve. Tube 10 is tightly inserted into the upper portion of tube 1; tube 10 is made of Teflon (internal diameter 1 mm). When mercury is fed into side valve 9 the drop appearing in the aperture of tube 10 indicates that the entire upper portion of the installation is filled with mercury. If tube 10 is absent, the mercury will not be retained in the upper portion of the tube, but will fall downward.

Brass sheath 11 with two apertures is mounted around tube 1. A scale is placed on the sheath with divisions of 1 mm. Nonius 12 is used to determine the level to an accuracy to 0.1 mm. The nonius has a control scale which moves in the rear aperture of the sheath, eliminating any error [parallax] in reading the level of mercury or liquid which might be caused by improper positioning of the eye during readout. The liquid and gas are agitated by bar 12 made of magnetic steel, moved by permanent magnet 14.

The amplitude of agitation of the magnet can be changed by fastening the rod to which the magnet ring is connected to the pulley wheel in various locations on the radial slot cut in the pulley wheel. The middle point about which the magnet moves can also be altered by raising or lowering the motor.

The plan of the entire installation for investigating phase equilibria by visual observation is shown in Fig. 232.

Mercury press 1 generates pressure in tube 2. The pressure on the standard and rough manometer is created by oil press 3, which is connected to mercury press 1 through manifold 4 and mercury equilibrator 5. The mercury equilibrator has two columns filled with mercury. The upper portion of the left column is connected to the mercury press, the upper portion of the right column being filled with oil and connected to press 3. The mercury level is checked by contact 6 and a lamp. The pressure in the installation is created by feeding mercury and oil alternately as the lamp is observed.

Before an experiment is begun, side collecting valve 8 is filled with mercury, using the mercury press, until the mercury can be seen through

---

* This installation was designed by the author.

the aperture of insert 9. Then, valve 8 is closed and the glass tube is filled with gas through the upper valve. Then, the gas volume is measured (from the level of the mercury in the tube) and its pressure is determined. Then, valve 8 is opened and the gas charge is inserted in it, feeding mercury in from the press through the lower capillary. Valve 8 is closed once more, the pressure in the tube is increased to the initial experimental pressure, and the mercury level is measured. The difference in levels, pressures, and temperature are used to determine the quantity of gas in the collecting valve. Then, the gas residue is evacuated from the tube, the upper valve is opened and a medical syringe is used to add a measured quantity of liquid through the hole in the insert. The valve is replaced and the air is removed from the installation by raising the liquid to the level of tube 2. The gas from valve 8 is transferred to the tube, the agitation is begun and, when equilibrium is established, measurement is performed.

This installation (see Fig. 232) can be used to determine the moment when the last gas bubble or liquid drop disappears, as well as to study the solubility of liquid in liquid, the compressibility of liquid and gas systems and critical phenomena in liquid-gas and liquid-liquid systems.

The range of pressures which can be used depends on the diameter and quality of the glass tubing (see Chapter I). A similar device of another design is shown in Fig. 233 [29].

Steel vessel 1 contains calibrated tube 2 made of pyrex glass, 360 mm long and 17/8 mm in diameter. The upper, sealed end of the tube is included in a separate chamber and compressed by two packing rings: rubber ring 4 and gold ring 5, compressed by nut 3. Nut 6 seals the upper chamber. The lower, open end of the tube is carefully polished and placed against copper packing 7. The lower chamber of the vessel has a capacity of 125 ml and is closed with ring 8 and Teflon packing 9.

The material to be investigated is introduced through connecting pipe tap 10 to the lower chamber. Mercury is introduced to the vessel through pipe taps 11 and 12, increasing the pressure in both chambers simultaneously. The tube structure permits the moment of disappearance of the last gas bubble in the upper end of the lower chamber to be observed.

After a measured quantity of liquid and gas is introduced to the chamber, the contents are agitated by pumping mercury into the vessel and are observed through a periscope placed within the glass tube to note the disappearance of one of the phases as the pressure is gradually increased.

The mercury in the upper chamber prevents leakage of gas through packings 4 and 5, packing 5 preventing adsorption of the gas by packing 4.

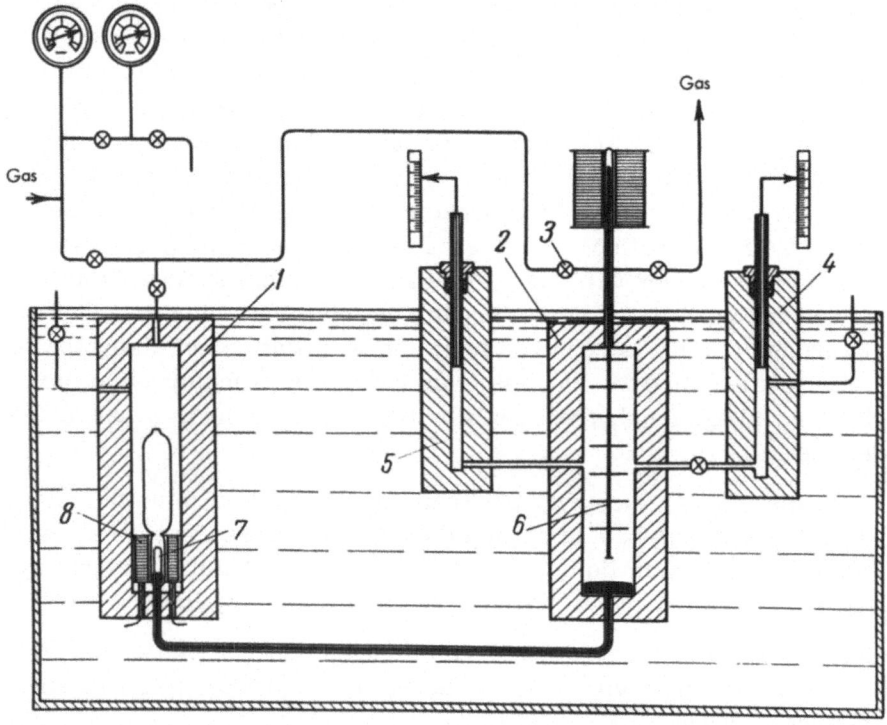

Fig. 235. Gamburg installation for determination of volume relations in solutions of liquids in compressed gases: 1, 2) high-pressure vessels; 3) valve; 4) meterer; 5) volumometer; 6) electromagnetic stirrer; 7) core; 8) coil.

Tube 2 is subjected to external pressure, which also acts on its sealed end. Thanks to its positioning, the tube can withstand rather high pressures.

A comparatively simple device allowing observation of the behavior of phases is shown in Fig. 234 [30]. Burette 1, made of quartz and carefully heat treated to remove residual stresses, is placed in support 2 and clamped between the body of valve 3 and nut 4. The sealing device is rubber tube 5 with sealing clamp 6. The apparatus (with a tubing diameter of 22-24 mm, wall thickness 4 mm) can withstand pressures up to 80 atm at temperatures up to 120°C.

The moment of phase conversion can be determined not only visually, but also by any other method allowing establishment of the dependence of a change in any property of the system on the change in temperature, pressure, volume, etc. For example, a measured quantity of liquid and gas may be introduced into the system, and the pressure may be increased without changing the quantity of material in the system, i.e., the volume of the system may be reduced. By noting the dependence of any of the properties of the system or of any phase in the system (volume, density, refractive index, etc.) on pressure, it is easy to note that this dependence will be a smooth curve until the number of phases existing in equilibrium changes. At this moment, a break appears on the curve of pressure vs. any given property. The pressure corresponding to this break in the curve as the gas dissolves in the liquid will be the pressure at which the last bubble of gas disappears. Similar results can be produced if the pressure is increased by introducing additional, precisely measured quantities of liquid or gas to the system. If the initial composition of the gas phase is known, it is possible, while increasing the pressure in the system, to determine the moment when the first drop of liquid is separated, i.e., the dew point can be determined.

If the quantities of the liquid and gas phases and their densities are known, the phase compositions can be determined by solving the following equations:

$$m'_l + m''_l = m_l, \qquad m'_g + m''_g = m_g$$
$$\frac{m'_l + m'_g}{v'} = d', \qquad \frac{m''_l + m''_g}{v''} = d''$$

where m'_l and m''_l are the number of moles of liquid in the gas and liquid phases; m'_g and m''_g are the number of moles of gas in gas and liquid phases; v' and v'', d' and d'' are the volumes and densities of gas and liquid phases, respectively.

Synthetic methods using these approaches have become widespread in the study of liquid-gas equilibria in oil-gas systems as well [31-34].

The Gamburg Installation. It was stated above that the dependence of a change in volume of a system at constant pressure and temperature on the quantity of liquid introduced to the system can be studied. An installation using this method of investigation has been created by Gamburg [35]. This installation consists of three parts: a system for compressing and purifying the gas, the experimental installation itself, and a control panel. The installation is designed and operated at 5000 atm, and therefore has two compressors, a 1000 atm compressor and a subsequent 5000 atm stage (see Chapter III). After compression, the gas is purified of oil and placed in the experimental installation (Fig. 235). The installation consists of two vessels 1 and 2 submerged in a thermostat and connected by a capillary filled with mercury. Both vessels are simultaneously filled with gas. Since the pressures in the vessels are identical, the mercury level does not change. If valve 3 is then closed, i.e., if the system of vessels is made into a mercury differential manometer, then the introduction of liquid from vessel 4 (the meterer) to primary vessel 2

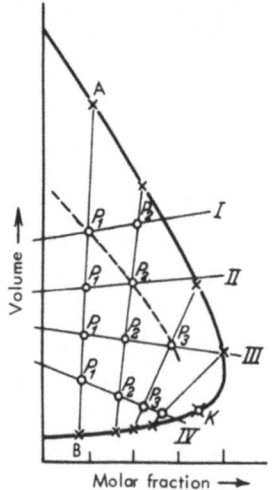

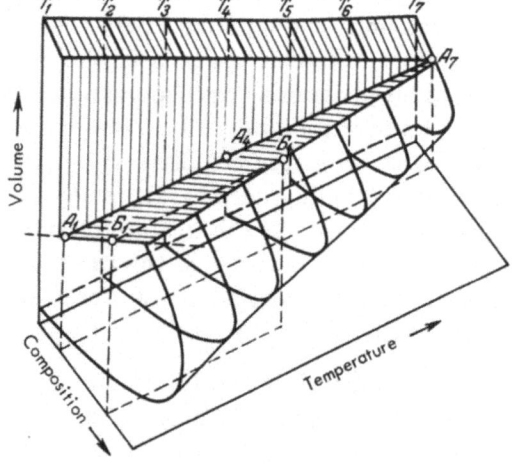

Fig. 236. Equilibrium of liquid
and gas in coordinates volume vs. composition.

Fig. 237. Equilibrium of liquid and gas in coordinates
volume vs. composition vs. temperature.

causes a change in the mercury level. By changing the volume (by raising
the piston) in vessel 5 (the volumometer), the pressure in vessel 2 can be
reduced and float (core) 7 can be returned to its earlier position. If we
know the quantity of liquid fed into vessel 2 from the reading of the meterer,
and the change in volume of the system after this quantity of liquid is dissolved (from indication of volumometer 5), we can perform all necessary
calculations. In order to hasten the onset of equilibrium, electromagnetic
stirrer 6 is placed in vessel 2.

The most important portion of the installation is the mercury differential manometer. The level of mercury in its columns is determined
from the position of the core which floats on the mercury. As the position
of core 7, consisting of a light, hollow float of pure iron, changes, the in-
ductive reactance of coil 8 also changes, unbalancing a bridge circuit, and
causing the arrow of a galvanometer to be deflected. The arrow is returned
to its initial position by raising the piston of volumometer 5; the change in
position of the piston is used to measure the change in volume of the system.

Cylinders 4 and 5, also designed for 5000 atm, are made of heat-
treated alloy steel, as are all other parts of the system. Each cylinder
has a bore which contains a piston, riding in the gland at its end. The
piston is moved by a reducing gear system, and its movement is measured by a height gauge. Since the volume of the piston is precisely known,
it is easy to produce exact data on the amount of liquid fed into the vessel.

This manometer can measure pressure drops to 0.025 mm Hg with total pressures varying from atmospheric to 5000 atm. This corresponds to measurement of a change in volumometer volume of ±0.005 ml.

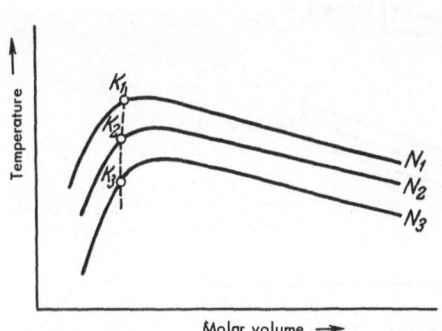

Fig. 238. Graph in coordinates molar volume vs. temperature.

The Polythermic Method. Any experimenter studying phase equilibria is well aware of the complexity and difficulty of investigations of multi-component systems (in comparison with two-component systems). At the present time, when the number of systems to be investigated and their complexity are both increasing constantly, it is extremely important to create investigative methods which allow savings of time and labor. In this respect, the synthetic methods, especially the polythermic method developed by I. R. Krichevskii and G. A. Sorina [36], are very effective. The method is based on the following facts, established by experiment.

1. On a graph drawn in the coordinates volume vs. composition (Fig. 236) the line of nodes connecting the points representing the composition of the corresponding gas (A) and liquid (B) phases are the geometric loci of the points of equal pressures (P_1, P_2, P_3, ...). Therefore, regardless of where points A and B are located on the graph — on the boundary or within this curve in the heterogeneous area — the pressures in mixtures whose composition and volume correspond to these points will be identical.

2. The isothermic boundary curves limiting the composition of heterogeneous mixtures on this graph will not intersect. This means that if a system of a given composition and volume (for example, A_7 on Fig. 237) is a two-phase system at a given temperature T_0, then at any temperature $T < T_0$, it will remain a two-phase system, A_4, A_1.

Investigation of liquid-gas equilibria in a binary system by the polythermic method consists of two parts. The first part of the experiment is performed by the sealed ampule method (or any other method allowing notation of the moment of transition of the system from a two-phase system to a single-phase system). As this is done, the connection between the molar volume of the system and temperature is noted. Several series of glass ampules are prepared for the experiment, each containing solutions of the components to be investigated identical in density but varying in composition; for example, four series of ampules containing solutions with 0.2, 0.4, 0.6, and 0.8 molar portions of one of the components may be

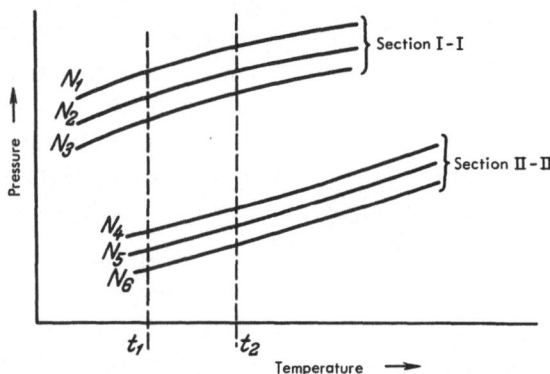

Fig. 239. Change in pressure in system as a function
of temperature.

prepared. Each series consists of four or more ampules containing solutions of identical composition but different density.

As the ampules are placed one after the other in the thermostat [26],
the temperature at which the last gas bubble or liquid drop disappears, or
the temperature at which the meniscus in the middle of the ampule disappears is noted; the results of observation are used to construct a graph in
coordinates molar volume vs. temperature, showing the points for each
series of ampules containing mixtures of identical composition (Fig. 238).
Then, a graph in the coordinates of volume vs. composition is constructed
using the data on volumes and critical points of the system being analyzed,
and the boundary curves for various temperatures are shown on the graph
(see Fig. 236).

The second portion of the investigation consists of determination of
the direction of the lines of nodes, i.e., the lines of constant pressure connecting the points corresponding to the volume and composition of coexisting liquid and gas phases. For this, smooth lines (perhaps straight lines),
so-called rays, or sections (I, II, III) are drawn on the graph (see Fig. 236),
on which certain figurative points (P_1, P_2, P_3, ...) are selected.

Then, the components of the solution are placed in a steel autoclave
supplied with a device for measurement of pressure in the system in a
quantity sufficient to produce the mixture fixed by one of the figurative
points of composition and volume. If this mixture is then heated in the
autoclave, the dependence of pressure on temperature in the heterogeneous
area can be produced. These measurements are performed for a number
of figurative points selected on the section; the results is a selection of
so-called polytherms (Fig. 239) (whence the name of the method).

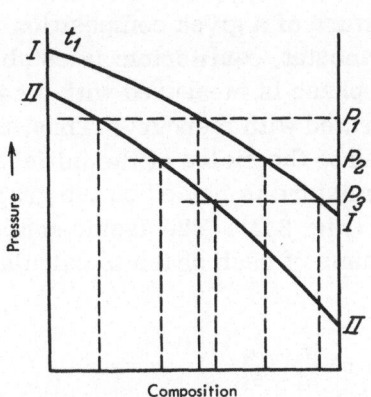

Fig. 240. Graph in coordinates pressure vs. composition: I, II) rays (sections).

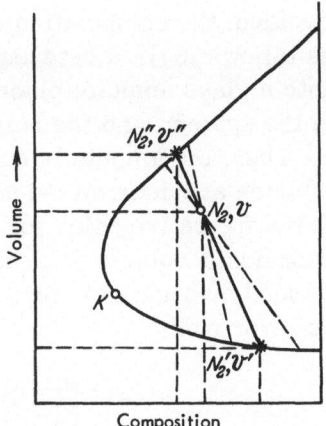

Fig. 241. Method of node rotation.

The results produced are used in constructing graphs in the coordinates pressure-composition at constant temperatures, equal to the temperatures on the volume-composition graph isotherms. This results in the production of smooth curves (Fig. 240) for each section. Then, values of pressure are selected and lines are drawn on Fig. 240 parallel to the abscissa. The intersections of these lines with the curves give values of the compositions of the phases over which the pressures are equal to the value selected. The volumes of these phases can be easily determined from the rays which were drawn above; in this way, points of identical pressures on the volume-composition graph are produced. These points are connected by a straight line, which is continued to intersection with the boundary curves; the points of intersection characterize the compositions of the coexisting phases. It can be seen from Fig. 237 that the volume and composition are identical at any temperature for each figurative point (A_{1-7}, B_{1-4}), as long as liquid-gas equilibrium exists in the system. Therefore, the inclinations of the nodes are also retained at any temperature. Only the pressure changes. This situation allows the new rays (sections) to be connected by studying the experimental dependence of pressure on temperature and, by constructing new dependences of pressure on composition, new points are located on the node lines, etc.

This method allows a considerable reduction in the volume of experimental investigations and allows experimental investigations to be replaced by graphic constructions and calculations. This becomes possible after the necessary minimum experimental data are produced.

In order to measure the pressures corresponding to the node lines, any high-pressure apparatus of known volume [37] supplied with a device for pressure measurement can be used. If the measurement of pressure

can be avoided, the composition of the corresponding phases can be determined as follows [37]: a heterogeneous mixture of a given composition is
sealed into a glass ampule, placed in a thermostat, equilibrium is established in the system, and the height of each phase is measured with a cathetometer. Then, the ampule is cut and calibrated with mercury. Thus, the
phase volumes are determined and converted to the moles of the initial system, and the figurative point for the given mixture is placed on the graph
in the coordinate volume vs. composition (Fig. 241). The assumed node
line is passed through this point and the volume of each phase is calculated
using the lever rule:

$$V' = v' \frac{N_2'' - N_2}{N_2'' - N_2'} \, , \qquad V'' = v'' \frac{N_2 - N_2'}{N_2'' - N_2'}$$

where V' and V" are the volumes of the liquid and gas phases per 1 mole
of the heterogeneous mixture; v' and v" are the molar volumes of the corresponding phases (determined from the coordinates of the node line); N_2,
N_2', and N_2'' are the molar portions of the component in the initial mixture
in the liquid and gas phases (defined from the coordinates of the node line).

If the experimental values of V' and V" do not correspond with the
calculated values, the node is rotated by a few degrees and calculation is
performed again.

SOLID-LIQUID SYSTEMS

Phase equilibria in solid-liquid systems are also studied analytically
and synthetically.

Analytic Methods. Direct measurement of the solubility of a
solid in a liquid has been performed for some time. The history of this
problem can be studied in the literature [38]. Here, as in all the section
above, methods for producing a saturated solution under the experimental
conditions, and for analyzing this solution, will be analyzed. The solution
is saturated at atmospheric pressure [39, 40] and placed in a glass vessel,
where the solid component is placed in a special container made of platinum
screen. The vessel is placed within the coil of an electromagnet so that the
core of the electromagnet moves the special platinum-screen container
within the solution as it [the electromagnet core] moves. As the container
is moved, the solid component is dissolved and the solution is agitated.

After equilibrium is established, the pressure is reduced, the apparatus is opened and the solution is analyzed. These operations should
be performed rapidly in order that the composition of the solution saturated under pressure will not change. Since the pressure-transmitting
liquid used is oil, the solution is generally covered with a layer of oil and

special pipettes must be used to remove a sample for analysis, allowing the solution to be filtered.

In some designs for this type
of apparatus [41] the solution and
oil are separated with mercury. This
is achieved using two connected vessels; the capillary connecting the
vessels is filled with mercury. The
pressure-transmitting liquid used
is oil placed in one vessel. The solution and solid component are placed
in the other vessel. The mercury
serves as a sealing liquid; the mercury also agitates the solution as the
vessel is rocked.

An apparatus of improved design [42] is shown in Fig. 242. It
consists of a bent, shaped steel tube.
One column of the apparatus contains special pipette 1, the other contains rod 2, designed to reduce the
amount of oil which must be fed into
the vessel. The apparatus is located
in a thermostat and is rocked back
and forth.

A pipette consists of two parts.
The upper portion is filled with benzene, the lower portion with an aqueous solution of the material being
investigated, covered with benzene.
The lower, capillary end of the

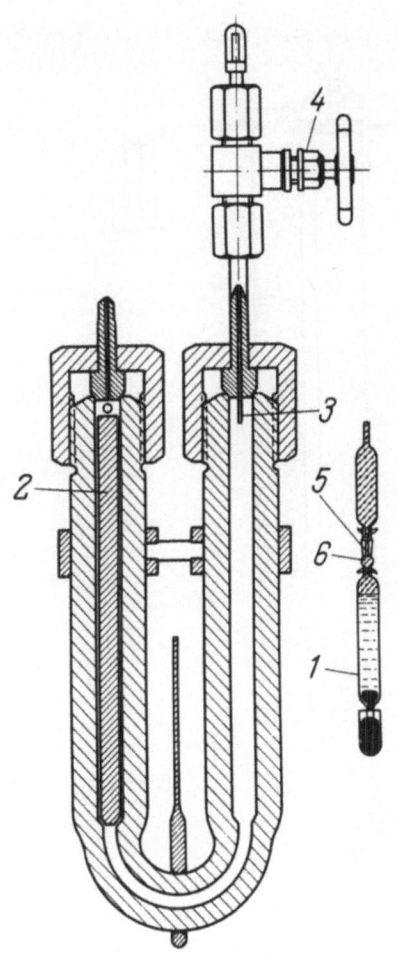

Fig. 242. Apparatus for determination of
solubility of solid in liquid: 1) pipette;
2) rod; 3) capillary; 4) valve; 5, 6) filters.

pipette is submerged in the mercury. The pipette assembled in this manner is connected to capillary 3 and placed in the column of the apparatus.
As the pressure increases, the oil forces the mercury into the glass pipette,
which is under biaxial pressure. As the apparatus rocks, the mercury, solution, and solid component placed in the pipette are agitated.

After the experiment is ended, valve 4 is opened and the benzine is
extracted from the upper portion of the pipette under constant pressure in
the apparatus. This causes the solution from the lower portion of the
pipette to move into the upper portion, after being transmitted through filters
5 and 6. Then the pressure is rapidly reduced and the pipette is removed.

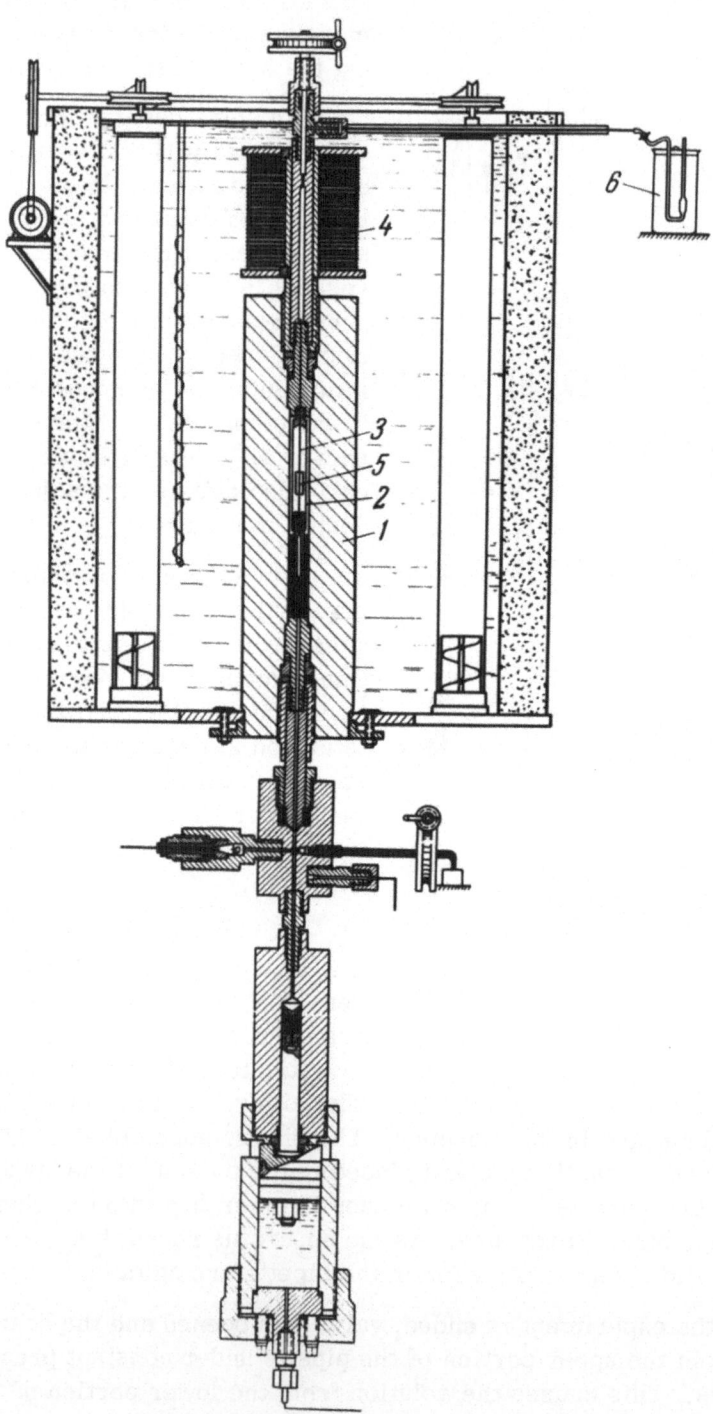

Fig. 243. Krichevskii–Lebedeva installation for studying solubility of solids in liquids: 1) high-pressure column; 2) piezometer; 3) electromagnetic stirrer; 4) coil; 5) container; 6) pycnometer.

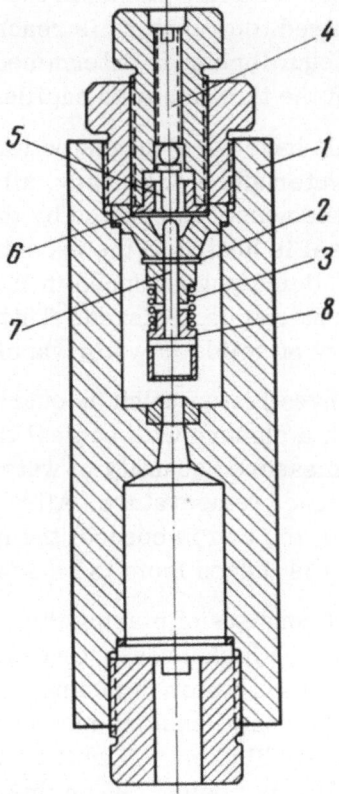

Fig. 244. Apparatus for determination of solubility of solids
in vapor at supercritical temperatures: 1) vessel; 2) pin; 3)
spring; 4) screw; 5) plunger; 6) gold diaphragm; 7) stem; 8)
lower portion of valve.

These methods are complex and are not universal, since only a limited number of solid materials and liquids can be investigated. A more
perfected and universal method has been developed by I. R. Krichevskii
and E. S. Lebedeva [38].

High-pressure column 1 (Fig. 243) contains piezometer 2, within
which is located electromagnetic stirrer 3, moved by coil 4. Either a pure
solvent or an unsaturated solution of the material being investigated is
placed within the piezometer. Container 5 with the material to be dissolved is attached to the shaft of the stirrer which moves within the piezometer. The piezometer is placed in a vessel on the bottom of which mercury is poured to act as a hydraulic seal separating the contents of the
piezometer from the pressure-transmitting liquid.

    After the solution and solid are placed in the column, the pressure is raised and, after the required temperature is reached, the agitation is begun. The achievement of equilibrium is determined by taking samples at constant pressure through the thermostated capillary in pycnometer 6.

    A number of methods have been suggested for investigating phase equilibria in systems of water and a solid body, all characterized by the fact that the pressure in the vessel is created by the evaporation of water as the high-pressure vessel is heated to the experimental temperature.* By varying the amount of liquid phase placed in the apparatus and the temperature, various pressures can be achieved. This same method allows investigation of the solubility of solids in water vapor.

    The mineral to be investigated (such as quartz) is placed in a bomb [43] of stainless steel with a closure with conical thread (similar to that shown on Fig. 73), and a measured quantity of water is added. Then, the bomb is heated to the assigned temperature. After equilibrium is reached, the bomb is rapidly cooled, the cap is opened, the mineral is extracted and weighed, and the solubility is judged from the mass loss.

    Another method [44] consists of placing the solid body in a loosely covered platinum crucible in a high-pressure vessel. The vessel is heated to the required temperature and water is pumped into it in order to create the required pressure. The conditions are maintained constant for 2 to 3 days for establishment of equilibrium. Then the heat source is rapidly removed from the vessel and it is cooled. When this is done, the water vapor condenses and the pressure drops. Then, the vessel is opened, and the liquid and solid material from vessel and crucible are weighed and analyzed, thus determining the composition of the coexisting phases.

    Similar investigations are performed by a dynamic method [45]. For this, water vapor at constant pressure is passed through a pulverized solid material in a high-pressure vessel placed in a furnace. The water vapor is collected at the bottom, through a throttling valve, in the form of a condensate. The vapor flow rate is determined from the number of drops of solution flowing through the valve. The condensate is collected, weighed, and analyzed. The pressure in the installation is maintained by feeding a measured quantity of water into the steam generator.

    Sometimes, a crucible with a filter is fastened to the cover of the high-pressure vessel in order to separate the solid material from the solution [46]. The bottom is made of a caked material like a shot filter. A known quantity of solid material is placed in the crucible, and a measured

---

*These methods are used to investigate the solubility of various mineral complexes, quartz, common salt, and other materials in water.

quantity of water is placed on the bottom of the vessel. The air is extracted from the vessel using a stream of argon, the vessel is closed and placed in a heated furnace. After the predetermined temperature is reached, the furnace is inverted. When this is done, the crucible is submerged in the water. The furnace and vessel are rocked in order to ensure establishment of equilibrium. After the experiment, the furnace is once more turned over, the vessel is removed from the furnace and rapidly cooled. The condensation of the water vapor creates a pressure difference, causing the solution to be forced out of the crucible through the filter into the vessel. The vessel is opened, the solution is poured off and analyzed.

The solubility of solids in vapor at supercritical temperatures can be studied using an improved method. The apparatus [47] (Fig. 244) suggested for this purpose allows investigation not only of the composition but also of the density of the vapor and liquid phases. The vessel 1 has two chambers, an upper valve chamber and a lower sample chamber. The chambers may be separated from each other by a valve. For investigation of the vapor phase, the apparatus is so rotated that the sample chamber is at the top. The aqueous solution of the solid material to be investigated is placed in the apparatus in a quantity sufficient that the sample chamber will contain only the vapor phase at the experimental temperature. For investigation of the liquid phase, a volume and composition of solution are selected such that when the sample chamber is turned downward it will be filled with liquid.

After charging, the vessel is placed in a furnace, heated to the experimental temperature and the contents of the vessel are agitated by rocking in order to ensure establishment of equilibrium between the phases. When a sample is removed, the vessel is placed vertically with the sample chamber above or below depending on whether a vapor or liquid phase sample is to be removed. After the sample is removed, the valve is closed. Up to this moment, the valve is open due to the fact that platinum-iridium pin 2 (0.9 mm in diameter) is passed through stem 7 of the valve, holding INCONEL spring 3 in the compressed position. In order to close the valve, screw 4 is turned, which presses on steel plunger 5. The plunger presses against gold diaphragm 6, 0.5 mm thick. The diaphragm bends and moves the stem, which shears pin 2. The spring expands, the lower portion of the valve 8 is moved and pressed against the valve chamber seat.

Even when expanded, the spring has an elastic force of about 12 kg, which presses the gold plate on the end of the stem into the seat. After the valve is closed, the vessel is cooled by placing it in cold water, the nut is removed and the solution is removed from the sample chamber with a calibrated pipette.

Synthetic Methods. It is known that when some salts dissolve in water or crystallize, a considerable change in the volume of the system

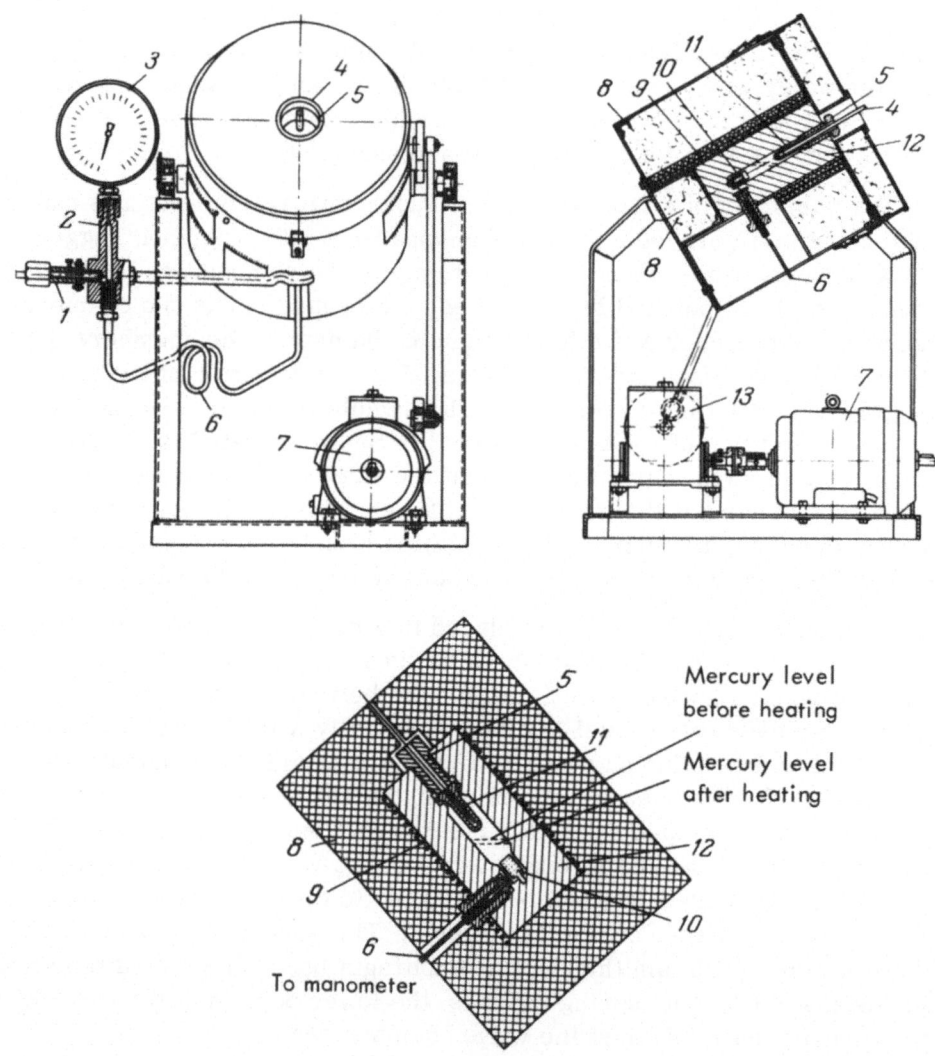

Fig. 245. Ravich-Borovaya installation: 1) mercury outlet valve; 2) divider; 3) manometer; 4) thermocouple; 5) pressure nut; 6) steel tube; 7) motor; 8) thermal insulation; 9) heater; 10) perforated shell; 11) thermocouple well; 12) body of autoclave; 13) reducing gear.

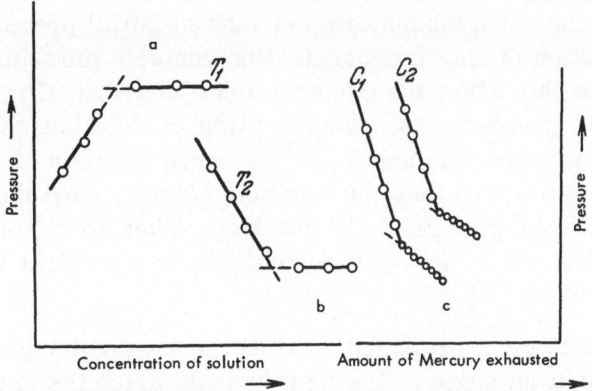

Fig. 246. Dependence of pressure on composition of solution
(a, b) and on quantity of mercury exhausted (c).

takes place. In consideration of this, M. I. Ravich and F. E. Borovaya [48]
developed the method described below for studying phase equilibria in a
solid-liquid system.

The salt to be investigated is pressed into tablets and placed in perforated steel shell 10 (Fig. 245) which is screwed into the bottom of autoclave 12. Then, a quantity of mercury is placed in the autoclave sufficient
to cover the shell and fill steel tube 6 connecting the autoclave with divider
2. The divider connects the autoclave with manometer 3 and is filled from
the bottom with mercury, from the top with oil, which also fills the arc of
the manometer.

An aqueous solution of this same salt, of known concentration, is
poured into the autoclave over the mercury. The autoclave is filled with
the solution until no air bubbles remain in it. Then, the autoclave is closed
with its closure. As these operations are performed, care must be taken
to assure that the shell containing the tablets is completely submerged in
the mercury and does not touch the aqueous salt solution. After the autoclave is loaded, it is placed in an electric furnace and heated to a predetermined temperature. During heating, the pressure begins to rise, and a
fraction of the mercury is allowed to escape through valve 1 to maintain
constant pressure. After the assigned temperature and pressure are
achieved, the autoclave and furnace are put in oscillating motion by motor
7 through reducing drive 13. When this happens, the solution contacts the
solid salt, and additional solutioning of the salt takes place, causing a
change in pressure in the autoclave. If the solution is supersaturated under
the experimental conditions, when the autoclave is heated, salt will precipitate and the pressure will remain constant during the process of agitation.

This part of the investigation produced information on the drop in pressure (with constant temperature and a given initial pressure) resulting when the solution of known concentration contacts the tablets of the solid salt. These data allow the construction of a graph (Fig. 246a and b) in the coordinate pressure vs. concentration of solution charged (for example for temperatures T_1 and T_2). The curve consists of two branches. If the solution is saturated then, as was noted above, when the solution contacts the solid salt the pressure will not drop. Therefore, the points corresponding to some saturated solutions will lie on a straight line parallel to the abscissa.

The second branch, produced in experiments involving unsaturated solutions, will be at an angle to the first branch, since the pressure over an unsaturated solution depends on the concentration of the solution. The point of intersection of the two branches corresponds to the solution concentration saturated at a given temperature and pressure. In this series of experiments, identical quantities of solid material and mercury must be charged in the autoclave.

If the solubility of the salt under the experimental conditions is greater than under normal conditions, the following procedure is used: in each series of determinations, the solution being investigated is charged into the autoclave at identical concentration (C_1 and C_2, Fig. 246c) and the shell contains various, precisely known quantities of tablets. The quantity of mercury is so selected that in one series of determinations the volumes of solution charged and salt charged are identical (under normal conditions). As the experiment is performed as described above, the pressure drop is noted; this drop will be greater, the higher the concentration of the solution formed. After the solution is saturated, a further increase in the quantity of solid salt in this shell will not cause any further drop in pressure. Then, a graph is constructed in the coordinates pressure vs. solution concentration (using the concentration which corresponds to solution of all the salt charged into the liquid solution), and once more two branches are produced, which intersect at a point corresponding to the concentration of a saturated solution under the given conditions of temperature and pressure.

Both these methods correspond to construction of a dependence of pressure on concentration at constant temperature. Investigations can be performed otherwise, constructing a dependence of pressure on volume at constant temperature for a solution of constant concentration. For this, a salt solution of known concentration is placed in the autoclave, it is heated to a fixed temperature and a pressure is set at which the solution is known to be unsaturated. Then, mercury is drawn out of the autoclave in small portions, the portions are weighed, and the pressure in the autoclave after

draining each portion of mercury is noted. On the basis of this information, curves are constructed in the coordinates pressure vs. quantity of
mercury exhausted, and the break on the curves corresponds to the pressure of beginning of crystallization (Fig. 246c). Finally, placing the solid
salt and concentrated solution of this salt in the autoclave, and heating the
contents of the autoclave, a predetermined (high) temperature is set and
the pressure is noted. Then, the autoclave is gradually cooled, noting the
drop in pressure with temperature. A graph is constructed in the coordinates pressure vs. temperature and curves are produced which will have
a break at the point where the new phase appears.

A combination of all these approaches allows exhaustive information
to be produced concerning the behavior of the system.

GAS-GAS SYSTEMS

I. R. Krichevskii and his colleagues [49, 50] showed that there are
conditions under which certain gas mixtures may be stratified into two
phases, i.e., their mutual solubility is limited. Figure 247 shows the plan
of an apparatus used to study the limited mutual solubility of gases in the
systems nitrogen—ammonia, methane—ammonia, etc.

High-pressure column 1 contains piezometer 2, submerged in mer-
cury-filled vessel 3, separating the contents of the piezometer from the
pressure-transmitting liquid (in this case a mixture of kerosene and oil).
The gas mixture to be investigated is prepared in rocking autoclave 4,
which contains a metal ball. One component (such as the ammonia) is
placed in the autoclave and heated to a temperature above the critical temperature. The pressure in the autoclave is brought up to the required pressure, the apparatus is cooled, the second component is introduced and the
apparatus is once more heated, while intense agitation is performed. The
prepared mixture is drained through heated tube 5 and valve 6 to a column
containing the working liquid, heated to a temperature over the critical
temperature of the less volatile component. If the pressure-transmitting
liquid is now forced into the column, this liquid will press on the mercury
in the vessel; the mercury will enter the piezometer and compress the gas.

When the required pressure is produced (at the experimental temperature), samples of the mixture are carefully drawn off through valve 6
for analysis. Stratification will be easily discovered from the results of
analysis of the mixture, in that the composition from sample to sample
will differ sharply. Then, while slowly forcing liquid into the apparatus,
the gas phase is exhausted at constant pressure until periodic analyses
show that the second layer of the gas-gas system is being removed from
the apparatus.

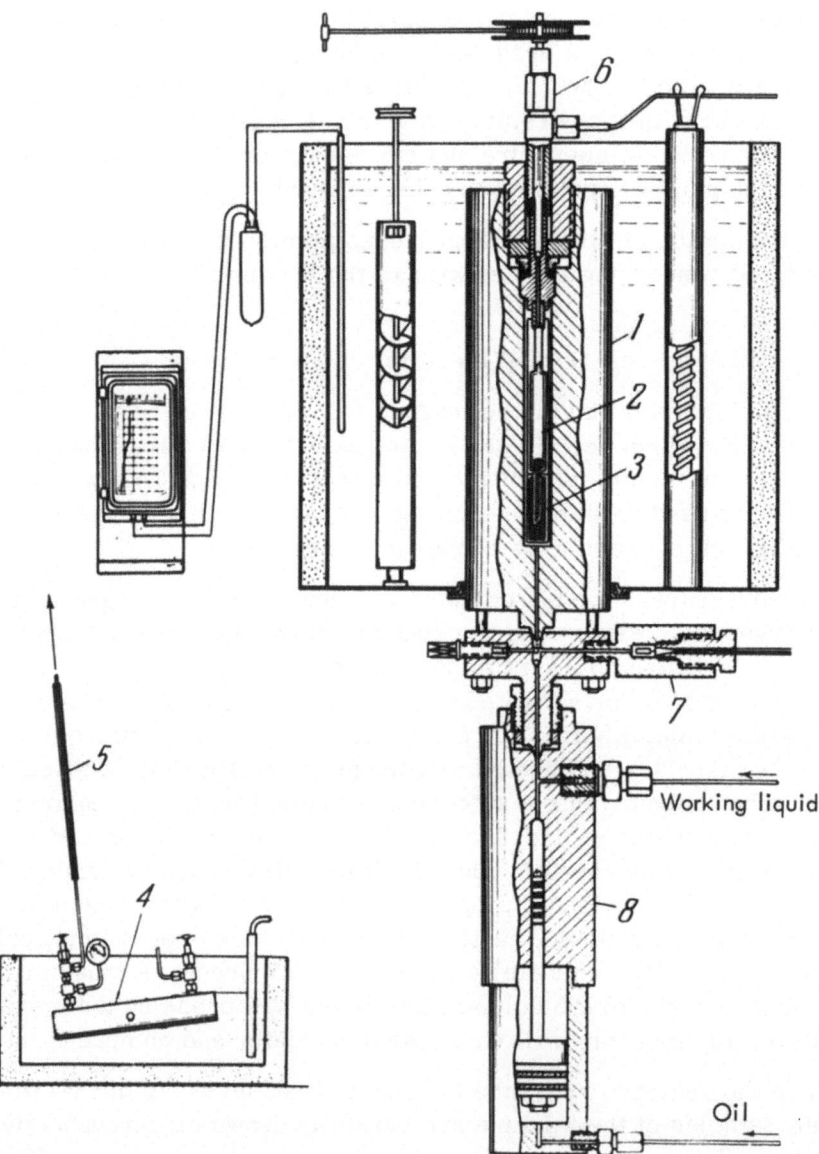

Fig. 247. Installation for studying gas-gas equilibria: 1) high-pressure column; 2) piezometer; 3) vessel containing mercury; 4) rocking autoclave; 5) heated tube; 6) reducing valve;
7) manganin manometer; 8) pressure intensifier.

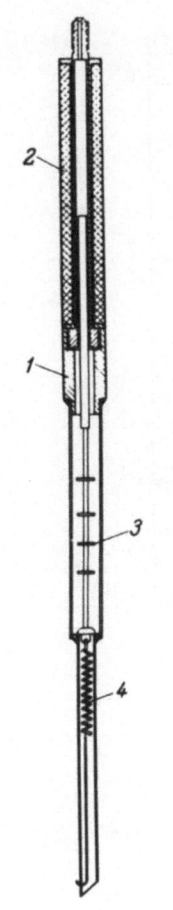

A deficiency of this method is the absence of any gas mixing, which greatly retards the investigation by not allowing, for example, subsequent experiments at lower pressure than the first experiment. This deficiency is not present in the installation developed by the author [51], in which a solenoid 2 is applied to the component metal piezometer 1 (Fig. 248), made of stainless, nonmagnetic steel. The piezometer and solenoid are placed in the high-pressure column, and electric leads are passed through the head of the high-pressure column to supply the solenoid.

The piezometer contains stirrer 3, whose tail is connected by spring 4 to the end of the piezometer. This spring returns the stirrer to its initial position when the current is cut off in the solenoid, since the stirrer would not penetrate into the mercury under its own weight.

A deficiency of both these installations is that the pressure-transmitting medium is kerosene with oil and mercury. At high temperatures, this makes operation with these installations dangerous.

In his investigations of mutual limited solubility of gases at high pressures and temperatures up to 500°C, the author has used an installation [52] (see also [53]) as shown in Fig. 249. It consists of thick-walled vessel 1, made of heat-resistant steel, containing electromagnetic stirrer 2. The vessel is connected to valve body 3, to which a pressure intensifier (not shown on the figure), press valve 4 for disconnection of the pressure intensifier if the back-flow valve fails, manganin manometer 5, and valve 6 for sampling are connected. All the parts of the installation, with the exception of the manganin manometer, pressure intensifier, and press valve are placed in an air thermostat. The manganin manometer is insulated from the medium whose pressure is being measured by a bellows filled with dry kerosene.

When the components are placed in the installation in a quantity as required to produce a mixture of a given total composition and the required pressure is produced, the installation is heated to a predetermined temperature. After equilibrium is established, the pressure is measured and samples are removed for analysis. The drop in pressure is compen-

Fig. 248. Piezometer with stirrer: 1) piezometer; 2) solenoid; 3) stirrer; 4) spring.

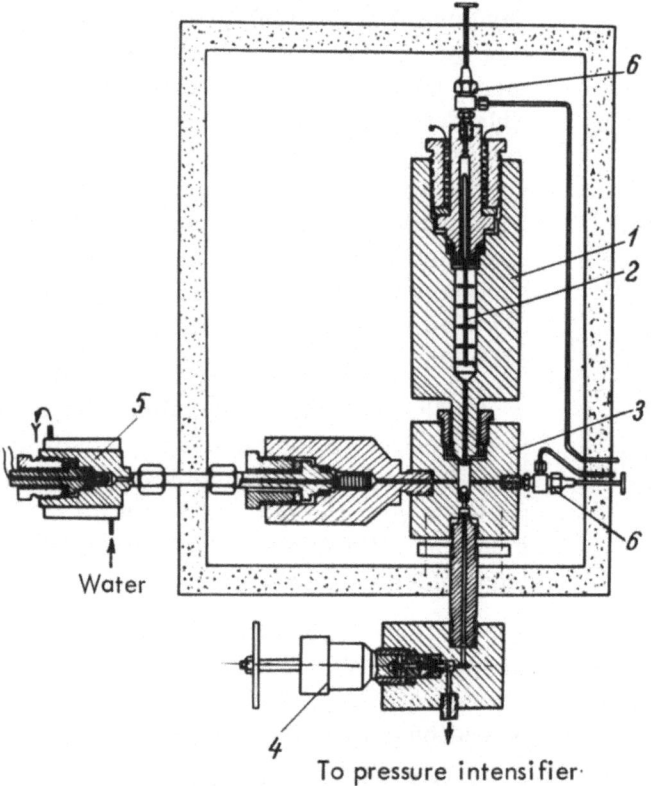

Fig. 249. Installation for investigation of gas-gas equilibria at high
temperatures: 1) high-pressure vessel; 2) stirrer; 3) valve body; 4)
press valve; 5) manganin manometer; 6) valves for removing samples.

sated by adding the liquid component of the mixture (under the given conditions), such as water, by using the pressure intensifier. The change in
the total composition of the mixture influences only the quantities of materials in phases (according to the lever principle), but does not influence
the composition of the phases so long as the total composition of the mixture is within the limits of the heterogeneous equilibrium.

The gas-gas equilibrium can be investigated by synthetic methods as
well in an installation with viewing windows similar to that used by Tsiklis
and Vasil'ev for measurement of the surface tension and volume of phases
in the systems helium–ethylene, etc., (see [12], Chapter XI). A similar
installation, with which the authors studied the gas-gas equilibrium in the
system helium–xenon at pressures up to 2000 atm is described in the
literature [54].

SOLID-GAS SYSTEMS

The solubility of solids in compressed gases may also be determined by static and dynamic methods. In essence, these methods do not differ from the methods of investigating phase equilibria in liquid-gas systems.

Static Methods. A deficiency of static methods in this case is the fact that very little saturated gas is present in the autoclave, so that a small quantity of this phase must be analyzed. The most convenient static method is the following [55, 56]: a high-pressure column of known volume contains a tablet of solid material which has been pressed and weighed on an analytic balance; the required quantity of compressed gas is introduced. At the experimental temperature, the contents of the column are agitated and after equilibrium is established, the pressure is measured. Then, the

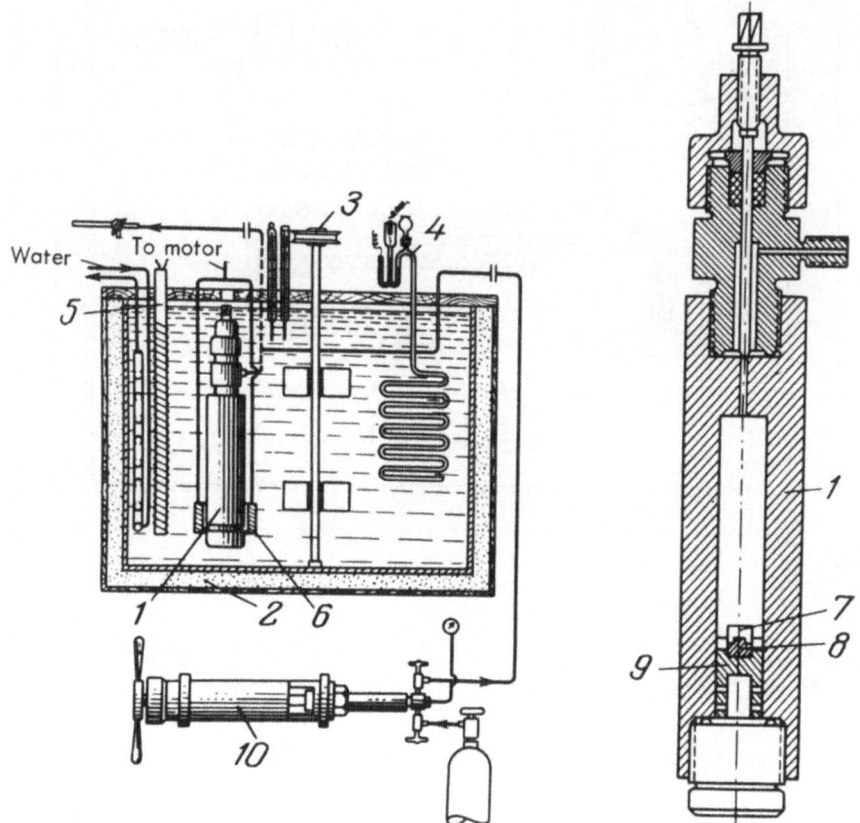

Fig. 250. Installation for studying solubility of solids in compressed gases: 1) high-pressure column; 2) thermostat; 3) stirrer; 4) thermoregulator; 5) heater; 6) permanent magnet; 7) tablet; 8) armature; 9) stirring cylinder; 10) press.

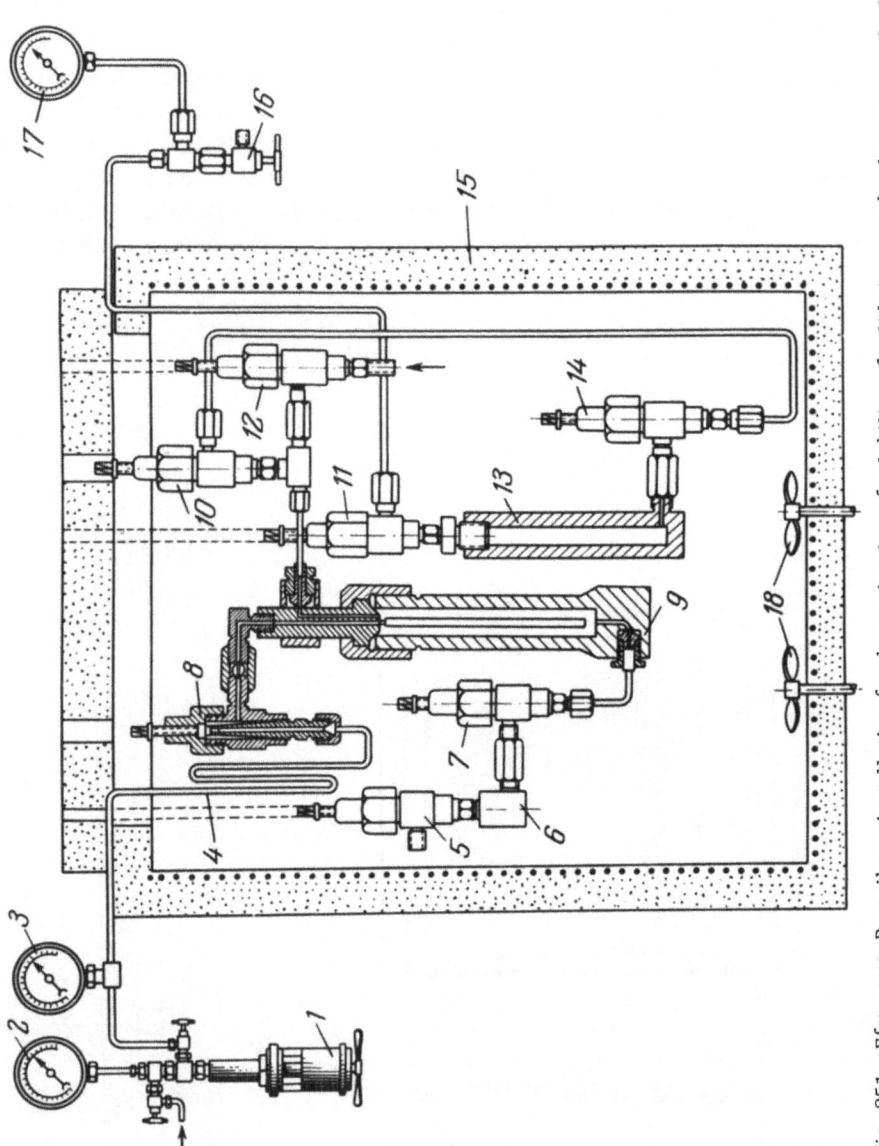

Fig. 251. Efremova–Pryanikova installation for determination of solubility of solids in gases by dynamic method: 1) press; 2, 3, 17) manometers (pressure gauges); 4) coil; 5, 7, 8, 10, 11, 12, 14) membrane valves; 9) saturator; 6, 13) sample collectors; 15) thermostat; 16) throttle valve; 18) fans.

gas from the column is drawn off in its entirety into calibrated flasks in order to determine the quantity of gas at atmospheric pressure.

The tablet is weighed before and after the experiment. If the reduction in mass of the tablet, the quantity of gas in the column, the temperature and pressure of the experiment, and the capacity of the column are known, the solubility of the solid material in the compressed gas and the volume of the solution thus produced can be determined.

An installation for studying the solubility of solid naphthalene in carbon dioxide is shown in Fig. 250. It consists of column 1, placed in thermostat 2 which is supplied with stirrer 3, thermoregulator 4, and heater 5. A permanent circular magnet 6 is placed around the column, which is made of stainless, nonmagnetic steel. Tablet 7, pressed from 1-1.5 g naphthalene is placed within the column. The tablet is pressed with armature 8 made of iron and placed in stirring cylinder 9, which prevents the tablet from dropping to the bottom of the column.

The tablet is weighed on an analytic balance and placed in the vessel; the air is evacuated from the vessel, hydraulic press 10 filled with carbon dioxide is connected to the vessel, and all the connecting capillaries are blown out. Then, the required quantity of carbon dioxide is placed in the column and the contents of the column are agitated by the permanent magnet at constant experimental temperature. As the solid material is dissolved, the pressure drops somewhat (since the volume of the column is increased due to the reduction of the volume of the tablet). Gas is added, and the phases are once more agitated until constant pressure is reached. At the end of the experiment, the gas is drawn off into an evacuated calibrated flask and its quantity is judged from the reading of a mercury manometer.

When the pressure is reduced, crystals of the solid material precipitate onto the surface of the tablet. They are removed with a soft cloth and the tablet is weighed. Then, if the volume of the tablet, stirrer, and column are known, the volume of the solution can be calculated and the composition of the solution can be determined from the reduction in mass of the tablet. Similar methods have been described in the literature [56].

Dynamic Methods. A dynamic method for determining the solubility of a solid in a compressed gas can be illustrated by a description of the work of an installation [57, 58] developed by G. D. Efremova and R. O. Pryanikova. Steel saturating autoclave 9 (Fig. 251) filled with melamine (whose solubility in ammonia is to be studied) is connected to two sample collectors 6 and 13. The saturator and sample collectors contain packingless membrane valves. The ammonia is fed into the installation using hydraulic press 1. The installation is placed in air thermostat 15, and the temperature in the air thermostat is maintained with an accuracy of 1°C.

The sequence of operations is as follows: the hydraulic press is used to force a small quantity of ammonia into the saturator, and the heat in the thermostat is turned on. Sample collector 13 and the capillary connecting this sample collector to throttle valve 16 are filled with ammonia. The thermostat is heated to the experimental temperature, the experimental pressure is created in the saturator, and the sample collector and the valves connecting the saturator to the sample collector 13 are opened. Then, ammonia is passed through the saturator and sample collector with constant velocity at constant pressure. The ammonia passes through coil 4, where it is heated to the experimental temperature, moves to the saturator, and travels through the tube to the bottom, where it passes through the layer of melamine and enters the sample collector 13. Throttle valve 16 is installed outside the thermostat, so that the melamine dissolved in the ammonia is precipitated in the cold capillary before arriving at valve 16, in order to avoid blocking the valve aperture.

The pressure in the installation during the course of the experiment is checked by manometers 2, 3, and 17. The experiments are performed at various gas flow rates, in order to determine the flow rate at which equilibrium is achieved between the melamine and the ammonia in the installation.

If the experiments are performed at temperatures greater than the melting temperature of melamine, a liquid-phase sample can be taken from sample collector 6. For this, valve 11 is closed and valve 12 is opened; nitrogen is forced into the installation under the experimental pressure. The nitrogen forces the liquid melamine out of the saturator into sample collector 6. After termination of the experiment, the sample collectors are disconnected from the saturator and their contents are analyzed.

METHODS OF SAMPLING FOR ANALYSIS

We noted above that the method of sampling is extremely important in the production of correct results. A sample should be taken so that equilibrium is undisturbed both in the sample and in the remaining system. A sample may be taken by two methods: it may be drawn directly from the apparatus or it may be separated from the total mass of matter being investigated.

Direct Extraction of Samples from Apparatus. This method is, of course, simpler. However, it is also the least perfect. When a liquid phase is analyzed, its total quantity must be rather large, otherwise extraction of even a small quantity of liquid from the apparatus will cause a considerable drop in pressure and will disrupt the equilibrium (if special measures are not taken to maintain the pressure). The liquid drawn from the high-pressure vessel must be protected from losses of the gas dissolved in it.

In analysis of a gas phase, direct sampling also requires a considerable vessel capacity, in order that the pressure will not drop. If there is little gas, special measures must be taken to prevent a pressure drop. In this case, the analysis itself may be performed under pressure (for example, freezing of a liquid) in order to avoid the error connected with throttling a gas.

The methods which use pressurized analysis include all indirect methods (determination of saturation pressure, dew points, etc.).

Method of Sample Separation. This method is much more accurate, since it allows a certain portion of the entire mass of material being investigated, which is at equilibrium, to be separated and analyzed.

Another advantage of this method is that, as a rule, the density of a sample can also be determined, since its volume is generally very precisely measurable.

In the following, in our description of methods of sample analysis, we will also indicate the method by which the samples are taken.

METHODS OF ANALYSIS

The analytic methods used in investigation of phase equilibria can be divided into two large groups: volumetric and gravimetric measurement of the quantities of liquid and gas. Of course, we will not mention the generally-used and well-known methods of analysis of gas mixtures, or those which can be taken from the existing literature. We would like in this section to acquaint the reader with an approach to the solution of the problem of analysis of samples. Naturally, it is not possible to give a universal solution to this problem.

Volumetric Measurements

Volumetric measurements consist of the most accurate possible measurement of the volumes of liquid and gas phases into which the (liquid or gas) sample drawn off from the apparatus is separated. As an example, let us describe in detail the analytic portion (Fig. 252) of the installation (see Fig. 218) for determination of the solubility of gases in liquids [6].

The sample is taken carefully, so as to avoid a loss of gas as the liquid is throttled. The sampling valve is opened very carefully, and a small quantity of liquid is drawn off from the apparatus to wash out all tubes leading to the analytic portion of the installation. Then, the installation is connected through three-way stopcock 1 to the analytic portion, which consists of glass sphere 2 plus the narrow graduated tube below it and manometer 3.

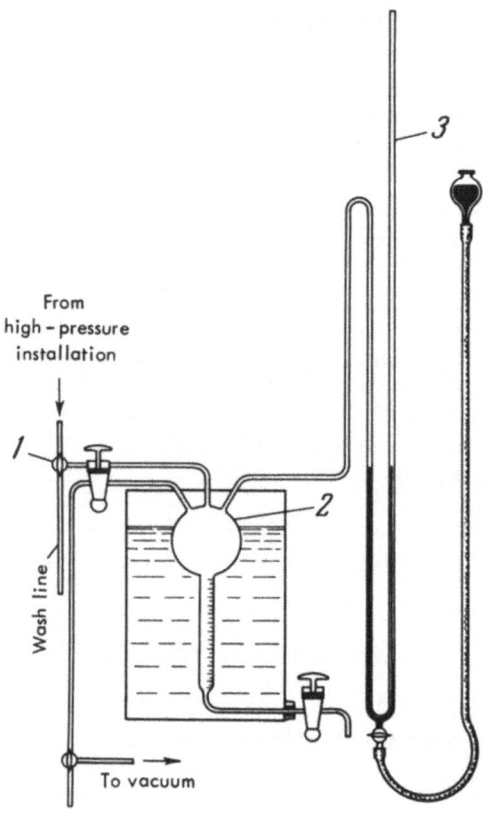

Fig. 252. Apparatus for analysis of liquid phase:
1) three-way stopcock; 2) glass sphere; 3) manometer.

The volume of the sphere must be precisely known. It is selected depending on the solubility of the gas, which may be approximately known. The capacity is usually selected such that, as the gas separates from the liquid, the pressure in the evacuated sphere will increase to not over atmospheric pressure. The amount of gas separated depends on the quantity of liquid drained off from the vessel. An attempt must be made to take the minimum possible amount of liquid for analysis (as little as accuracy of determination allows), in order to maintain as much pressure as possible in the vessel as the liquid is drained off for analysis.

A residual pressure of 1-2 mm Hg is created in the sphere before the sample is drawn off into it. The mercury in the manometer is filled up to a mark made on the manometer column connected with the vacuum pump and the liquid from the apparatus is slowly drained into the sphere. The dissolved gas in the liquid, which is evolved after the liquid is placed in the sphere, increases the pressure in the sphere. This increase in pres-

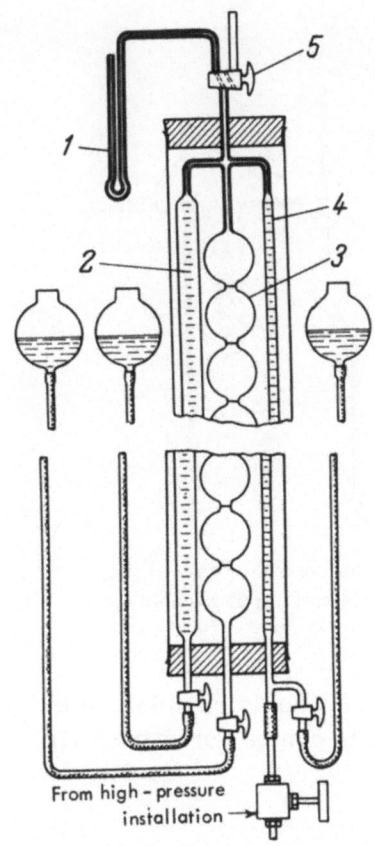

Fig. 253. Burette for analysis of liquid phase: 1) water manometer; 2) 50 ml burette; 3) 50 ml sphere; 4) 25 ml burette; 5) stopcock.

sure can be used to determine the quantity of gas separated from the liquid. If the quantity of liquid is known and a solubility correction factor is introduced for the solubility of the gas and the liquid at atmospheric pressure and at the temperature in the thermostat containing the sphere, the solubility of the gas under pressure can be calculated.

An analogous method of analysis was used in the apparatus of Zel'venskii [59]. However, the analytic portion of the installation differs from that described above. The analysis of the liquid is performed in an apparatus (Fig. 253) which consists of three burettes interconnected at the top. Two of the burettes are ordinary burettes of 50 and 25 ml capacity, graduated each 0.05 ml; the third burette is a ball burette with a total capacity of 500 ml. The volume of gas separated from the liquid is measured in the 25 ml burette. Each burette is connected to an equilibrating bulb and filled with mercury or a solution of sodium sulfate (where a gas phase is being sampled). The burettes are contained in water jackets. Before a sample is taken, the burettes for containing gas samples are filled with a sealing fluid; the burettes for liquid sampling are filled with the gas from the preceding analysis.

Another installation is shown in Fig. 254. The air is evacuated from a portion of the device (from stopcock 3 to mercury seal 2) by pump 1. Coil 4 is submerged in a Dewar flask with liquid air. Then, the line from the pump is blown through with the mixture to be analyzed and connected to the instrument. The ammonia condenses in coil 4, and the nitrogen—hydrogen mixture evolved is pumped by pump 1 into burette 5.

The installation is evacuated to the same vacuum which was present before the sample was taken. Then the ammonia is pumped into sulfuric acid and titrated, and the nitrogen—hydrogen mixture is analyzed in gas analyzer 6 for hydrogen content.

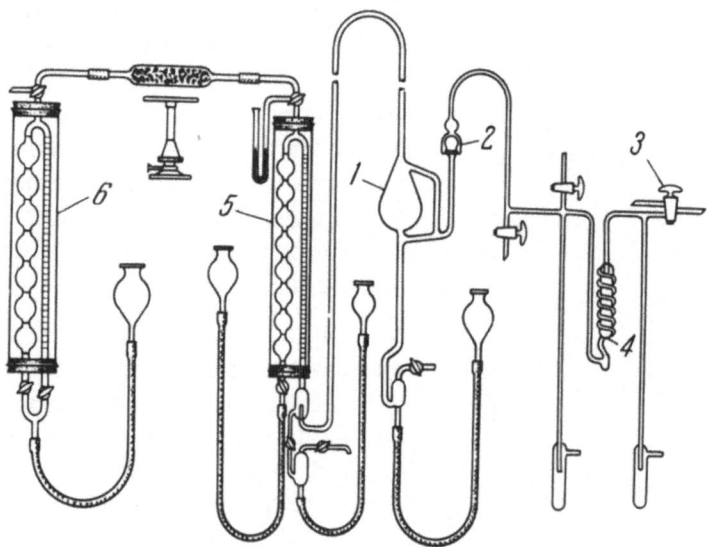

Fig. 254. Installation for analysis of liquid phase for content of gas: 1) mercury pump; 2) mercury seals; 3) stopcock; 4) coil; 5) burette; 6) gas analyzer.

It should be noted that this method is unjustifiably complex, since analysis of the gas mixture containing ammonia can be performed with great accuracy in the gas analyzer.

Gravimetric Measurements

A considerably more accurate, though more complex, method is that of weighing a sample which has been divided into phases. A gravimetric method, in combination with the method of separating the sample in the apparatus, allows production of precise and complete data. As an example, we can present a portion of an installation which we have already described [20], shown in Fig. 255. Since the installation was designed for investigation of methane—hydrocarbon systems, the possibility of freezing the liquid phase is included in the apparatus.

During analysis of the liquid phase, the entire amount of liquid is collected in removable condensor 1 and its quantity is determined by weighing. If we know the mass (amount) of liquid, its volume under pressure (considering expansion of piezometers under pressure) and the amount of gas dissolved in the liquid, we can determine the density of the liquid.

The gas which is separated as the liquid is throttled is pumped by pump 2 to the right portion of the installation, where its volume is measured in flask 3. The density of the gas is determined in flask 4. During analysis of a gas phase containing considerably less liquid, the installation

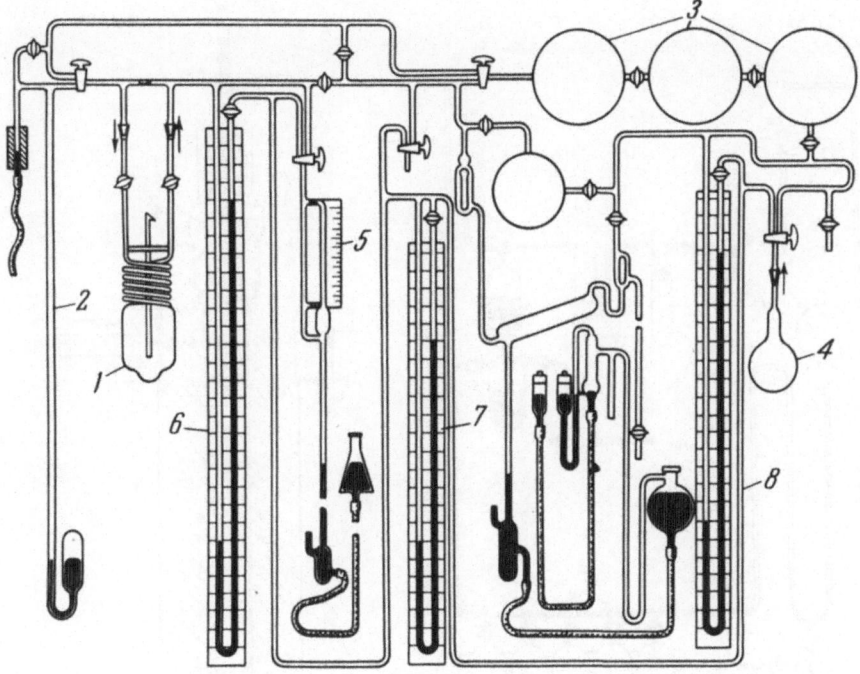

Fig. 255. Instrument for analysis of liquid phase by gravimetric method: 1) removable condensor; 2) mercury pump; 3) flask; 4) density-measurement flask; 5) vacuum meter; 6, 7, 8) manometers.

is evacuated to low residual pressure, measured by vacuum gauge 5. Then, the gas is allowed to enter the instrument, passing through the condensor, where its liquid content is condensed. The gas is pumped into the collectors until the pressure in the left portion of the installation drops to its initial vacuum level. Manometers 6, 7, and 8 measure the pressure of the gas in the flask, which makes it possible to determine the quantity of gas involved.

Another device used for analysis using this method is the Krichevskii-Efremova installation [24], shown on Fig. 256. This is a part of a larger installation (connected to the remainder of the installation by a copper—glass seal); this portion of the installation is maintained under a vacuum. It consists of receptor 1, two removable ampules 2 (in which the liquid is frozen out of the sample), control trap 3, mercury pump 4, gas receptors 5, and closed mercury manometer 6. Before filling the manometer with mercury, it is evacuated to 1 mm Hg, and the remainder of the air is separated with the mercury pump. A liquid sample is placed in piezometer 2, shown on Fig. 226 (see page 310).

This installation combines the most accurate method of analysis known with the most accurate method of sample separation. Before analy-

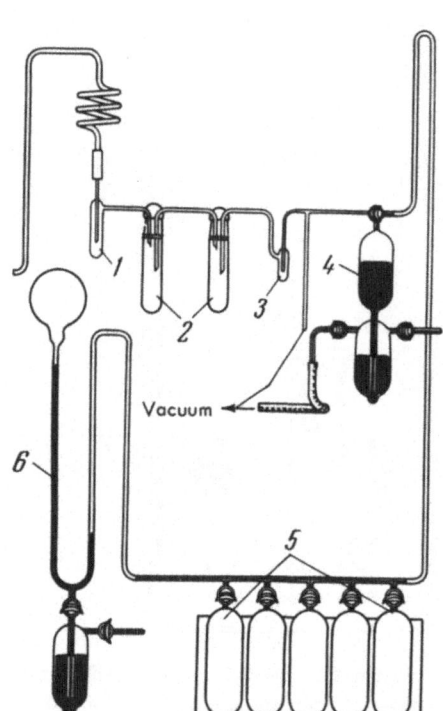

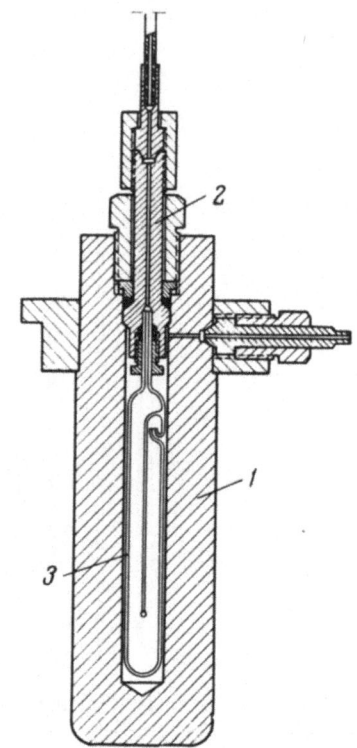

Fig. 256. Krichevskii-Efremova apparatus for analysis of liquid phase: 1) receiver; 2) removable ampules; 3) control trap; 4) mercury pump; 5) gas receivers; 6) manometer.

Fig. 257. Condenser: 1) stainless steel vessels; 2) plug; 3) ampule.

sis, the air is pumped from the installation and the liquid from the piezometer is drained off into the ampules, where it is frozen at the temperature of liquid nitrogen. The gas separated from the liquid as the temperature is lowered is pumped by pump 4 into receptors 5. When the entire amount of liquid has been frozen, the gas is once more drawn over into the ampules to freeze out any liquid vapor which might be carried away by the rapid gas flow. After this, the gas, completely free of liquid, is once more pumped by pump 4 into receptors 5. The volume of gas is determined by the increase in pressure measured by the mercury manometer, considering the calibrated volume of the collectors used. The quantity of liquid is determined by weighing the ampules.

The volumes of the collectors and connecting tubes must be precisely known. It is extremely important to determine the volume of each piezometer as connected into the installation, i.e., taking into account all the specific features of the assembly. This volume is determined by calibration with the working liquid, i.e., a special run is made in which the ampules are pumped full of pure solvent, containing no gas. If the mass of the ampules and the density of the liquid are known, the capacity of the piezometer can be determined (in work [24], the capacity was approximately 25 ml). At pressures of 500 to 1000 atm, the capacity of the piezometer increases by only 0.005 ml, so that no correction for increased volume of piezometer need be made at these pressures. At higher pressures, failure to consider expansion of the vessel under pressure may lead to considerable error.

The volumes of liquid and gas and the volume of the piezometer can be used to calculate the composition, density, and molar volume of the solution.

Analysis of the gas phase during determination of the solubility of liquids in gases is usually rather difficult. The problem is that analysis of the gas phase, especially if it is connected with freezing of material from the gas, has always been performed at atmospheric pressure in all devices of this type. Gas is analyzed after the drops or particles of material are separated from it by reducing it and passing it rather slowly (2-3 liters/h) through condensers or absorbers. If the liquid is poorly soluble in the gas (for example, 0.0001 or 0.00001 g/liter at normal temperature and pressure), the accumulation of a sufficient quantity of material to be weighed in the condenser requires an extremely long experiment. Therefore, a method suggested by I. R. Krichevskii and D. Yu. Gamburg [15] is worthy of attention; the essence of the method is that the condensation of the liquid from the gas is also performed under pressure. This allows a considerable increase in the quantity of gas passed through the condensation system per unit time, while simultaneously reducing the flow rate through the condenser.

The condenser (Fig. 257) consists of vessel 1, made of stainless steel, closed with plug 2, in which glass condensation ampule 3 is enclosed, passing through an asbestos gland. The lower portion of the condenser is cooled, the upper portion is heated to avoid condensation of the liquid outside of the glass ampule. The quantity of gas passed through the condenser is determined from a gas meter into which the gas is passed after it is throttled at the output of the condenser. The quantity of liquid is determined by weighing the glass ampule.

Absorption can also be used in this case in place of freezing.

BIBLIOGRAPHY

1.   Alekseev, V. F., Zh. Russ. Fiz. Khim. Obshchestva, 8:329 (1876).
2.   Basset, J., and Dode, M., Compt. Rend., 203:275 (1876).
3.   Sander, W., Z. Phys. Chem., 78:513 (1912).

4. Wiebe, R., Gaddy, V., and Heins, H., Ind. Eng. Chem., 24:823 (1932).

5. Saddington, A., and Krase, N., Ind. Eng. Chem., 23:401 (1931).

6. Krichevskii, I. R., Zhavoronkov, N. M., and Tsiklis, D. C., Zh. Fiz. Khim., 8:317 (1937).

7. Bartlett, E., J. Am. Chem. Soc., 49:65 (1927).

8. Saddington, A., and Krase, N., J. Am. Chem. Soc., 56:353 (1934).

9. Miller, B., and Dodge, B., Ind. Eng. Chem., 32:434 (1940).

10. Larson, A., and Black, G., J. Am. Chem. Soc., 47:1015 (1940).

11. Krichevskii, I. R., and Kazanova, N. E., Zh. Fiz. Khim., 13:1690 (1939).

12. Krichevskii, I. R., and Kal'varskaya, R. S., Zh. Fiz. Khim., 14:748 (1940).

13. Krichevskii, I. R., and Koroleva, M. V., Zh. Fiz. Khim., 15:327 (1941).

14. Braune, H., and Strassman, F., Zh. Phys. Chem., 143:225 (1929).

15. Krichevskii, I. R., and Gamburg, D. Yu., Zh. Fiz. Khim., 17:215 (1943).

16. Ipat'ev, V. V., Teodorovich, V. P., Brestkin, A. P., and Artemovich, V. S., Zh. Fiz. Khim., 22:834 (1948).

17. Schen-Wu-Wan, and Dodge, V. F., Ind. Eng. Chem., 32:95 (1940).

18. Ipat'ev, V. V., and Teodorovich, V. P., Zh. Obshchei Khim., 4:395 (1934).

19. Wiebe, R., and Tremearne, T., J. Am. Chem. Soc., 55:975 (1933); 59:1984 (1937).

20. Boomer, E., Johnson, C., and Argue, D., Can. J. Res., B15:367 (1937).

21. Bol'shakov, P. E., and Linshits, L. P., Tr. Gos. Inst. Azo. Prom., No. 3, Goskhimizdat, 1954, p. 18.

22. Tsiklis, D. S., Mushkina, E. V., and Shenderei, L. I., Inzh. Fiz. Zh., 1(8):3 (1958).

23. Gonikberg, M. G., Fastovskii, V. G., and Gurvich, I. G., Zh. Fiz. Khim., 13:1669 (1939).

24. Krichevskii, I. R., and Efremova, G. D., Zh. Fiz. Khim., 22:116 (1948).

25. Krichevskii, I. R., Shurmovskaya, N. A., and Kal'varskaya, R. S., Zavodsk. Lab., 1:112 (1947).

26. Lebedeva, E. S., and Khodeeva, S. M., Zh. Fiz. Khim., 35(11):2602 (1961).

27. Cook, D., Trans. Faraday Soc., 49:716 (1953).

28. Krichevskii, I. R., and Efremova, G. D., Zh. Fiz. Khim., 30:877 (1956).

29. Wells, F. W., and Roof, I. G., Rev. Sci. Instrum., 26:403 (1955).

30. Bogdanova, V. A., and Martynenko, A. G., Zavodsk. Lab., 27(9):1159 (1961).

31. Sage, B., and Lacey, W., Ind. Eng. Chem., 26:106 (1934).

32. Sage, B., and Lacey, W., Trans. Am. Inst. Mining Met. Engrs., Petr. Div., 136 (1940).

33. Exline, P., and Dean, H. E., Oil Gas J., 45:82 (1947).

34. Khiteev, A. M., Dokl. Akad. Nauk AzSSR, 13:117 (1957).

35. Gamburg, D. Yu., Zh. Fiz. Khim., 24:272 (1950).

36. Krichevskii, I. R., and Sorina, G. A., Zh. Fiz. Khim., 33:1151 (1959).

37. Krichevskii, I. R., Efremova, G. D., Pryanikova, R.O., and Polyakov, E. V., Zh. Khim. Prom., 7:498 (1961).

38. Lebedeva, E. S., Dissertation, Moscow, 1948.

39. Kohen, E., and Sinnige, L., Z. Phys. Chem., 67:432 (1909).

40. Kohen, E., Incuge, R , and Euwen, C., Zh. Phys. Chem., 75:258 (1910).

41. Still, A. E., J. Am. Chem. Soc., 38:2632 (1916).

42. Kohen, E., et al., Z. Phys. Chem., 104:3203 (1923).

43. Kenndy, G. S., Econ. Geol., 45:629 (1950).

44. Morey, G. W., and Hesselgesser, J. M., Am. J. Sci., Bowen, 2:343 (1952).

45. Morey, G. W., and Hesselgesser, J. M., Econ. Geol., 46:821 (1951).

46. Booth, H. S., and Bidwell, R. M., J. Am. Chem. Soc., 72:2567 (1950).

47. Copeland, C. S., Silverman, J., and Benson, S. W., J. Chem. Phys., 21:12 (1953).

48. Ravich, M. I., and Borovaya, F. E., Zh. Prikl. Khim., 9(4):952 (1964).

49. Krichevskii, I. R., and Bol'shakov, P. E., Zh. Fiz. Khim., 15:184 (1941).

50. Krichevskii, I. R., and Tsiklis, D. S., Zh. Fiz. Khim., 17:126 (1943).

51. Tsiklis, D. S., Dokl. Akad. Nauk SSSR, 86:993 (1952).

52. Tsiklis, D. S., and Maslennikova, V. Ya., Dokl. Akad. Nauk SSSR, 57(2):417 (1964).

54. Tödheide, K., and Frank, E. M., Z. Phys. Chem., N. F., 37:387 (1963).

54. J. de Swaan Arons, Ontmenging in deGasfase, Thesis, Delft, 1963.

55. Tsekhanskaya, Yu. V., Iomtev, M. B., and Mushkina, E. V., Zh. Fiz. Khim., 36(10):2187 (1962).

56. Iomtev, M. B., Dissertation, Moscow, 1962.

57. Krichevskii, I. R., and Efremova, G. D., Zh. Fiz. Khim., 33(6):1328 (1959).

58. Efremova, G. D., and Pryanikova, R. O., Tr. Gos. Inst. Azo. Prom., No. 15, Goskhimizdat, 1960, p. 193.

59. Zel'venskii, Ya. D., Zh. Priklad. Khim., 12:1312 (1939).

Chapter X

Determination of Compressibility
of Gases and Liquids

Measurement of compressibility consists of the determination of volume occupied by a known quantity of gas (one mole, one kilogram, etc.) at a given pressure and temperature.* This problem may be solved by various methods. For example, the required experimental pressure of the material being investigated may be created in a vessel of known volume at a given temperature; then the quantity of the material in the vessel may be measured by its mass or by its volume at atmospheric pressure. Another method consists of compressing a known quantity of gas in a vessel whose volume can be changed during the process of the experiment. The volume of the material under pressure is measured indirectly in this case.

The compressibility of gases and liquids can be calculated from their density which in turn is measured by hydrostatic weighing. There are several varieties of these methods.

The many methods of measuring compressibility can be divided into the following groups:

1. Methods of measuring compressibility using piezometers of constant volume. The quantity of material placed in a piezometer is determined from the weight of the piezometer and the reduction in mass of the material in the vessel from which the charge is drawn or (after the experiment) by the mass or volume of investigated material at low pressure.

2. Methods of measuring the compressibility using variable-volume piezometers with a constant quantity of sample material.

---

* Certain methods of determination of molar volumes of solutions were described in Chapter IX.

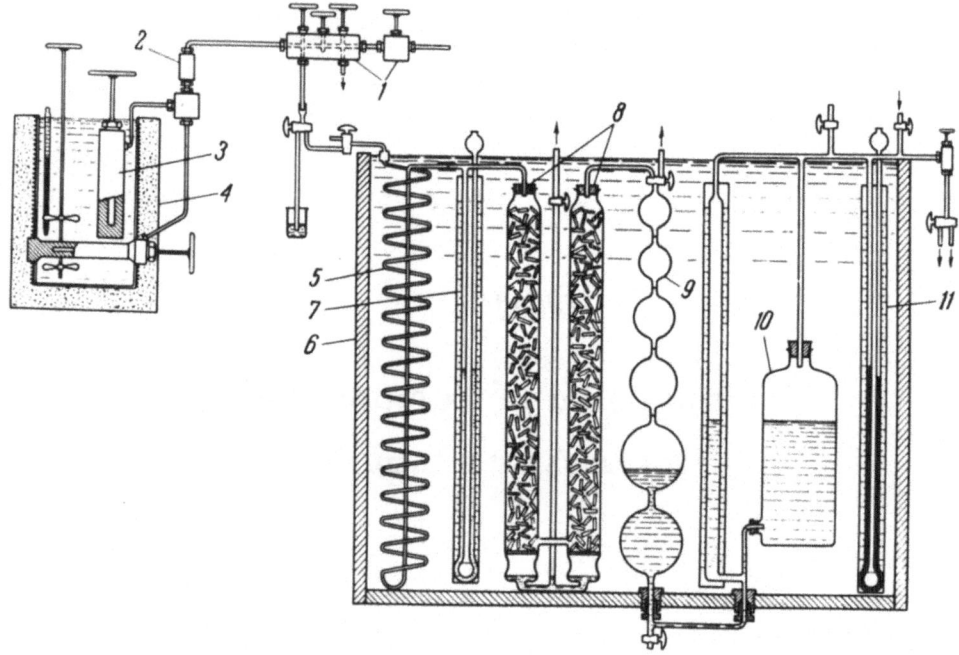

Fig. 258. Barlett installation: 1) valve system; 2) filter; 3) piezometer; 4, 6) thermostats; 5) coil; 7, 11) manometers; 8) humidifiers; 9) burettes; 10) bottle.

3. Methods of determining density of materials from weight loss of a solid body placed in the compressed medium (hydrostatic weighing methods).

4. Mixed methods.

COMPRESSIBILITY OF GASES

Measurement with Constant-Volume Piezometers

The Bartlett Installation [1] shown in Fig. 258 consists of two parts: the high-pressure portion (including the system for compressing the gas in the piezometer) and the measuring portion. The compressed, purified gas is placed in piezometer 3 through a system of valves 1 and filter 2; the piezometer is placed in thermostat 4. Depending on the pressure, piezometers of various volumes are used (in work [1], about 28 ml for pressures up to 100 atm and 2 ml for pressures up to 1000 atm).

After the piezometer is filled with gas, and a certain amount of time has passed for equilibration of temperature, the pressure of the gas is measured. Then, closing the stop valve, the gas which has not entered the piezometer is allowed to escape through the input valve, and the gas

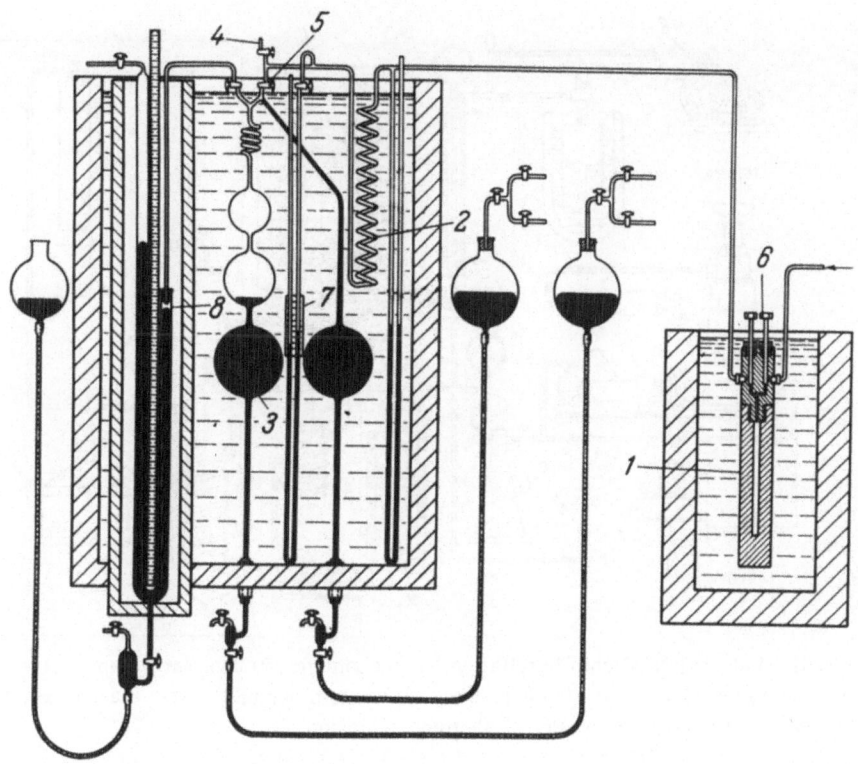

Fig. 259. Wiebe-Gaddy installation: 1) piezometer; 2) coil; 3) burette; 4, 5) stopcocks; 6) valves; 7, 8) manometers

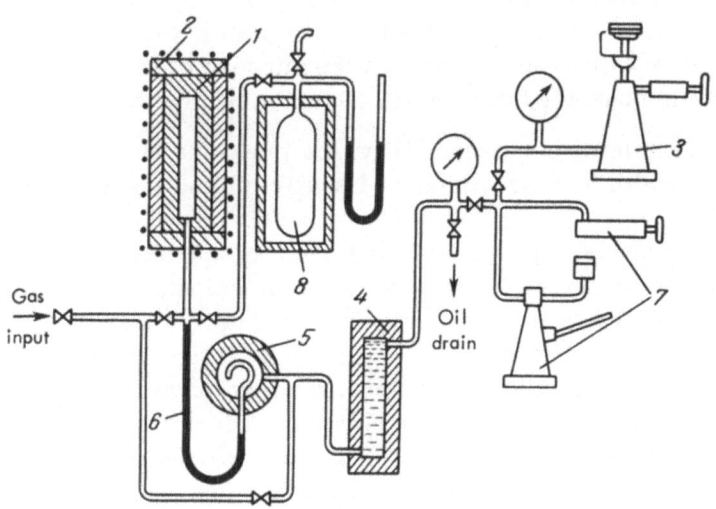

Fig. 260. Kirillin-Ulybin installation: 1) piezometer; 2) furnace; 3) piston manometer; 4) receiver; 5) divider; 6) mercury plug; 7) presses; 8) gas meter.

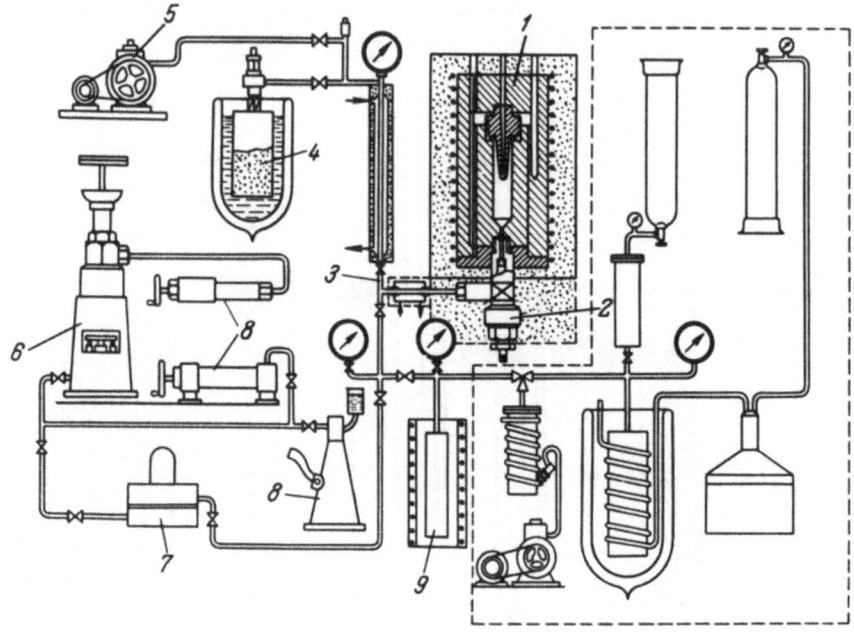

Fig. 261. Vukalovich-Altunin installation: 1) piezometer; 2) constant-volume valve; 3) connecting line; 4) vessel containing carbon; 5) vacuum pump; 6) piston manometer; 7) differential manometer; 8) presses; 9) thermocompressor.

conduits are disconnected. After this, the gas is let out of the piezometer to the measuring system. This measuring system consists of copper coil 5 (during passage through the coil, the gas takes on the temperature of thermostat 6), oil manometer 7, two humidifiers 8, and gas burette 9. Since the gas in burette 9 is collected over water, the gas must first be saturated with water vapor, which is the purpose of the humidifiers. The gas burette is connected with bottle 10 and mercury manometer 11.

In determining compressibility, the pressure in the piezometer must be measured with high precision.* Usually, piston manometers are used

---

*Data on compressibility or PVT are used to calculate the thermodynamic properties of materials. The equations for calculation of a number of thermodynamic properties contain the first and second derivatives of the volumes at pressure and temperature, the coefficient of compressibility with respect to temperature, etc. It is known that during differentiation, the accuracy of all initial data is reduced by an order of magnitude. Therefore, if an error in calculation is to be less than 1%, the error in determination of volume or coefficient of compressibility must be not over 0.1%. If an equation with a second derivative is used in calculating a thermodynamic property, this error should not exceed 0.01%.

for this purpose, which make it possible to measure pressures with accuracies to hundredths of a percent.

Wiebe and Gaddy Installation.   Figure 259 shows an installation [2] which differs from the preceding installation only in the method of measuring the volume of the expanded gas, which is collected over mercury, considerably simplifying the entire measurement portion. The gas from piezometer 1 passes through coil 2 for equilibration of the temperature and arrives at burette 3. The burette is filled with mercury and connected by stopcock 4 to the atmosphere. Before the gas is admitted, stopcock 4 is closed, stopcock 5 is opened, and the gas is drained from the piezometer by valve 6. After the gas is drained, atmospheric pressure is created in the line from valve 6 to stopcock 4 by letting out some of the mercury (pressure measured by manometer 7) and, after stopcock 5 is closed, the gas volume is measured. For this, the mercury is set at the nearest division between the spheres in burette 3 and, determining the pressure from manometer 8, the volume of gas is measured.

Kirillin-Ulybin Installation [3].   The gas to be investigated is collected in piezometer 1 (Fig. 260) made of stainless steel and located in electric furnace 2. The pressure in the piezometer is measured with piston manometer 3 connected to the installation through oil receiver

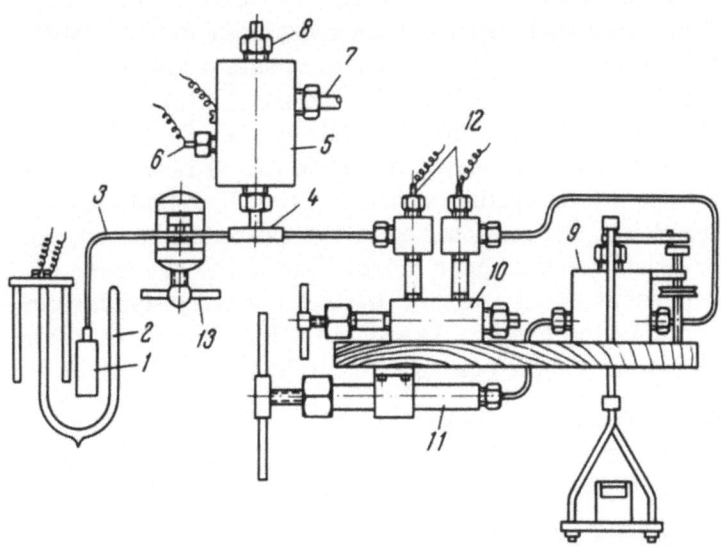

Fig. 262. Installation for measurement of compressibility by gravimetric method: 1) piezometer; 2) thermostat; 3) capillary; 4) T-joint; 5) block; 6) manganin manometer; 7, 8) leads; 9) piston manometer; 10) mercury plug; 11) oil press; 12) electrical contacts; 13) vise.

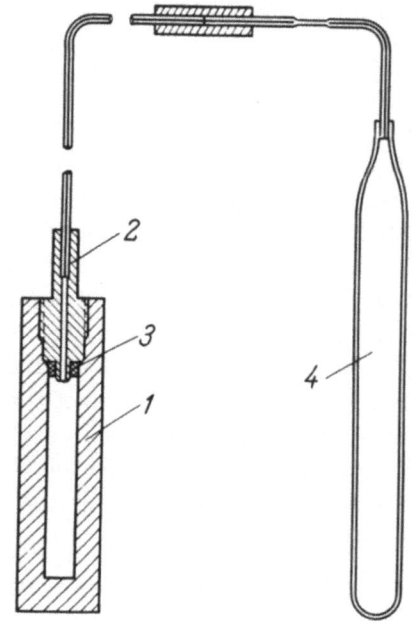

Fig. 263. Piezometer with bottle: 1) body;
2) plug; 3) packing; 4) bottle.

4 and divider 5. The divider is a differential manometer whose sensing element is a tubular manometer filled with water. In order to avoid solution of the gas being investigated in water, mercury plug 6 is placed between gas and water. The pressure in the piston manometer is created by presses 7. After the required temperature is reached in the furnace, the pressure in the piezometer is measured. Then the contents of the piezometer are drained into thermostatted gas meter 8 where the quantity of gas is determined at atmospheric pressure.*

Vulkalovich-Altunin Installation [5] (Fig. 261). This installation differs from those described above in the method of measuring quantity of gas. Here also piezometer 1 is filled with the gas being investigated.† After equilibrium has been established, constant-volume valve 2 is closed, separating the piezometer filled with gas from the remainder of the system. Then the gas located in line 3 connecting valve 2 to vessel 4 is opened, the line is evacuated using pump 5 and, after this, the gas (in this case carbon dioxide) is exhausted from piezometer 1 into vessel 4. The vessel, filled with an adsorbent (such as activated carbon AG-2) is weighed and placed in a medium with a temperature of about −100°C. At this temperature, the gas is adsorbed onto the carbon. After the gas is exhausted, vessel 4 is closed, heated, and weighed once more. Due to the presence of the adsorbent, the pressure in the vessel, even at ordinary temperature, is not great. This allows the use of a vessel with thin walls, i.e., a very light vessel which can be weighed on an analytic balance.

The pressure created by thermocompressor 9 is measured by piston manometer 6 connected to the systems through differential manometer 7.

---

*A similar method was used by D. S. Tsiklis and A. I. Kulikova [4] for measurement of molar volumes of oxygen at pressures up to 10,000 atm and temperatures up to 400°C.

†The right portion of the figure shows the installation for gas purification.

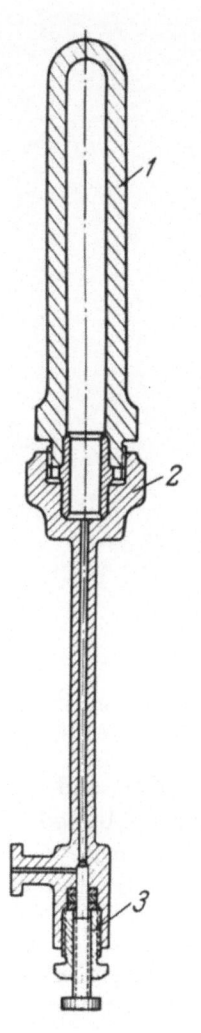

Fig. 264. High-pressure gravimetric piezometer: 1) ampule; 2) head; 3) valve.

Benedict Installation. The quantity of gas in a piezometer at a fixed pressure and temperature can be determined by direct weighing. These methods are called weighing or gravimetric methods [6]. The gravimetric method was first used by D. I. Mendeleev [7]. The method consists of determining the quantity of gas in a piezometer by the difference in masses between the full and empty piezometer. For this, piezometer 1 (Fig. 262), which is a thick-walled nickel vessel, located in thermostat 2, is connected through nickel capillary 3 and T-joint 4 with block 5 and piston manometer 9. The piezometer is filled with gas at high pressure through leads 7 and 8. Block 5 contains manganin manometer 6, which is used for measurement of pressures over 1000 atm.

After the piezometer is filled, it is allowed to equilibrate to the temperature of the thermostat and the pressure is measured. The piston manometer is connected to the installation by mercury plug 10 which is required to separate the gas being investigated from the oil filling the manometer; after this, the plug allows leakage of oil from beneath the piston of the manometer to be checked. Press 11 is used to create a pressure approximately equal to the measured pressure, and the mercury heights in the columns of trap 10 are set at identical levels, which is judged by closing of electrical contacts 12.

After the mercury level is set, the pressure is measured and capillary 3 is immediately clamped by vise 13. Then the capillary is twisted and sealed with silver solder. After this, the vise is released, the piezometer is weighed, the capillary is cut (without loss of metal), and the piezometer is weighed once more after the gas has been removed from it.

At pressures corresponding to densities over 550 amag units,* the cut end of the capillary is not soldered, but is rather connected to nickel bottle 4 (Fig. 263). The gas is allowed to enter the bottle, and the piezo-

---

* An amag density unit is the ratio of the density of a compressed gas to the density of the same gas at atmospheric pressure.

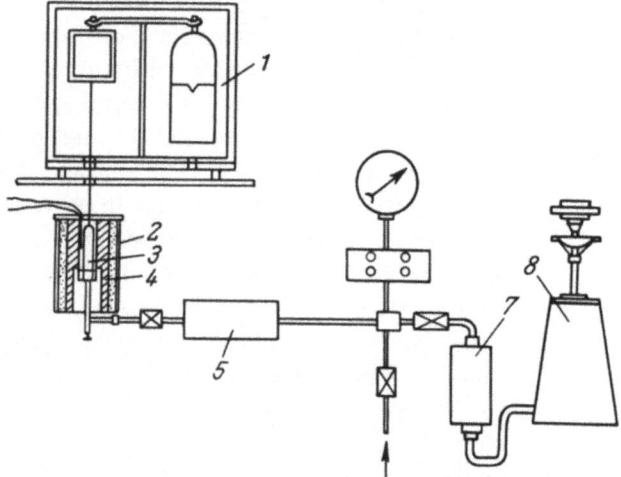

From high-pressure compressor

Fig. 265. Installation for measurement of molar volumes using
gravimetric piezometer: 1) balance; 2) electric furnace; 3)
piezometer; 4) copper block; 5) oil separator; 6) thermocouple;
7) divider; 8) piston manometer.

meter and bottle are weighed together. Then the gas is exhausted and both
vessels are weighed once more.

The structure of the piezometer is shown in Fig. 263. Body 1 and
plug 2 of the piezometer are made of pure nickel, the body being polished
on the inside. Copper packing 3 is placed beneath the plug, the plug is
tightened and, by heating the piezometer in hydrogen, the packing is melted
to provide a reliable seal.

The piezometer is autofrettaged to strengthen the metal. The piezometer is calibrated in water, allowing no bubbles of air to enter the
piezometer.

Bilevich Installation. The gravimetric method was used by
A. V. Bilevich, L. F. Vereshchagin, and Ya. A. Kalashnikov [8]. The piezometer (Fig. 264) consists of three parts: ampules 1, head 2, and vertical
capillary and valve 3. The piezometer is suspended on balance 1 (Fig. 265)
in copper block 4 which is located in electric furnace 2, and is connected
through divider 7 to piston manometer 8. The furnace containing the piezometer is heated to the experimental temperature, the piezometer is filled
with the gas being investigated under the defined pressure and, after thermal equilibrium is achieved, the pressure in the piezometer is measured
by the piston manometer. Then the valve is quickly closed and the piezometer is separated from the gas feed line and weighed; the gas is exhausted
and the piezometer is weighed once more.

Piezometers of constant volume with electromagnetic agitation are often used in installations for simultaneous study of phase and volume behavior of solutions [9]. This group of methods includes the polythermal method (see Chapter IX). There are many other methods [10], which amount to variations of those described above; we will not describe them in detail.

The change of capacity of piezometers with pressure and temperature must be considered. Therefore, in order to produce reliable results, piezometers must be calibrated at the conditions of the experiment, i.e., at the corresponding pressures and temperatures. The calibration should be repeated. Baric and thermal deformation of piezometers can be taken into consideration using the equation presented below (see page 385).

If experiments are performed carefully taking a good deal of time, errors can be reduced to 0.2%.

Krichevskii-Markov Installation. As was noted above, it is very important to be able to measure pressure precisely. However,

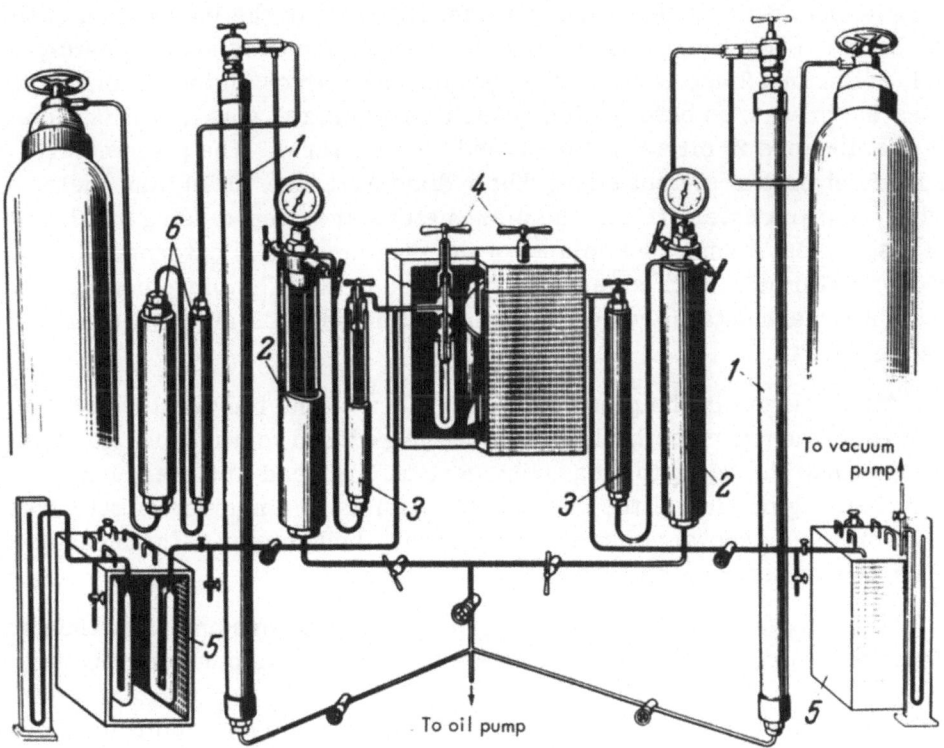

Fig. 266. Krichevskii-Markov installation: 1) low-pressure collector; 2) high-pressure collector; 3) oil separator; 4) piezometers; 5) measurement portion; 6) vessels for gas purification.

the investigator may not always have a piston manometer available. I. R. Krichevskii and V. P. Markov [11] used a method for measuring compressibility without using a piston manometer [12]. Their installation (Fig. 266) consists of two parts, the vessels of one part of which are connected to the vessels of the other portion through an oil line,* which maintains identical pressure in the corresponding vessels.

The principle of operation of the installation is as follows: the compressibility of the gas or gas mixture being investigated is compared with the known compressibility of any other gas. The left portion of the installation is designed for determination of the pressure from the compressibility of nitrogen, known from the literature; the right portion of the installation is designed for determination of the compressibility of the gas being investigated.

The investigated gas or gas mixture† is taken from a high-pressure cylinder into the entire right-hand portion of the installation. Then, by feeding oil into low-pressure collector 1, the pressure of the gas in the installation is increased to 150 atm. After the low-pressure collector is disconnected, oil is fed into the high-pressure collector 1, increasing the pressure to the limit pressure for which the vessels and oil pump are designed. The same operation is performed with the nitrogen in the left portion of the installation. When the pressure in both portions becomes approximately equal, collectors 2 are connected by opening the valves on the oil line. This causes the pressures between the vessels to equalize. Gas now fills a portion of collectors 2, oil separator 3, and piezometer 4. The piezometers are made of heat-resistant steel. Their diameters are 40/10 mm, their capacity approximately 10 ml; the piezometers are opened using a valve with a spherical stem. The volume of the piezometers is determined by mercury calibration with an error of not over 0.1%. The piezometers are placed in a thermostat in which the temperature is maintained with an accuracy to 0.1°C.

After the gas in the piezometer has taken on the temperature of the thermostat, the pressure in the feed lines is reduced to atmospheric in order to exhaust the gas into measuring portion 5, placed in a thermostat. The measuring portion of the instrument consists of a number of glass vessels of precisely known volume and a mercury manometer. The measuring portion is evacuated before the gas is let into it.

If the pressure is known in the glass portion before and after the gas is drained into it, and the capacity of the vessels in this portion is known,

---

*A study of the change in composition of a gas mixture due to different solubilities of its components in oil [11] has shown that the composition of the mixture practically does not change.

†Gas mixtures were studied separately on the installation.

it is not difficult to calculate the quantity of gas in the piezometer.* Then, using the known volume of the piezometer, quantity of nitrogen (or other gas) which it contains, and compressibility of the gas, the pressure can be calculated, after which the compressibility of the investigated gas can be determined.† After the piezometer is emptied, the subsequent experiment can be performed at lower pressure.

This method allows us to avoid the introduction of corrections for change in volume of the piezometer with pressure and temperature, since the correction is made unnecessary by the presence of two piezometers under identical conditions.

Measurement with Variable-Volume Piezometers

As was noted above, this method consists of compressing the gas in a vessel by changing the volume of the vessel. The volume is changed by various methods − by pumping mercury into the vessel, by forcing a piston into the vessel, etc. During the course of the experiment, the experimenter immediately produces the relation between volume and pressure at the given temperature.

Installation with Visual Observation of Meniscus.
Figure 267 shows one such installation [13], used to investigate the compressibility of various gases to 3000 atm. The piezometer consists of a tube with spheres 1 and expanded portion 3, terminating in a tip submerged in mercury. The entire piezometer is placed in high-pressure vessel 2. The gas in the piezometer is compressed by the mercury, which is introduced under oil pressure. The level of the mercury is determined by window 4 to which one of the spheres of the piezometer is set by screw 5; the mercury meniscus is then observed. A similar principle was used in the work of D. L. Timrot [14].

Another method of determining the mercury level is the introduction of platinum contacts between the spheres, the level of mercury being judged by the number of contacts which it touches. This method is being used by many investigators at the present time [15].

Kazarnovskii-Simonov-Aristov Installation. Ya. S.
Kazarnovskii, G. B. Simonov, and G. E. Aristov [16] also used a contact piezometer (Fig. 268). The wide, open end of glass piezometer 1 is placed in metal vessel 2, filled with mercury. Millivoltmeter 3 shows the potential differences between the ends of platinum spiral 4. The capacity of the spheres between the platinum contacts is carefully calibrated by mercury.

---

*It should be noted that this method for determining the quantity of gas included in a piezometer is the most convenient and precise.

†The principle of measurement of pressure was borrowed from a work by Witkowskii [12].

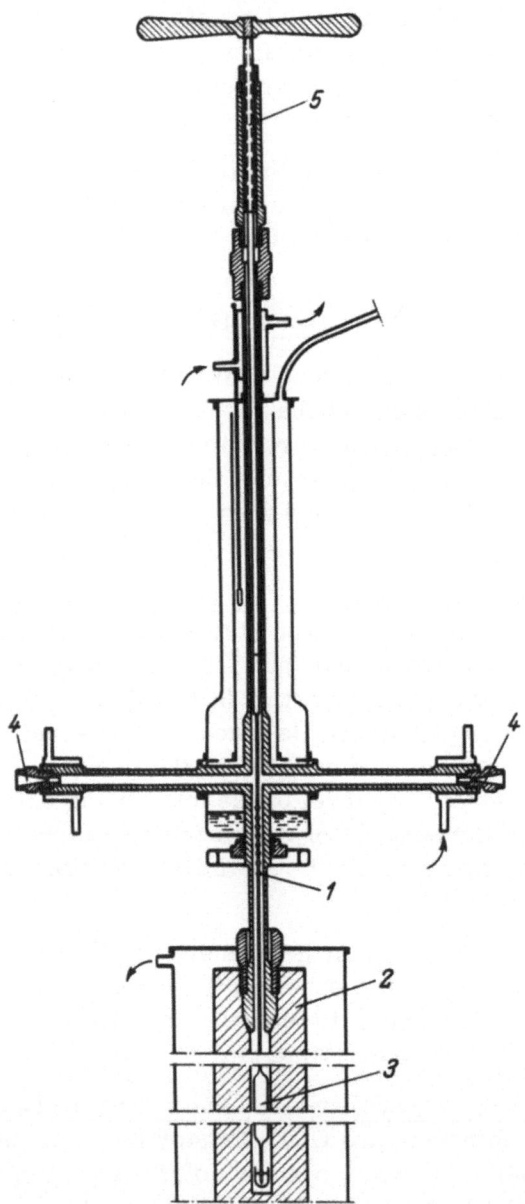

Fig. 267. Amagat installation: 1) piezometer spheres;
2) high-pressure vessel; 3) wide portion of piezometer;
4) viewing windows; 5) screw.

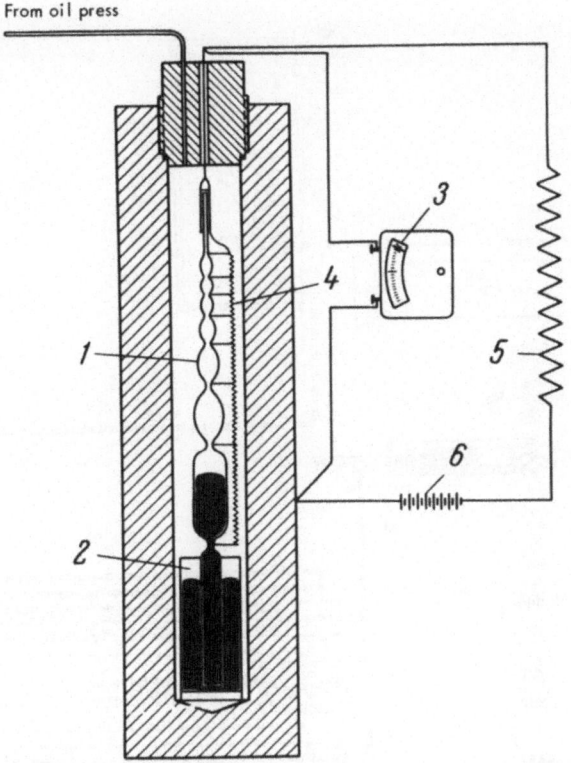

From oil press

Fig. 268. Plan of glass piezometer installation: 1) glass piezometer; 2) vessel; 3) millivoltmeter; 4) spiral; 5) resistance box; 6) battery.

As oil is pumped into the apparatus, the mercury enters the piezometer and, as it rises, makes contacts to connect different portions of the platinum spiral. The contacting is noted using millivoltmeter 3. Resistance box 5 is used as a shunt, the values of resistance being selected for each experiment.

After noting the pressure corresponding to connection of a new section of the spiral, the compression is continued, thereby producing a dependence between pressure and volume at the given temperature, the volumes being fixed by the distance between platinum contacts of the piezometer.

This plan permits the compressibility of a gas to be investigated at various pressures and temperatures in successive experiments without disassembling or emptying the apparatus.

The Kazarnovskii–Simonov–Aristov apparatus plan is shown in Fig. 269. The gas mixture is prepared in mixer 1 as described in Chapter IX (page 337).

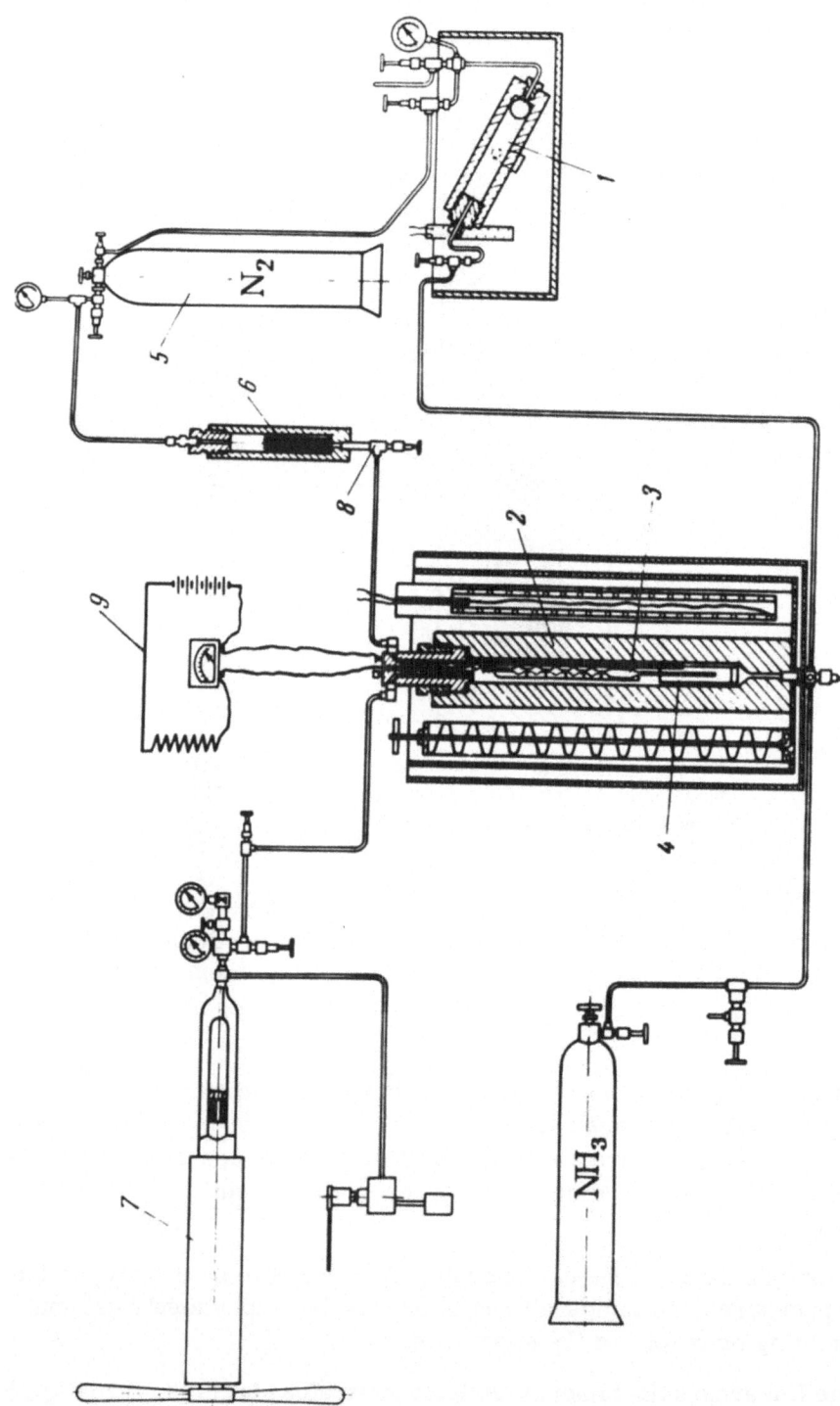

Fig. 269. Kazarnovskii-Simonov-Aristov installation for determination of compressibility of nitrogen-hydrogen-ammonia mixtures: 1) mixer; 2) column; 3) piezometer; 4) vessel; 5) gas cylinder; 6) mercury column; 7) oil press; 8) valve; 9) electrical circuit of measurement portion of installation.

A residual pressure of 10^{-2} mm Hg is created in column 2 containing the piezometer, after which the gas mixture is transferred from mixer 1 to column 2 through a heated tube (to prevent condensation of such gas components as ammonia). The mixture fills the piezometer, vessel 4, and all the space in the column. Then nitrogen is fed from cylinder 5 through column 6 to transfer all the mercury which it contains into vessel 4, thus sealing a portion of the gas in the piezometer. Press 7 is used to fill column 2 with oil, displacing the gas mixture from the space around the piezometer. The oil and gas are allowed to take on the temperature of the thermostat, after which the compression of the gas is begun. The closing of the contacts is noted as the pressure is increased, then again as it is reduced. The closer the values of pressure produced in the two series of measurements, the more accurate the results of the experiment. A pause of ten to fifteen minutes is necessary before each contact is closed in order to allow the heat of compression of the gas to be dissipated and to allow the portion of liquid just fed into the apparatus to take on the experimental temperature.

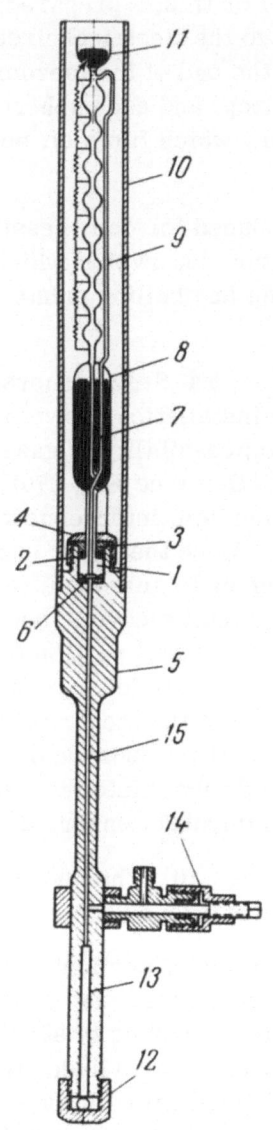

Fig. 270. Kazarnovskii-Sidorov piezometer for determination of gas compressibility at high pressures and low temperature: 1) glass piezometer; 2) leather ring; 3) metal collar; 4) nut; 5) metal portion of piezometer; 6) lead packing; 7) tube; 8) reservoir; 9) calibrated tube; 10) metal cover; 11) mercury vessel; 12) nut; 13) cylinder; 14) valve; 15) bore.

The total volume of the piezometer depends on the final pressure: in this installation it was 8 ml. The diameter of the capillary between contacts is 0.6-0.7 mm. The contacts are made of platinum wire, 0.3 mm in diameter, and the diameter of the connecting wire wound in the form of a spiral (in order to increase the resistance) was 0.2 mm. The piezometer is made of glass whose coefficient of volumetric expansion is between $2.6 \cdot 10^{-5}$ and $2.7 \cdot 10^{-5}$.

The piezometer is calibrated with mercury. For this, a prepared heat-treated ("tempered") piezometer is connected to the electrical circuit (see Fig. 268). A narrow capillary is connected to the end of the piezometer, so that the mercury will exit slowly, drop by drop, and each sphere of the piezometer is calibrated by weighing the mercury which flows off between closings of contacts.

At high pressures, a correction must be introduced for compressibility of the material of the piezometer. The temperature interval at which the installation can be used is limited by the freezing and boiling points of the oil and mercury under pressure.

Kazarnovskii-Sidorov Installation. Ya. S. Kazarnovskii and I. P. Sidorov [17] varied the construction of the installation shown in Fig. 269 to make it suitable for measurement of compressibility of gases at low temperatures. The piezometer in this installation (see Fig. 270) consists of two parts — a glass portion and a metal portion, interconnected by a method described in Chapter VI (see page 253). Here the lower end of the glass portion is a cylinder 12 mm in diameter and 8 mm high. At the upper end of the cylinder we find leather ring 2, metal sleeve 3, and nut 4. Lead packing 6 is applied to the upper end of the metal portion 5 of the piezometer and supports the conical shelf of the glass cylinder, compressed by nut 4. The mercury located in reservoir 8 of the piezometer rises through the tube 7 into the glass spheres under oil pressure and closes the contacts; the gas passes through calibrated tube 9 into the metal portion of the piezometer, which is kept at the experimental temperature.

The glass piezometer is protected by metal cover 10. The upper portion 11 is used to connect the piezometer to the electrial lead.

The mercury in reservoir 8 is connected to the metal portion of the piezometer by a conductor. Metal portion 5 of the piezometer also consists of two parts. Cylinder 13, which is a piezometer, has a volume of about 5 ml. It is closed by a steel sphere, held down by nut 12. The cylinder terminates in bore 15 (approximately 3 mm in diameter) lined with a hollow conductor, the conductor is placed in the bore to reduce the free space in the channel. The volume of the bore is precisely determined by calibration with mercury, and the volume of the conductor is determined by its mass and density.

Calibration of this piezometer differs from that described above and is performed in two stages: 1) before reservoir 8 is sealed, all the spheres and a portion of tube 9 are calibrated; 2) after reservoir 8 is sealed, the remaining portion of tube 9 is calibrated right up to the conical shelf at the cylinder. A special glass sleeve through which the level of off-flowing mercury can be seen is used for calibration.

Determination of compressibility of gases using this installation (Fig. 271) consists of the following operations: upper column 1 containing the glass portion of the piezometer is filled with oil using press 2. The piezometer is washed several times, then filled with the gas to be investigated, the initial pressure being selected such that at maximum compression the mercury will rise to the last contact. The glass piezometer and column 1 are held at a temperature of 0°C. The lower metal portion of the piezometer 3 is placed in a cryostat, held at the experimental temperature. After the gas in both portions of the piezometer takes on the proper temperatures, oil is fed into column 1 compressing the gas, and the piezometer contacts are closed. The pressure and temperature are noted as the contacts close.

In order to determine the molar volume of a gas at low temperatures, the quantity of gas in the metal portion of the piezometer must be known, which requires a measurement of the total number of moles of gas used in the experiment and subtraction of the number of moles of gas located in the upper glass portion of the piezometer and in the connecting tube. The measurement of the total quantity of gas is performed by passing all the gas through valve 4 into the glass calibrated vessels 5 located in the thermostat. The vessels are preliminarily evacuated by pump 6, and the quantity of gas is determined from the indications of the mercury manometers.

The compressibility of the gas located in the ring-shaped space of the connecting tube to the metal piezometer is calculated by relating it to the average temperature of both thermostats. The value of PV at this temperature is found by extrapolation of the data at the higher temperatures (if such data are available). The values of gas compressibility and the known volume of the connecting tube and part of the glass piezometer containing no mercury are used to calculate the number of moles of gas in the upper glass portion of the piezometer and the connecting tube.

The error in measurement performed by the authors [17] was evaluated by them as 0.3-0.5%.

Sidorov-Kazarnovskaya Installation. A further improvement of this method allowed I. P. Sidorov and D. B. Kazarnovskaya [18] to create an installation for measurement of compressibility of gases at high temperatures.

The apparatus (Fig. 272) consists of two steel columns located one above the other. The upper column 1, made of heat-resistant steel, is placed in thermostat 2 containing melted tin, the temperature in which may reach 500°C. The lower column 3 is held at room temperature.

Glass piezometer 4, consisting of two reservoirs in the upper and lower columns plus a connecting capillary, is placed in both columns. The lower, narrow end of the piezometer enters container 8, filled with mercury.

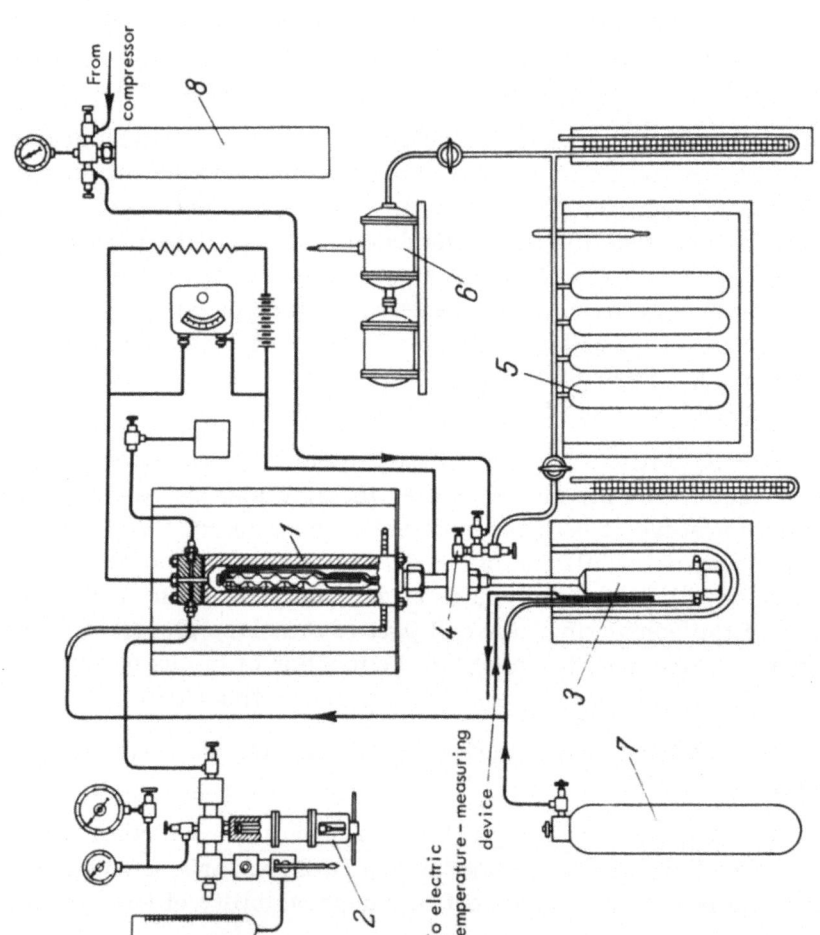

Fig. 271. Kazarnovskii-Sidorov installation for measurement of compressibility of gases at low temperatures: 1) column; 2) press; 3) metal portion of piezometer; 4) valve; 5) glass vessel; 6) vacuum pump; 7, 8) gas cylinders.

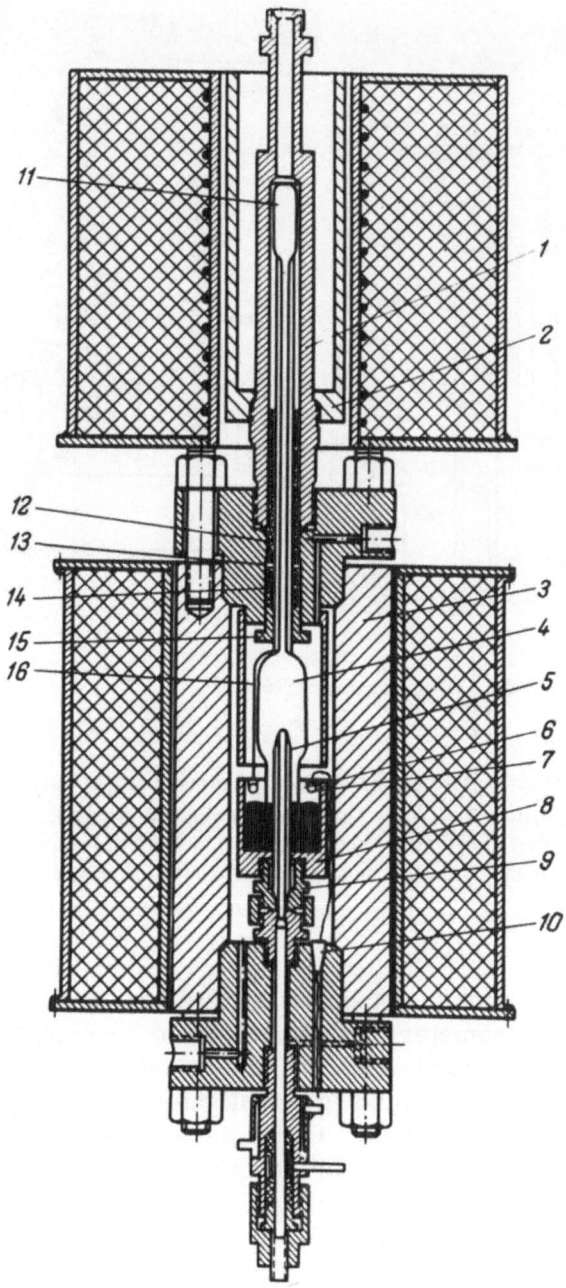

Fig. 272. Sidorov-Kazarnovskaya apparatus for measurement
of gas compressibility at high temperatures and pressures: 1)
upper column; 2) thermostat; 3) lower column; 4) glass piezometer; 5) capillary; 6) container; 7, 16) contacts; 8) outer container;
9) nipple; 10) electrical lead; 11) upper piezometer cylinder; 12) asbestos insulation; 13, 14, 15) glands.

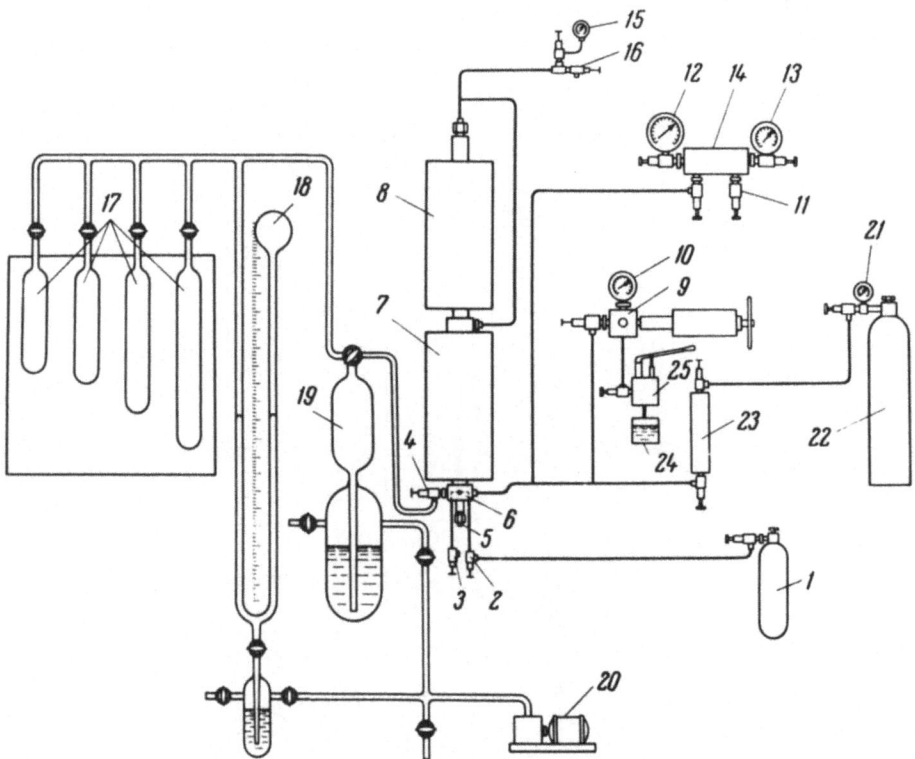

Fig. 273. Plan of installation for determining compressibility of gases at high pressures and temperatures: 1, 22) gas cylinders; 2, 3, 4, 11, 16) valves; 5) stop valve; 6, 14) blocks; 7) lower
column; 8) upper column; 9) hydraulic press; 10, 12, 13, 15, 21) manometers; 17) thermostatted
cylinder; 18) mercury manometer; 19) mercury pump; 20, 25) pumps; 23) compression chamber;
24) oil reservoir.

In the center of the container there is a glass capillary 5, the upper end of
which enters the piezometer. The lower end of the capillary has a plate
which is lapped to fit metal nipple 9. The container and capillary act as a
mercury plug to hold the gas being investigated in the piezometer. The
level of mercury in the container is maintained using lower contacts 7
through the mercury in glass inner-container 6, and electrical lead 10, connected to the electrical circuit. The mercury in the piezometer is maintained
at a level to contact upper platinum contact 16.

The total capacity of cylinder 11 and the connecting capillary to the
platinum contact is the final volume of gas after compression, the volume
of the capillary making up only 0.2% of the volume of cylinder 11. Therefore, the change in gas temperature in the capillary can not cause any great
error in the compressibility values determined. Still, measures are taken

to reduce the heat flow between the hot and cold portions of the installation. This is achieved by asbestos insulation 12, held between the capillary and glands 13, 14, and 15.

The method for performing the experiment is as follows: first, the glass piezometer is washed by the gas being investigated from glass cylinder 1 (Fig. 273) through block 6 and stop valve 5. In order to maintain the mercury level of the piezometer at the height of the lower contact, oil is fed into the piezometer by hydraulic press 9 simultaneously with the input of the gas. The required pressure is created by forcing the gas and oil into the piezometer; the pressure is measured by manometer 10.

After the piezometer has been washed and filled, valve 5 is closed and mercury is forced into the piezometer to the level of the upper contact. Then, the thermostat is heated to the required temperature. The gas in the piezometer is heated and, as it expands, presses on the mercury. The mercury is maintained at the upper level by forcing oil into the installation.

When the gas takes on the temperature of the thermostat, the level of mercury stops changing and the pressure becomes constant. At this time, the level of mercury is finally set in the piezometer (at the upper contact) by draining excess oil through valve 11; then the pressure is measured. Then, a portion of the gas from the piezometer is exhausted to the vacuum portion of the installation. This is done by draining a small quantity of oil from the installation through valve 16, and adding mercury to the level of the lower contact. When valve 5 is opened, the gas is vented to thermostatted calibrated cylinders 17.

Before the gas is exhausted, valve 4 is opened and valves 2 and 3 are closed, and the entire vacuum installation and metal block 6 are evacuated to a residual pressure of 0.5 mm Hg using vacuum pump 20. Then, with valve 4 closed, valve 5 is opened and, by carefully opening control valve 4, the gas in the glass installation is reduced, bringing the pressure to 400 mm Hg. After this, valve 5 is closed and the gas exhausted from the high-pressure installation is pumped by mercury pump 19 into calibrated vessels 17.

By raising the mercury level in the piezometer to the upper contact, the remaining gas is transferred to the working portion of the piezometer and once more, after the temperature is established, the pressure is measured. Then, the gas is once more fed into vessels 17. The number of such operations depends on the initial pressure. When the last portion of gas is fed into the vacuum installation, the gas pressure in the piezometer should be approximately 1 atm as the mercury reaches the level of the lower contact. When the mercury is subsequently raised to the upper con-

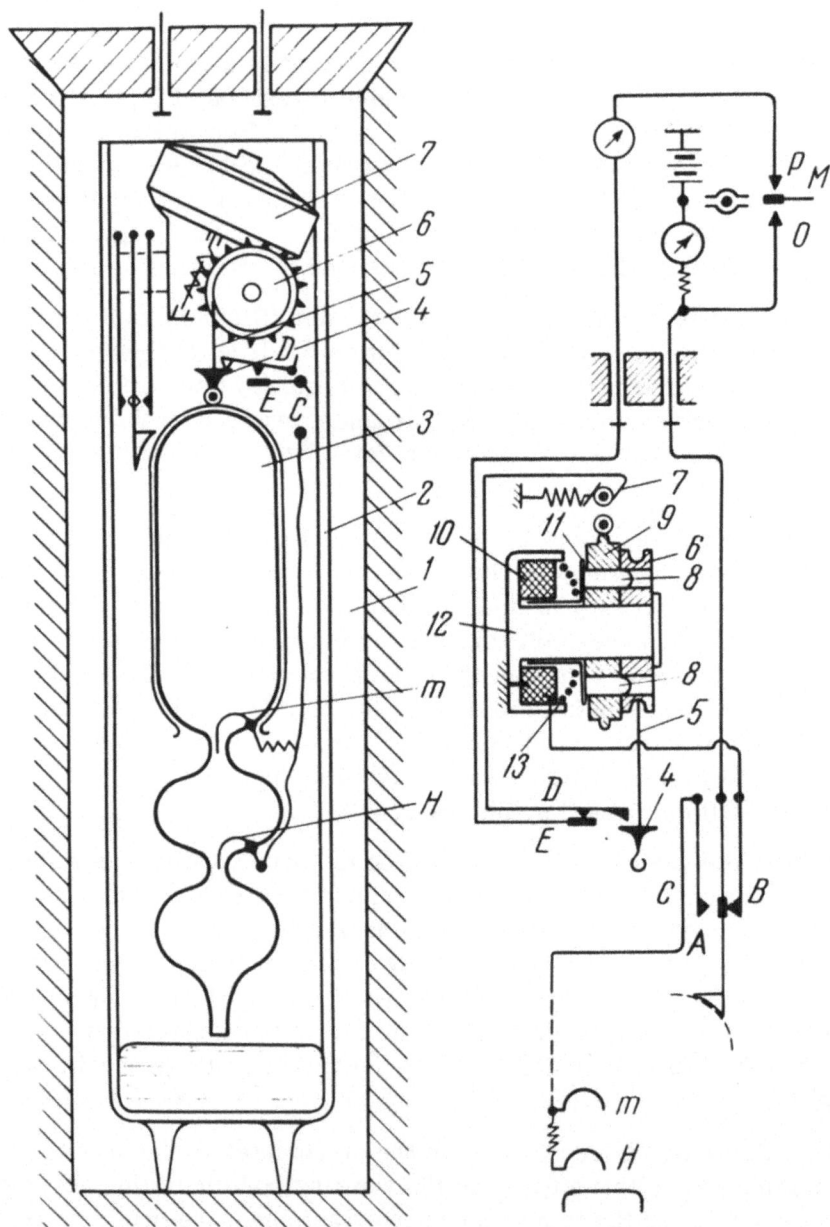

Fig. 274. Apparatus for measurement of compressibility at pressures up to 10,000 atm:
1) high-pressure vessel; 2) cylinder; 3) piezometer; 4, 11) plates; 5) cable; 6) pulley;
7) motor; 8) pin; 9) worm drive; 10) electromagnet; 12) axis; 13) spring.

tact, the pressure is increased to approximately 10 atm. The quantity of gas at this pressure in the piezometer at a temperature of about 300°C can be calculated from the equation for an ideal gas.

This method can be used to produce data with high accuracy. The errors introduced by unevenness of heating of the connecting capillary can be ignored, since its volume amounts to only 0.2% of the volume of the cylindrical portion of the piezometer. The volume of the capillary through which the gas is introduced from the nipple to the valve pin is only 0.04% of the volume of the cylindrical portion.

Contact piezometers can be used at pressures up to 10,000 atm. The method of attaching the platinum contacts must be carefully developed. A type of glass must be selected which will not cause fissuring of piezometer seals at high pressures due to varying values of coefficient of compression of the glass and platinum under pressure.

Basset Installation. Basset [19] suggested another method for measuring compressibility using glass piezometers (Fig. 274). The principle of this method is that the gas being investigated, located in the piezometer, is sealed in by mercury at the initial pressure by placing the open end of the piezometer in mercury. Then, as the pressure is increased in the apparatus, the gas is compressed in the piezometer by the mercury until the pressure is reached at which the mercury closes the contact.

The high-pressure vessel 1 contains cylinder 2, into which the mercury is poured. The cylinder contains piezometer 3 made of pyrex glass. The piezometer is suspended on cable 5 so that its open end is not far from the surface of the mercury. The upper portion of cylinder 2 contains a device which is used for movement of the piezometer. Cable 5 is wound around pulley 6. Electric motor 7 rotates worm gear 9 which is connected by bar 8 to pulley 6.

The gas being investigated is compressed in vessel 1 to the initial pressure of about 4000 atm. Under this same (carefully measured) pressure, the gas is allowed to fill the piezometer. Then, by rotating switch M to position O, electromagnet 10 is connected which, overcoming the resistance of spring 13, pulls plate 11. The plate pulls pin 8 out of pulley 6, which then turns freely on axis 12 and drops the piezometer into the mercury.

As it drops, the piezometer frees contact A, which is swept from position B to position C. As this occurs, the circuit of contacts m and H is automatically switched on; this circuit is used for measurement of the mercury level of the piezometer. The electromagnet circuit is disconnected, as a result of which spring 13 moves plate 11 back to the right. Pin 8 enters the aperture in pulley 6. At the same time, plate 4 frees contact DE, which is closed.

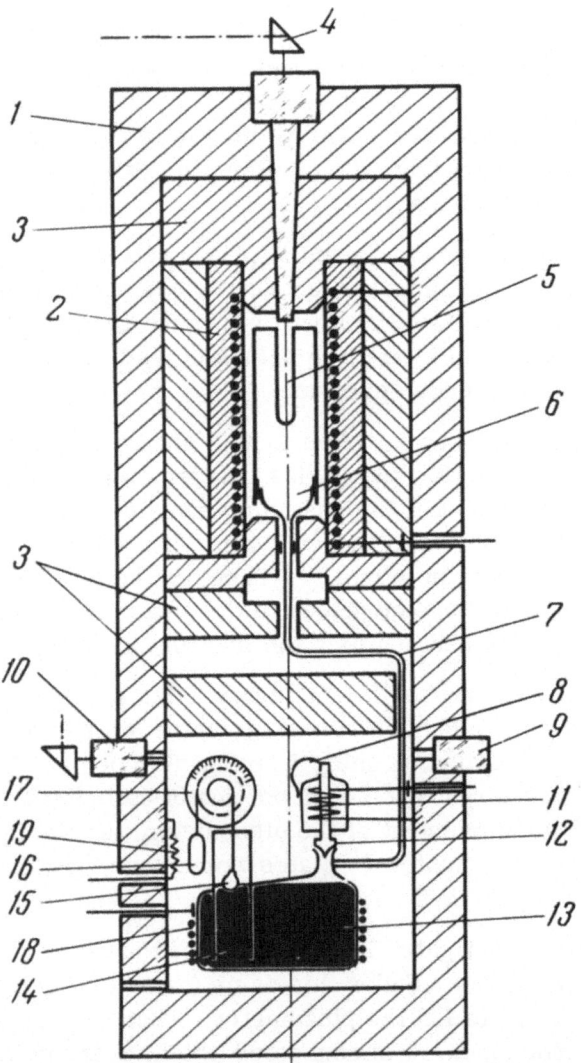

Fig 275. Installation for measurement of compressibility at high pressures and temperatures up to 1200°C: 1) high-pressure vessel; 2) electric furnace; 3) radiation screens; 4) pyrometer; 5, 7, 14) tubes; 6) piezometer; 8) spring; 9, 10) viewing windows; 11) electromagnet; 12) pin; 13) measurement vessel; 15) plumb; 16) counterweight; 17) disc; 18, 19) resistance thermometers.

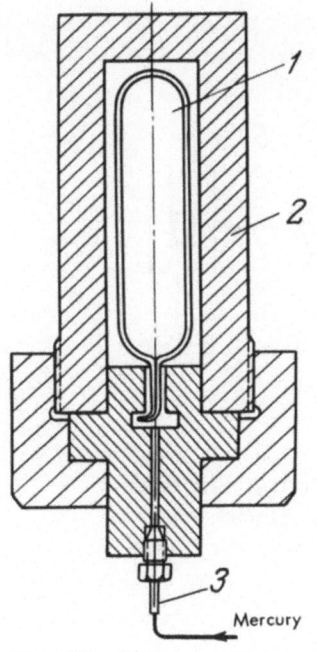

Fig. 276. Glass piezometer: 1) glass cylinder; 2) steel vessel; 3) capillary.

Now, switch M is placed in the central position and the pressure in the apparatus is increased by pumping in the gas being investigated. The mercury enters the piezometer and closes contacts H and m one after the other.

When the contacts are closed the compression is stopped and, after a constant temperature is attained, the pressures at which the contacts open and close are measured.

Having measured the pressure corresponding to the position of the mercury at the level of contact m, the pressure is reduced again to the initial pressure* and switch M is placed in position P. This turns on electric motor 7 (see Fig. 274). The piezometer is returned to the initial position, contact A is switched to B once more, and contacts DE are opened. The motor stops. Then, the quantity of gas in the piezometer is measured, increasing or decreasing the pressure, and the procedure is repeated. The installation is designed for measurement of compressibilities at pressures up to 10,000 atm and temperatures up to 200°C.

The measurement of compressibility at temperatures on the order of 1200°C has been performed using a method [19] based on direct measurement of the expansion of a gas located in a platinum piezometer of known volume. High-pressure vessel 1 (Fig. 275) is cooled with water. Vessel 1 contains electric furnace 2, carefully screened by radiation screen 3 made of foamed magnesite. The furnace contains platinum piezometer 6 with internal tube 5, which is used for measurement of the temperature by optical pyrometer 4. The piezometer is connected to capillary platinum tube 7 which is in turn connected to vessel 13 which is used for measurement of the increase in volume of the gas as the piezometer is heated. Steel vessel 13, the volume of which is precisely known, is carefully insulated from the thermal radiation of the piezometer. The vessel is equipped with a conical aperture which is closed by pin 12, retained by spring 8. This valve is opened using electromagnet 11. Measurement tube 14 contains plumb 15, suspended on a cable. The cable passes over a pulley and is connected to counterweight 16 (Fig. 275).

---

*The pressure is reduced somewhat, so that the mercury goes out of the tip of the piezometer.

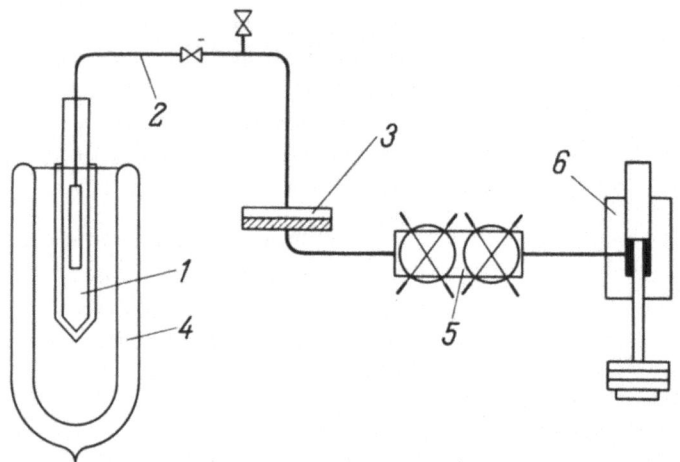

Fig 277. Plan of installation for measurement of compressibility at low temperatures with membrane differential manometer: 1) steel vessel; 2) capillaries; 3) differential manometer; 4) cryostat; 5) oil press; 6) weight manometer.

The device is used as follows: by raising pin 12, the vessel is filled with the gas to be investigated at a pressure of up to 6000 atm. The temperature is allowed to equalize and the pin is closed. Then, the furnace is turned on. The gas in the piezometer expands and exits through capillary 7 into vessel 13, driving the mercury into tube 14. As this occurs, float plumb 15 rises and disc 17 rotates. This disc carries a scale which is observed through window 10 and illuminated through window 9.

Gas under pressure P and at temperature T is located in a piezometer of a known volume. The temperature is changed to T_1 at the same pressure and the change in gas volume is measured. Knowing the temperature of vessel 13, measured by resistance thermometer 18, and of the lower portion of the high-pressure vessel (measured by thermometer 19), we can perform all necessary calculations to determine the compressibility of the gas at the high temperature.

A simpler device [20] similar to that shown on Fig. 275 is shown in 276; it consists of gas cylinder 1 containing a known quantity of gas. The cylinder has a thin sealed tail and is placed in thick-walled vessel 2. The air is evacuated from around the cylinder, and vessel 2 is filled with mercury through capillary 3. Then mercury is placed into cylinder 1 under pressure. The tail of the cylinder is broken and the mercury, entering the cylinder, compresses the gas. If we know the quantity of mercury fed into the cylinder we can determine the volume of gas under the experimental conditions.

Using a membrane differential manometer as a zero instrument, Michels simplified the method for measurement of the compressibility of gases at temperatures considerably different from room temperature [21]. The gas being investigated (Fig. 277) is contained in steel vessel 1 and also fills capillaries 2 and the upper portion of differential manometer 3. Vessel 1 is placed in cryostat 4,* the temperature of which can be reduced to −180°C. The differential manometers contained in an aluminum block whose temperature (25°C) is maintained to an accuracy of 0.01°C. The lower portion of the differential manometer, filled with oil, is connected to oil press 5 and weight manometer 6. The gas is separated from the oil by a membrane made of beryllium bronze; the diameter of the membrane is 64 mm, the thickness is 0.4 mm. Bending of the membrane is measured by an electrical micrometer. The membrane reacts to a change in pressure of 0.001 atm, which corresponds to a change in gas volume in the space of 1.2 mm^3.

Measurement consists of a series of experiments each of which consists of determination of the pressure of the gas in the piezometer at various temperatures of the cryostat. At first, the gas pressure is measured at 25°C and the known volume of the piezometer (27 cm^3) and data on compressibility are used to determine the density of the gas. As the piezometer is cooled, a portion of the gas from the differential manometer enters it. The density of the gas in the piezometer can be determined if we know the pressure and the ratio of volumes of vessel 1 and capillary 2 (see Fig. 277). This ratio is determined by calibration.

Corrections must be introduced for the change in capacity of the piezometer with temperature and pressure. For the piezometer in question,

$$v_t = v_0 (1 + 48.65 \cdot 10^{-6} t + 3.82 \cdot 10^{-8} t^2)$$
$$v_P = v_0 (1 + 1.80 \cdot 10^{-6} P)$$

where v_t and v_p are the volumes at the given temperatures t and pressure P.

Installation for Measurement of Compressibility by Displacement of Piston. This method [22] consists of compressing the gas filling a high-pressure vessel with a piston moving in a cylinder; the position of the piston is used to determine the volume of the gas. The installation [23] (Fig. 278) consists of three parts: upper cylinder 1 is connected by tube 3 to lower cylinder 2. The volume of gas is measured from the movement of piston 4, whose position is determined using rod 5 on scale 6. Before beginning the experiment, piston 4 is placed at its uppermost position and the apparatus is filled with the gas for the investigated through-tube 7. The initial pressure of the gas is established on the basis of the

---

*Vessel 1 may be placed in a furnace.

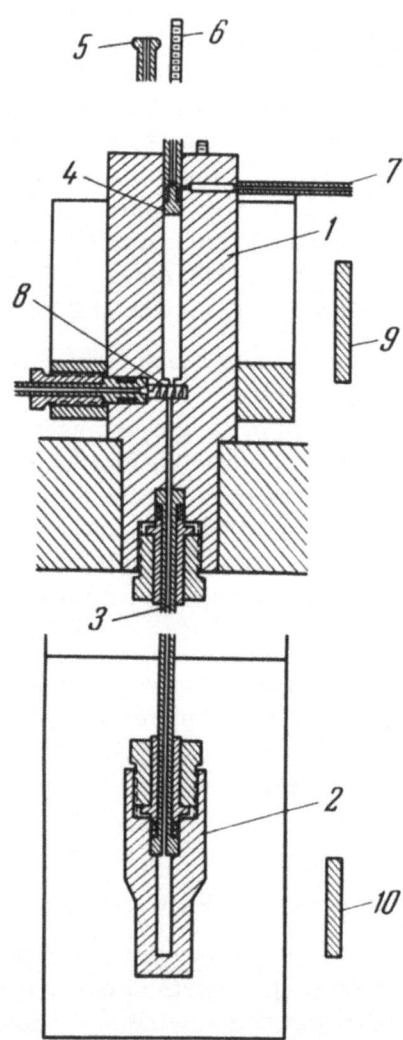

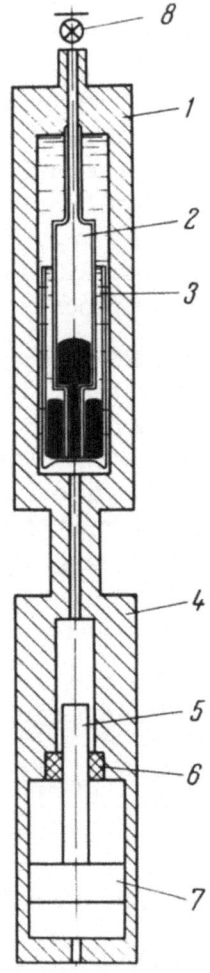

Fig. 278. Installation for measurement of
compressibility by piston displacement method:
1) upper cylinder; 2) lower cylinder; 3) tube;
4) piston; 5) rod; 6) scale; 7) tubes; 8) manganin manometer; 9, 10) inserts.

Fig. 279. Plan of Krichevskii-Tsiklis apparatus: 1)
high-pressure vessel; 2)
piezometer; 3) container;
4) pressure intensifier; 5)
small piston; 6) gland; 7)
large piston; 8) valve.

final compression desired. Then the piston is dropped until it passes the inlet aperture. After this, a temperature of 25°C is created in the entire installation and the piston is set in the position corresponding to the so-called standard pressure, i.e., a pressure at which the compressibility of the investigated gas is known. From this, the quantity of gas located in the installation can be calculated.

Then, while the temperature in cylinder 1 is held constant (25°C), the temperature in cylinder 2 is varied over the required range.* At the assigned temperature, piston 4 is moved in cylinder 1, noting the displacement of the piston required to create the pressure desired, as measured by manganin manometer 8.

The change in volume of the vessels at high pressure can be checked by measuring the compressibility of the gas while steel inserts 9 and 10 are in cylinders 1 and 2; the volumes of inserts 9 and 10 are precisely known at 25°C and atmospheric pressure. If the volumes of the inserts are known under pressure, the change in volume of the insert can be calculated and a correction can be introduced for the change in cylinder capacity under pressure. The volume of the inserts under pressure can be calculated from the cross-sectional area. This calculation is rather complex and introduces an error of no less than 0.3%.

Errors also appear due to inaccuracy of assembly and disassembly of the apparatus when the inserts are placed within it. The entire idea of the control experiment is to exclude the total capacity of the apparatus which should be identical in both cases, from calculation. Also, at high pressures, the residual deformation of the apparatus increases so that the check should be made immediately after an experiment. The method also requires introduction of corrections for unevenness of the distribution of temperature in the walls of the cylinder.†

It should be noted that manufacture of apparatus for this method is difficult. Absolute hermetic sealing of the packing on the piston must be achieved; the piston bore must be carefully polished and it must be absolutely cylindrical.

A more convenient method has been developed by I. R. Krichevskii and D. S. Tsiklis [24]. The high-pressure vessel 1 (Fig. 279) contains metal piezometer 2, placed in vessel 3 containing mercury. Pressure intensifier 4 is connected to the high-pressure vessel. The high-pressure piston 5 of the pressure intensifier is sealed by unsupported-area gland 6. The

---

*When working in a narrow temperature range and when there is no danger of leakage past the seal on the piston, one cylinder can be used alone.
†This problem is investigated in article [23].

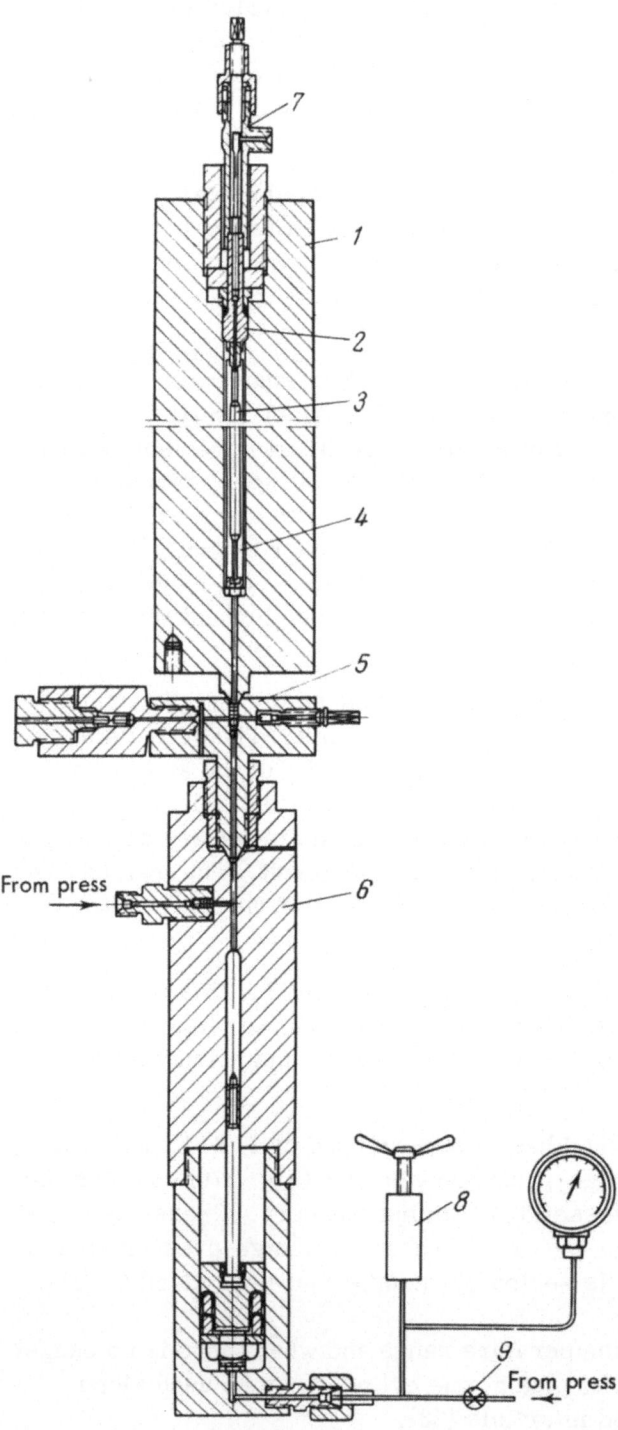

Fig. 280. Krichevskii-Tsiklis apparatus for determination of compressibility at high pressures: 1) high-pressure column; 2) plug; 3) piezometer; 4) vessel containing mercury; 5) valve body; 6) pressure intensifier; 7, 9) valves; 8) press.

piezometer is filled with the gas being investigated through valve 8. The gas is compressed with the intensifier by feeding oil to the low-pressure piston. Mercury, forced into the piezometer by the pressure-transmitting fluid, compresses the gas.

After the assigned pressure and temperature are achieved under piston 7, a calibrated press is used to feed a precisely known volume of oil, increasing the pressure in the piezometer by several hundred atmospheres. Gas is then extracted from the piezometer through valve 8 into a vessel with calibrated volume until the pressure in the installation is reduced to the experimental pressure again.

$$V_2 = \frac{V_1}{K}, \qquad V = \frac{V_2 V_0}{V_3}$$

where V_2 is the volume of mercury which enters the piezometer; V_1 is the volume of liquid fed beneath piston 7; K is the ratio of areas of the large and small pistons in the pressure intensifier; V is the molar volume of the gas under experimental conditions; V_0 is the molar volume of the investigated gas (at 0°C and 760 mm Hg); V_3 is the volume of the gas extracted from the installation (reduced to 0°C and 760 mm Hg).

The change in volume of piston 5 under the experimental pressure due to compression can be easily calculated using well-known methods.

The use of a gland rather than the ordinary seal at the end of the piston allows us to avoid the necessity of introducing a correction for the change in bore diameter of the pressure-intensifier cylinder. Without this gland, a piston seal would adhere to the walls of the bore and the area of the piston is determined not by its diameter, but rather by the diameter of the bore. The increase in diameter of the bore of the cylinder from this deformation causes an increase in piston area. The change in cylinder bore dimensions can be evaluated. Calculations show that with a piston diameter of 12 mm the increase in area amounts to about 0.016 cm^2, or 1.3%.

The installation (Fig. 280) consists of high-pressure column 1 with plug 2, which is connected to piezometer 3, which is placed in vessel 4 containing mercury. The lower portion of the column is connected through valve block 5 to pressure intensifier 6. The gas is fed to the assembled apparatus through valve 7. The initial gas pressure should be not less than 300-400 atm. If a dense gas is being investigated, the pressure may be less. Valve 7 is closed, and a capillary for exhausting the gas into the measuring system is connected to it. Then, after the required pressure and temperature are created in the installation, the measurement itself is begun. For this, valve 9 is closed in the oil line from the hydraulic press to the intensifier and a defined volume of oil is forced under the low-

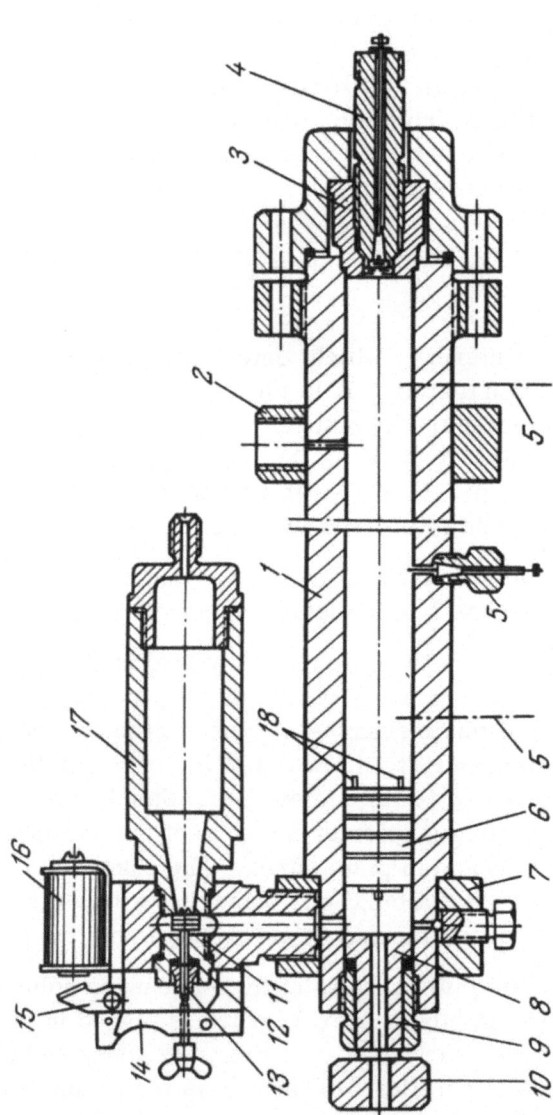

Fig. 281. Ryabinin-Tsiklis device for measurement of compressibility by adiabatic compression: 1) column; 2, 7) rings; 3) sleeve; 4) manometer body; 5) electrical lead; 6) piston; 8) plug; 9) shank of fly-out valve; 10) fly-out valve; 11) valve; 12) packing; 13) shaft; 14) lever; 15) armature; 16) solenoid; 17) accelerating-gas collector; 18) lead protuberances to be crushed.

pressure piston using calibrated press 8. This causes the low-pressure piston, and therefore also the high-pressure piston, to move upward and a portion of the pressure-transmitting liquid from the high-pressure cylinder of the pressure intensifier enters the column, forcing a portion of the mercury into the piezometer. The pressure in the piezometer thus increases.

Then gas is exhausted through valve 7 to the calibrated volume until the gas pressure in the installation is reduced to the initial gas pressure. The pressure intensifier is kept at room temperature and the column is kept in a thermostat at a temperature other than room temperature. The liquid, entering the column, is heated which causes a further increase in pressure in the column. This, it would seem, should lead to error. However, the quantity of liquid forced into the column is so small that only the content of the bore in the valve body (below the valve) actually enter the column; this liquid has taken on a temperature near the temperature of the thermostat. The liquid actually moved by the high-pressure piston of the pressure intensifier first enters the bore of the valve body, where it is heated. Also, the coefficient of thermal expansion of organic fluids (such as kerosene or oil) is equal at atmospheric pressure to 10^{-3} cm^3/deg, at high pressures decreasing by a factor of 2-2.5. Therefore it can be considered that the error caused by temperature gradient is not great, and will not exceed 1-2%.

The accuracy in measurement of pressure is determined by the accuracy in measurement of the electrical resistance of the manganin coil (see Chapter IV). If a measuring microscope is used for determination of position of the arrow of a null-type galvanometer, pressure can be read with an accuracy of ±10 atm. The increase in pressure in the installation caused by addition of oil from press 8 (see Fig. 280) is about 500 atm. Consequently, the accuracy of measurement of pressure is ±2%.

Methods of determination of compressibility in piezometers of variable volume include the entire group of methods used for determination of parameters of stratal petroleum [25, 26].

This group of methods includes measurements of PVT ratios using adiabatic-compression installations. We recall that in these installations the gas is compressed by a piston flying rapidly through a column closed at one end. During compression, the gas is heated to very high temperatures. By measuring the pressure and volume of the compressed gas and estimating its temperature, the compressibility of the gas at very high pressures and temperatures can be determined.

An installation designed by the author (Fig. 281) is based on the principle of single compression, developed and checked by Yu. N. Ryabinin [27]. Column 1, made of heat treated 18KhNVA steel, contains piston 6. Ring 7,

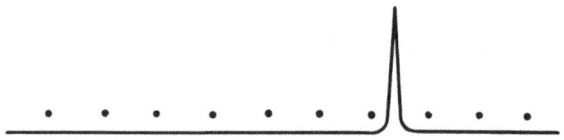

Fig. 282. Oscillogram of compression of nitrogen.

which holds the collector filled with the accelerating gas 17 and valve 11, encircles the column. Valve 11 is used to admit the accelerating gas into the channel of the column. The end of the column is closed with plug 8, employing the unsupported-area principle. This plug contains shank 9, to which load 10 is applied.

When current is passed through coil 16, armature 15 is pulled upward and moves lever 14. The compressed gas in collector 17 moves shaft 13 and rushes into the bore of the column. Piston 6 is forced forward by the accelerating gas and flies toward the closed end of the column, compressing the gas being investigated. When it reaches its terminal position, having expended all its energy received from the accelerating gas, the piston stops and then flies back, under the influence of the expanding gas being investigated.

The weight of valve 10 is so calculated that at this moment it flies out of cap 8 along with shank 9, making an opening for the accelerating gas to leave the column as it is forced back by the returning piston. Piston 6 returns to its previous location and stops.

The column is evacuated through the valve in ring 2 and filled with the gas to be investigated. Packing 12 prevents gas leakage. The second end of the column is closed with sleeve 3; a brass membrane 0.1 mm thick is soldered to the bottom of sleeve 3. The sleeve contains manganin manometer 4 with a flat coil. The space between the sleeve and the coil is filled with kerosene. As it moves in the column, the piston compresses the investigated gas, and its pressure is transmitted through the brass membrane to the kerosene and coil. This causes the resistance of the manganin to be changed, and unbalances the balance bridge in which the coil is connected. The unbalanced current is fed to the input of an electronic oscillograph and recorded as a curve of pressure vs. time.

Nipples with electrical leads to which tin contacts are soldered, directed toward the interior of the column, are screwed into the body of the column. They protrude within the column by 0.1 mm. As it moves through the column, the piston closes these contacts one after the other, the moment of passage being marked on a second line of the same oscillograph. This recording allows the rate of movement of the piston through the column to be established.

Before an experiment is performed, the installation is carefully cleaned, and two lead protuberances 18, 1 mm in diameter, are placed in the piston; the piston is placed in the column, plug 8 is closed and the installation is filled with the gas being investigated. Then, the accelerating gas is fed into collector 17 and, after the required pressure is created, valve 11 is opened. The "shot" is fired, after which an oscillogram is produced with a recording of the pressure and piston velocity. After the piston is extracted, the lengths of the lead protuberances 18 are measured, i.e., the distance from the piston to the end of the cylinder at the moment of piston halt is determined and, if now the initial volume of gas at atmospheric pressure is known, its volume at the final pressure can be calculated.

Knowing the volume and pressure of the gas at the moment that the piston stopped and assuming that the gas obeys the adiabatic law, we can calculate the compression temperature and thereby produce the PVT data at pressures up to 10,000 atm and temperatures on the order of 5000–7000°C.

Figure 282 shows an oscillogram of compression of nitrogen, from which we see that the entire process of movement of the piston takes about 10 msec, the sharp increase in pressure about 1 msec.

The necessity of calculating the gas temperature according to the law for an ideal gas is a great deficiency of this method. This necessity is a result of the great difficulty which arises in attempts to measure the temperature of a compressed gas.

The temperature can be calculated by an elegant method suggested by I. R. Krichevskii. The piston moving in the column receives its energy

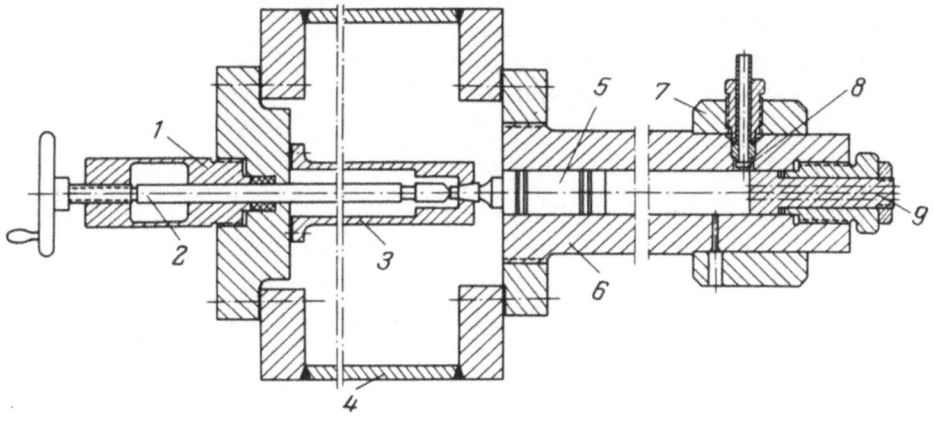

Fig. 283. Price-Lalos adiabatic installation: 1, 3, 7) sleeves; 2) rod; 4) accelerating gas reservoir; 5) piston; 6) column; 8) window; 9) plug.

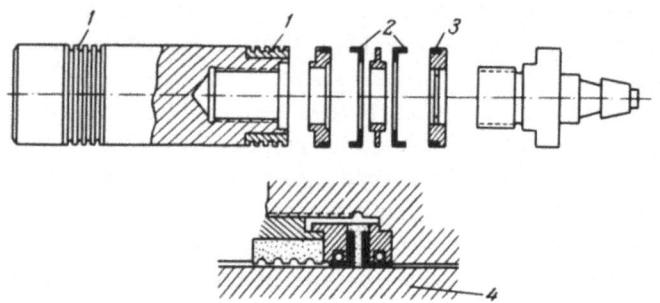

Fig. 284. Piston with sealing ring: 1) phosphor-bronze ring; 2) Teflon; 3) O-ring; 4) wall of cylinder.

from the accelerating gas and transmits this energy to the compressed gas. If we ignore friction, which is very slight in a polished column, the following energy balance equation can be set up:

$$\Delta E_1 = \Delta E_2 + \Delta E_3$$

where ΔE_1 is the change in energy in the accelerating gas; ΔE_2 is the change in energy in the piston; ΔE_3 is the change in energy in the compressed gas.

The equation balance is correct under one condition: the accelerating gas must expand isentropically. If data are available on the dependence of the energy of the accelerating gas on its volume and temperature, calculated from the PVT data on this gas, and if we know the velocity W of movement of the piston of mass m in the column of the installation, we can calculate ΔE_2, since $\Delta E_2 = mW^2/2$. We know now how the energy of the compressed gas changes as its volume changes. The pressure of the gas can be determined and it can be checked whether the calculated data conform with experimental data using the equation

$$(\partial E/\partial V)_S = -P$$

where P is the pressure of the compressed gas; S is the entropy.

By filling the column with the investigated gas at various pressures, the entropy of the gas can be measured and a dependence between energy change and volume at various constant entropies can be determined. If we know the values $(\partial E/\partial V)_{S_1}$, $(\partial E/\partial V)_{S_2}$, $(\partial E/\partial V)_{S_3}$, etc., we can calculate the derivative $(\partial E/\partial S)_V$, i.e., the temperature of the compressed gas, since

$$(\partial E/\partial S)_V = T$$

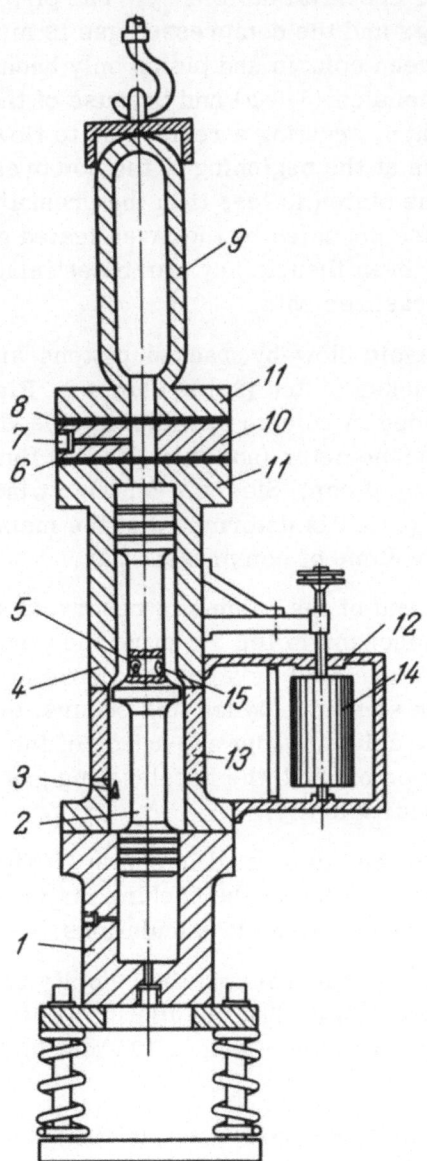

Fig. 285. Graham-Maas adiabatic installation: 1) column; 2) piston; 3) lead crusher
[protuberance]; 4) guiding cylinder; 5) lamp;
6, 8) membranes; 7) hole for inlet and outlet
of supporting gas; 9) collector for accelerating
gas; 10) ring; 11) flanges; 12) cassettes; 13)
windows; 14) drum; 15) lens.

The installation we have described, like other installations of the same type, has one essential deficiency. The piston moving in the column has no sealing rings and the compressed gas is minimized in escape through the clearance between column and piston only because of the small size of this clearance dimension (5-6 μ) and because of the presence of grooves on the body of the piston, creating a resistance to flow. However, a portion of the accelerated gas at the beginning of the compression cycle, when the pressure before the piston is less than the pressure of the accelerating gas, does enter the space occupied by the investigated gas, and vice versa. These phenomena, even though they can be estimated, considerably reduce the accuracy of measurements.

In order to avoid blow-by, packed pistons are used [28]. One installation with a packed piston [29] is shown in Fig. 283. Piston 5 with its sealing ring is placed in column 6, which is closed with plug 9. The plug contains a quartz manometer (not shown in the figure) for measurement of the pressure in the column. Sleeve 7 contains window 8, through which the position of the piston is determined at the moment when it stops, for calculation of the volume of compressed gas.

At the other end of the column is reservoir 4 containing the accelerating gas and the mechanism for mounting the piston. As can be seen from the figure, the piston is held by a polished cone in sleeve 3. It is freed by rotation of rod 2 in sleeve 1. When this occurs, the rod extracts the piston, which flies into the column under the action of the accelerating-gas pressure. With this type of feed, the accelerating gas is not throttled, but rather expands isentropically.

The design of the piston seal is shown in Fig. 284. When using a piston with packing, friction of the packing material on the wall of the column must be considered in all calculations.

Another method of measuring the velocity of the piston was used in the work of reference [30]. The installation (Fig. 285) consists of high-pressure column 1, in which piston 2, 125 mm in diameter, weighing 50 kg, moves.

The piston has a seal (aviation [piston] rings) and consists of two parts: one moves in column 1, the other in directing cylinder 4. Ring 10 with two membranes 6 and 8 is compressed between flanges 11. The upper flange ends with collector 9 for the accelerating gas, capacity 40 liters. When the collector is filled, the gas is simultaneously fed through aperture 7, creating a supporting pressure under membrane 8 which, however, cannot break through membrane 6.

The installation is fired by exhausting the gas through aperture 7. Then, membrane 8 is broken, as is membrane 6, and the gas from the col-

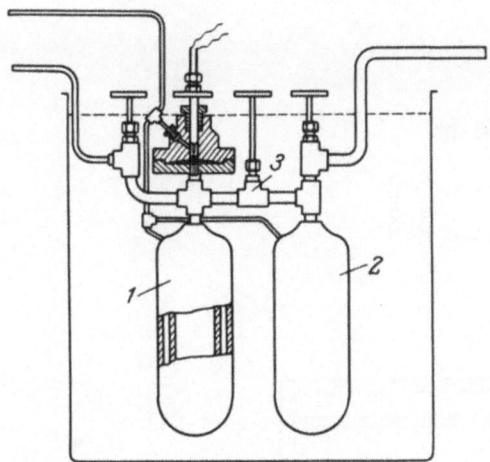

Fig. 286. Plan of Burnett installation: 1, 2)
piezometers; 3) constant-volume valve.

lector enters cylinder 4. The rate of movement of the piston is recorded
by drum 14 which rotates at constant velocity in cassette 12. The drum is
covered with light-sensitive paper, which is exposed through window 13 by
light from zirconium arc lamp 5 placed in the piston. The beam is focused
by lens 15. Lead crusher [protuberance] 3 is used to determine the volume
of the compressed gas.

Measurement with Variable-Volume Piezometer with Variable Quantity of Material

Burnett [31] suggested a method for determining the coefficient of
compressibility of gases* which differs from those described above in that
it does not require measurement of the volume of the piezometer. We know
that it is much easier to measure pressure with high accuracy than it is to
measure volume; therefore Burnett's method has been widely used in recent years [34, 35].

The installation (Fig. 286) consists of two piezometers 1 and 2 — high-
pressure vessels with double walls placed in a thermostat. The piezometers
are connected by constant-volume valve 3. The pressure in the space between the walls is maintained equal to the experimental pressure so that
there is not need to correct for deformation of the piezometer walls. The
gas being investigated is fed into vessel 1 and, when the temperature is
equilibrated, the pressure is measured. Then, valve 3 is opened and the

---

*This method is also used for measurement of compressibility of the li-
quid phase of multi-component systems [32, 33].

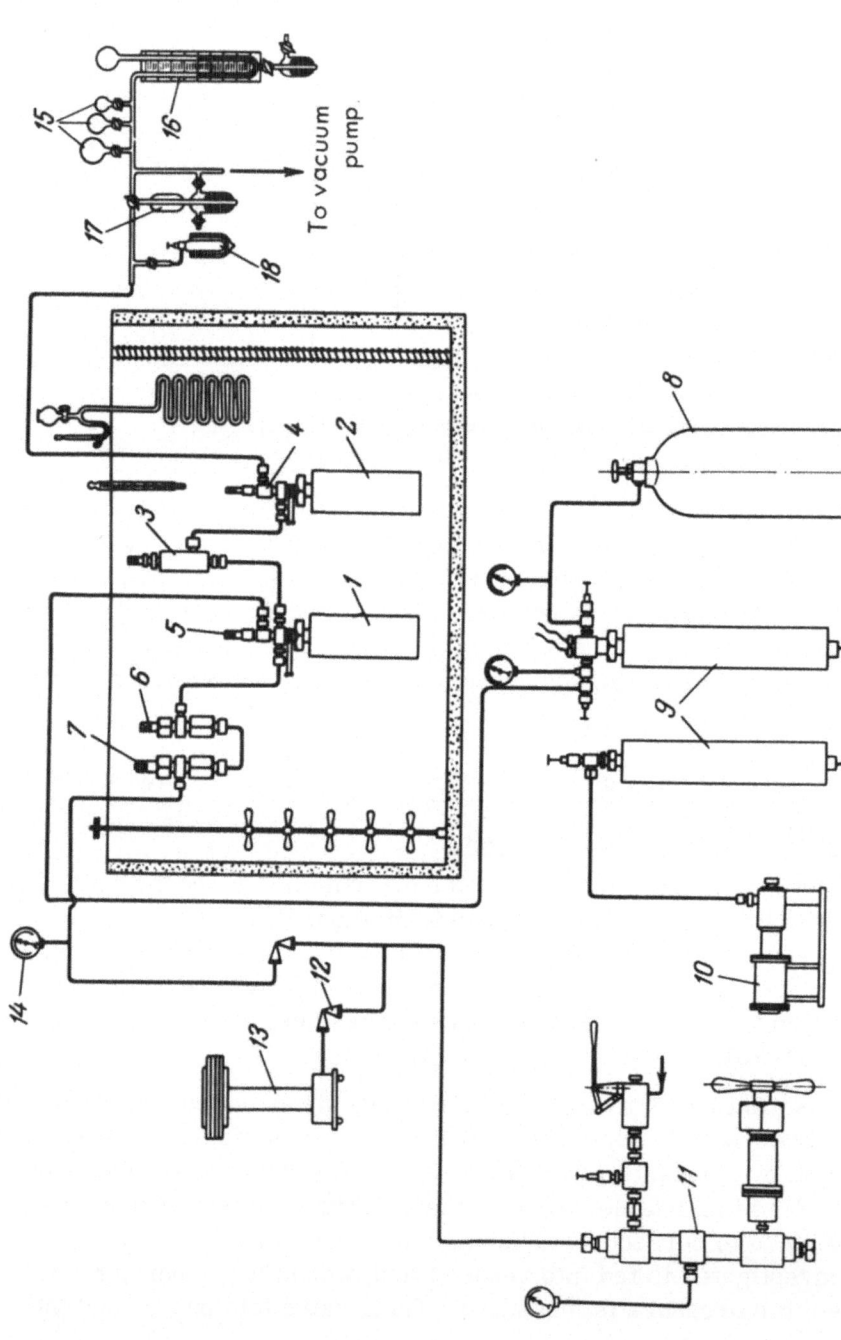

Fig. 287. Tsiklis-Linshits installation: 1, 2) piezometers; 3) constant-volume valve; 4, 5, 6) shut-off valves; 7) membrane null instrument; 8) cylinder; 9) mercury press; 10) oil pump; 11) oil press; 12) valve; 13) piston manometer; 14) tubular manometer; 15) calibrated flasks; 16) mercury manometer; 17) mercury pump; 18) metal ampule.

gas is allowed to flow from vessel 1 into vessel 2 from which the air has been evacuated. After the pressure has been measured in this vessel, valve 3 is closed, the gas is pumped from vessel 2 and the operation is repeated. The experiment is continued until it is no longer possible to measure the pressure of the gas in the vessels precisely.

The result of this experiment is the production of a number of values of pressure: P_0, P_1, P_2, ..., P_n (where P_0 is the pressure of the gas in vessel 1 before the first transmission of the gas; P_1 is the pressure after the first movement of gas; P_2 is the pressure after the second movement of gas, etc.).

The ratios P_0/P_1, P_1/P_2, ..., P_{n-1}/P_n are determined and a graph of the dependence of these ratios on the values of P_0, P_1, ..., P_n is constructed. The points (when all operations are performed properly) should lie on a smooth line. This line is continued to its intersection with the ordinate. As $P \to 0$, the relation $PV = $ const becomes correct and the sector cut off by the line on the ordinate, i.e., P_{n-1}/P_n, will be equal to V_n/V_{n-1}. In other words, if we represent the volumes of vessels 1 and 2 by V, and V'', then

$$(V' + V'')/V' = N$$

where N is a constant.

Thus, without knowing the volumes of the piezometers, it is possible to determine the ratio of the volumes.

Next, the product of (1) this constant to a power numerically equal to the number of the operation, and (2) the pressure measured during the operation, is calculated ($N^n P_n$), and a graph of this product vs. pressure is drawn. This produces a straight line which cuts off sector $N^s P_s$ on the ordinate which Burnett calls the basis sector. Then, all products $N^n P_n$ are divided by $N^s P_s$ and the graph of these values vs. P is constructed once more. The result is a straight line passing through unity where $P = 0$. The equation of this straight line is

$$K = 1 + bP$$

where K is the isothermal coefficient of compressibility of the gas being investigated.

If we determine the constant N for helium or hydrogen (for which the P_{n-1}/P_n ratios vs. P fall on a straight line over a rather great pressure range) the value of N can be determined to a high degree of accuracy.

Improvement of this method has permitted an installation to be assembled which can be used to measure the coefficient of compressibility

and molar volumes of gases and gas mixtures. This installation, assembled
by D. S. Tsiklis and L. R. Linshits, is shown in Fig. 287. Two thick-walled
vessels 1 and 2 made of ÉI437BU steel, designed for pressures of 1000 atm,
are placed in a thermostat. The vessels are closed by plugs with type I unsupported-area packing and supplied with electromagnetic agitation. This
permits the formation of mixtures to be studied directly in the piezometers,
and also allows phase behavior of systems to be studied using this installation. Constant-volume valve 3 is located between vessels 1 and 2. The gas
(or gas mixture) being investigated is fed from cylinder 8 to mercury press
9 where it is compressed to the required pressure by pump 10. From here,
the gas is fed through valve 5 into vessel 1. After the gas has assumed the
temperature of the thermostat, valve 6 is opened, the gas is let into the
membrane null instrument 7, and tubular manometer 14 is used to measure
the pressure in the system. Then valve 12 is opened and the pressure is
measured precisely with piston manometer 13.

Next valve 3 is opened and the gas is exhausted from vessel 1 into
evacuated vessel 2. After equilibrium is established, the pressure is measured once more and, after closing valve 3, the gas is exhausted from column
2 to the atmosphere through valve 4. Then, the gas is pumped from vessel
2 once more with the vacuum pump. All operations are repeated until the
pressure can no longer be measured precisely with the piston manometer.
During one gas transfer, the gas from vessel 2 is extracted to the volumetric portion of the installation, consisting of calibrated flask 15, manometer 16, mercury pump 17, and metallic ampule 18. When this is done, the
number of moles of gas in vessel 2 at the given temperature and pressure
is determined.

If the number of moles of gas in vessel 1 at the beginning of the experiment is represented by n_0, and if the number of moles of gas in vessel
1 after each gas movement is represented by n_1, n_2, ..., n_n, and if the number of moles of gas in vessel 2 after each gas movement is represented by
n', then it is not difficult to see that

$$n_0 = \frac{n' N^{n-1} N^n}{N^n - N^{n-1}}$$

After determining n' with any gas movement, it is possible to calculate n_0; knowing the volume of vessel 1 (V'), it is possible to calculate
the molar volumes of gas at all pressures. If a condensable gas is investigated, the gas can be exhausted from vessel 2 to a gravimetric ampule
18 cooled with liquid nitrogen rather than to the volumetric portion of the
installation, so as to determine n' gravimetrically. It should be noted that
if vessels 1 and 2 are made of the same material, coefficient N will not depend on the temperature and there will be no need to introduce a correction
for thermal expansion of piezometers during calculation of compressibility.

COMPRESSIBILITY OF LIQUIDS

Methods for determination of the compressibility of liquids also may be divided into three main groups: 1) methods using constant-volume piezometers containing various quantities of liquid; 2) methods using piezometers (pycnometers) of various volume, containing constant quantities of liquids, and 3) methods of hydrostatic suspension.

Measurement with Constant-Volume Piezometers

One of the most widely used methods [36] is that in which a piezometer as shown on Fig. 288 is used. The piezometer is made of glass; two contacts 1 and 2 are sealed into the glass. Before the experiment, a known quantity of mercury is poured into the piezometer, sufficient to close both contacts.

A measured quantity of the liquid being investigated is poured into the wide portion of the piezometer and the piezometer is closed with a tight plug. Then the piezometer is placed in the high-pressure vessel which has an insulated electrical lead. After the vessel is closed, the pressure-transmitting fluid is pumped into it. This causes the mercury level in the tube containing contact 1 to drop in proportion to the compressibility of the liquid in the wide portion of the piezometer. The pressure is noted at the moment when contact 1 is opened. Then the pressure is reduced in the apparatus until the contact is closed. By repeating this procedure several times, the pressure where the contact is opened can be determined precisely. Next, the pressure is reduced, the apparatus is opened, a portion of the liquid in the piezometer is replaced with a weighed quantity of mercury, and the entire procedure is repeated until finally the piezometer contains only mercury. Since the volume of the piezometer to the level of contact 1 is known precisely, and the quantity of mercury and liquid are precisely known, it is easy to determine the volumes occupied by various quantities of liquid at pressures corresponding to the opening of the contact. Data on the compressibility of pure mercury under these conditions, i.e., in the same piezometer, allows a determination of the volume of the mercury to be made and makes it unnecessary to introduction a correction for the compressibility of the glass of which the piezometer is made.

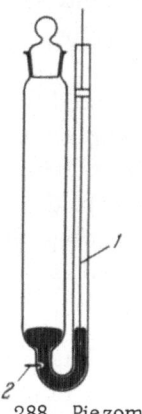

Fig. 288. Piezometer for measurement of differential coefficient of compression of liquids (1, 2, contacts).

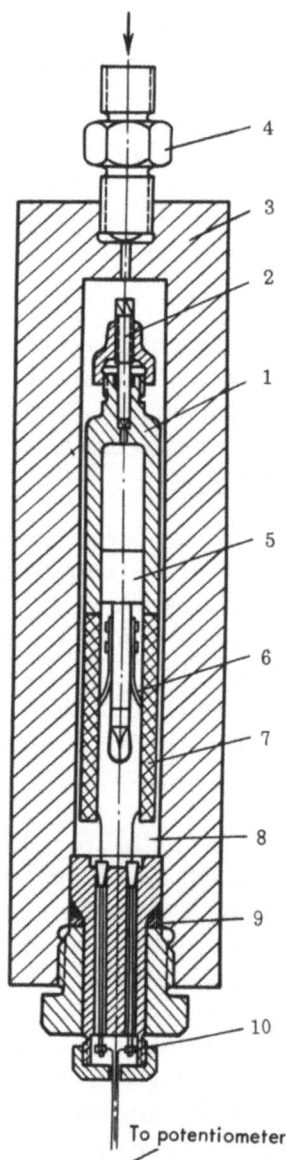

Measurements with Variable-Volume Piezometers

Variable-volume piezometers are used for determination of the compressibility from the displacement of a piston [37]. This method differs little from the method described above for gases. The liquid being investigated is separated from the compressing medium by mercury, so that the joint compressibility of the investigated material, mercury, and compressing medium is actually determined. Corrections must be introduced for the compressibility of the medium and of the mercury, and for expansion of the pressure container. The correction for compressibility of the mercury and the medium is found by performing a control experiment with a metal insert for which the coefficient of compression is known.

Another method [38] consists of determining the compressibility from the displacement of a piston directly contacting the liquid being investigated. The piston must be very tightly sealed to the cylinder and the material must be compressed under the action of a pressure-transmitting fluid pumped into the apparatus on the other side of the piston. The position of the piston in the cylinder is determined by using a contact firmly fixed to the body of the apparatus and slipping along a platinum wire which moves together with the piston. The wire is insulated from the piston. Its resistance changes depending on the position of the contact (piston) and is measured potentiometrically.

L. F. Vereshchagin and V. A. Galaktionov [39] placed a piston piezometer in a high-pressure vessel (Fig. 289). Piezometer 1, into which the investigated liquid is placed through valve 2, is placed in thick-walled cylinder 3 which contains the same liquid. Then the investigated liquid is pumped into cylinder 3 through nipple 4. When this occurs,

To potentiometer

Fig. 289. Vereshchagin and Galaktionov piezometer for compressibility measurement: 1) piezometer; 2) valve; 3) high-pressure chambers; 4) nipple; 5) piston; 6, 10) contacts; 7) insulator; 8) wire; 9) packing.

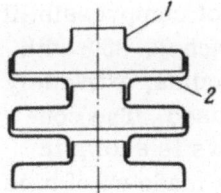

Fig. 290. Metal bellows
(1, 2, sections)

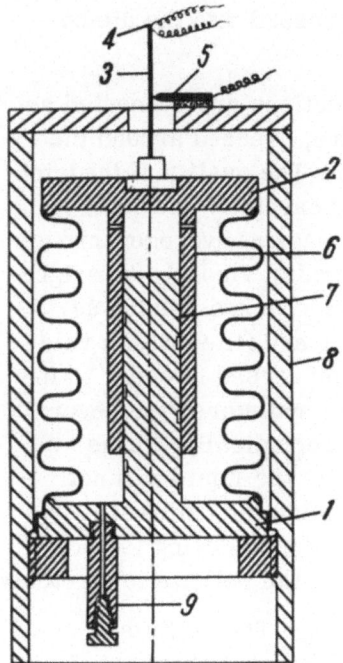

Fig. 291. Piezometer with metal
bellows: 1, 2) covers; 3) manganin
wire; 4, 5) contacts; 6, 7) directing
cylinders; 8) cover; 9) tube.

piston 5, carefully lapped to the walls of the piezometer, compresses the liquid contained in the piezometer. Two contacts 6 are fastened to the back of the piston, slipping along a manganin-wire rheostat placed on base 7 made of insulating material. As the piston moves, the resistance of the rheostat changes and is measured by a potentiometer connected to contacts 10 in plug 9. If the position of the piston in the piezometer is known, the volume of liquid at the given pressure can be determined.

This method reduces the correction for deformation of the piezometer which is placed under complete surrounding pressure.

Piezometer with Metallic Bellows. A piezometer of this type [40] is a specially made metal bellows filled with the liquid to be investigated and placed in a high-pressure apparatus. The bellows is compressed under the influence of the hydrostatic pressure of the fluid pumped into the apparatus until the internal pressure becomes equal to the external pressure. By measuring the decrease in length of the bellows, the change in volume, and consequently the change in volume of the liquid which the bellows contains, can be determined.

This method has a number of advantages over methods involving piston displacement. The liquid in the bellows is reliably insulated from the pressure-transmitting liquid, which eliminates the danger of contamination; the data produced by this method are more precise, since there is no leakage of the investigated liquid through the sealing. Finally, the bellows can be placed rather far from the manganin manometer, which makes it possible to measure the compressibility at high temperatures. The bellows is under surrounding pressure so that no plastic deformation occurs.

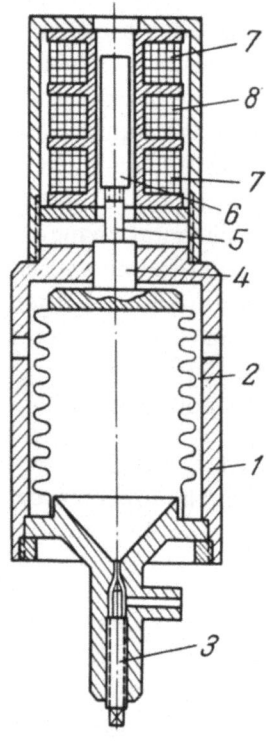

Fig. 292. Bellows piezometer:
1) container; 2) bellows; 3)
valve; 4) directing sleeve; 5)
rod; 6) core; 7) secondary winding; 8) primary winding.

For measurement of compressibility of liquids, which may reach 30–35% with strongly compressible liquids, extremely elastic bellows must be used. The construction of such a bellows is shown in Fig. 290. The bellows is made up of nine shaped sections 1 and 2, each 2.5 cm long. The sections are stamped of sheet (gasket) brass, 0.04 mm thick. For this, disks of the corresponding diameters are punched out,* annealed, and formed in a special press. The press is supplied with a shaped die; a lead plug is pressed into the die by a steel piston.

The formed sections are connected as is shown in the figure, pressed around the edges, and soldered. The quality of the joint should be especially carefully checked; joints are tested by passing hydrogen through the bellows submerged in alcohol. The tightness of the seams is checked finally as follows. The bellows are filled with a volatile liquid,† then compressed in order to increase the pressure, and stored for several days, being weighed periodically. If the seams are well-made, the liquid will not evaporate. Before using the bellows, it must be certain that the soldering is secure and the position of the sections does not change when the bellows is stretched

The location of the bellows is shown in Fig. 291. The bellows is attached to covers 1 and 2. Cover 2 carries a device for measurement of the compression of the bellows, consisting of manganin wire 3 to which contact 4 is soldered and which slips along contact 5, insulated from the body. The degree of compression can be judged by measuring the resistance of the wire between contacts 4 and 5.

The bellows is filled with the specimen liquid through tube 9. During filling the air must be pumped out of the bellows and bubbles of air

---

*Attempts to use phosphor bronze, soft iron, and even platinum as a material for manufacture of the disks have been unsuccessful.
†The volatile fluid used was carbon disulfide [40].

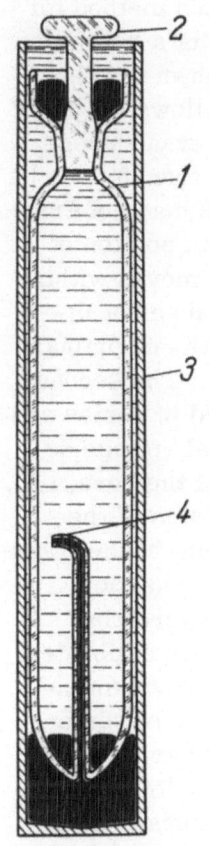

Fig. 293. Pycnometer for determination of compressibility of liquids: 1) pycnometer; 2) plug; 3) metal container; 4) capillary.

must be removed from the liquid. The tube is sealed with a plug made of argentan, and the quantity of liquid is determined by weighing. The bellows can contain up to 5 cm³ of liquid.

Cylinders 6 and 7 are guiding cylinders and are used to protect the bellows from bending. They are lapped to each other. In order to avoid digging in, the cylinders are made of different materials — brass and copper. A groove is cut around the outside of cylinder 7 to collect dirt and contaminants and thus to minimize friction between the cylinders. The bellows after assembly is placed in cover 8, and the electrodes are connected to a three-pole electrical lead, after which the entire device is placed in the high-pressure chamber. Readings are taken during forward and reverse movement (increase and decrease of pressure).

It should be noted that viscous liquids deform the bellows, and if a liquid freezes under pressure, the bellows is completely destroyed.

In order to check the dependence of the cross-section of the bellows on length, a calibration is made, which consists of the following: the bellows is filled with kerosene and placed between the plates of a height gauge. The bellows is compressed with a micrometric screw and the kerosene which is forced out of the bellows is weighed.

The manganin wire must also be calibrated.

For production of precise data, corrections are also introduced for the compressibility of the metal. For example, the correction for brass in cylinder 7 (see Fig. 291) can be calculated from the following formula (at 30°C):

$$\frac{\Delta V}{V_0} = 9.207 \cdot 10^{-7} P - 6.42 \cdot 10^{-12} P^2$$

where P is the pressure, kg/cm².

Also, the change in resistance of the manganin with pressure should be considered.

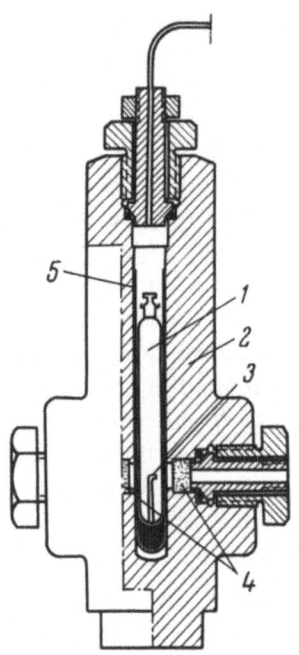

Fig. 294. Installation with visual observation for study of compressibility of solutions: 1)pycnometer; 2) high-pressure apparatus; 3) capillary; 4) viewing windows; 5) glass container.

A more highly perfected method for measuring the compression of a bellows is used in the device [41] shown in Fig. 292. Here, the change in bellows length is measured by a transformer transducer [linear variable differential transformer]. Brass rod 5 carrying core 6 made of Armco iron is fastened to the bottom portion of bellows 2. The brass rod 5 moves within guiding sleeve 4. The displacement indicator is a transformer with one primary and two secondary windings. The secondary windings 7 are connected in series and counter-connected. The total voltage depends on the displacement of the core, i.e., is a function of the compression of the liquid in the bellows. The emf of the transducer is measured by a compensation method using a secondary differential transformer which is carefully calibrated, and by constructing a graph of volume of liquid in bellows vs. indication of scale of instrument. This method permits measurement of the compressibility of a liquid to be made with an accuracy of up to 0.08% at pressures up to 2000 atm.

The literature contains descriptions of other methods for measuring the density of liquids using electromagnetic compensation [42], or compression by mercury, whereby the position of the mercury is determined by the position of an iron core which changes the magnetic field of a differential transformer [43].

Measurement of Compressibility using Pycnometers

These methods [44] consist of placing a piezometer which is in itself a pycnometer in a high-pressure apparatus. Pycnometer 1 (Fig. 293) contains the liquid; the pycnometer is closed with plug 2 and placed in metal container 3 with mercury. The container is placed in the high-pressure apparatus. As the pressure is increased, the mercury enters through capillary 4 and drops to the bottom of the pycnometer. As the pressure is reduced, the excess liquid leaves the pycnometer through the capillary. If we know the mass of the mercury in the pycnometer, and the compressibilities of the mercury and of the glass, we can calculate the compressibility of the liquid being investigated.

This method requires great care in operation and its accuracy is determined by the size of a single drop of mercury which will break off the tip of the capillary.

An improved apparatus of this type is shown in Fig. 294 [45]. Here pycnometer 1 is placed in high-pressure vessel 2 so that the tip of capillary 3 is located opposite two windows 4. The capillary is illuminated through one window and observed using a microscope through the opposite window. The mercury is located in glass container 5 in which the pycnometer floats.

It has been established that the dimensions of drops falling from the capillary change over wide limits. Thus, for example, if the pycnometer contains water, a drop of mercury has a volume of about 0.002-0.003 cm^3; if the pycnometer contains a 10% solution of lithium bromide, drop volume will be 0.0015 cm^3.

Pycnometers designed for operation under high pressure should be made particularly carefully and should be specially heat-treated. A detailed description of methods of operation with pycnometers for determination of the compressibility of solutions can be found in a work by E. S. Lebedeva [46].

Determination of Compressibility by Hydrostatic Weighing*

Methods of hydrostatic weighing, used for determination of the density of gases and liquids, are based on the fact that a body submerged in the medium being investigated changes its weight with a change in density of the medium. If the change in weight of a float is determined as the density of the medium around the float is changed, and if the volume of the float is known, the density of the medium can be calculated at the experimental temperature and pressure [47, 48]. The apparatus used for determination of density by hydrostatic weighing differ primarily in the device used for weighing the float.

Figure 295 shows an instrument for determination of the density of liquids designed by V. N. Razumikhin [49]. Thick-walled steel cylinder 1 contains bath 2 with a balance. The balance consists of equal-arm-length bar 4, 150 mm long, with prism fulcrum and two bodies: steel body 3, volume 1.4 cm^3, and duraluminum body 5, volume 5.4 cm^3. Steel body 3 can be shifted along the balance arm.

The balance is equilibrated in the liquid being investigated which fills the bath at atmospheric pressure, by adding weight to the depression in

---

*The methods described below can also be used to determine the compressibility of gases.

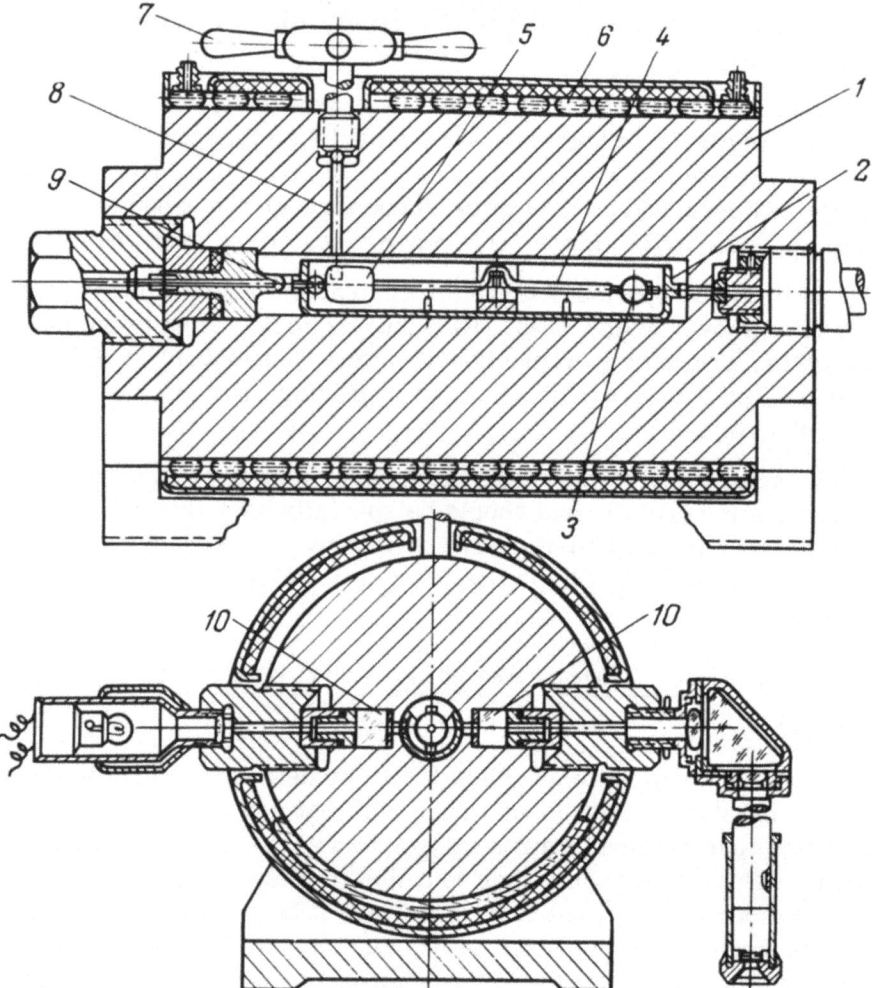

Fig 295. Razumikhin installation for measurement of liquid densities: 1) high-pressure vessel; 2) bath; 3) iron body; 4) balance arm; 5) duraluminum body; 6) coil; 7) valve; 8) aperture for addition of weight; 9) thermocouple cover; 10) window.

body 5. In equilibrating the balance, care must be taken that there are no air bubbles in the liquid. Then bath 2 is placed in the apparatus, it is more horizontal, and the cylinder is filled with the liquid being investigated. Next, the cylinder is closed and, by passing water at the desired temperature through coil 6, the temperature of the apparatus is set. Then, an excess weight is placed on the balance, aperture 8 for addition of weights is closed, and the pressure in the apparatus is increased until the balance is once more equilibrated. Metal spheres which drop into the depression in

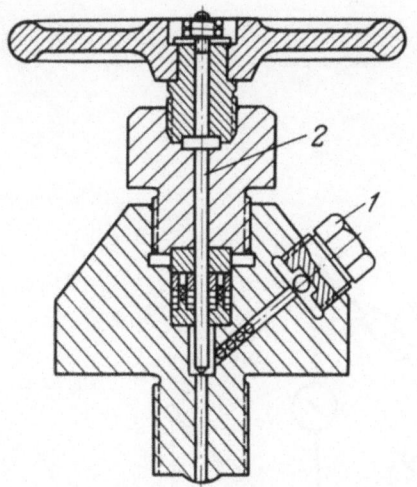

Fig. 296. Valve for addition of load under pressure: 1) plug; 2) pin.

the duraluminum body through aperture 8 are used as the additional weights. The position of the balance indicator is observed through two windows 10, located perpendicular to the horizontal axis of the cylinder.

In order to avoid a reduction in pressure each time additional weights are added, the device shown in Fig. 296 is used. It consists of a high-pressure valve whose pin is closed during the experiments. Weights are added by opening plug 1 and placing a sphere in the aperture. Then, the pin 2 is opened, and the sphere rolls into the cavity of the body. After the load has been added, the pressure is once more increased and the balance is equilibrated.

This device can be used to perform measurements by absolute and relative methods. During measurements by the absolute method

$$\rho = \rho_0 \left[1 + \frac{C}{(B-A)\rho_0} + \frac{A}{B-A} \right]$$

where ρ_0 is the density of the liquid at atmospheric pressure;

$$A = V_1 K_1 - V_2 i K_2 + \frac{m_3}{\rho_p} K_3$$

$$B = V_1 - V_2 i + m_3/\rho_p$$

$$C = m_3 (\rho_p - \rho_0)$$

V_1 and V_2 are volumes of bodies 3 and 5 (see Fig. 295); i is the ratio of lengths of the balance arm; K_1, K_2, and K_3 are the coefficients of volumetric compressibility of the materials of bodies 3 and 5 and of the additional weights; ρ_p is the density of the material of the additional weights; m_3 is the mass of the additional weights equilibrating the balance arm under pressure.

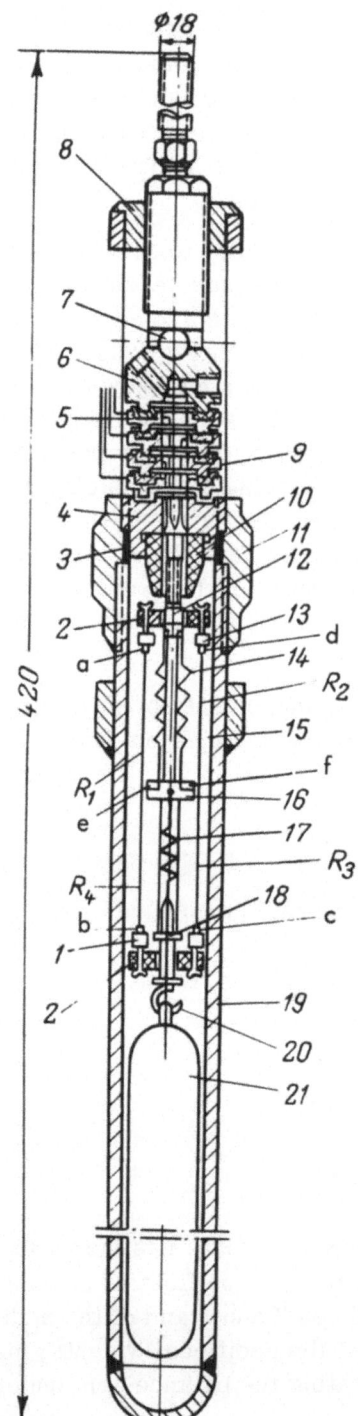

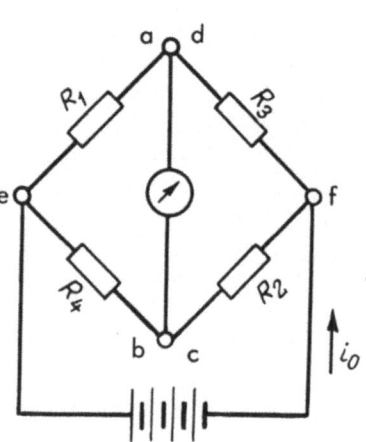

Fig. 297. Tensimetric balance: 1) lower regulating screw; 2) rings; 3) packing; 4) centering nut; 5) insulating ring; 6) pressure cap; 7) metal sphere; 8) internal nut; 9) removable bronze ring; 10) plug for fastening weights; 11) cover nut; 12) tension regulating screw; 13) upper regulating screws; 14) constantan spring; 15) constantan wires; 16) textolite plate; 17) spring for supporting load; 18) limiting stop; 19) operating tube; 20) hook; 21) hollow quartz float.

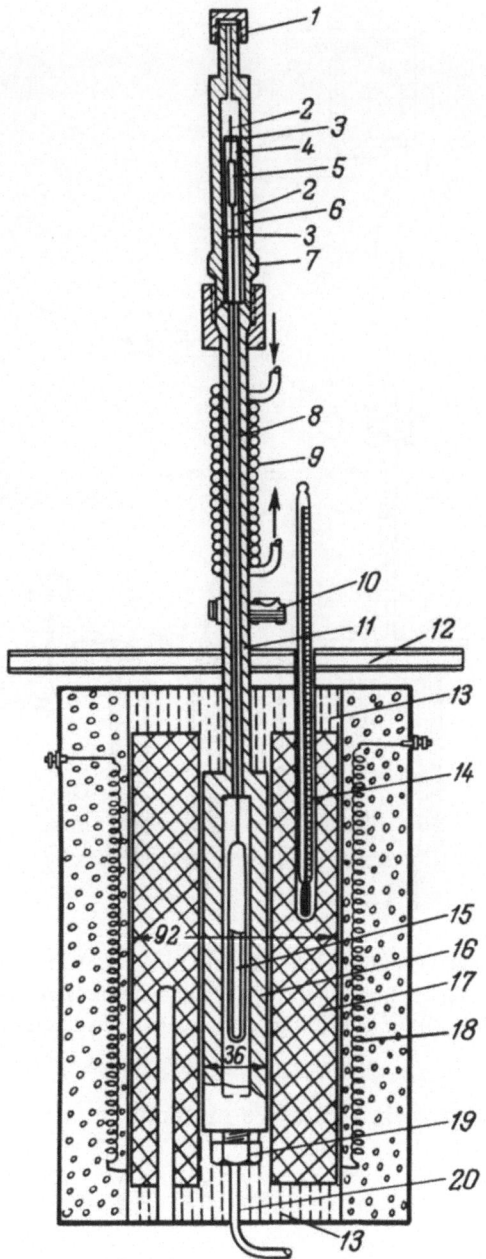

Fig. 298. Golubev instrument for determination of compressibility by the hydrostatic weighing method: 1) sealing nut; 2) control needle; 3) Teflon plate; 4) brass insert; 5) iron core; 6, 16) upper and lower vessels; 7) mounting insert; 8) manganin wire; 9) cooling coil; 10) level; 11) steel tube; 12) thermal insulating ring; 13) insulation; 14) thermometer; 15) quartz float; 17) aluminum (copper) block; 18) electric furnace; 19) nut; 20) capillary.

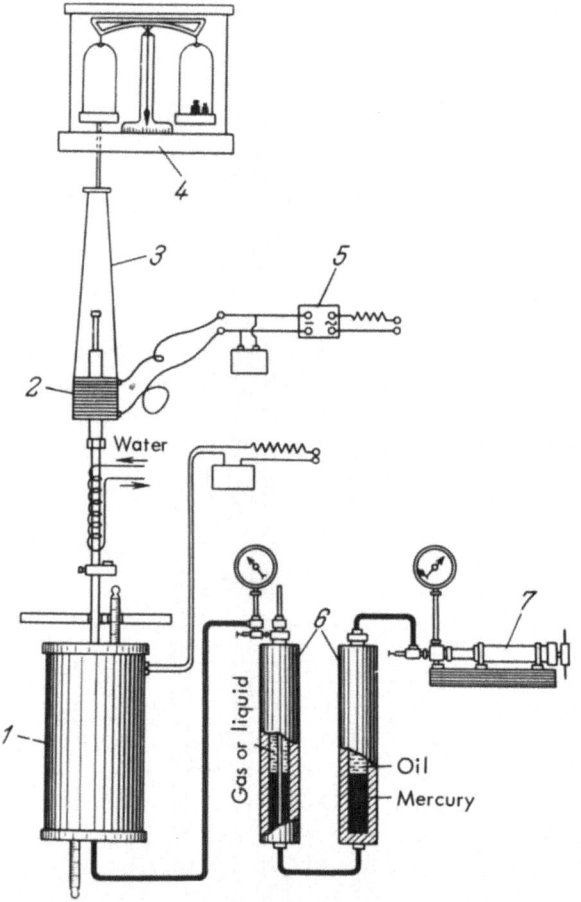

Fig. 299. Plan of Golubev installation: 1) electric furnace with lower vessel; 2) solenoid; 3) support system; 4) analytic damper balance; 5) rectifier; 6) mercury press; 7) hydraulic press.

When working with a relative method*

$$\rho = \rho' \left(1 + \frac{A}{B} \right)$$

where ρ' is the density of the standard liquid at atmospheric pressure.*

A method for determining the density of a liquid under compressed-gas pressure [51] consists of the following: a float is placed in a liquid poured into a high-pressure vessel with a window. Gas is used to create pressure over the liquid and the height of an indicator in the gas phase over

*See also [50].

the level of the liquid is determined. In order to determine the density of
the liquid, the density of the gas over the liquid must be known, as must the
surface tension at the liquid-compressed gas boundary and the wetting angle
of the walls of the indicator by the liquid.

Tensimetric Balance. The tensimetric balance, designed by
N. V. Pavlovich and D. L. Timrot [52] (Fig. 297) consists of an equal-arm-
length electrical bridge. All the arms are made of thin constantan wire whose
resistance changes with a change in tensile stress. The tensile stress depends on the weight of a float suspended on the wires.

The device consists of high-pressure tube 19 in which hollow quartz
float 21 is suspended on constantan wires 15. The sensitivity of the balance is increased by supporting the primary part of the weight of the float
on constantan springs 14, the tension of which can be controlled. When the
density of the medium is increased, the float rises and the tension of the
two opposite arms of the bridge is decreased, that of the two other arms
being increased. The device is calibrated using a medium of known density
(such as nitrogen) to determine the instrument constant, i.e., the coefficient of proportionality between the change in resistance and load.

Golubev Installation. An interesting method of hydrostatic
weighing was developed by I. F. Golubev [53]. The device (Fig. 298), which
can be used to determine the density of liquids and gases, consists of two
high-pressure vessels 6 and 16 made of nonmagnetic steel and connected
by steel tube 11. The device is placed absolutely vertically using level 10.
The lower vessel contains quartz float 15, the upper iron core 5, connected to the float by manganin or platinum wire 8. The lower vessel is
placed in copper block 17, which is used for equilibration of the temperature and is located in electric furnace 18. The ends of the lower vessel
are closed with thermal insulation 13. Screen 12 and cooling coil 9 protect the upper portion of the instrument from heating.

Core 5, located in vessel 6, is placed in brass insert 4, the ends of
which hold Teflon guide-rings 3. Stainless-steel medical needles 2 pass
through these directing rings, centering core 5 in insert 4. This careful
centering is necessary in order to be sure the core does not touch the walls
during its vertical movement.

Solenoid 2 (Fig. 299) is applied to the upper vessel, suspended from
balance 4. The balance is equilibrated. If a direct current is passed
through the winding of the coil, the magnetic field will pull core 5 (see
Fig. 298) and the coil will become heavier. The balance will be moved off
the middle point. After this, the balance is once more equilibrated, thus
determining the weight of the support system in the medium being investigated at the given temperature and pressure. If the weight when empty is

known as well as the volume of the float, core, and connecting wire, the density (in g/cm^3) of the investigated medium can be determined.

The results of investigations using this method must be corrected for change in volume of core and float caused by a change in temperature and pressure. Also, the influence of lateral thrust on the core, as well as on the entire support system, must be considered.

* * *

In conclusion to this chapter, we would like to discuss two general problems concerning accuracy of measurements:

1. It was noted above that under experimental conditions the walls of a piezometer are deformed resulting in a change in its volume. The deformation of walls takes place under the influence of pressure and temperature. In those cases when the piezometer is located under unilateral pressure, its volume V_p under pressure P will be

$$V_P = \frac{V_0 P}{E(R^2 - r^2)} \left[3(1 - 2\mu) r^2 + 2(1 + \mu) R^2 \right]$$

where V_0 is the volume of the piezometer at normal temperature and pressure taken as the zero point of calculation; E is Young's modulus; R and r are the external and internal radii of the piezometer; μ is Poisson's ratio.

The deformation of a piezometer under pressure may be elastic; however, under extreme loads, residual [plastic] deformation may appear. Therefore, from time to time the piezometers must be calibrated in order to determine the value of V_0. It is most accurate to calibrate piezometers at the experimental conditions under which they will be used.

Sometimes, in order to reduce deformation from pressure, conditions are created under which the piezometer is exposed to surrounding pressure. In this case, the volume of the piezometer is also changed. It is reduced the same amount by which the volume of a solid bar of metal of the same type used for manufacture of the piezometer is reduced under the same conditions. For an iron bar [22],

$$V_P = V_0 (1 - 5.826 \cdot 10^{-7} P - 0.80 \cdot 10^{-12} P^2)$$

It must be noted that the calculated correction is not always exact. Thus, checking of the volume of a piezometer under pressure has shown [8] that the change in external volume of a piezometer at a pressure of 3000 atm is greater than the calculated change by a factor of almost 2.

The change in volume of a piezometer with temperature is calculated from well-known formulas (see Chapter IV).

2. In many methods of investigating phase equilibria and volume relations, the plugging- and pressure-transmitting liquid is mercury. Although mercury has certain valuable properties, its use in many cases is undesirable. Mercury is dangerous for the health and at high temperatures any leakage from high-pressure apparatus is immediately evaporated and poisons the atmosphere. Also, mercury dissolves in compressed gases [54]. Investigations have shown that the solubility of mercury in butane compressed to 400 atm and at temperatures from 200 to 300°C is greater than that calculated from the saturated vapor pressure by approximately a factor of 4. This must be considered in performing precise measurements under high-temperature and pressure conditions when the mercury concentration in the gas phase may be considerable.

BIBLIOGRAPHY

1. Bartlett, E., J. Am. Chem. Soc., 46:687 (1927); 50:1275 (1928).
2. Wiebe, R., and Gaddy, V., J. Am. Chem. Soc., 60:2800 (1928).
3. Kirillin, V. A., and Sheindlin, L. E., Investigations of the Thermodynamic Properties of Materials, Gosénergoizdat, 1963.
4. Tsiklis, D. S., and Kulikova, A. I., Zh. Fiz. Khim., 39, No. 7 (1965).
5. Vulkalovich, M. P., and Altunin, V. V., Teploénerg., No. 11 (1959).
6. Benedict, M., J. Am. Chem. Soc., 59:224 (1937).
7. Mendeleev, D. I., Zh. Russ. Fiz. Khim. Obshchestva, 4:309 (1872).
8. Bilevich, A. V., Vereshchagin, L. F., and Kalashnikov, Ya. A., Pribory i Tekhn. Éksperim., 3:146 (1961).
9. Tsiklis, D. S., Kulikova, A. I., and Shenderei, L. I., Dokl. Akad. Nauk SSSR, 131:887 (1960).
10. Holborn, L., and Schultze, H., Ann. Phys., 47:1089 (1915).
11. Krichevskii, I. R., and Markov, V. P., Zh. Fiz. Khim., 14:101 (1940).
12. Witkowski, Phys. Mag., 41(5):228 (1896).
13. Amagat, E., Ann. Chem. Phys., 29(6):68 (1873).
14. Timrot, D. L., Izvest. Vsesoyuz. Teplotekh. Inst., I (1949).
15. Michels, A., and Gibson, R., Ann. Phys., 87:850 (1928); 12:562 (1932).
16. Kazarnovskii, Ya. S., Simonov, G. B., and Aristov, G. E., Zh. Fiz. Khim., 14:774 (1940).
17. Kazarnovskii, Ya. S., and Sidorov, I. P., Zh. Fiz. Khim., 21:1363 (1947).
18. Sidorov, I. P., and Kazarnovskaya, D. B., Tr. Gos. Inst. Azo. Prom., No. 3, Goskhimizdat, 1954, p. 200.
19. Basset, J., and Basset, J., J. Phys. Radium, Suppl. 15, No. 1, 47A (1954).

20.   Douslin, D. R., Moore, R. T., Dauson, J. P., and Waddington, G.,
      J. Am. Chem. Soc., 80:2031 (1958).
21.   Michels, A., Wassenaar, T., and Zwietering, T. N., Physica, 18:67 (1952).
22.   Bridgman, P. W., High-Pressure Physics, Otdel. Nauk Tekhn. Inst.,
      1935.
23.   Benedict, M., J. Am. Chem. Soc., 59:2224 (1937).
24.   Krichevskii, I. R., and Tsiklis, D. S., Dokl. Akad. Nauk SSSR, 78:1169
      (1951).
25.   Cherney, B., Marchman, H., and York, R., Ind. Eng. Chem., 41:2653
      (1949).
26.   Trivus, N. A., and Vinogradov, K. V., Investigation of Petroleum and
      Gas in Stratal Conditions, Aznefteizdat, Baku, 1955.
27.   Ryabinin, Yu. N., Gases at High Densities and Temperatures, GIFML,
      1959.
28.   Longwell, P. A., Reamer, H. H., Wilbourn, N. P., and Sage, B. H.,
      Ind. Eng. Chem., 50(4):603 (1958).
29.   Price, D., and Lalos, G. T., Ind. Eng. Chem., 49(12):1987 (1957).
30.   Graham, G. T., and Maas, O., Can. J. Chem., 38:2482 (1960).
31.   Burnett, E. S., J. Appl. Mech., 3(4):A-136 (1936).
32.   Burnett, E. S., U.S. Bureau of Mines, R 16267, Pittsburg (1963).
33.   Elington, R. T., and Eakin, E. B., Chem. Eng. Progr., 59(11):80
      (1963).
34.   Schneider, W. G., Can. J. Res., B27:339 (1949); Beer, H., Chem. Eng.
      Techn., 31:784 (1959).
35.   Silberberg, I. H., Kobe, K. A., and McKetta, J. J., J. Chem. Eng.
      Data, 4:314 (1959).
36.   Richards, T. W., Z. Phys., 4:393 (1926).
37.   Bridgman, P. W., Proc. Am. Ac., 49:I (1913).
38.   Bridgman, P. W., Z. Krist., 67:363 (1928).
39.   Vereshchagin, L. F., and Galaktionov, V. A., Pribory i Tekhn.
      Éksperim., 1:98 (1957).
40.   Bridgman, P. W., Proc. Am. Ac., 66:185 (1931).
41.   Shakhovskoi, G. P., Lavrov, I. A., Pushkinskii, M. D., and Gonikberg,
      M. G., Pribori Tekhn. Éksperim., 1:181 (1962).
42.   Hilcter, R. G., Bull. Soc. amic. sci. et lett., Poznan, B, 16:201 (1962).
43.   Doolitle, A. K., Simon, I., and Courish, R., AICHE J., 6(1):150 (1960).
44.   Adams, L., J. Am. Chem. Soc., 53:3769 (1931).
45.   Gibson, R., J. Am. Chem. Soc., 59:1521 (1937).
46.   Lebedeva, E. S., Dissertation, Moscow, 1948.
47.   Ekhlakov, A. D., and Podionov, I. L., Fiz. Metal. i Metalloved.,
      9:982 (1960).
48.   Mirinskii, D., Jubilee Collection of Moscow Higher Technical School,
      1955.

49. Razumikhin, V. N., Tr. Inst. Kom Standartov, Mer i Izmerit. Prib., 46(106):96 (1960).

50. Borzunov, V. A., and Razumikhin, V. N., Tr. Inst. Kom. Standartov, Mer i Izmerit. Prib., 75(135):134 (1964).

51. Goldman, Brit. J. Appl. Phys., 9(1):40 (1958).

52. Pavlovich, N. V., and Timrot, D. L., Teploénerg., No. 4, p. 69 (1958).

53. Golubev, I. F., Tr. Gos. Inst. Azo. Prom., No. 7, Goskhimizdat, 1957, p. 47.

54. Roulinson, I. S., Trans. Faraday Soc., 55:1333 (1959).

Chapter XI

Methods of Measuring
Surface Tension and Wettability*

The methodology of the measurement of surface tension and wettability at high pressures does not differ in principle from methods of analogous measurements at atmospheric pressure [1-3, 13]. Therefore, the performance of these measurements at high pressures consists of reproduction of known methods under these conditions. The difficulties which arise in performing these measurements are much more closely connected with the solution of common problems of investigation at high pressures than with the specific nature of the measurements of surface tension and wettability being performed.

On the assumption that the reader is familiar with the theory of surface phenomena and methods of measuring surface tension and wettability at atmospheric pressure, we will use several examples to show the possibility of using these methods for measurement of surface tension and wettability at high pressures in liquid-gas-solid systems, as well as in gas-gas systems.

MEASUREMENT OF SURFACE TENSION AT
LIQUID-GAS BOUNDARY

Capillary-Rise Methods

The measurement of surface tension by the capillary-rise method consists of determining the height of a column of liquid in a capillary

---

*This chapter uses materials kindly presented to the author by Candidate of Chemical Sciences P. E. Bol'shakov.

placed in the liquid being investigated. If we know the radius of the capillary, we can calculate [2] the surface tension. The height of rise of the liquid in the capillary should be measured with great accuracy; at high pressures, only the visual method is convenient for these observations.

An apparatus for determining surface tension by the capillary–rise method (Fig. 300) consists of a rectangular steel chamber 1 with two longitudinal viewing windows on the sides, pressed into packing flanges. The packing is insulated from the action of the medium by aluminum foil. Within the chamber, a glass tube with an internal diameter of about 0.5 mm is mounted on two springs. The steel chamber is mounted in air thermostat 2, equipped with a device for rocking the chamber within limits of ±30° from its horizontal axis for agitation of the liquid and gas phases. The

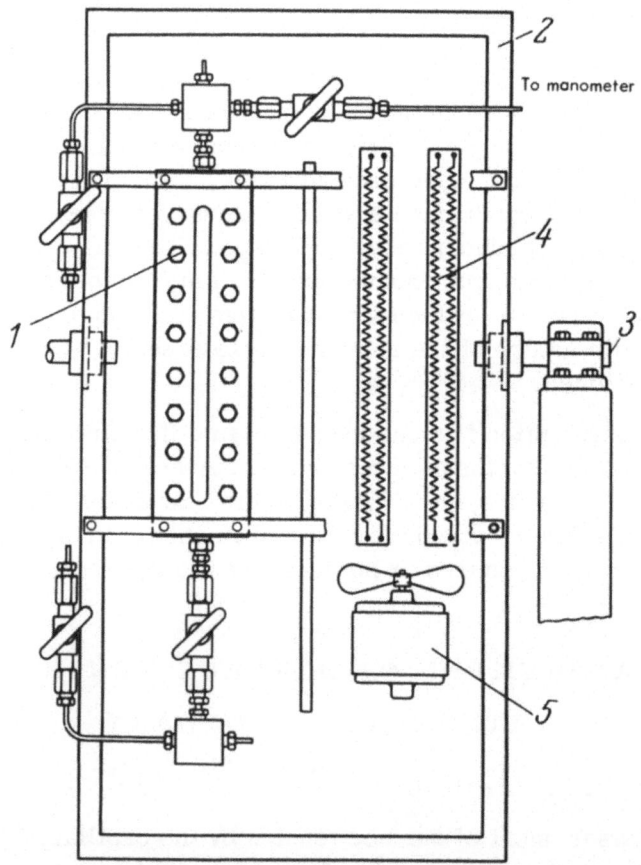

Fig. 300. Installation for measurement of surface tension on liquid–gas boundary by capillary–rise method: 1) steel chamber; 2) thermostat; 3) axle; 4) heaters; 5) fan.

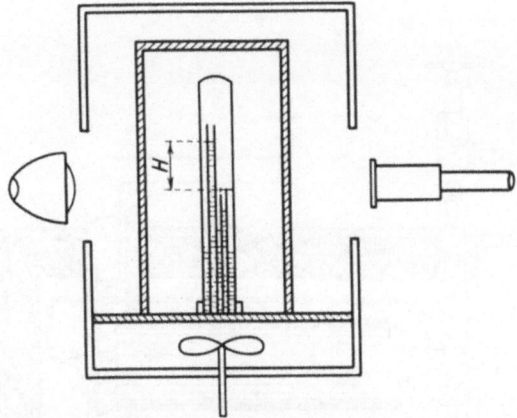

Fig. 301. Installation for measurement of surface tension using difference in heights of liquids in capillaries of different diameters.

mixing of the phases is necessary, since the amount of surface tension depends not only on the pressure, but also on the composition of the phases, which changes with pressure due to solubility of one phase in the other.

The capillary tube should have uniform cross-section throughout its entire length, which can be checked by measuring the length of a mercury column as it moves through the tube. The diameter of the capillary can be determined from the mass of the mercury. The bore of the capillary need not be strictly circular. If the difference between the largest and smallest radii of the bore is 6%, the error in measurement of the height of capillary rise will be approximately 0.1% [4].

The air is pumped out of the assembled apparatus and the apparatus is washed through by the gas being investigated or by vapors of the liquid being investigated. Then, a known quantity of liquid is introduced into the chamber and, after the experimental temperature is reached, the surface tension at the boundary between the liquid and its vapor is determined by the height of rise of the liquid in the capillary. Then, the gas being investigated is fed in through the bottom of the column, the pressure required is generated, and the phases are mixed. The level of the liquid in the capillary is allowed to become stable and the height of the liquid column is measured once more.* Then, the gas pressure is increased, the phases are mixed once more, and the measurement is performed again. After the measurement has been performed at the maximum pressure, the pressure is reduced step by step in the apparatus and a second series is performed.

---

* The level of liquid in the capillary is measured by a cathetometer with an accuracy of 0.1 mm.

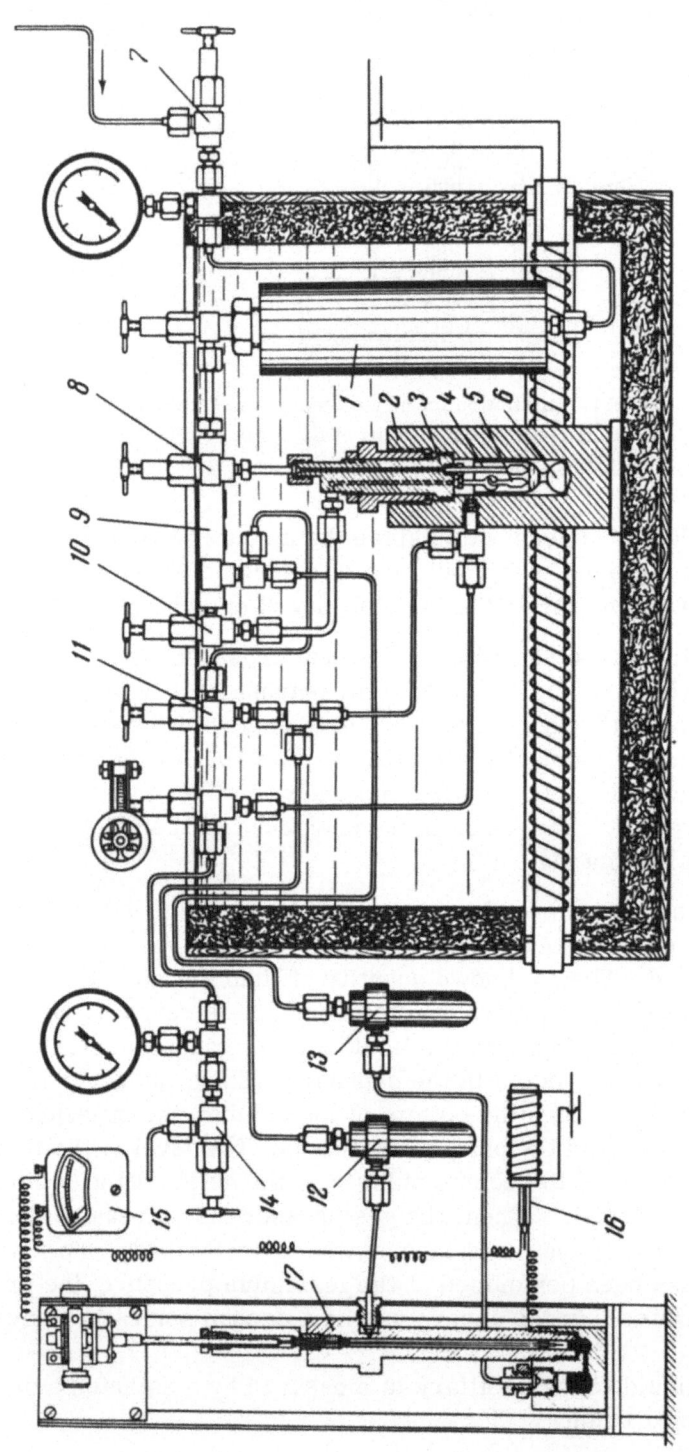

Fig. 302. Bol'shakov installation: 1) saturator; 2) column; 3) head; 4, 5) capillaries; 6) container; 7, 8, 10, 11, 14) valves; 9) distributing block; 12, 13) spray trap; 15) galvanometer; 16) thermocouple; 17) differential manometer.

The structure of the optical portion of this installation can be considerably simplified if the difference in heights in capillaries of different diameters is measured rather than the height of the column of liquid. This allows a considerable reduction in the dimensions of viewing windows. An installation of this type (Fig. 301) [5] consists of a quartz vessel, a thick-walled aluminum block with apertures for windows capable of withstanding high pressure and an air thermostat. Two capillaries of different diameters are located in the quartz vessel. A steel vessel with windows can be used in place of the quartz vessel (see Chapter VI).

Method of Maximum Pressure in Bubble

This method consists of measurement of the pressure required for forcing a gas bubble through the end of a tube submerged in the liquid being investigated. Once this pressure is determined, if the radius of the aperture at the end of the tube is known, the surface tension can be calculated.

The use of this method at high pressures causes certain difficulties. The pressure required to force the gas bubble out depends on the depth to which the capillary is placed in the liquid. This depth can be determined at high pressures through viewing windows.

It is impossible to set the capillary so that it touches only the surface of the liquid, since the level of the liquid changes with temperature and pressure and solution of gas in the liquid. In order to exclude the depth to which the tube is submerged from calculations, two capillaries of different diameters are used, submerged to identical depth in the liquid; the pressure required to force a bubble through first one, then the other, capillary is measured.

Strictly speaking, an error is made in this case since the bubbles blown from the different capillaries have different dimensions. The larger diameter bubbles are actually submerged to a greater depth in the liquid and the pressure required for formation of a larger bubble is greater by the difference in diameters of the large and small bubbles. However, the error which this introduces is less than the accuracy of measurement of the pressure difference, so it can be ignored.

The advantage of the maximum-pressure method is that there is no need to know the precise values of liquid density and gas-phase density at high pressures in order to calculate the surface tension using this method.* Also, visual observations are unnecessary, which greatly simplifies the design of apparatus.

---

* The difference in densities is included in the denominator as a component whose value is low in comparison to the value of the entire denominator.

An installation for measurement of surface tension by the maximum pressure in bubble method designed by P. E. Bol'shakov [6] consists (Fig. 302) of saturator 1, columns 2, distributing block 9 with valves, and differential manometer 17.

The compressed gas enters the installation through valve 7 and passes through a layer of the liquid being investigated in saturator 1. At this point, the liquid is dissolved in the compressed gas. Then, the gas enters capillaries 4 and 5 through block 9 and valves 8 and 10; these capillaries are installed with asbestos glands in head 3. The gas bubbles through the liquid being investigated in container 6 and is dissolved in it. Also, the gas passes through valve 11 directly to the column and fills the entire installation.

The capillaries are submerged in the liquid to identical depths. When the required pressure has been created in the installation and the thermostat containing the installation is heated to the experimental temperature, valve 7 is closed and, by opening valve 13, the gas is begun to be released into the atmosphere. Then, a flow of gas from the saturator is begun through the wide capillary. The gas bubbles through the liquid in container 8. This creates a more complete equilibrium between the gas and liquid at the experimental pressure.

Then a slight excess pressure is created in the system and the exhausting of gas to the atmosphere is continued at a rate such that the number of bubbles passing through the capillary does not exceed 1 or 2 per minute. The bubbles are counted by observing the deflection of the arrow of galvanometer 15. At the moment when a bubble leaves the capillary, the level of mercury in the narrow column of the differential manometer increases due to the increased pressure in the capillary. The mercury touches the platinum tip of manometer 10 and closes the thermocouple-galvanometer circuit. After the bubble leaves the liquid, the pressure in the capillary decreases and the circuit is opened again.

The gas is exhausted until the pressure in the system becomes equal to the experimental pressure. Then, the mercury level in the differential manometer is measured, i.e., the maximum pressure in a bubble is measured. Measurement consists of establishing the platinum tip at a level at which the contact of the mercury with the tip is brief. This is done by raising and lowering the manometer pin. At each position where the pin is stopped, the indication of the differential manometer micromometer (see Chapter IV) is observed, and the height of the mercury column is determined. The manometer in the Bol'shakov installation allows readings to be made to an accuracy of 0.005 mm Hg.

In order to produce reliable data, carefully manufactured capillaries are a necessity. In the manufacture of such capillaries, the end of the

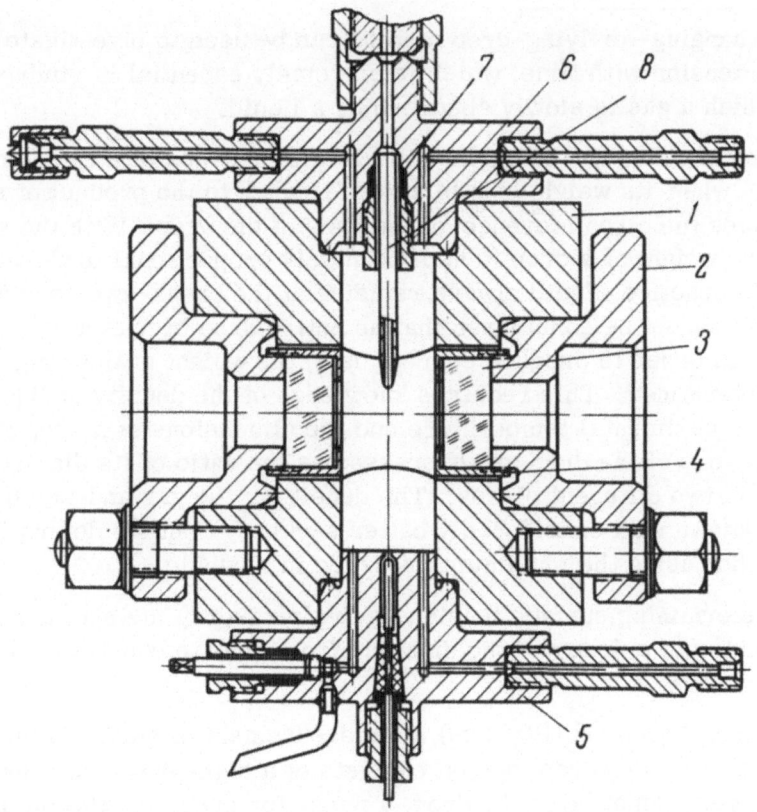

Fig. 303. Installation for measurement of surface tension by hanging-drop method:
1) body; 2) flanges; 3) viewing windows; 4) cover; 5, 6) heads; 7) gland; 8)
dropper.

capillary tube is blown out to a diameter of 6 mm and sharply pulled. Then,
the end of the capillary is gripped in the chuck of a watchmaker's lathe, a
glass-cutting diamond is used to make a scratch around the capillary, and
the tip is broken off clean. A microscope is used to select capillaries with
the proper curvature of apertures and slight depth of cut. The capillaries
selected are carefully washed with a warm chromic acid mixture, with
water, and with alcohol, then calibrated at atmospheric pressure. The
diameter of the large capillary is determined by measurement of the
maximum pressure in a bubble blown in a liquid whose surface tension is
known (such as double-distilled water or methyl alcohol). If the surface
tension and maximum pressure in the bubble are known, the radius of
curvature of the surface and radius of the capillary can be calculated. For
thin capillaries, the radius of curvature differs little from the radius of the
capillary.

Hanging-Drop Method

The hanging- or lying-drop method can be used to investigate changes in surface tension with time, which is extremely essential in studying systems in which a gas is slowly absorbed by a liquid.

A drop formed on the horizontal circular surface of a capillary tip breaks off when its weight equals a force equal to the product of surface tension times the circumference of the base of the drop. With the same surface, the weight of a drop is approximately proportional to the surface tension. In practice, liquid always remains at the break-away surface; therefore it cannot be considered that an entire drop breaks away from the capillary. In order to produce correct data, the weight of the hanging drop must be determined. This requires knowledge of the density of liquid and gas at the experimental temperature and the dimensions of a drop [7, 8, 9]. The dimensions of the drop are expressed as the ratio of its diameters measured in two different planes. The density of the gas and liquid phases should be known with considerably better accuracy than for determination by the method using the maximum pressure in a bubble.

In determining surface tension by the lying-drop method, the maximum diameter of a drop and the distance from peak to equatorial diameter are measured.

An installation [7] (Fig. 303) for measurement of surface tension by the hanging- or lying-drop method consists of a high-pressure chamber, a glass tip for formation of the drop, a press for supplying the liquid under pressure, an optical system and a measuring apparatus. In body 1 of the chamber, which consists of a rectangular vessel, two apertures are drilled straight through for viewing windows 3, mounted on flanges 2. Optically flat glass (40 mm diameter and 20 mm thick) is lapped to fit the internal surface of the flange and pressed in lightly by cap 4. The top and bottom of the column are covered with heads 5 and 6. A thermocouple enters through the lower head; glass dropper 8, in which the drop is formed, is placed in the column through asbestos gland 7. The heads have apertures for supplying and exhausting liquid and gas.

The liquid being investigated is saturated with gas in a steel vessel, from which it is transmitted to the press and fed into dropper 8. Liquid and gas phase densities at high pressures are measured using an upper head of another design with an aperture 25 mm in diameter for a spring made of steel wire 0.25 mm in diameter. A thick-walled glass sphere 20 mm in diameter is suspended on the spring.

This device makes it possible to measure the change in density with increased pressure. The density of the liquid at atmospheric pressure and the experimental temperature should be known. The glass sphere is sub-

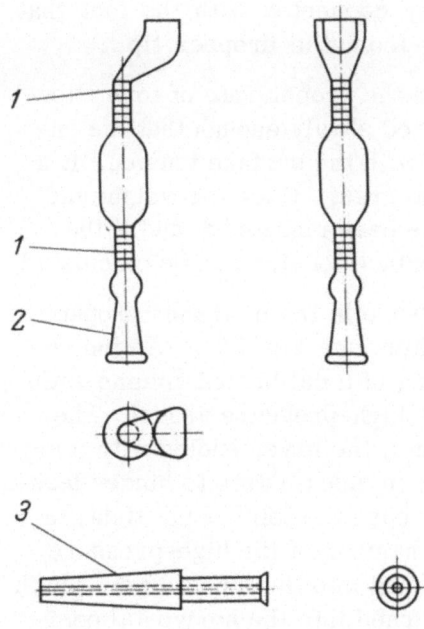

Fig 304. Instrument for measuring surface tension by drop-weight method: 1) graduated tube; 2) joint; 3) brass body.

merged in the liquid poured into the chamber. The extension of the spring is determined from a scale mounted with the sphere, by measuring the position of the scale with a cathetometer to an accuracy of 0.001 mm.

In this work [7], changes in sensitivity of the spring with pressure and the change of the weight of the spring in the compressed-gas medium were not taken into consideration.

The image of the meridional section of the drop and the outline of the dropper is produced on a screen or photographic film using an optical system consisting of a series of lenses. Measurement is performed as follows: the top of the chamber is closed with the head, which carries the carefully washed and dried dropper tip. The hydraulic press is disconnected from the installation and the gas is fed into the chamber to the required pressure. The hydraulic press is filled with the liquid being investigated, which is saturated by the gas. An excess pressure (10 atm) is created in the press. The image of the tip is focused on the screen and photographed. This photograph serves as a scale determiner for determination of the degree of increase in drop size. Then, the valve connecting the dropper to the press is opened. The liquid is fed to the dropper under excess pressure by gradual rotation of the crank. A drop (when ready for measurement) is held several minutes at the tip of the dropper for establishment of equilibrium, then photographed. The average value from several measurements is used for determination of surface tension.

Method of Drop Weight (Volume)

This method consists of measuring the volume or weight of a drop gradually breaking away from the tip of the vertical tube. It is based on the fact that the weight of a drop is proportional to the radius of the tube and depends on the physical and chemical properties of the liquid. In essence, this method is a variety of the method of hanging or lying drop described above, which avoids the difficulty connected with the fact that a portion of the drop does not break away from the dropper tip.

It is extremely important to determine the proper rate of formation of the drop on the tip. Drops should be formed slowly enough that the entire liquid flowing into drop will equilibrate with the surface tension. If a drop forms too rapidly, its weight will be too great. Once the weight and volume (or density) of a number of drops has been measured, and if the radius of the tip of the tube is known, the surface tension can be calculated.

This method was used [3] at high pressure in the methane—propane system, using an installation described earlier (see Fig. 300). An instrument for counting drops (Fig. 304), consisting of a calibrated volume with two graduated tubes 1, was placed within the high-pressure vessel. The upper graduated tube was formed into a funnel, the lower ending with joint 2 into which brass body 3, which terminates in sharp edges to hinder leakage of the liquid along the walls of the tip in case the joint is not tight, is placed. The instrument is installed in the chamber of the high-pressure vessel and, rotating the vessel, liquid is poured into the funnel, from which it flows into the capillary as the vessel is turned into the horizontal position. Then, the drops are counted and their weight and volume are determined.

MEASUREMENT OF CONTACT ANGLES
(OF WETTABILITY) AT HIGH PRESSURES

The method of measuring contact wettability angles of solid bodies by liquids in the presence of compressed gases was developed by P. E. Bol'shakov [10], who used the method of P. A. Rebinder [11] for this purpose.

The principle of the method consists of applying a drop of liquid to the investigated surface and projecting an image of this drop on a screen. The projection can be photographed or the contact angle can be directly measured on the screen.

The installation designed by P. E. Bol'shakov (Fig. 305) allows visual observation of the drop under pressure. The gas, compressed and purified of traces of oxygen* is fed into humidifier 1, in which it is saturated with vapors of the liquid being investigated. Then, the gas is fed into column 2 in which there are two nipples 3 with high-pressure viewing glasses.

The column for measurement of the contact angles in the Bol'shakov installation is shown in Fig. 306. The top of the column 1 is closed by head 2, the upper portion of which contains press 3. Glass dropper 4 is in-

---

*The presence of oxygen may cause formation of an oxide film on the surface being investigated and considerably change the wettability.

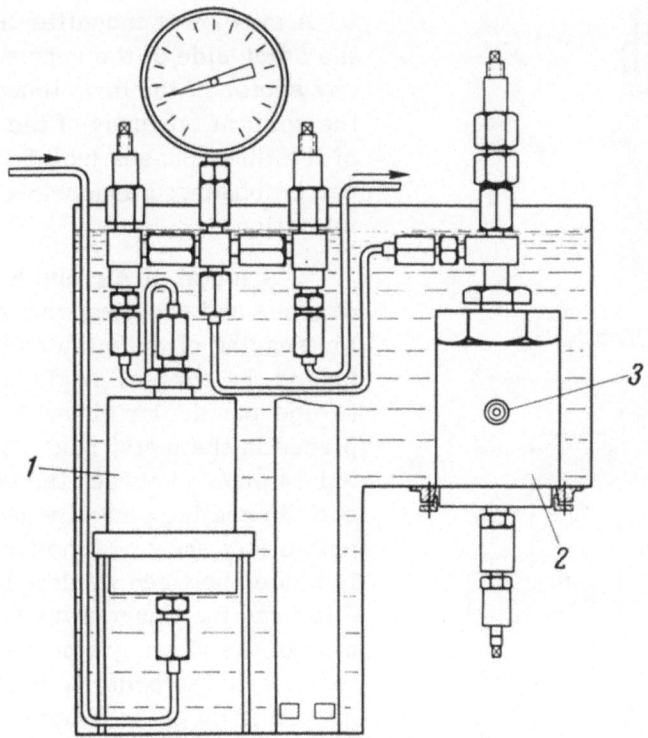

Fig. 305. Bol'shakov installation for measurement of
contact angle under pressure: 1) humidifier; 2) column;
3) nipples.

stalled in an asbestos gland, and the liquid is forced by press 3 through the
dropper onto the surface being investigated. The bottom of the column is
closed by plug 7 in which rod 6 is set in gland 8. The rod is rotated in plug
7 on screw thread 9 and rotates disk 5 attached to its bottom. The disk is
divided into five sectors by grooves. In four of these sectors, samples of
the metals being investigated are placed. By rotating wheel 10, any of the
four samples can be moved to the window and placed under the dropper.
The fifth sector is left opened and used for formation of test drops. Wheel
10 is indexed like disk 5. This makes it easy to determine which sample
is next to the viewing glass.*

The optical system of this installation (Fig. 307) consists of point
light source 1 whose light passes through condenser 2 and falls as a parallel beam on viewing window 3 of the column. Then, the light passes through
the drop, through window 4, and enters lens 5. The image of the drop, mag-

*The method of manufacture and preparation of nipples and attachment of
viewing glasses is described in Chapter VI.

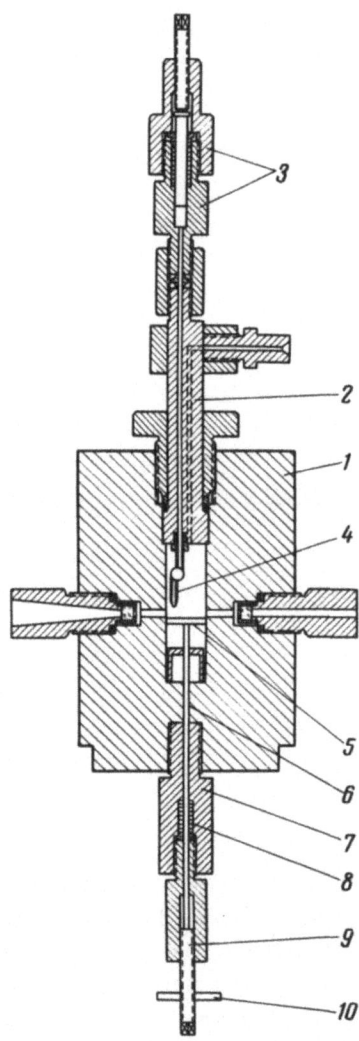

Fig. 306. Column for measurement of contact angle in Bol'shakov installation: 1) body of column; 2) head; 3) press; 4) dropper; 5) disk; 6) rod; 7) plug; 8) gland; 9) screw; 10) ring.

nified 8 times, is projected on screen 6. A six-power magnifier is placed on the other side of the screen. Thus, any sector of the drop image, such as the point of tangency of the boundary of the three phases (solid-liquid-gas) can be observed under 48-power magnification.

A metal circle divided into 360 degrees and supplied with a nonius for reading of the minutes is installed behind the screen. A glass disk with two perpendicular cross hairs is placed in the metal ring. A wooden bar is placed between the screen 6 and the sealing plug of window 3. A prism (not shown in the illustration) is placed between window 4 and lens 5 to bend the image away from the axis of the viewing window. The image is deflected in order to protect the observer from glass fragments.

Before an experiment, metal samples are polished with emery powder and chromium oxide (GOI paste) in a closed chamber under a current of nitrogen. The polished samples are placed on the disk and held down with a thin metal band, washed with pure benzene, and dried in a current of nitrogen. The disk carrying the samples is placed in the column, which is preliminarily washed with the gas being investigated; then the head is closed. The compressed gas from the humidifier is drawn into the column through the dropper at low speed, the dropper being half filled with liquid. The gas enters the tip of the dropper, bubbles through the liquid, is dissolved in the liquid and fills the space between the rod and liquid.

After the required pressure and temperature are established, the press piston is slowly lowered. This causes the liquid to begin flowing

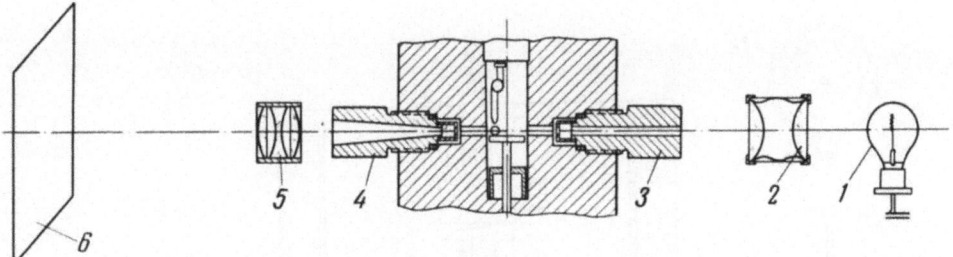

Fig. 307. Optical circuit of Bol'shakov installation: 1) point light source; 2) condenser; 3, 4) viewing windows; 5) lens; 6) screen.

from the dropper, drop by drop. Several drops are allowed to fall on the section of the disk which does not carry a sample. When a drop of the required dimensions is formed, the disk is rotated to place one drop of liquid on each sample; the contact angle is measured immediately after formation of each drop. The drop can be applied at maximum pressure, then the pressure can be lowered, performing measurement at intermediate pressures. However, if the surface under the drop is corroded (by water or carbon dioxide) the measurements are performed at the given pressure and the samples are polished once more, repeating all operations.

In measuring the contact angle, the cross hair of the glass disk is oriented along the projection of the solid surface, then moved to the tangent to the projection of the drop at the point where the drop touches the surface of the solid.

MEASUREMENT OF SURFACE TENSION

AT GAS-GAS BOUNDARY

An installation (Fig. 308) designed for measurement of surface tension between two gas phases formed by stratification of gas mixtures [12] consists of high-pressure vessel 1 placed in thermostat 2. In the vessel, frame 3 carrying matte glass 5 and two capillaries 6 with bores of different diameters is placed opposite windows 4. When the frame connected with armature 7 of an electromagnetic stirrer is moved, the phases are mixed in the vessel and in the bores of the capillaries.

The gas being investigated is fed into the installation in the quantity necessary to produce a mixture of the required composition and mercury is forced into vessel 8 with a press (not shown in the figure) to compress the mixture to the required pressure. The quantity of mixture is selected such that the level of the phase division is located opposite the viewing windows, and the ends of the capillaries are submerged in the heavier

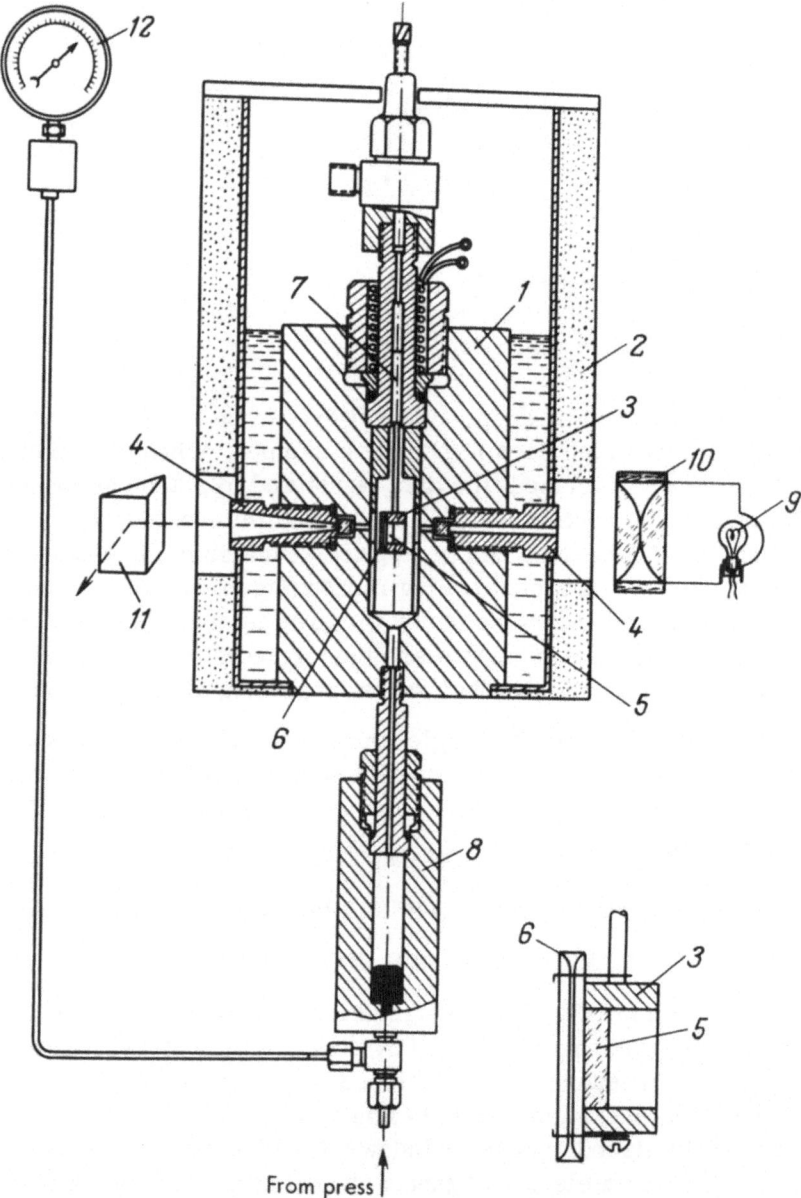

From press

Fig. 308. Tsiklis-Vasil'ev installation for measurement of surface tension at boundary between two gas phases: 1) high-pressure vessel; 2) thermostat; 3) frame; 4) window; 5) matte glass; 6) capillary; 7) armature; 8) mercury press; 9) lamp; 10) condenser; 11) prism; 12) manometer.

phase. Therefore, the heavy phase rises in the capillaries. Since the diameters of the capillary bores are different, the height of capillary rise is also different. The difference in height is measured with a cathetometer. Thus, the installation uses the Volyak principle [5].

The capillaries are illuminated by point light source 9 through condenser 10. The light beam is fed through ground glass 5 and the second window to total internal reflection prism 11, which rotates the beam for observer safety. The pressure in the installation is measured with Bourdon manometer 12.

This installation can also be used to measure the density of coexisting phases (density values are required for calculation of surface tension) and the contact wetting angle of the solid phase of the capillary wall. For this, the ground glass is removed and the surface of the phase division boundary is photographed in transmitted light with the capillary. The contact angle can be measured on enlarged photographs.

BIBLIOGRAPHY

1. Ostval'd-Lyuder-Druker, Physical and Chemical Measurements, Otdel. Nauk Tekhn. Inst., 1935.
2. Adom, N. N., Surface Physics and Chemistry, Goskhimizdat, 1947.
3. Weinaug, C. E., and Katz, D. L., Ind. Eng. Chem., 35:229 (1943).
4. Richards, I. E., and Carver, E. K., J. Am. Chem. Soc., 43:827 (1921).
5. Volyak, L. D., Dokl. Akad. Nauk SSSR, 74:307 (1950).
6. Bol'shakov, P. E., and Levchenko, G. T., Tr. Gos. Inst. Azo. Prom., No. 1, Goskhimizdat, 1953, p. 30.
7. Kusakov, M. M., Lubman, N. M., and Koshevnik, A. Yu., Dokl. Akad. Nauk SSSR, 74:319 (1950); Koshevnik, A. I., Dissertation, Inst. Nefti, Akad. Nauk SSSR, 1953.
8. Hough, E. W., Rsasa, M. J., and Wood, B. B., Trans. Am. Inst. Min. a Met. Eng., Petrol. Div., 192:57 (1951).
9. Hough, E. W., Wood, B. B., and Rsasa, M. J., J. Phys. Chem., 56:996 (1952).
10. Bol'shakov, P. E., Tr. Gos. Inst. Azo. Prom., No. 1, Goskhimizdat, 1953, p. 43.
11. Rebinder, P. A., Physics and Chemistry of Floatation Processes, Metallurgizdat, 1933.
12. Tsiklis, D. S., and Vasil'ev, Ya. N., Dokl. Akad. Nauk SSSR, 136:349 (1961); Zh. Fiz. Khim., 37:1355 (1963).
13. Adamson, A. W., Physical Chemistry of Surfaces, London, 1960.

Chapter XII

Optical, X-Ray, and Electrical Measurements

This chapter presents individual examples of the use of optical, x-ray, and electrical methods for investigation of the refraction coefficients, absorption spectra and Raman-effect spectra, rotation of polarization plane, compressibility, and many other properties of a material under high pressure.

OPTICAL MEASUREMENTS

Observation of Raman-Effect Spectra. The apparatus used for this purpose [1] (Fig. 309) consists of a two-layered high-pressure vessel 1 with four glass windows. Sealing of glass 2 is done in the same way as described in Chapter VI. Windows 3 are designed for illumination of the material. Window 4 is used for measurement. Diaphragms 5 prevent the light from striking the walls.

The Raman spectra are excited by the blue line of mercury ($\lambda = 4538\,\text{Å}$). The light of a mercury-quartz lamp is directed into the vessel through the condenser. The vessel, the type LSP-PI spectrograph which is installed coaxially with it, and the pressure intensifier connected to it are filled with the liquid being investigated which is then compressed by the pressure intensifier to the required pressure. The spectra of compressed gases can be investigated in this same installation.

Observation of Absorption Spectra. In many cases, sapphire windows are used for visual observation [2]; this allows measurement to be performed in the infrared area. For investigations in wider spectral areas, diamond windows and high-pressure vessels made of diamond are used. In one such installation [3] designed for investigation using

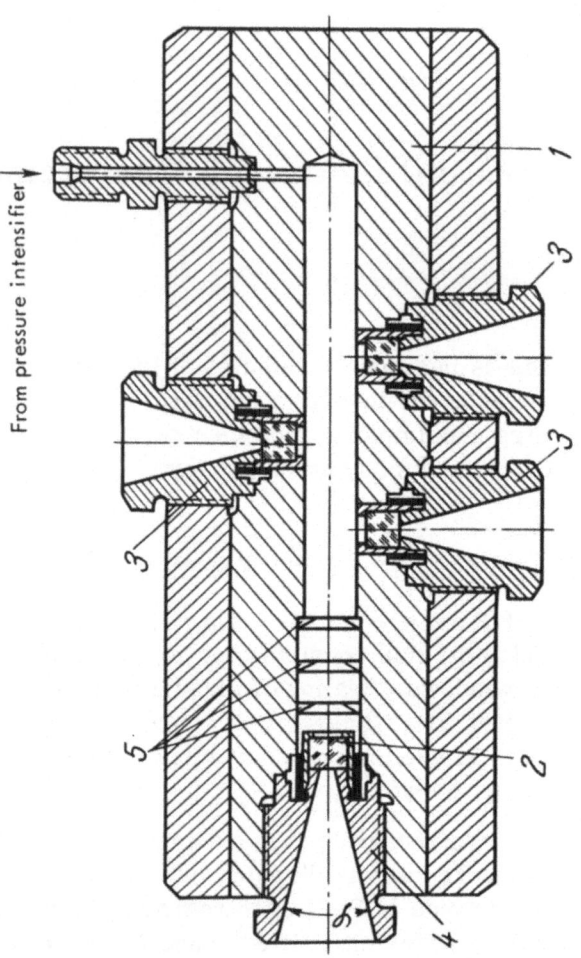

Fig. 309. Device for observation of Raman spectra: 1) high-pressure vessel; 2) glass; 3) illuminating windows; 4) observation window; 5) light filters.

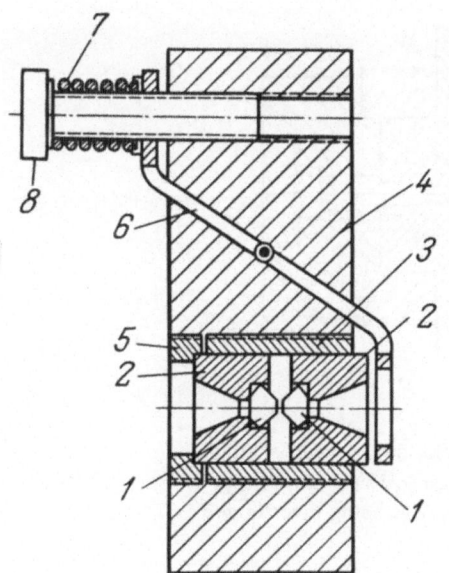

Fig. 310. Diamond high-pressure apparatus: 1) diamond; 2) piston; 3) support; 4) block; 5) rests; 6) lever; 7) spring; 8) screw.

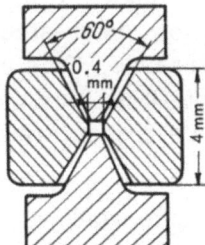

Fig. 311. Diamond vessel.

wavelengths in the visible, infrared, and ultraviolet spectral areas (Fig. 310), type II diamond is used [4], the absorption spectrum of which allows measurement to be performed in these wavelength areas.*

The primary portion of the apparatus — the high pressure — is made of two diamonds, each 0.036 g in mass. They form two anvils, the compressing surfaces of which are ground until their planes (0.0013 mm^2 area) are parallel. The material being investigated is placed between the diamonds. Each diamond has conical form. The large base of the diamond cone rests on piston 2 made of stainless steel, into which an aperture 1.5 mm in diameter is drilled. The material being investigated is placed in the focus of a beam passing through the system. Pistons 2 can move freely in bronze support 3 screwed into steel block 4. One of the pistons rests against part 5, also screwed into the block. The other piston can move in the support under the influence of lever 6, which is pressed upon by spring 7. A degree of compression of the spring is controlled by screw 8. This device can be used to create a pressure between the diamonds of up to 160,000 atm. The entire installation is placed in a thermostat. The temperature-controlling medium is heated air or the vapors of liquid nitrogen.

An apparatus of another design [5] uses a diamond vessel (Fig. 311) consisting of a diamond cylinder approximately 6 mm in diameter and 4 mm high in which a central aperture 0.4 mm in diameter is drilled between two conical broader areas. This diamond is a reduced copy of the

---

* Heating to 800°C and compression to 160,000 atm do not change the optical properties of this type of diamond.

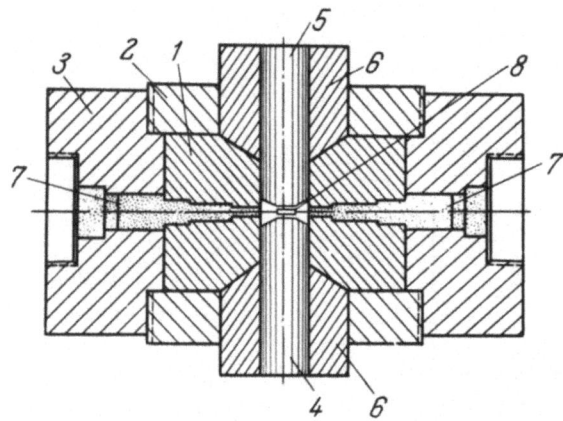

Fig. 312. Apparatus with sodium chloride windows: 1) high-pressure vessel; 2, 3) reinforcing rings; 4, 5) pistons; 6) supporting ring; 7) salt windows; 8) material being investigated

apparatus described in Chaper III. The conical portions of the diamond carry pistons which compress the material being investigated when a force is applied to the pistons using a spring, screw, or other method.

Sodium chloride is a very convenient material for manufacture of high-pressure-vessel windows. It has a low shear-strength limit, so that compression creates quasihydrostatic stresses. This material is transparent for wavelengths from 0.25 to about 10 μ. Due to these properties, sodium chloride can be used as a pressure-transmitting medium and a medium for manufacture of windows.

One apparatus [6] with sodium chloride windows is shown in Fig. 312. It consists of a high-pressure cylinder 1 made of steel with hardness 58-60 R_c. Reinforcing rings 2 and 3 of softer steel are placed around the cylinder. The cylinder contains two cemented-carbide pistons. Piston 4 is stationary, and piston 5 can move under the influence of the force applied. Guiding bronze cylinders 6 encircle the pistons.

Cylinder 1 has telescopic apertures which are filled with sodium chloride. The tremendous pressure acting on the small area of the sodium chloride (diameter 6 mm) is therefore transmitted to an ever increasing area and finally the end of the window is subjected to stresses so slight that it need not even be supported by a plug.

The telescopic tube should be thoroughly cleaned of rust, dust, and moisture. For this, it is polished with rouge and carborundum powder, cleaned with wood chips, and blown out with dry air.

A sodium-chloride monocrystal is placed in the smallest-diameter aperture, heated to 500°C, and pressed through the bronze tube which seals the salt with a carboloy piston carefully lapped to the aperture (clearance not over 0.001 mm). The apertures of larger diameters are filled in the same manner. Then, the salt is sealed by pressing it up to a pressure of 15,000 atm. The pressure is dropped and the salt is compressed once more to 12,000 atm. Subsequent compression cycles are performed at 9000, 6000, and 3000 atm. The pressure is applied rapidly and maintained for one minute. In this way, a rather transparent window is produced. After both windows are filled, a crystal approximately 1 mm thick is formed, heated and placed in the central portion of the apparatus. The crystal is compressed to 30,000 atm with flat pistons. During this operation, the windows are closed with plugs with polished surfaces screwed into apertures in ring 3. These same plugs should be kept in place when the apparatus is not being used.

After the windows are formed, the salt is removed from the crystal portion of the apparatus and replaced with fresh salt, containing the material being investigated 8. Windows made in this manner can withstand repeated pressures up to 160,000 atm. However, they are rapidly destroyed by moisture and dirt.

A chamber with conical windows for optical investigations under pressures up to 30,000 atm (Fig. 313) [7] consists of die 1, made of ShKh15 steel (58-60 R_c) pressed into cone 6 (45KhNMFA steel) and compressed by nut 7. Conical apertures are drilled into the matrix for windows filled with

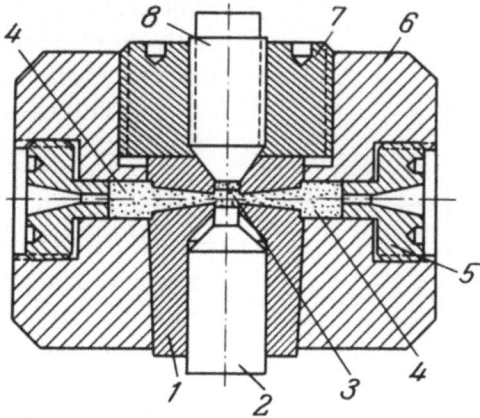

Fig. 313. Chamber for optical investigations under pressures up to 30,000 atm: 1) die; 2, 8) pistons; 3) steel packing; 4) viewing window; 5) obturators; 6) mount; 7) bracing nut.

sodium chloride. The windows are reinforced by obturators 5. The material being investigated is placed under plate 3. It is compressed with piston 2. Piston 8 is stationary. The conical form of the windows aids in reducing the stresses on their external surfaces. For improved transparency, the salt is compressed under a pressure of up to 25,000 atm.

X-RAY MEASUREMENTS*

X-ray measurements yield extremely important information on the behavior of materials under pressure. Determination of the crystal-lattice constants allows us to judge the structure of a compressed material, to establish the number of phase transitions, and to decide which transitions have taken place: a phase transition, or a transition of an electron within the atom. † Also, the data produced in X-ray investigations enable calculation to be made of the compressibility of materials and the coefficients of thermal expansion. This method excludes possible errors caused by porosity of a material and eliminates the necessity of introducing corrections for deformation of the high-pressure vessel.

Since only a very slight quantity of material is required for performance of X-ray investigations, this method can be used to investigate materials which are not available in large quantities.

The difficulties arising in performance of X-ray investigations [8, 9] are primarily connected with the fact that the material of the vessel also absorbs the X-rays. Therefore, the vessel must be as thin as possible and the material used to manufacture the X-ray chamber of the high-pressure apparatus or the walls which transmit the X-rays should be properly selected. Materials transparent to X-rays include Be, C, LiH, BN, LiF, B_4C, BeO, etc. However, some of these materials have low mechanical strength. Many organic materials which are transparent to X-rays also have low strength.

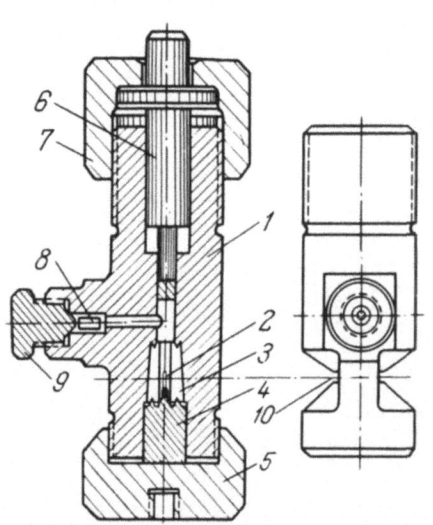

Fig. 314. X-ray rotation chamber for investigation of monocrystals: 1) high-pressure vessel; 2) crystal; 3) beryllium cone; 4) steel plug; 5, 7) nuts; 6) piston; 8) manganin manometer; 9) electrical lead; 10) slit for X-rays.

*X-ray investigations at high pressures have been described by L. F. Vereshchagin [8].
†Determination of the break on the PV curve in this case will not allow the difference to be established.

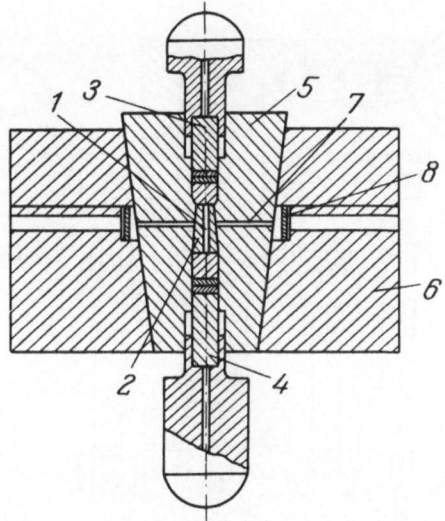

Fig. 315. Chamber for X-ray investigatiions
at pressures up to 30,000 atm: 1) material; 2)
beryllium cone; 3, 4) pistons; 5) cone; 6) mount;
7) slit; 8) photographic film holder.

In one installation designed for X-ray investigations [10], the window
is made of beryllium, the target and film being placed within the high-
pressure vessel and the pressure (up to 5000 atm) being created by compressed helium. An inconvenience of this method of creating pressure is
the fact that the helium may erode the photographic layer away from the
film if the pressure is dropped rapidly.

Investigation of monocrystals under pressures up to 7000 atm can
be performed using an X-ray rotation chamber [11] (Fig. 314)* consisting
of a steel high-pressure vessel 1 which contains the investigated crystal
2 in beryllium cone 3. The axis of rotation of the steel vessel should correspond to one of the primary crystallographic planes of the crystal being
investigated. The aperture in the cone is sealed at the bottom with steel
plug 4, at the end of which there is a conical rod which closes the opening.
Nut 5 serves to press the beryllium cone. The pressure is created by steel
piston 6, which presses on the benzine filling the bore of the beryllium cone
and the vessel. The piston is moved by a hydraulic press, and after the
required pressure is created, is fixed by nut 7. After this, the sample

---

* A rotation X-ray diagram requires several strata of lines; therefore, a
wide photographic film is required. The X-ray diagram of a polycrystal-
line sample may be limited as to height without loss of measurement accuracy.

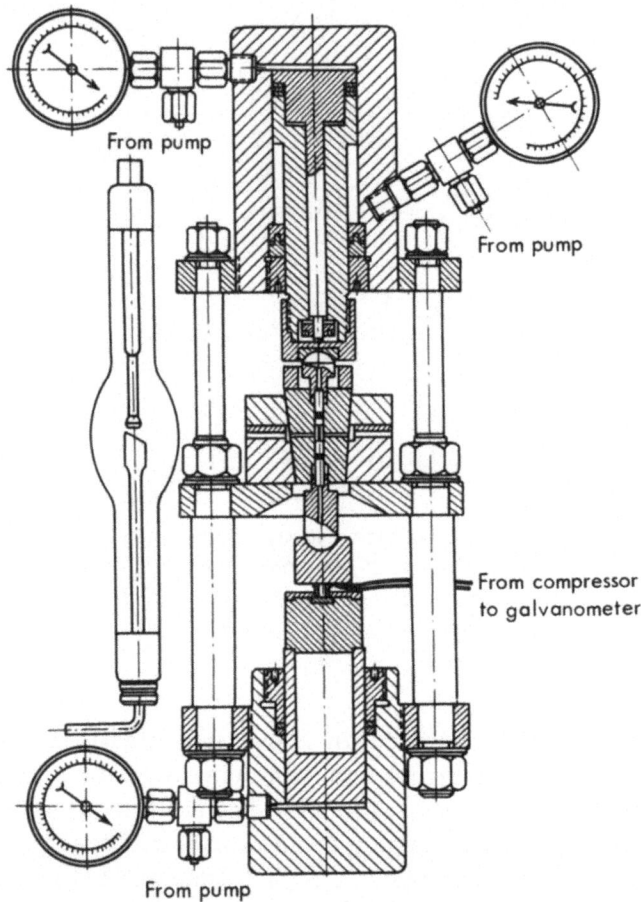

Fig. 316. Location of chamber and high-pressure presses.

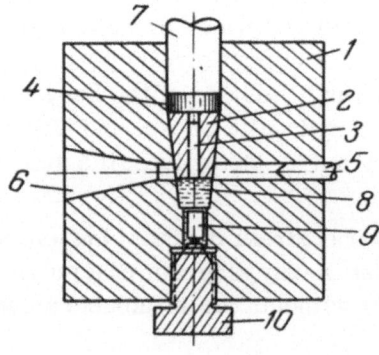

Fig. 317. X-ray chamber for work at pressures up to 16,000 atm: 1) high-pressure vessel; 2) beryllium cone; 3) sample; 4) plug; 5) diaphragm; 6) slit; 7) piston; 8) liquid; 9) manganin manometer; 10) electrical lead.

chamber, now under pressure, is placed in the X-ray installation and the exposure is made. The compression pressure is measured with manganin manometer 8 and electrical lead 9. Slit 10, 3 mm high and conical in form, is used for entrance and exit of the X-rays, allowing the diffraction cones, forming an angle up to 25° with the horizontal plane, to fall on the film.

A chamber [12] for X-ray investigations of materials at up to 30,000 atm is shown in Fig. 315. The investigated material 1 is located within beryllium cone 2. The pressure on the sample is transmitted through lithium from pistons 3 and 4, made of VK-8 alloy. The cone is reinforced mechanically (see Chaper III). Cone 2 is placed in steel cone 5, which is located in mount 6. Slit 7, 2 mm in diameter, in steel cone 5 passes the X-rays. The exiting ray exposes photographic film held in mount 8. The chamber is placed in a hydraulic press and the X-ray tube is placed near it (Fig. 316).

A considerably simpler chamber [13] for investigations at pressure up to 16,000 atm (Fig. 317) consists of a high-pressure vessel 1, containing beryllium cone 2. The investigated material 3 is placed in the bore of the cone. A bore 0.4 mm in diameter is closed with steel plug 4. The rays enter the chamber through diaphragm 5, exiting through slit 6. The force of the press is transmitted to steel piston 7 and beryllium cone 2. When the cone is forced into the vessel, liquid 8 is compressed. The pressure which is then created in the vessel is measured by manganin manometer 9, connected to electrical lead 10.

It was stated above that materials capable of transmitting X-rays include carbon. Therefore, an ideal material for construction of X-ray chambers is diamond. Many X-ray investigations have been performed in the diamond installations described above [14].

ELECTRICAL MEASUREMENTS

Polarographic Measurements. An installation for polarographic investigations at high pressures [15] is shown in Fig. 318. Polarographic cell 3 is fastened on frame 8 in high-pressure vessel 1. The side wall of the vessel contains triple electrical lead 4, which is used for connection of the electrical leads of the cell to the measuring circuit. The cell consists of a glass cylinder containing the capillary of dropping electrode 9 and the salt bridge of reference electrode 5. The lower (narrow and notched) open end of the cell is submerged in mercury, poured into container 2. The mercury in the container forms a plug separating the liquid being investigated from the pressure-transmitting liquid, and also serves as the floor of the cell and as its anode.

The capillary of the dropping electrode has a spoon to force breakaway of the drop, which allows the time between drop falls to be maintained

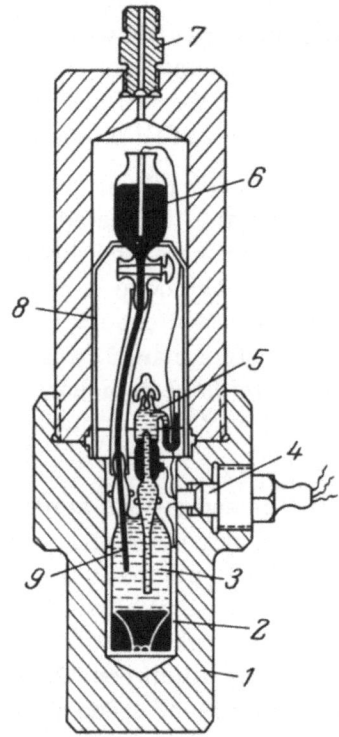

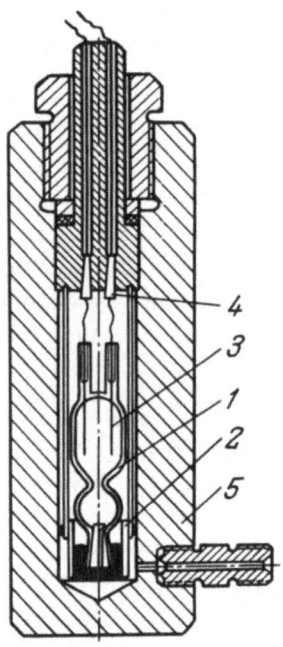

Fig. 318. Cell for polarographic investigations under pressure: 1) high-
pressure vessel; 2) container with mercury; 3) polarographic cell; 4) electrical lead; 5) reference electrode; 6)
reservoir; 7) nipple; 8) frame; 9)
dropping electrode.

Fig. 319. High-pressure vessel
with cell for measurement of
electrical conductivity of solutions: 1) cell; 2) steel container
with mercury; 3) electrodes; 4)
electrical leads; 5) high-pressure vessel.

constant during measurement of the electrode potential. The capillary of
the electrode is connected by vacuum tube with the glass reservoir 6, filled
with mercury. Since the quantity of mercury expended in an experiment is
slight and the capacity of reservoir 6 is great, the height of the mercury
column remains practically unchanged during the course of a single experiment (2-3 h).

The reference electrode 5 is a saturated calomel electrode. The tip
of the salt bridge (containing potassium-chloride solution) is closed with
filter paper and a plug of agar-agar. The electrode contains a mercury-
filled syphon, which serves as a plug and equalizes the pressure within
and without the cell. The entire vessel is placed in a thermostat.

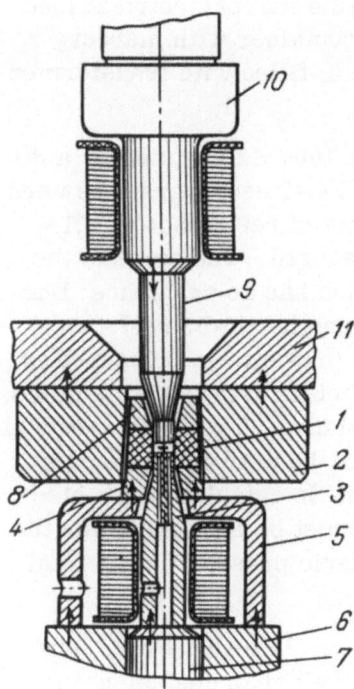

Fig. 320. Installation for investigation of galvanomagnetic phenomena under pressure: 1) microlite die; 2) mount; 3, 9) dies; 4, 8) nonmagnetic conical rings; 5) brace; 6) lower-press piston; 7) rod; 10) upper-press piston; 11) intermediate platen.

The measurements are performed as follows. Mercury is poured into container 2 and the cell is set in place. Reservoir 6 is placed in holder 8. Dropping electrode 9 is placed in the cell and blown through with nitrogen to remove oxygen from the air. Then, the specimen solution is placed in cell 3 in a current of nitrogen and the reference electrode 5 is set in place. The holder and cell are placed in the lower portion of vessel 1 and the conductors from the electrodes are soldered to the terminals of electrical lead 4. Then, the stopcock is opened under reservoir 6 and the top portion of the vessel is set in place and is filled with oil, which serves as the pressure-transmitting liquid; the entire vessel is placed in the thermostat. After the temperature is established, a control polarogram is taken at atmospheric pressure. Then, the temperature is increased and the measurements are performed.

Polarography is performed by the usual method. A potential difference is created between the mercury in vessel 2 and dropping electrode 9 and the current flowing in the circuit is measured with a microammeter (type M-95, sensitivity 5-10 μA). The potential of the dropping electrode is measured in relation to the reference electrode by a potentiometer P-4 with expanded measurement scale.

Measurement of Electrical Conductivity under Pressure [16]. The electrical conductivity of a solution under pressure is measured in the apparatus shown in Fig. 319. Cell 1 is made of pyrex glass. The lower portion of the cell is closed with a fitted plug whose bore is bent in order to prevent the mercury from entering glass container 2 in the electrode space when the pressure is increased. Two electrodes 3 made of blackened platinum are soldered at the top of the cell. The one-sided surface area of the electrode is 4 cm^2, the distance between the electrodes is

0.9 cm. The electrodes of the cell are soldered to the ends of electrical lead 4, insulated from the body by mica. Cell 1 and container with mercury 2 are fastened into the steel high-pressure vessel 5, filled with transformer oil, and placed in a thermostat.

The electrical conductivity is measured at 1000 Hz supplied by a 3G-10 oscillator with a Shidlovskii circuit [17]. An EO-7 oscillograph is used as a null instrument. The error in measurements of resistance is 0.1-0.4%, depending on the amount of resistance measured. The arms of the bridge are balanced by resistance box R-58, which has no reactance. During measurements of resistance over 30,000 Ω, another R-58 resistance box is used as a shunt. At the moment when the pressure is increased, a certain increase in temperature of the solution being investigated is noted, which leads to a reduction in its electrical resistance. Therefore, several measurements at intervals of 10-15 minutes must be performed, until a constant value of solution resistance is achieved. In calculating the electrical conductivity of the solution, a correction must be introduced for the electrical conductivity of the solvent at atmospheric pressure. The total error in measurements is 1.5-2%.

Investigation of Galvanomagnetic Phenomena. The investigation of galvanomagnetic phenomena can be performed using the apparatus [18] shown in Fig. 320. Conical microlite* die 1, 20 mm high with an external diameter of 45 mm, is placed in the conical aperture of mount 2, made of 40 Kh steel (40-42 R_c). Two conical parts 4 and 8 made of 1Kh18N9T nonmagnetic steel enter the mount above and below the die. Die 1 and parts 4 and 8 are fitted to enter the conical aperture in mount 2 as a single whole. During assembly, a lead plate 1 mm thick is placed between the die and the steel parts and the parts are wrapped in lead foil, up to 0.3 mm thick; graphite lubricant is applied. Compressive force is applied to lower part 4 through mount 5 (also made of nonmagnetic steel) as supplied by piston 6 of the lower press. The supporting force which arises not only around the external surface of the microlite die, but also at its ends, creates almost completely surrounding support. The support force is transmitted from mount 2 by intermediate plate 11 of the press.

The internal pressure in the die is created by two anvils 3 and 9, made of R18 steel. Lower anvil 3 carries four electrical leads which are fastened to the electrodes from the sample for simultaneous measurement of the Hall effect, resistance, and the dependence of resistance on the magnetic-field intensity. The pressure-transmitting medium is silver chloride. The force on the anvils is transmitted from the upper press. The lower anvil 3 rests against rod 7 which passes through piston 6 of the lower press

---

* Microlite is a ceramic material suggested by I. I. Kitaigorodskii.

and rests against the bottom of the press cylinder. Piston 10 and anvil 3 carry magnetizing coils which create a magnetic field in the compression chamber. If the distance between anvils is 4 mm, the mean magnetic field intensity at the location of the sample will be about 4 kOe. During an experiment, the pressure is increased in the lower and upper press cylinders simultaneously, increasing the support pressure of the die as the internal pressure increases.

BIBLIOGRAPHY

1. Gonikberg, M. G., Stern, Kh. E., Ukholin, S. A., Opekunov, A. A., and Aleksanyan, V. T., Zh. Obshchei Fiz., 6:108 (1959).

2. Waggener, W. C., Rev. Sci. Instrum., 30(9):788 (1959); Warschauer, D. M., Op. Cit., 29(8):13 (1958).

3. Weir, C. E., Lippincott, E. R., and Bunting, F. N., J. Res. Nat. Bur. St., A63:55 (1959).

4. Bunting, E..N., and van Valkenburg, A., Am. Min., 43:102 (1958); Lansdall, K., Nature, No. 892:669 (1944); Lippincott, E. R., et al., Spectrochim. Acta, 16:59 (1960).

5. Kasper, J. S., Hillard, J. E., Caht, I. W., and Phillips, V. A., W.A.D.C. Techn. Rept. 59-747. Gen. Elec. Co., New York, 1960.

6. Fitch, R. A., Slykhouse, T. E., and Drickamer, H. G., J. Opt. Soc. Am., 47:1015 (1957).

7. Klyuev, Yu. A., Dokl. Akad. Nauk SSSR, 144:538 (1962).

8. Vereshchagin, L. F., X-ray Structural Investigations of Materials at High Pressures. Supplement to book: Svenson, K., High Pressure Physics, IIL (Translation edited by L. F. Vereshchagin), 1963.

9. Jamieson, I. C., and Lawson, A. W., Debye-Shearer X-Ray Techniques for Very High Pressure Studies. In collection: Modern Very High Pressure Techniques, London, 1962.

10. Jacobs, R. W., Phys. Rev., 54:325 (1938).

11. Vereshchagin, L. F., Kabalkina, S. S., and Ebdokimova, V. V., Pribory i Tekhn. Éksperim., No. 3:90 (1958).

12. Vereshchagin, L. F., and Brandt, I. V., Dokl. Akad. Nauk SSSR, 108 (1956).

13. Kabalkina, S. S., and Vereshchagin, L. F., Dokl. Akad. Nauk SSSR, 131:300 (1960).

14. Jamieson, J. C., J. Geol., 65:334 (1957).

15. Mairanovskii, S. G., Gonikberg, M. G., and Opekunov, A. A., Dokl. Akad. Nauk SSSR, 123:312 (1958).

16. Ershov, Yu. A., Gonikberg, M. G., Neiman, M. B., and Opekunov, A. A., Dokl. Akad. Nauk SSSR, 128:759 (1959).

17. Weisberger, A., Phys. Meth. Org. Chem., 2:1022 (1949).

18. Likhter, A. I., Pribory i Tekhn. Éksperim., No. 1:131 (1960).

APPENDICES

APPENDIX I

Properties of Some Heat-Resistant Metals

Properties	Beryllium	Titanium	Zirconium	Hafnium	Niobium	Tantalum	Tungsten
Ultimate strength, kg/cm^2	3500–5700	4800–8000	4300	3500	4800	4800	16,000
Elongation, %	0–20	15–25	30	35	49	30	1–4
Young's modulus × 10^{-6}, kg/cm^2	3.0	1.1–1.3	0.8	1.06	1.06	2.0	3.5–4.2
Melting point, °C	1283	1660	1860	2130	2485	2950	3410
Specific heat, cal/(g · degree)	0.42	0.129	0.066	0.034	0.065	0.036	0.034
Coefficient of linear expansion × 10^{-6} degree^{-1}	11.5	8.5	5.4	5.9	6.86	6.5	4.0

APPENDIX II

Conversion from Brinell and Rockwell Numbers to Tensile Strength

Brinell numbers		Rockwell numbers		$\sigma_b \cdot 10^{-2}$, kg/cm^2		
sphere, 10 mm diameter, load 3000 kg		H_{R_C}	H_{R_B}	Carbon steel	Chrome steel	Nickel and chrome — nickel steel
d_B —indentation diameter, mm	H_B	150 kg	100 kg			
2.44	632	62	—	227	220	214
2.46	621	61	—	223	217	210
2.48	611	60	—	220	213	207
2.50	601	—	—	216	210	204
2.51	597	59	—	214	208	202
2.52	592	—	—	212	206	200
2.53	587	—	—	210	204	199
2.54	582	58	—	208	203	197
2.56	573	—	—	—	201	—
2.57	569	57	—	205	200	194
2.58	564	—	—	203	198	192
2.59	560	—	—	—	—	—
2.60	555	56	—	200	195	189
2.61	551	—	—	198	193	187
2.62	547	—	—	196	191	185
2.64	538	55	—	194	189	184
2.65	534	—	—	192	187	182
2.67	526	—	—	—	—	—
2.68	522	53	—	185	180	175
2.70	514	—	—	183	178	173
2.71	510	—	—	—	—	—
2.72	507	52	—	180	175	170
2.74	499	—	—	178	173	168
2.75	495	51	—	—	—	—
2.76	492	—	—	175	170	165
2.78	486	—	—	173	168	163
2.79	481	50	—	171	—	—
2.81	474	—	—	169	165	160
2.82	471	49	—	167	163	158
2.84	464	—	—	165	—	—
2.85	461	48	—	—	160	—
2.87	454	—	—	162	158	153
2.88	451	—	—	[161]	[157]	[152]
2.90	444	47	—	160	156	—
2.91	441	—	—	158	154	150
2.93	435	46	—	—	—	—
2.95	429	—	—	155	150	146

Appendix II (Continued)

Brinell numbers		Rockwell numbers		$\sigma_b \cdot 10^{-2}$, kg/cm^2		
sphere, 10 mm diameter, load 3000 kg		H_{R_C}	H_{R_B}	Carbon steel	Chrome steel	Nickel and chrome – nickel steel
d_B – indentation diameter, mm	H_B	150 kg	100 kg			
2.96	426	45	–	153	–	–
2.98	420	–	–	151	147	143
3.00	415	44	–	149	145	141
3.02	409	–	–	147	143	139
3.04	404	43	–	145	141	137.7
3.06	398	–	–	143	138	135.5
3.08	393	42	–	141	137	133.5
3.10	388	–	–	139.5	136	132
3.12	383	41	–	138	134	130
3.14	378	–	–	136	132	128.5
3.16	373	–	–	134	130.3	127
3.18	368	40	–	132	128.5	125
3.20	363	–	–	130.5	127	123.5
3.22	359	39	–	129	125.5	122
3.24	354	–	–	127.5	124	120.5
3.26	350	38	–	126	122.5	119
3.28	345	–	–	124	121	117.5
3.30	341	37	–	122.5	119	116
3.32	337	–	–	121	118	114.5
3.34	333	–	–	120	117	113 5
3.36	329	36	–	118.5	115.5	112
3.38	325	–	–	117	114	110.5
3.40	321	35	–	115.5	112	109
3.42	317	–	–	114	111	108
3.44	313	34	–	112.5	109.5	106.5
3.46	309	–	–	111	108	105
3.48	306	33	–	110	107	104
3.50	302	–	–	108.5	105.5	102.5
3.52	298	–	–	107	104	101.5
3.54	295	32	–	106	103	100.5
3.56	292	–	–	105	102	99.5
3.58	288	31	–	103.5	101	98
3.60	285	–	–	102.5	100	97
3.62	282	30	–	101.5	98.5	96
3.64	278	–	–	100	97.5	94.5
3.66	275	–	–	99	96.5	93.5
3.68	272	29	–	98	95.5	92.5
3.70	268	–	–	97	94	91.5

Appendix II (Continued)

Brinell numbers		Rockwell numbers		$\sigma_b \cdot 10^{-2}$, kg/cm^2		
sphere, 10 mm diameter, load 3000 kg		H_{R_C}	H_{R_B}	Carbon steel	Chrome steel	Nickel and chrome-nickel steel
d_B−indentation diameter, mm	H_B	150 kg	100 kg			
3.76	260	—	—	93.5	91	88.5
3.80	255	—	—	92	89	86.5
3.86	246	25	—	88.5	86	83.5
3.90	241	24	100	87	84.5	82
3.96	234	23	99	84	82	80
4.00	229	22	98	82.5	80	77.5
4.06	222	21	97	80	77.5	75.5
4.10	217	—	—	78	76	74
4.16	211	—	—	76	74	72
4.20	207	—	—	74.5	72.5	70.5
4.26	200	—	93	72	70	68
4.30	197	—	—	71	69	67
4.40	187	—	—	67.5	68.5	63.5
4.52	177	—	88	63.5	62	60
4.72	161	—	84	58	56.5	55
5.00	143	—	—	51	49.5	48.5
5.20	131	—	72	47	45.5	44.5
5.40	121	—	68	43.5	42.5	41
5.61	111	—	63	40	39	38
5.80	103	—	57	37	—	—
6.04	94	—	51	34	—	—

APPENDIX III

Pressure Unit Conversion Table

Unit	Barye, dyn/cm^2	Bar	kg/cm^2	kg/m^2	Normal [standard] atmosphere
Barye, dyn/cm^2	1	$1 \cdot 10^{-6}$	$1.01972 \cdot 10^{-6}$	$1.01972 \cdot 10^{-2}$	$0.98692 \cdot 10^{-6}$
Bar	$1.0000 \cdot 10^6$	1	1.01972	$1.01972 \cdot 10^4$	0.98692
kg/cm^2	$98.0665 \cdot 10^4$	$98.0665 \cdot 10^{-2}$	1	$1 \cdot 10^4$	0.96784
kg/m^2	98.0665	$98.06665 \cdot 10^{-6}$	$1 \cdot 10^{-4}$	1	$0.96784 \cdot 10^{-4}$
Normal [standard] atmospheres	$1.01325 \cdot 10^6$	1.01325	1.033227	$1.033227 \cdot 10^4$	1
Centimeters Hg at 0°C	$1.33322 \cdot 10^4$	$1.33322 \cdot 10^{-2}$	$1.3591 \cdot 10^{-2}$	$1.3591 \cdot 10^2$	$1.3158 \cdot 10^{-2}$
Inches Hg at 0°C	$3.38639 \cdot 10^4$	$3.38639 \cdot 10^{-2}$	$3.3453 \cdot 10^{-2}$	$3.3453 \cdot 10^2$	$3.3421 \cdot 10^{-2}$
cm of water at 4°C	$9.80638 \cdot 10^2$	$9.80638 \cdot 10^{-4}$	$0.99997 \cdot 10^{-3}$	9.9997	$9.6781 \cdot 10^{-4}$
Inches of water at 4°C	$2.49082 \cdot 10^3$	$2.49082 \cdot 10^{-3}$	$2.5390 \cdot 10^{-3}$	$2.5399 \cdot 10$	$2.4582 \cdot 10^{-3}$
Pounds per square inch	$6.8947 \cdot 10^4$	$6.8947 \cdot 10^{-2}$	$7.0306 \cdot 10^{-2}$	$7.0306 \cdot 10^2$	$6.8045 \cdot 10^{-2}$

APPENDIX IV

Chemical Composition of Steels Used for Manufacture of High-Pressure Apparatus (Weight %)

Type of steel	All-Union State Standard	C	Cr	Mn	Mo
St. 5	380-50	0.28-0.37	—	0.50-0.80	—
18KhNVA	4543-48	0.14-0.21	1.35-1.65	0.25-0.55	—
30KhMA	4543-48	0.25-0.33	0.8-1.10	0.40-0.70	0.15-0.25
OKhNZM	4543-48	0.30-0.40	0.80-1.0	0.30-0.60	0.20-0.30
30KhGS	4543-48	0.25-0.35	0.8-1.1	0.8-1.1	—
45KhNMFA	4543-48	0.42-0.50	0.8-1.1	0.50-0.80	0.20-0.30
1Kh18N9	5632-51	≤ 0.12	17-20	≤ 0.2	—
1Kh18N9T	5632-51	≤ 0.12	17-20	0.3-0.7	—
2Kh13	5632-51	0.15-0.25	12-14	≤ 0.50	—
ShKh13	801-58	0.95-1.10	1.05-1.40	0.20-0.40	—
ShKh15	801-58	0.95-1.10	1.30-1.65	0.20-0.40	—
KhVG	5950-51	0.9-1.05	0.9-1.20	0.8-1.1	—
Kh20N77T2YuR (ÉI437B)†	5632-61	0.08	19-23	0.5	—

*W may be replaced by Mo in the proportions of 3 parts W per 1 part Mo (18KhNMA steel).

† The steel contains 0.4-1.1% Al and 0.005% B.

Appendix III (Continued)

Centimeters Hg at 0°C	Inches Hg at 0°C	cm of water at 4°C	Inches of water at 4°C	Pounds per square inch
$0.75006 \cdot 10^{-4}$	$0.2953 \cdot 10^{-4}$	$1.01974 \cdot 10^{-3}$	$4.0147 \cdot 10^{-4}$	$0.145039 \cdot 10^{-4}$
$0.75006 \cdot 10^{2}$	29.5300	$1.01974 \cdot 10^{3}$	401.474	14.5039
73.5561	28.9590	1000.03	$3.93712 \cdot 10^{2}$	14.2235
$73.5561 \cdot 10^{-4}$	$28.9590 \cdot 10^{-4}$	0.100003	$3.93712 \cdot 10^{-2}$	$1.42235 \cdot 10^{-3}$
76.0002	29.9212	$1.03326 \cdot 10^{3}$	$4.06794 \cdot 10^{2}$	14.6961
1	0.393700	13.5955	5.35253	0.193369
2.54	1	34.5325	13.5955	0.491158
$7.3554 \cdot 10^{-2}$	$2.89582 \cdot 10^{-2}$	1	0.393701	$1.42231 \cdot 10^{-2}$
$1.8683 \cdot 10^{-1}$	$7.35538 \cdot 10^{-2}$	2.54	1	$3.61266 \cdot 10^{-2}$
5.17147	2.03600	70.3083	27.6804	1

Appendix IV (continued)

Ni	P	S	Si	Ti	W	V
	not over					
—	0.050	0.055	0.17-0.35	—	—	—
4.0-4.5	0.035	0.03	0.17-0.37	—	0.8-1.2*	—
≤ 0.40	0.035	0.03	0.17-0.37	—	—	—
2.25-3.0	0.035	0.035	0.18-0.35	—	—	—
≤ 0.40	0.04	0.04	0.9-1.20	—	—	—
1.3-1.8	0.035	0.03	0.17-0.37	—	—	0.1-0.2
8-11	0.035	0.03	0.3-0.8	—	—	—
8-11	0.035	0.03	0.3-0.8	≤ 0.8	—	—
≤ 0.6	0.035	0.035	≤ 0.7	—	—	—
≤ 0.20	0.027	0.020	0.15-0.35	—	—	—
≤ 0.20	0.027	0.020	0.15-0.35	—	—	—
≤ 0.25	0.025	0.020	0.15-0.35	—	1.2-1.6	—
base	≤ 0.02	≤ 0.015	≤ 1.0	2.0-2.9	—	—

APPENDIX V

Mechanical and Physical Properties of Certain Steels

| Type of steel | Mechanical properties | | | | | | Heat |
| | Ultimate strength $\times 10^{-2}$, kg/cm² | Yield strength $\times 10^{-2}$, kg/cm² | Unit Elongation % | Unit reduction in area, % | Hardness (heat treated) | | Normalization |
					Brinell	Rockwell	
St. 5	50-62	28	17-15	—	—	—	—
18KhNVA	120	100	10	45	340-400	—	—
30KhMA	90-130	80-105	5-12	40-50	300-400	—	800-900
OKhNZM	100-160	84-138	10-13	45-47		48-53	—
45KhNMFA	160-200	150-165	5-10	20-40	477-500	—	850-870
30KhGS	90-190	70-140	8-28	40-60	250-500	—	870-890
1Kh18N9	60-65	20-25	45	60	—	—	—
1Kh18N9T	54	22	40	60	140-170	—	—
2Kh13	90-125	70-95	7	45	285-341	—	—
ShKh12	—	—	—	—	—	59-63	—
ShKh15	—	—	—	—	—	59-63	780-800
KhVG	—	—	—	—	—	62-65	780-800
ÉI437B	92-109 90*	66 58*	11-24 22-29*	10-21 19-27*	— —	— —	1080°C-8 h air-cooled

* At 500°C

†Of linear expansion.

Appendix V (Continued)

treatment, °C		Physical properties					
Annealing	Temper-ing	Density γ, g/cm^3	Heat capacity C, cal/ g·degree	Thermal coefficient† $\alpha \cdot 10^6$ (0-100°C)	Coefficient of heat con-ductivity λ (20°C), cal/cm·sec	Electrical resistivity ρ, $\Omega \cdot$mm^2/ m	Magnetic character-istics
–	–	–	–	–	–	–	–
870 (oil)	200 (air)	7.94	–	14.5	0.057	–	Magnetic
880 (oil)	300-600 (oil)	7.82	–	12.3	0.102	–	Magnetic
840-880 (oil)	200-600 (oil)	–	–	–	–	–	Magnetic
860-880 (oil)	350-420 (air)	7.81	–	12.4	–	–	Magnetic
860-880 (oil)	200-550 (air)	7.75	–	11	0.09	–	Magnetic
1100 (water)	–	7.9	0.12	16	0.039	0.73	Nonmag-netic
1100 (water)	–	7.9	0.12	16	0.04	0.73	Nonmag-netic
1050 (air)	500	7.75	–	10.5-11.0	0.0595	0.57-0.60	Magnetic
820-850 (oil)	180-200 (air)	–	–	–	–	–	Magnetic
820-840 (oil)	180-200 (air)	–	–	–	–	–	Magnetic
820-840 (oil)	180-200 (air)	–	–	–	–	–	Magnetic
aged 700°C-16h air-cooled		8.2	12.6	0.03	–	–	Nonmag-netic

APPENDIX VI

Resistance of Copper Conductors

Conductor diameter, mm	Section, mm^2	Resistance of 1 m, Ω	Mass of 1 m, g
0.05	0.00196	3.95	0.018
0.08	0.0050	3.50	0.045
0.10	0.0079	2.22	0.070
0.11	0 0095	1.84	0.085
0.12	0 0113	1.55	0 101
0 13	0 0133	1.32	0.118
0.14	0.0154	1.14	0.137
0 15	0 0177	0 99	0.158
0 16	0.0201	0.87	0.178
0.17	0 0227	0.772	0.202
0.18	0.0255	0.685	0.227
0.19	0.0284	0.617	0.253
0.20	0.0314	0.557	0.280
0 22	0.0380	0.460	0.339
0.25	0.0491	0.357	0.457
0.30	0.0707	0.248	0.630
0.35	0.0962	0.182	0.857
0.40	0.1260	0.139	1.130
0.45	0.1590	0.110	1.417
0.50	0.1960	0.0895	1.750
0.60	0.2830	0.0618	2.520
0.70	0.3850	0.0455	3.43
0.80	0.5030	0.0348	4.48
0.90	0.6360	0.0275	5.67
1.00	0.7850	0.0223	7.07
1.20	1.1310	0.0155	10.98
1.50	1.7670	0.00992	15.75

APPENDIX VII

List of Literature Containing Information on Experimental Techniques at High and Ultrahigh Pressures

Korndorf, B. A., High Pressure Techniques in Chemistry, Goskhimizdat, 1952.

Zhokhovskii, M. K., Theory on Design of Instruments with Unsealed Pistons, Mashgiz, 1959.

Golubev, I. F., The Viscosity of Gases and Gas Mixtures, Fizmatgiz, 1959.

Kirillin, V. A., and Sheindlin, L. E., Investigation of Thermodynamic Properties of Materials, Gosénergoizdat, 1963.

Superhigh Pressure Apparatus, An annotated list of domestic and foreign literature, 1950-1960, VNII p'ezoopt. i mineral. syr'ya, 1961.

Lloyd, E. K., and Johnson, L. P., Static and Dynamic Calibration of Instruments for Measurement of Pressures at the National Bureau of Standards, USA, Izd. Akad. Nauk SSSR, 1960.

Swenson, C., High Pressure Physics, Translation edited and with supplement by L. F. Vereshchagin (X-ray Investigations of Materials at High Pressures), IIL, 1963.

Nemets, Ya., Calculating the Strength of Vessels Operating Under Pressure, Translated from Czechoslovakian, Mashinostroenie, 1964.

Modern Very High Pressure Technique [translated from English], Izd. Mir, 1964.

Comings, E. W., High Pressure Technology, New York, 1956.

Paul, W., and Warshauer, D. M., (Eds.) Solids Under Pressure, London, 1963.

Giardini, A. A., and Lloyd, E. C., (Eds.), High Pressure Measurements, London, 1963.

Bradley, R. S., (Ed.), High Pressure Physics and Chemistry, 1, 11, 1963.

Bundy, F. P., Hibbard, W. R., and Strong, H. M., Progress in Very High Pressure Research, New York, 1961.

Author Index

Aarons, A. B., 183, 195
Abbot, L. H., 182, 195
Adams, L. H., 170, 194, 402, 411
Adamson, A. W., 415, 429
Adom, N. N., 415, 416, 429
Afanas'ev, P. A., 19, 22, 36
Aibinder, S. B., 22, 36
Alaeva, T. A., 260, 269
Aleksandrov, B. S., 151, 152, 176, 194
Aleksanvan, V. T., 431, 443
Alekseev, E. N., 181, 195
Alekseev, K. A., 178, 195
Alekseev, V. F., 299, 351
Altunin, V. V., 358, 360, 411
Amagat, E. H., 149, 150, 193, 264, 270, 365, 411
Anderson, O. L., 34, 37, 76, 78
Andreatch, P., 34, 37, 76, 78
Anson, D. J., 233, 269
Argue, D., 307, 348, 352
Aristov, G. E., 188, 195, 200, 207, 285, 290, 365, 368, 369, 411
Arons, J. de Swaan, 340, 353
Artemovich, V. S., 304, 352
Ashcroft, K., 116, 134, 174, 195
Ashmarin, N. V., 15, 36
Astrov, D. N., 222, 268

Babb, S. E., 30, 33, 37, 230, 261, 269, 270
Bakhvalova, V. V., 192, 196
Balchan, A. S., 174, 195
Baldin, 267, 270

Barcas, W., 183, 195
Barnet, S. S., 198, 207
Bartenev, G. M., 26, 37
Bartlett, E., 303, 352, 356, 411
Bashkirov, A. N., 197, 199, 207
Basset, J. J., 12, 35, 130, 132, 135, 300, 351, 377, 379, 411
Beard, C. S., 222, 243, 268
Beck, R., 285, 290
Belyaev, A. M., 49, 76
Belyaev, N. N., 46, 76
Benedict, M., 361, 383, 411
Benson, S. W., 333, 353
Bersenev, B. I., 14, 15, 35, 94, 134
Bett, K. E., 33, 37, 147, 193
Bewilogua, L., 127, 135
Bidwell, R. M., 332, 353
Bilevich, A. V., 362, 410, 411
Bishnevskii, 282, 290
Björkman, 273, 289
Black, G. N., 19, 22, 36, 303, 352
Blosser, L., 261, 270
Bobrowsky, A., vi, 6
Bogdanova, V. A., 322, 352
Boksha, S. S., 87, 130, 131, 133, 246, 269
Bol'shakov, P. E., 23, 36, 187, 188, 195, 265, 270, 306, 309, 337, 352, 415, 418, 420, 424, 425, 426, 427, 429
Bomberg, H. I., 169, 194
Bomshtein, E. I., 222, 268
Boomer, E., 307, 348, 352
Boon, E. F., 233, 269
Boon, F. F., 238, 269

Booth, H. S., 332, 353
Borodina, M. D., 88, 133, 180, 184, 195
Boronely, A. V., 222, 268
Borovaya, F. E., 334, 335, 353
Borzov, V. A., 175, 194
Borzunov, V. A., 96, 111, 134, 217, 241, 261, 262, 268, 408, 413
Boulestreau, M., 200, 207
Bovenkerk, H. P., 69, 77
Bowen, I. S., 298
Bowen, J. C., 85, 133
Boyd, F. R., 75, 78
Bradley, 453
Brandt, I. V., 439, 443
Brandt, N. B., 127, 135
Braude, G. E., 202, 203, 207, 276, 289
Braune, H., 303, 352
Brestkin, A. P., 304, 352
Bridgman, P. W., v, 13, 16, 35, 43, 48, 55, 63, 64, 71, 76, 77, 104, 109, 112, 123, 130, 134, 135, 169, 180, 194, 205, 207, 222, 225, 247, 263, 264, 268, 269, 270, 381, 398, 399, 400, 410, 412
Browman, H., 22, 36
Brown, B., 206, 207
Bruns, B. P., 202, 203, 207, 276, 289, 290
Buckhoff, F. J., 19, 36
Bundy, F. P., 69, 77, 121, 122, 134, 205, 207, 453
Bunting, E. N., 433, 443
Bunting, F. N., 431, 443
Burnett, E. S., 393, 412
Burova, L. L., 170, 173, 178, 194
Bush, H. W., 183, 195
Butterman, H. I., 149, 193
Butuzov, V. P., viii, 67, 77, 110, 116, 130, 134, 135
Byerly, I. I., 3, 4, 34

Caht, I. W., 433, 443
Carson, I. A., 183, 195
Carver, E. K., 417, 429
Charters, A. C., 183, 195
Chegodaev, D. D., 22, 36
Cherney, B., 387, 412
Cherry, J., 183, 195
Chipizhenka, A. P. 49, 76
Christiansen, E. W., 76, 78

Class, J., 15, 17, 22, 36, 60, 77
Coffin, D. O., 164, 194
Coffin, L. F., 51, 77
Cole, R. H., 183, 195
Comings, E. W., 15, 36, 60, 77, 79, 83, 133, 194, 222, 243, 268, 453
Compton, L. A., 206, 207
Consins, E., 183, 195
Cook, D., 319, 352
Copeland, C. S., 333, 353
Cornish, R. H., 261, 270
Courish, R., 402, 412
Crookes, W., 130, 135

Dachille, F., 174, 195
Dadson, R. S., 148, 193
Daniels, W., 75, 78, 218, 268
Darling, H. E., 170, 194
Dauson, J. P., 380, 412
Davis, P. C., 255, 269
Dean, H. E., 323, 353
De la Shez, 55, 77
Demyashkevich, B. P., vii, 90, 134
Devis, R., 256, 269
Dew, I. E., 295, 298
Diets, A. G. H., 22, 36
Diment, J., 183, 195
Dode, 300, 351
Dodge, B. F., 15, 36, 303, 305, 352
Domanskaya, P. I., 206, 207
Doolittle, A. K., 402, 412
Douslin, D. R., 380, 412
Dricamer, H. G., 174, 194, 434
Drozd, M. S., 8, 35
Duns, H. S., 22, 36
Dyment, J., 21, 36

Eakin, E. B., 393, 412
Ebookimova, V. V., 437, 443
Edwards, W. A. M., 87, 133
Efremova, 255, 269, 289, 290, 310, 311, 319, 327, 328, 342, 343, 349, 351, 352, 353
Eiring, H., 5, 34
Eitel, W., 130, 135, 234, 269
Ekhlakov, A. D. 403, 412
Elington, R. T., 393, 412
England, J. L., 75, 78
Ermilov, P. I., 298

Ershov, Yu. A., 441, 443
Ershova, T. P., 5, 35
Euwen, C., 328, 353
Exline, P., 323, 353

Fastovskii, V. G., 284, 285, 290, 308, 310, 352
Fateeva, K. S., 130, 135
Faupel, J. H. 55, 77
Fedorovskii, A. E., 34, 37
Fey, R. F. 222, 268
Fisher, J. C., 51, 77
Fitch, R. A., 434, 443
Flom, D. G., 271, 289
Fradkov, A. B., 12, 35
Frank, E. M., 339, 353
Freindlin, L. Kh., 104, 134
Fuchs, F. J., Jr., 67
Furman, M. S., 88, 133
Furucawa, G., 22, 36

Gaddy, V., 301, 307, 352, 357, 359, 411
Gadolin, A. V., 57
Gaida, A. V., 213, 268
Galaktionov, V. A., vii, 73, 77, 113, 119, 120, 134, 398, 412
Gamburg, D. Yu., 15, 79, 133, 189, 196, 289, 290, 303, 322, 323, 351, 352, 353
Ganz, S. N., 222, 230, 268
Garbar, M. I., 19, 36
Garrison, J. B., 206, 207
Garber, R. I., 4, 34
Gavrilovskii, 267, 270
Gebhart, I. E., 218, 268
Gibson, R. F., 170, 194, 365, 403, 411
Giardini, 453
Gielessen, J., 167, 194
Gill, 255, 269
Gindin, I. A., 4, 34
Gladkovskii, V. A., 51, 77
Glikman, L. A., 8, 35
Gogarty, W. B., 76, 78
Goldman, 408, 413
Golubev, I. F., 407, 408, 409, 413, 453
Golydman, A. M., 216, 268
Gonikberg, M. G., vii, 66, 77, 110, 117, 130, 134, 284, 285, 290, 308, 310, 352, 402, 412, 431, 439, 441, 443

Goranson, R. W., 130, 133, 135, 170, 194
Gore, T. L., 255, 269
Gourlag, J. S., 19, 36
Govze, M. N., 17, 36
Graham, G. T., 391, 392, 412
Graig, L. E., 295, 298
Granvill, I. I., 222, 268
Grazhdankina, N. P., 206, 207
Griggs, D. T., 130, 135, 240, 269
Grishenko, A. P., 255, 269
Gross, M., 181, 195
Gubkin, S. I., 6 .
Gudstov, N. T., 17, 36
Gulyaev, A. P., 6, 35, 41, 76
Gunn, D. A., 26, 37
Gurvich, I. G., 284, 285, 290, 308, 310, 352

Hall, H. T., 69, 73, 77, 119, 134, 174, 195, 207
Hashman, U., 255, 269
Hayes, P. H., 33, 37, 193
Heins, H., 301, 307, 352
Hesselgesser, 332, 353
Hibbard, W. R., 453
Higgs, P. M., 183, 195
Higrave, P., 285, 290
Hilcter, R. G., 402, 412
Hillard, J. E., 433, 443
Hiller, 23, 36
Hinde, J., 35
Hitton, H., 113, 134
Hodgson, G., 26, 37
Holborn, L., 363, 411
Hough, E. W., 422, 429
Hyde, G. R., 27, 37

Ibser, H. W., 138, 193
Incuge, R., 328, 353
Indrik, P. V., 193
Inoue, K., 30, 31, 37
Iomtev, M. B., 341, 353
Ipat'ev, V. V., 304, 307, 352
Isaikov, V. K., 34, 37, 123, 134, 234, 269
Ivanov, V. E., 51, 77, 84, 90, 94, 112, 133, 134, 217, 235, 236, 268, 269

Jackson, T. W., 198, 207
Jacobs, R. W., 437, 443

Jamieson, I. C., 4, 36, 439, 443
Jellinek, K., 138, 193
Jenkins, R. L., 298
Jennett, E., 297, 298
Johnson, C., 307, 348, 352
Johnson, D. P., 113, 134, 153, 173, 194, 453
Johnson, L. P., 174, 194
Jolly, 183, 195
Jones, M., 19, 36, 75, 78
Jonson, D. P., 222, 230, 268

Kabalkina, S. S., 437, 439, 443
Kalashnikov, Ya. A., 133, 135, 206, 207,
    259, 269, 362, 410, 411
Kal'varskaya, R. S., 303, 312, 340, 352
Kamzolkin, V. V., 197, 199, 207
Kan, L. S., 124, 126, 135
Kantarovich, Z. V., 45, 47, 65, 76
Kantorovich, Z. B., 12, 35
Kasper, J. S., 433, 443
Kats, G. S., 67, 77, 116, 134
Katsnel'son, O. G., 279, 289
Katz, D. L., 415, 424, 429
Kazarnovskaya, D. B., 371, 373, 411
Kazarnovskii, Ya. S., 216, 253, 268, 269,
    365, 368, 370, 371, 372, 411
Kedi, U., 183, 195
Kenndy, G. S., 332, 353
Kennedy, G. G., 12, 35, 127, 130, 135, 173,
    174, 194
Kent, E. R., 273, 289
Key, W. B., 26, 36
Khalilov, Kh. M., 25, 36, 253, 269
Khariton, Yu. B., 88, 133
Khazanova, N. E., 15, 35, 243, 269, 303, 352
Khimushin, F. F., 12, 35
Khiteev, A. M., 323, 353
Knodeeva, S. M., 314, 318, 352
Khrapovitskii, Yu. S., 37
Kiebler, M. W., 271, 289
Kikoin, A. I., 206, 207
King, G., 22, 36
Kirillin, V. A., 357, 359, 411, 452
Kiselev, D. V., 189, 196
Kistler, S. S., 76, 78
Kiyama, R., 30, 31, 37
Klyuev, 435, 443
Knop, 295, 298
Knowles, J. K., 22, 36

Kohen, E., 328, 329, 353
Kondrat'eva, T. F., 297, 298
Kondrat'ev, V. N., 7, 35
Konstantinov, A. A., 197, 207
Konyaev, Yu. S., 160, 194
Korndorf, B. A., 60, 77, 79, 83, 133, 452
Kornilov, I. I., 12, 35
Koroleva, M. V., 303, 352
Koshevnik, A. Yu., 422, 423, 429
Kousmine, E., 88, 133
Krase, N., 301, 303, 352
Krichevskii, I. R., vii, 15, 35, 50, 76, 101,
    134, 243, 245, 249, 255, 269, 289, 290,
    301, 302, 303, 310, 311, 312, 319, 325,
    327, 328, 330, 331, 337, 340, 343, 345,
    349, 351, 352, 353, 363, 364, 382, 383,
    384, 411
Kristal', M. M., 12, 35
Kryukov, I. M., 4, 34
Küchler, 277, 289
Kulikova, A. I., 87, 133, 360, 363, 411
Kurata, E., 255, 269
Kurdyumov, G. V., 5, 34
Kusakov, M. M., 422, 423, 429
Kuzin, N. N., 119, 120, 134
Kuzlovskii, D. E., 223, 268
Kuznetsov, L. F., 260, 269

Lacey, R., 156, 194
Lacey, W., 281, 289, 323
Lakes, M. G., 22, 36
Lalos, G. T., 392, 412
La Mory, 173, 174, 194
Lancaster, I. F., 2, 34
Lane, P. H., 181, 195
Lansdall, K., 433, 443
Larson, A., 303, 352
Lauson, A. W., 206, 207, 436
Lavergne, A., 99, 134
Lavrov, A. S., 55
Lavrov, I. A., 402, 412
Lazarev, V. G., 124, 135
Lebedeva, E. S., 314, 318, 328, 330, 331,
    352, 353, 403, 411
Lees, J., 116, 134, 174, 195
Legallis, 183, 195
Leipunskii, 23, 36
Leonidov, G. G., 195
Lessels, D., 64, 77

Levchenko, G. T., 420, 429
Levchenko, V. G., 160, 194
Liberman, L. Ya., 2, 34
Likharev, K. K., 43, 76
Likhter, A. I., vii, 112, 134, 442, 443
Linshits, L. P., 306, 309, 352, 394
Lippincott, E. R., 431, 433, 443
Livshits, L. D., 13, 14, 15, 34, 35, 75, 77, 94
Lloyd, E., 113, 134, 174, 194, 453
Lobo, P. A., 295, 298
Lok, H. H., 233, 269
Longwell, P. A., 392, 412
Lowenstein, I. G., 297, 298
Lubman, N. M., 422, 423, 429

Maas, O., 391, 392, 412
Maccary, R. R., 222, 268
Maiors, N. Yu., 22, 36
Mairanovskii, S. G., 439, 443
Makargv, V. V., 12, 35
Makhnev, G. A., 232, 269
Makhonina, T. A., 51, 77
Maksimov, N. M., 8, 35
Mal'tsev, B. K., 288, 290
Marchman, H., 387, 412
Markov, V. P., 363, 364, 411
Marschall, 255, 269
Martynenko, A. G., 322, 352
Martynov, E. D., 14, 35
Maslennikova, V. Ya., 280, 289, 339, 353
Mayers, A. F., 48, 60, 77
McCoskey, R., 22, 36
McGeer, P. L., 22, 36
McGregor, G. W., 51, 77
Mech, M. J., 181, 195
Mendeleev, D. I., 25, 36, 138, 193, 361, 411
Merkel, E., 22, 36
Merret, F., 173, 194
Michels, A., 169, 194, 255, 269, 365, 381, 411
Mikhailov-Mikheev, P. B., 2, 34
Miller, B., 303, 352
Mil'vitskii, R. V., 67, 77
Mirinskii, D. S., 67, 77, 116, 134, 403, 412
Molstad, M. C., 15, 36
Molyneax, F., 20, 36
Montgomery, L. H., 183, 195
Moore, R. T., 380, 412

Morey, 25, 36, 332, 353
Morley, J. G., 26, 37
Moskvin, N. N., 12, 35
Munday, G., 297, 298
Mushkina, E. V., 225, 268, 309, 341, 352
Myers, M. B., 174, 195

Nadai, A., 39, 46, 49, 76
Naidich, I. M., 51, 54, 77
Neiman, E. Ya., 60, 77
Neiman, M. B., 441, 443
Nelson, B., 206, 207
Nemets, 453
Ness, H. C., 15, 36
Nevyazhskaya, L., 31, 37
Newhall, D. H., 86, 133, 170, 173, 194
Newitt, D. M., 33, 37, 183, 193, 232, 253, 269, 297
Nikolaeva, O. I., 2, 34
Nisman, L. N., 197, 207
Nogatkin, A. G., 167, 194
Norrish, R., 194
Novikov, L., 31, 37
Novmanov, N. K., 104, 134

Oglobin, A. N., 209, 268
Oleinik, I. I., 260, 269
Opekunov, A. A., 66, 77, 117, 134, 431, 439, 441, 443
Osipyan, Yu. A., 5, 34
Ostrovskii, I. A., 263, 270

Parker, C. M., 12, 35
Paul, W., 15, 35, 173, 194, 453
Pavlovich, N. V., 409, 413
Payne, I. W., 273, 289
Pearson, G. H., 222, 243, 268
Peisikhis, M. I., 2
Pereverkin, S. M., 37
Pertsov, N. V., 17, 36
Peters, E., 3, 4, 34
Phillips, V. A., 433, 443
Pimshtein, P. G., 60, 77
Plachenov, T. G., 235, 269
Platen, B. von, 118, 134
Podionov, I. L., 403, 412

Polandov, I. N., 195
Ponyatovskii, E. G., 5, 35
Polyakov, E. V., 327, 328, 353
Polyakovo, R. S., 12, 35
Porill, N. T., 271, 289
Porter, R. L., 295, 298
Poulter, T. C., 264, 270
Preobrazhenskii, V. A., 86, 133
Price, D., 392, 412
Pryanikova, R. O., 327, 328, 342, 343, 353
Pryanishnikov, V. P., 26, 37
Pugh, H. L. D., 26, 37
Pushkinskii, M. D., 402, 412

Rakovskii, V. S., 4, 34
Ramsauer, S., 14, 133
Ranbo, M., 285, 290
Ratner, S. I., 14, 35
Ravich, M. I., 334, 335, 353
Razumikhin, V. N., 173, 175, 194, 403, 404, 408, 412, 413
Reamer, H. H., 156, 192, 194, 392, 412
Rebinder, P. A., 17, 36, 424, 429
Reeves, L. E., 30, 37
Richards, I. E., 417, 429
Richards, T. W., 397, 412
Robin, J., 133, 135
Robin, S., 133, 135
Roebuck, I. R., 138, 193
Rogosinski, M., 255, 269
Roof, 321, 352
Roöner, R., 127, 135
Rose, W., 267, 270
Roters, 279, 289
Roth, R. F., 19, 36
Roulinson, I. S., 411, 413
Roy, R., 174, 195
Rozen, A. N., 39, 51, 214
Rozhanskii, V. N., 13, 36
Rozhdestvenskii, L. A., 4, 34
Rsasa, M. J., 422, 429
Rubinshtein, A. M., 104, 134
Rudenko, N. S., 87, 133
Ruheman, M., 87, 133
Ruoff, A. L., 261, 269
Rusching, 218, 268
Ryabinin, Yu. N., vii, 13, 14, 15, 34, 35, 75, 77, 88, 123, 134, 184, 185, 195, 264, 270, 387, 412

Saddington, A., 301, 303, 352
Saechtling, H., 22, 36
Sage, B. H., 392, 412
Sage, B. N., 156, 192, 194, 281, 289, 323
Sakhin, S. I., 5, 35
Sander, 300, 351
Sarrak, V. I., 4, 34
Schen-Wu-Wan, 305, 352
Schneider, W. G., 393, 412
Scholl, W., 256, 269
Schultz, E. A., 169, 194
Schultze, H., 363, 411
Schuster, M., 149, 193
Schwaigerer, S., 60, 77
Scott, G. I., 30, 33, 37, 261, 270
Semendyaev, K. A., 279, 289
Semerchan, A. A., 34, 37, 73, 77, 119, 120, 123, 134, 234, 269
Semin, V. P., 96, 111, 134, 217, 241, 261, 262, 268
Sensil, E., 285, 290
Seren, L., 267, 270
Shakhovskoi, G. P., 110, 134, 246, 264, 270, 402, 412
Shakovskii, M. G., 59, 132
Shapiro, M. B., 12, 35
Shapochkin, V. A., 13, 35, 123, 135
Shatenshtein, A. I., 256, 269
Shchukin, E. D., 17, 36
Sheindlin, 359, 411
Shelaputin, K. N., 189, 195
Shenderei, L. I., 309, 352, 363, 411
Shevelkin, B. N., 12, 35
Shishkov, N. Z., 234, 269
Shorr, 255, 269
Shteinman, E. B., 2, 14, 34
Shurmovskaya, N. A., 312, 352
Shumov, K. M., 181, 195
Shvarts, G. L., 12, 35
Sidorov, I. P., 99, 134, 200, 201, 203, 207, 213, 216, 243, 253, 258, 268, 269, 285, 288, 290, 369, 370, 371, 372, 373, 411
Siebel, E., 60, 77
Silberberg, 393, 412
Silverman, J., 333, 353
Simon, F., 87, 133, 175, 195
Simon, I., 402, 412
Simonov, I. P., 365, 368, 369, 411
Sinitskii, A. K., 49, 76
Sinnige, L., 328, 353

Sirota, A. M., 288, 290
Slavin, D. O., 2, 14, 34
Sliepcevich, C. M., 295, 298
Slesarev, V. N., 34, 37, 73, 77, 119, 120, 134
Slutskii, A. B., 130, 135
Slykhouse, T. E., 434, 443
Smirnov, F. F., 4, 34
Smirnov-Alyaev, G. A., 55, 77
Sokolov, O. G., 5, 35
Sokolov, S. N., 51, 76
Solov'ev, P. V., 298
Sorina, G. A., 325, 353
Speransov, N. N., 31, 37
Sreed, C. W., 273, 289
Stepanov, V. A., 217, 262, 268, 270
Stephens, H. R., 297, 298
Stern, Kh. E., 431, 443
Stewart, J. W., 20, 36
Still, A. E., 329, 353
Strassman, F., 303, 352
Straub, R. A., 99, 134
Strel'chuk, N. A., 298
Strong, H. M., 69, 77, 205, 207, 453
Sullivan, J. W. W., 12, 35
Swenson, C., 22, 36, 205, 207, 436, 453

Tait, P. G., 168, 194
Tait, T., 255, 269
Tal, S. E., 238, 269
Teranishi, H., 30, 31, 37
Teodorovich, V. P., 304, 307, 352
Thomson, M., 183, 195
Thomson, T., 235, 239, 243, 269
Thygeson, J. R., 15, 36
Tikhomirov, N. A., 264, 270
Tiller, F. M., 138, 193
Timoshenko, S. P., 64, 77
Timrot, D. L., 365, 409, 411
Tödheide, K., 339, 353
Tomashchik, A. K., 127, 135
Toropov, V. N., 12, 35
Tremearne, T., 307, 352
Trivus, N. A., 387, 412
Tsekhanskaya, Yu. V., 341, 353
Tsiklis, D. S., v, vii, 37, 50, 66, 76, 77, 82,
    87, 88, 101, 117, 133, 134, 163, 173,
    180, 184, 194, 218, 221, 225, 245, 249,
    268, 269, 277, 280, 284, 289, 301, 302,
    309, 314, 337, 339, 345, 352, 360, 363,
    382, 383, 384, 394, 411, 427, 428, 429

Ukholin, S. A., 431, 443
Ulybin, 357
Ushakov, I. P., 233, 269
Ustinovich, V. V., 203, 207
Uzhik, G. V., 12, 35

Van Valkenburg, A., 433, 443
Vasil'ev, A. V., 203, 207
Vasil'ev, Ya. N., 427, 428, 429
Vereshchagin, L. F., vii, 13, 15, 34, 35, 51,
    73, 77, 84, 86, 88, 90, 94, 104, 112,
    119, 120, 123, 130, 133, 134, 135, 151,
    152, 176, 194, 206, 207, 216, 217, 233,
    250, 260, 268, 269, 362, 398, 410, 411,
    436, 437, 439, 443
Vinogradov, K. V., 387, 412
Voitovich, M. V., 189, 195
Volarovich, M. P., 28, 31, 37
Vollbrecht, H., 233, 242, 269
Volyak, L. D., 419, 429
Voronel, A. V., 140, 176, 193
Vulkalovich, M. P., 358, 360, 411
Vyaznikov, N. F., 2, 4, 34

Waddington, G., 380, 412
Waggener, W. C., 267, 270, 431, 443
Walker, K. E., 297, 298
Ward, J. W., 183, 195
Warschauer, D. M., 15, 35, 173, 194, 453
Wassenaar, T., 381, 412
Weale, K. E., 37
Weber, G. G., 297, 298
Weinaug, C. E., 415, 424, 429
Weir, C. E., 173, 194, 431, 443
Weisberger, A., 442, 443
Welbergen, H. G., 23, 36
Wells, F. W., 321, 352
Wentorf, R. H., 68, 69, 77, 118, 122, 134
Whalley, E., 99, 134
Whitsides, R. H., 198, 207
Wiebe, R., 301, 307, 352, 357, 359, 411
Wilbourn, N. P., 392, 412
Willey, B. J., 183, 195
Wilson, R. O., 264, 270
Wilson, W. B., 75, 77
Witkowski, 364, 365, 411
Wolf, R. C., 85, 133
Wood, B. B., 422, 429

York, R., 387, 412
Young, H. S., 261, 270

Zaikov, M. A., 12, 35
Zakharchenko, E. S., 222, 268
Zelikman, A. N., 4, 34
Zelinskii, 216, 268
Zel'venskii, Ya. D., 347, 353
Zelyaev, A. F., 181, 195
Zhavorankov, N. M., 301, 302, 345, 352
Zhokhovskii, vii, 29, 140, 143, 147, 148,

Zhokhovskii, (cont.), 149, 150, 155, 156,
    157, 159, 160, 164, 173, 175, 176,
    192, 193, 452
Zhuravlev, V. N., 2, 34
Ziebland, H. J., 21, 36
Zil'berg, G. A., 197, 207
Zlunitsyn, S. A., 87, 133
Zolotykh, E. V., 28, 31, 33, 37, 170, 173, 194
Zubkov, V. M., 119, 120, 134
Zubova, E. V., 13, 35, 123, 135
Zukowsky, 36
Zwietering, T. N., 381, 412

Subject Index

Absolute manometers, 137-164
Absorption-spectra installation, 431-436
Accuracies of
  intensifiers, 210
  manometers, 148, 150, 156, 159, 162, 164, 166, 168, 170-172
Action of gases and liquids, 15-17, 20
Activated carbon as drying agent, 79, 293
Adiabatic-compression compressibility installations, 386-393
  Graham-Maas, 391-393
  Price-Lalos, 389, 390, 392
  Ryabinin-Tsiklis, 386-390, 392
Adiabatic press, 87, 88
Adsorption on steel, 1
Aging of
  manganin coil, 177, 178
  ShKh12 Steel, 266
Agitation, see: Mixing
Agitator-pump, 289
Air compressor, 82
Alcohol, see specific kind
Alloy steel
  compositions, 10
  mechanical properties (T), 9, 10
  thermal pressure, 52
Alumel
  electrical properties (T, P), 205
  in thermocouples, 205
Aluminum
  alloy, see: Duraluminum
  cast, 18
  chemical properties of, 20

Aluminum (continued)
  hard, 18
  in joints, 251
  mechanical properties of, 18
  in packings, 20, 222, 251
  seizing property, 20
  soft, 18
  in steel, 12, 448
  temperature limit of, 18
  thermal properties of, 18
  in titanium alloy, 3, 8, 256
Aluminum hydrosilicate, friction (P), 27
Aluminum oxide
  in Bridgman anvils, 71
  as electrical insulation, 277
  in stirrers, 277
  as thermal insulation, 71
Amagat
  compressibility installation, 365, 367
  density apparatus, 361
  manometer, 149, 150
  window, 263, 264
Amalgamation, 16
AMK-2 (Magnesium alloy), mechanical properties, 14
Ammonia, solubility
  in gases, 303, 337
  of solids in, 343
Ampules
  of glass, 25, 313, 314
  Lebedeva-Kodeeva, 315
  with neck, 311, 312
  in phase-equilibria work, 311-315, 318, 319, 325

Note: In this index we use the following abbreviations. T) over a range of temperatures; P) over a range of pressures; LT) low temperature; RT) room temperature; ET) elevated temperature.

Ampules (continued)
  in phase-transformations work, 129, 130
  of porcelain, 124, 129, 130
  Tsiklis, 315
  tubing in, 25
  with valves, 315, 318
Analysis
  gas-phase, 351
  gravimetric, 348-351
  liquid-phase, 347-350
  methods of, 345-351
  volumetric, 345-348
Analytic methods, phase equilibria
  circulation methods, liquid—gas, 301,
    310, 311
  definition of, 299
  dynamic methods, liquid—gas, 300-305
  dynamic methods, solid—gas, 343, 344
  in gas—gas systems, 337-340
  limitations, 300, 301, 303, 339
  in liquid—gas systems, 301-311
  pressure limits, 304
  in solid—gas systems, 341-344
  in solid—liquid systems, 328-333
  static methods, liquid—gas, 301, 305-309
  static methods, solid—gas, 341-343
Anhydrone, 79
Anisotropy, effect on ductility, 2
Annealing
  of manganin, 178
  of steels, 451
  under pressure retains ductility, 5
Anson seal, 233
Antimony in babbitt, 20
Anvils, 70-72, 123-125
  in belt, 72, 120
  in installations, 112-116, 122-125
  in Institute of High Pressure Physics device, 73
  with shear force, 123-125
  in tetrahedral press, 69, 70
  with thermal insulation, 71
Apparatus
  assembly of, 209-211
  construction of, 55-76
  design of, 39-55
  installation of, 293-297
Aristov manometer, 188, 189
Aristov-Sidorov
  gas-flow meter, 200, 202
  pump, 285

Armco iron, mechanical properties (P), 14
Arons phase-equilibrium apparatus, 340
Arsenic in babbitt, 20
Asbestos
  with boron nitride, 23
  chemical properties, 23
  in clingerite cement, 253
  in electrical leads, 259
  in glands, 22, 23, 211, 253, 254
  with graphite, 23, 211
  in joints, 253-255
  mechanical properties (T), 22
  with molybdenum disulfide, 23
  with oil, 23
  in packings, 22, 23
  pressure limit, 22
Astrov-Boronely valve, 222
AT-4, 6, 8, 9, 10, 12 (titanium alloys), me-
    chanical properties (T), 8
Atmosphere, 137, 448, 449
Austenitic steel, compositions and mechani-
    cal properties (T), 10, 11
Austenite—martensite conversion during
    thermomechanical treatment, 5
Autoclaves, 272, 273
  Björkman autoclave, 273
  flexible capillary in, 273
  in installations, 305, 307, 334, 335
  with internal window, 266
  Payne-Sreed-Kent autoclave, 273
  rocking autoclave, 272, 273
  rotating autoclave, 273
  windows for, 266, 267
Autofrettaged pressure chambers
  die autofrettage, 55
  equivalent stress in, 56
  intensifier bores, 246
  machining allowance for, 56
  piezometers, 360
  pressure limits for, 60
Autofrettage pressure, 56
Average normal stress
  effect on ductility, 2, 5, 6
  during tensile necking, 43
  in unsupported-area seals, 1, 2
Aviation oil, mechanical properties
    (T,P), 28
Avtol (oil), viscosity of, 28
Axial stress, 44
AZh-6 alloy, mechanical properties, 14

B-70 (gasoline), mechanical properties (P), 32
B 90 (babbitt), properties, 18
    thermal properties, 18
Babbitt
    in bearings, 20
    composition, 20
    in glands, 20
    in installations, 83
    mechanical properties, 18
    in packings, 20
    thermal properties, 18
Babb seal, 230
Balance
    decimal manometric, 154, 156
    manometric, 154, 156
    Shtyukrat, 154, 156
    tensimetric, 406, 409
Balanced-bridge circuit, 171
Balanced-pressure item, see: Diaphragm item
Baldin-Gavrilovskii window, 267, 268
Ball valves
    in installations, 90, 96, 100, 104
    limitations (P), 31
Bar, conversion table, 448, 449
Barium, reference point of resistance
        change, 175
Bartlett compressibility installation, 356,
        358, 359
Barye, 137, 448, 449
Barnet-Jackson-Whitesides rheometer, 198,
        199
Basset compressibility installations, 376-380
Bearings in pumps, 288
Bearings, babbitt, 20
Beck-et al. pump, 285, 286
Behavior of material under load, 39-44
Bellows
    compressing devices, 85-87
    compressor, 87
    installations, 339
    manometers, 166, 168, 179
    piezometers, 399-402
    press, 85-87
    pump, 285, 286
    valves, 222
Belt apparatus, 72, 73, 120, 122
Bending strength, see: Mechanical properties,
        solids
Benedict compressibility instruments, 359-
        361, 381-383

Benzine, solubility of gases in, 304, 305
Beryllium
    mechanical properties (RT), 444
    pressure chambers, 436-439
    thermal properties, 444
    X-ray transparent, 436-439
Beryllium bronze, kinds and properties, 18
Beryllium—copper alloy, see: Beryllium
        bronze
Beryllium oxide, 436
BF-6 (glue), 263
Biaxial stress, effect on ductility, 2, 6
Bilevich compressibility installation, 361, 362
Binding-ring apparatus, 72, 73, 120, 122
Bishnevskii stirrer, 282-284
Bismuth
    electrical properties (P), 174, 175
    mechanical properties (P), 13
    polymorphic transformation, 174
    in solder, 256
Björkman autoclave, 273
Blosser-Young-Cornish electrical lead, 261
Bodies, see: Valve bodies
Boksha-Shakhovskii heated press, 131, 132
Boksha thermo-compressor, 87
Bol'shakov-Linshits phase-equilibrium appa-
        ratus, 306, 309
Bol'shakov
    manometer, 187, 188
    surface-tension instrument, 418, 420,
        421
    wettability installation, 424-427
Bolts, seizing of, 209
Boomer-Johnson-Argue phase-equilibrium
        apparatus, 305, 307-309, 348, 349
Boon-Lok seals, 233, 234
Boon-Tal seal, 237-239
Boric acid, friction (P), 27
Boron carbide
    friction (P), 27
    as lapping compound, 105
    as X-ray transparent, 436
Boron in
    steel, 448
    titanium alloy, 8, 12
Boron nitride
    in asbestos, 23
    friction (P) of, 27
    as lubricant, 23, 27
    as X-ray transparent, 436

Borosilicate glass, mechanical properties
    (P,T), 24
Borzunov-Semin
    electrical lead, 261, 262
    gland, 239-242
    valve, 217, 218
Boulestreau gas flow meters, 200
Bourdon gauges, 164-167, 186, 296
Boyd-England supported-piston and cylin-
    der, 75
BrA7 (bronze), mechanical properties (T), 11
Bracket cap, 223-225
Brass
    as crusher, 185
    as diaphragm, 163, 180
    in installations, 90, 95, 100, 256
    mechanical properties (LT), 11
    mechanical properties (RT), 11, 18, 401
    in packings, 222
    in sealing rings, 240
    in seals, 243-246
    thermal properties, 18
BrB2 beryllium bronze, kinds and proper-
    ties, 18
Break-away wall, 295
Bridgman anvils, 70
    with aluminum-oxide insulation, 71
    of cemented carbide, 70
    with graphite packing, 71
    with thermal insulation, 71
Bridgman-Benedict compressibility instal-
    lation, 381-383
Butuzov-Boksha-Gonikberg heated press,
    129, 130, 132
Bridgman
    compressibility installations, 398-401
    compressimeter, 180, 181
    double-support intensifier, 103, 104, 109
    electrical lead, 263
    lipped anvils, 71, 72
    piezometers, 399, 400
    shear anvils, 123
    submerged intensifier, 110, 112
    unsupported-area seal, 222
    variable single-support intensifier, 63
Brinell hardness, see also: Mechanical prop-
    erties, solids
    conversion to Rockwell hardness and to
      tensile strength of steels, 7, 445-447
    errors, 9

Brittleness, see: Ductility
Bronze
    in installations, 99, 109
    mechanical properties (RT), 11, 14
    mechanical properties (LT), 11
    as saturator, 311
    as sealing rings, 240, 242
    as seals, 248, 249
BrS-30 (bronze), mechanical properties, 14
Bruns-Braude
    dispenser (meterer), 202, 204
    proportioning pump, 202, 204
Bubbling, 301, 311
Buildings, 291-292
    shielded room, 293, 294
Bulges, 15
Bulk-modulus manometer, 182
Burnett compressibility installations, 393,
    395
Burette
    for analysis of liquid phase, 347
    in installations, 319, 322
Bursting strength, see: Pressure limit, compo-
    nents
Bush-Barcas manometer, 183
Butuzov-Boksha-Gonikberg installation for
    investigation of polymorphic transi-
    tions, 129, 130, 132
Butuzov-Mirinskii-Kats
    Cubic clamp (restrained) intensifier, 114,
      116
    six-anvil clamp (restrained) intensifier,
      115, 116
Butuzov-Shakhovskii-Gonikberg intensifier,
    106, 110, 111

Cadmium in materials
    babbitt, 20
    solder, 256
Calcium, electrical properties (P), 174, 175
Calcium chloride, drying agent with gas
    compressors, 79
Calcium hydroxide, friction (P), 27
Calibration of
    intensifiers, 99, 101
    manometers, 139, 140, 156, 166, 172-
      176, 181, 186
Canadian balsam, 266
Caoutchouc, see: Rubber

Capacitive manometers, 182, 183
Capillary, see also: Tubing
    of glass, 24, 25, 186
    in installations, 312, 320
    joints for, 250-255
    in manometers, 150, 167-169
    pressure limit, 23, 24
    of quartz, 130, 132
    in rheometers, 197-199
    of steel, 197-199
    as windows, 267
Capillary-rise surface-tension methods, 415-
    421
Caproamide, see: Nylon
Carbides, see also: Cemented carbides
    titanium−tungsten carbide, 3
    tungsten carbide, 3
Carboloy, see: Cemented carbides
Carbon, see also: Graphite, Diamond
    chemical properties, 15
    in liquid-level gauge, 86
    in steels, 10, 12, 15, 448
    as X-ray transparent, 436
Carbon dioxide
    solubility in liquids, 301, 305
    solubility of solids in, 343
    vapor pressure of, 147
Carbon monoxide
    chemical properties, 15
    compressor for, 82
    solubility in liquids, 301
Carbon steel
    classification of, 10
    effect of mercury, 17
    hardness-tensile strength conversion,
    445-447
    mechanical properties (T), 9, 10
Cathetometer, with manometer, 139
Castor oil, mechanical properties (P, T),
    28, 31
Cast iron
    mechanical properties (RT), 14
    in seals, 249
Cellulose nitrate (varnish), as electrical in-
    sulation, 176, 177
Cemented carbides
    in belt, 72
    in Bridgman anvils, 70, 71
    classification of, 3
    compositions, 3

Cemented carbides (continued)
    in hydraulic-supported pressure cham-
    ber, 64
    mechanical properties (LT), 13
    mechanical properties (P), 14, 64
    mechanical properties (RT), 3, 4, 105
    in pistons, 4, 63, 72, 105, 109, 155, 244,
    247
    in pressure chamber, 74
    pressure limits, 4
    temperature limits, 3
    type numbers, 3
    in wedge pressure chambers, 66
Certified heat treatment, 7
Cesium, polymorphic transformation, 174
Chamber, see: Pressure chamber
Chamfer rings
    in installations, 59, 63, 98, 102, 106,
    107, 111
    of 18 KhNVA steel, 106
    in seals, 229, 230, 245
Check valves, 249, 250
    limitations, 249, 250
    machining, 250
Chemical properties, gases
    carbon monoxide−steel, 15
    nitrogen−iron (P, ET), 15
    nitrogen−steel (P, ET), 15
    oxygen−manganin, 87
    resistance of steels to hydrogen pressure,
    17
Chemical properties, liquids
    design, 30
    liquid metals, 17
    mercury−steel, 16, 17, 20
Chemical properties, solids
    action of gases and liquids, 15-17, 20
    asbestos, 23
    corrosion resistance of steels, 4
    manganin, 87
    polymers, 22
    resistance of steels to hydrogen pressure,
    17
Chemical reactions, see also: Chemical
    properties, solids (or liquids or gases)
    in adiabatic compressors, 88
Chromel
    electrical properties (P, T), 205
    in thermocouples, 205
Chrome−moly steel, 52

Chrome—nickel steel
  hardness-tensile strength conversion,
    445-447
  hydrogen resistance, 15
  thermal pressure, 52, 53
Chrome steel
  hardness-tensile strength conversion, 445-7
  hydrogen resistance, 15
Chrome-tanned leather, see: Leather
Chromium oxide, friction (P), 27
Chromium in
  steel, 9, 10, 12, 15, 448
  titanium alloy, 3, 8, 12
Circulating pumps, 284-289
  agitator-pump, 289
  Aristov-Sidorov pump, 285
  Beck-et al. pump, 285, 286
  bellows pump, 285, 286
  bellows in, 285
  electrical lead into, 287
  electromagnetic pump, 286, 288
  glandless pumps, 282
  glands in, 284, 286
  Gonikberg-Fastovskii-Gurvich pump,
    284, 285
  high-pressure pump, 287, 288
  in installations, 308, 310
  Katsnel'son-Bruns-Gamburg pump, 289
  Krichevskii-Efremova pump, 289
  packing in, 286
  pistons in, 284-286, 288
  resonant-circuit pump, 289
  shaft in, 285
  Sirota-Mal'tsev pump, 287, 288
  valve body in, 286
  valves in, 285, 286, 288
  vane pumps, 284, 285
Circulation, 284-289, see also: Mixing
Circulation methods, phase equilibria, see:
    Analytic methods, phase equilibria
Circumferential stress, see: Tangential stress
Clean compression, 85, 87
Clearance
  in compressors, intensifiers, presses, and
    pumps, 84, 88, 90, 94, 106, 108
  in glands, 235, 236
  in manometers, 140, 141, 145, 147-149,
    152, 155, 159, 181, 182
  in seals, 230, 246
  of threads, 209

Clingerite (cement), 253, 255
  composition, 253
  as packing, 255
Closures, 222, 235, see also: Seals
Cobalt, in
  cemented carbides, 3, 4
  steels, 12
Coefficients of
  friction of solids, 27
  thermal stress, 53, 54
Coil, electrical
  design, calculations for, 277-280
  internal, 277
  power supply for, 274, 275
Coil, glass, for solubility determination,
    301, 302
Coil manometers, 170-172, 176-180
Collar for metal—glass joints, 255, 256
Collars of
  leather, 21
  rawhide, 21
  titanium alloy, 256
Combustion, see also: chemical properties,
    solids (or liquids or gases)
  in compressors, 87
Compensated-area seals, 222-225
Components, 209-268
Composite glass, 26
Compressibility, gases, 355-411
  accuracy of, 358, 359, 363, 364, 383, 387
  adiabatic-compression methods, 386-393
  Amagat installation, 365, 367
  Bartlett installation, 356, 358, 359
  Basset installation, 376-380
  Benedict installation, 359-361
  Bilevich installation, 361, 362
  Bridgman-Benedict installation, 381, 383
  Burnett installation, 393, 395
  corrections to, 363, 365, 370, 371, 381,
    410, 411
  definition of, 355
  Douslin-et al. installation, 379, 380
  Golubev installation, 407-410
  Graham-Maas installation, 391, 393
  gravimetric method, 361
  hydrostatic weighing, 403-410
  Kazarnovskii-Sidorov installation, 369-
    372
  Kazarnovskii-Simonov-Aristov installa-
    tion, 365, 367-370

Compressibility, gases (continued)
   Kirillin-Ulybin installation, 357, 359
   Krichevskii-Markov installation, 363
   Krichevskii-Tsiklis installation, 249, 382-386, 387
   methods of measurement, 355, 356
   Mitchels installation, 380, 381
   pressure limits, 356
   Price-Lalos installation, 389, 390, 392
   Ryabinin-Tsiklis installation, 386-390, 392
   Sidorov-Kazarnovskaya installation, 371, 373-375, 377
   Tsiklis-Kulikova method, 360
   Tsiklis-Linshits installation, 394-396
   Vukalovich-Altunin installation, 358
   Wiebe-Gaddy installation, 357, 359
Compressibility, Liquids, see also: Mechanical properties, liquids
   by Bridgman installations, 398-401
   corrections to, 397, 401
   by Doolittle-Simon-Courish installation, 401-403
   by Gibson installation, 402-403
   by Goldman method, 408-409
   by Golubev installation, 407-410
   by hydrostatic weighing, 403-410
   measurement of, 101
   method of measurement of, 397
   by optical methods, 431
   by pycnometer installation, 401-403
   by Razumikhin installation, 403-405, 408
   by Richards installation, 397
   by Shakhovskii-et al. installation, 400, 402
   by tensimetric balance, 406, 409
   by Vereshchagin-Galaktionov installation, 398, 399
Compressibility of solids, see: Mechanical properties, solids
Compression stages, 83, 85
Compressive strength, see: Mechanical properties, solids
Compressimeters, 180, 181
   in installations, 110, 112
Compressors, see also: Intensifiers, Presses, Pumps
   bellows compressor, 87
   Boksha compressor, 87
   in buildings, 291, 293

Compressors (continued)
   diaphragm (membrane) compressor, 82, 85-87
   glycerin-water-transfer, 86, 311
   hand pumps, 96
   hydraulic, 88-92, 94
   Ivanov hydraulic compressor, 90-92, 94
   Korndorf compressor, 80, 83, 84
   laboratory compressors, 79, 82, 83
   no lubrication in, 80
   mercury-transfer compressor, 84, 86
   NZhR compressor, 99
   for oxygen, 86, 87
   single-stage compressors, 80, 81, 83-85
   thermo-compressor, 87
   Tsiklis hydraulic pump, 93, 95
   Tsiklis-Kulikova bellows compressor, 87
   Tsiklis-Radle hydraulic pump, 94, 99
   variable-stroke compressor, 94, 99
   Vereshchagin hydraulic compressor, 88-90
   Verschchagin-Ivanov compressor, 81, 84, 85
   Vereshchagin-Preobrazhenskii oxygen compressor, 86
   Whalley-Lavergene compressor, 99
   Wolf-Bowen compressor, 84, 85
Concentricity, 210, 211
Concrete, 292, 295
Condenser, 350, 351
Cones, 251, 252, 256-263
   inverted, 257, 258
Conical
   joints, 251
   packing, 71, 72
Conical dies
   apparatus with, 71
   Bridgman anvils, 70
   design, 71
   in Institute of High Pressure Physics device, 73
   pyrophyllite packing, 71
Conical stem-end valve, 211-213
Conicity, 211
Constant
   electrical properties (P), 181, 183
   electrical properties (P,T), 205
   in manometers, 181, 183
   in thermocouple, 205
Constant-clearance manometer, 153, 155-6

Constant-volume valve, 219, 220
Construction materials, 1-34
Contact angles, 424-427
Contactless manometer, 188, 189
Contact piezometer, 365, 367
Contacts, see: Electrical contacts
Container for specimens for presses, 121
Controlled-clearance manometer, 152, 153, 155
Controls
    pressure, 163, 191, 192
    proportioning, 202-204
Control valves, 211-213, 218
Convection in heated pressure chambers, 133
Cooling
    in compressors, 80, 83
    in presses, 120, 121, 128, 130
    of pressure chambers, 80
    in pumps, 95, 99
    of valve glands, 215, 216
Cooled-gland valve, 215, 216
Copeland-Silverman-Benson phase-equilibria apparatus, 331, 333
Copper
    in babbitt, 20
    chemical properties, 15, 20
    electrical properties, 452
    electrical properties, (P,T), 205
    hard, 18
    in installations, 95, 105, 118
    mechanical properties (LT), 11
    mechanical properties (P), 13
    mechanical properties (RT), 11, 18
    as packing material, 20
    in packings, 20, 202, 251, 255, 318
    in sealing rings, 246
    soft, 18
    in steels, 12
    temperature limit, 18
    in thermocouples, 205
    thermal properties, 18, 20
    in unsupported-area seals, 20
Corded asbestos, 23
Corrections to manometers, 141, 149, 150, 168, 169
Corrosion-resistant metals, 4
Corrosion, see also: Chemical properties, solids (or liquids, or gases)
    action of gases and liquids, 15-17, 20
Cotton wool, 293

Cracks, 209
    propagation, 2
Crankshaft drive in installations, 79, 82, 85
Creep-resistant steels, 9
Creep, 9, see also: Mechanical properties, solids
Crusher
    of brass, 185
    in compressibility installation, 386, 391
    in manometer, 185, 186
Crusher manometer, 184-186
Crystallization under pressure, elimination of defects, 5
Cryostats in installations, 308, 310, 311
Cubic presses, 114-116, 118, 119
Current interrupters, 274-276
Cylindrical sleeves, elastic, 232, 233, 235, 236
Cylindricity, 211

D1 (duraluminum), mechanical properties (T), 11
Daniels-Jones supported-piston and cylinder, 75
Dead-weight
    manometers, see: Piston manometers
    testers, see: Piston manometers
Decarburization, 7, 15
Decimal manometric balance, 154, 156
Defect-free crystals, 5
Defects
    due to design, 6
    elimination by crystallization under pressure, 5
    in forging, 5, 6
    in glass, 25
    due to heat treatment, 6
    self-healing by deformation under pressure, 14
    in solids, 5
Deformation, see also: Mechanical working, Plastic deformation
    of crystal lattice, 40, 41
    ductility, 13
    elastic, 40-42
    plastic, 41-43
    under pressure, 13
    residual, 41
    self-healing, 14

Deformation coefficients of manometers, 141, 143, 144
Deformations in assemblies
  intensifiers, 246
  manometers, 140, 141, 145, 146, 153, 155, 161, 179
  piezometers, 363, 385
Delta-ring seal, 233, 234
Density
  determination of, 356, 429
  units, 361
  of liquids, see: Mechanical properties, liquids
  of solids, see: Mechanical properties, solids
Design
  of apparatus, 39-55
  ductility of steels, 5
  of electrical leads, 23
  of manometers, 141, 176-180
  of manganin manometers, 176, 177
Design of components
  glass capillary, 25
  glass tubing, 25
  monobloc pressure chamber, 44-51
  seals, 228, 231, 232
  unsupported-area seal, 1, 2
  valve stems, 211
Design, shape
  angles, 6, 210
  blind holes, 6
  sharp edges, 6, 210
  surfaces, 209
  threads, 9
  through-holes, 6, 210
Diamond
  apparatus containing, 431, 433, 434, 439
  graphitization of, 5
  optical properties (P,T) of, 433
  sintering under pressure with metal, 5
  synthesis of, 119
  as windows, 263
  as X-ray transparent, 436
Diaphragm(s)
  of brass, 163, 180
  compressors, 82, 85-87
  in compressors, intensifiers, presses, and pumps, 82, 85
  design of, 221
  manometers, 179, 181, 183, 184

Diaphragm(s) (continued)
  in manometers, 162-164, 179-181, 192
  in null instruments, 85
  in valves, 85, 220-222, 331, 333
  of stainless steel, 163, 164
  of steel, 183
Diaphragm-mercury
  divider, 163
  null instrument, 163, 164
Diaphragm null instruments, 159, 161-164
  in installations, 307, 309
Diaphragm valves, 220-222
  in installations, 307, 309
Die autofrettage, 55
Differential manganin manometers, 192, 193
Differential manometers, 186-193
Differential-piston manometers, 140, 142-144, 146, 156-161
Differential pistons
  dual, 140, 143
  single, 140, 142
Differential-pressure regulator, 191
Differential thermocouples, 205, 206
  in presses and intensifiers, 130-132
Diffusion, 300
Diment-Carson-Charters manometer, 182, 183
Dislocations, 5
Dispenser
  Bruns-Braude dispenser, 202-204
  packing in, 202
  pressure chamber in, 202, 203
  pump in, 202, 204
  rod in, 202, 204
Distortion-energy theory, 46
Distortion-free steel, 4
Divider, diaphragm, 163
Doolittle-Simon-Courish compressibility installation, 401-403
Double-cone packed presses, 72, 73, 75, 120, 122
Douslin-et al. compressibility installation, 379, 380
Drain valve, 218, 219
Drop stirrers, 280, 281
Drop-weight surface-tension methods, 423, 424
Drying column, 79
Dual differential-piston manometer, 140, 143, 144, 146

Ductility, see also: Mechanical properties,
      solids
    action of gases and liquids, 15-17, 20
    effect of anisotropy, 2
    effect of average normal stress, 2, 6
    effect of heat treatment, 5
    effect of pressure, 13, 14, 75
    effect of size, 2
    effect of stress state, 2, 5, 6
    limit for steels, in design, 5
    nitrogen embrittlement, 15
    requirement for steels, 2
Duraluminum, mechanical properties (T), 11
Dynamic methods, phase equilibria, 300-
      305, 343, 344
Dynamic presses, 87, 88
Dynamic-pressure manometers, 182-186
Dynamic seals, 235-249

É16 (steel), see: 18KhNVA steel
Ebonite as electrical insulator, 177
Edged-sleeve seal, 232, 234
Efremova-Pryanikova installation for deter-
      mining solubility of solids
    in gases, 342-344
ÉI437 steel, mechanical properties (P), 13
ÉI437B (non-magnetic stainless steel)
    composition, 448, 449
    heat treatment, 450, 451
    magnetic properties, 4, 451
    mechanical properties (ET), 9, 450
    mechanical properties (RT), 450, 451
    as pressure chamber, 280
    thermal properties, 451
ÉI698 non-magnetic steel, as pressure cham-
      ber, 280
ÉI183 non-magnetic steel as pressure cham-
      ber, 280
Elastic
    deformation, 41
    distortion, 40, 41
    limit, 41, 42
Elastic-electrical manometers, 180-184
Elastic-plastic deformation, 48, 49
Elastic-property manometers, 164-169
Elastic-sleeve gland, 235, 236
Electrical
    arc heaters, in installation, 130
    circuits for manometers, 171-3, 190-3

Electrical (continued)
    manometers, 169-184
    relay in installations, 87
    resistors of platinum, 169, 186
Electrical coils
    in agitators, 283
    design of, 277-280
    internal coils, 277
    power supply of, 274
    in pumps, 286, 288, 289
    in stirrers, 280, 281, 283, 284
Electrical-conductivity installations, 440-442
Electrical conductivity of solids, see also:
      Electrical properties, solids
    pressure scale, 175
Electrical-contact manometers, 187, 188
Electrical contacts, see: Electrical leads
    of mercury, 17, 33
    of platinum, 169, 186, 188, 200, 203
Electrical heaters
    of graphite, 118, 121, 130, 132
    of platinum, 132
    of platinum—rhodium, 130
    of tungsten, 132
    in presses, 118, 121, 128-133, 187
Electrical insulators, 102
    of aluminum oxide film, 277
    cellulose nitrate varnish, 176, 177
    classification of, 27
    of ebonite, 177
    of enamel, 176
    of fibra, 23
    friction coefficient (P) of, 27
    of glasses, 23-26, 277
    of ivory, 23
    of mica, 27, 130
    of organic materials, 23
    of plastics, 277
    of Plexiglas, 177
    of polymethylmethacrylate, 23
    of porcelain, 24, 27, 277
    of quartz glass, 24, 25
    of silk, 176, 177
    of steatite, 177
    of wax paper, 176, 177
Electrical leads, 257-263, see also: Electri-
      cal contacts
    Blosser-Young-Cornish lead, 261
    Borzunov-Semin lead, 261, 262
    Bridgman lead, 263

leads (continued)
pressing devices, 85
, 23
illations, 317
d-cone leads, 257-263
nikoff lead, 259, 260
essure leads, 257
iometers, 162, 164, 181, 184, 188,
  192
nsulated lead, 258, 259
wire leads, 259-262
 lead, 260
essure chambers, 83, 85, 86, 102,
  110-112, 128, 130-132, 205, 287
e limits, 257-259, 263
yllite-cone lead, 262, 263
imeters and flow meters, 200, 201,

ead, 261
3abb lead, 261
 lead, 258, 259
ov lead, 262, 263
orted-area leads, 257-263
chagin-Kuznetsov-Alaeva lead, 260
and magnetic properties, solids
 (P,T), 205
 (P), 174, 175
h (P), 174, 175
n (P), 174, 175
el (P,T), 205
itan (P,T), 205
, 205, 452
20, 23
23
chromium alloy (P,T), 170, 171
), 174, 175
23
'), 174, 175
nese-silver alloy (P), 169
nin (P), 169
nin (T), 169, 170, 178
—molybdenum alloy (P,T), 205
 (P,T), 205
im (P,T), 205
im—rhodium alloy (P,T), 205
ethylmethacrylate, 23
im (P), 174, 175
?), 180, 181
 451
measurements, 439-443

Electrical measurements (continued)
  electrical conductivity, 440-442
  Ershov-et al. electrical-conductivity
    unit, 440-442
  galvanomagnetic phenomena, 441-443
  Likhter phenomena installation, 441-443
  Mairanovskii-Gonikberg-Opekunov in-
    stallation, 439-441
  polarography, 439-441
Electrical-resistance liquid-level gauge, 86
  in installations, 305, 308, 365, 367
  in manometers, 169, 186
Electrical-resistance
  manometers, 169-182
  thermometers, 204
Electric-power loss as temperature measure-
    ment, 206, 207
Electromagnetic
  pump, 286
  transducers, 163, 164, 180-182, 189-192,
    203, 216
Electromagnetic stirrers, 274-284
  current interrupters, 276
  drop stirrer, 280, 281
  external-rotor stirrer, 281, 282
  power supply, 274, 275
  pressure limits, 274
  rotating-reciprocating stirrer, 281
Electromagnetic-transducer manometers,
    163, 164, 180-182, 189-192
Electropneumatic manometer, 189
Elongation, see also: Mechanical properties
    of solids
  definition, 42
  limiting value in steels, for design, 5
Embrittlement, see also: Action of gases and
    liquids
  carbon monoxide, 15
  helium, 15
  hydrogen, 15
  mercury, 16, 17, 20
  nitrogen, 15
Emulsification, 82
Enamel as electrical insulation, 176, 261
End windows, 265-268
Epoxy resin, in electrical leads, 261
Equivalent stress, 46-51
  in autofrettaged pressure chamber, 56
  equivalent thermal stress, 53, 54
  in heated pressure chamber, 54

Equivalent stress (continued)
    in monobloc pressure chamber, 48
    in multilayered pressure chamber, 57
Ershov-et al. electrical-conductivity instal-
        lation, 440-442
Etched glass, mechanical properties, 26
Etching of glass, 26
Ether, see: Petroleum ether
Ethyl alcohol, mechanical properties (T,P), 29, 32
    as pressure-transmitting medium, 124,
        126, 127, 264
Ethylene, see also: Polyethylene
    solubility in gases, 340
Evaporated-gas press, 87
Expanding-section plug, 234
Explosion, see also: Chemical properties,
        gases; Chemical properties, liquids
    to generate pressure, 82
    precautions, 291, 293-295, 297
Explosives, as heat source, 130
External-motor stirrer, 271
External-rotor stirrer, 281, 282
External-support pressure chamber, see: Vari-
        able external-support pressure chamber

Factor of ignorance, 46, see also: Strength
        reserve
Factor of safety, 47, see also: Strength reserve
Fiber as packing, 187
Fibra (a leatheroid)
    chemical properties, 20
    design limitations, 20
    in electrical leads, 257
    electrical properties, 20, 23
    in joints, 251, 253
    mechanical properties (RT), 21
Filler for hydraulic valves, 33
Filters in installations, 329, 332, 333
Fissures, 6, 14, 15, 16, 210, 247, 256
Fitch-Slykhouse-Drickamer absorption-spectra
        installation, 434, 435
Flat-packing seals, 222-225
Flow measurement, see: Measurement of flow
        of fluids
Flow meters, see also: Rheometers
    Aristov-Sidorov gas flow meter, 200, 202
    Boulestreau gas flow meters, 200
    electrical contacts in, 200, 203

Flow meters (continued)
    with external heating, 200
    Kamzolkin-Bashkirov gas flow meter, 199, 200
    pressure chamber in, 200
    rotameter, 203
    valve in, 200, 203
    window in, 199
Flow of solids, see: Plastic flow
Flow stress, 43
Fluoroplast-3, see: Gostaflon
Fluoroplast-4, see: Teflon
Fly-out valve, 386
Fracture
    in belts, 75
    due to defects, 210
    of pistons, 75, 105, 244
    in pressure chamber, 48
    of supports, 105
    tensile, 40, 41, 49
Free-cone valves, 213, 214
Friction determination, supported pressure
        chamber, 62, 63
Friction in
    compressors, 89
    intensifier, 101, 180
    manometers, 140, 144, 145, 152, 156
    stirrers, 271
Friction properties of materials (P), 27
    asbestos—graphite, 27

Galaktionov tetrahedral press, 113, 116
Galvanomagnetic installation, 441-443
Gamburg-Katsnel'son manometer, 189-191
Gamburg phase-equilibria apparatus, 322-325
Ganz seal, 223-225
Gases
    action on construction materials, 15-17
    compressibility of, 355-396
    compression of, 79-88, 124-127
    determination of solubility of and in,
        301-305
    control of flow of, 211-213
    purification of, 293
Gas holders, 292
Gaskets, see: Packing materials
Gasoline
    to calibrate manganin manometer, 174
    chemical properties, 21
    gasoline—oil (P), 31

Gasoline (continued)
    mechanical properties (P), 31, 32
    as pressure-transmitting medium, 106,
        126, 178
Germanium, thermal properties (P), 174
Gibson compressibility installation, 402, 403
Gielessen manometer, 165, 167, 168
Girdle, 73, 75
Glandless pumps, 282
Glandless valves, 220-222
Glands, 235-243
    Boon-Tal seal, 237-239
    Borzunov-Semin gland, 239-242
    clearance in, 235, 236
    in compressors, intensifiers, presses, and
        pumps, 20, 80, 81, 83-86, 88-96, 98-
        101, 105, 284-286
    cooled gland, 215, 216
    design of, 238
    elastic sleeve gland, 217, 235, 236
    Griggs gland, 237, 240, 241
    hydrodynamic seal, 237-239
    in installations, 315
    in joints, 253, 254
    labyrinth gland, 236, 238
    in manometers, 162
    natural-sealing glands, 235-239
    packed glands, 235, 237, 239-243
    for pistons, 22, 156, 235
    pressure limits, 236, 238
    in rheometer, 197, 198
    self-sealing gland, 237, 240, 241
    self-tightening gland, 239-243
    for shafts and rods, 235
    slipping glands, 242, 243
    in stirrers, 271
    surfaces of, 209
    unsupported-area gland, 238, 240, 241
    in valves, 20, 21, 211, 212, 214, 215, 235
    Vollbrecht gland, 242, 243
Glands, materials
    asbestos, 22, 23
    babbitt, 20, 83
    leather, 21, 187, 188
    rawhide, 21
Glass, 23-26
    in agitators, 313
    in ampules, 25, 313, 314
    in capillaries, 186
    capillary, see: Glass capillary

Glass (continued)
    classification of, 24
    in coil, 301, 302
    in electrical insulators, 23, 277
    etched, 26, 256
    in joints, 253-257
    mechanical properties, etched, 26
    mechanical properties (P,T), 23-26
    in optical windows, 23, 198, 199, 263-
        265, 316, 318, 431, 432
    in piezometers, 23
    pressure limit, 24
    tubing, see: Glass tubing
    in tubing, 24-26, 316, 318-321
    in valve plates, 202, 204
    water attack on, 263, 264
Glass capillary, see also: Capillary
    design, 25
    pressure limit, 24, 25
    in rheometers, 197, 198
Glass fibers, 26, 259
Glass tubing, see also: Tubing
    design, 25
    pressure limit, 24, 25, 26
Glucose, as pressure transmitting medium, 152
Glue, 263
Glycerin
    as lubricant, 83, 86, 105, 107
    in lubricating paste, 63
    mechanical properties (T,P), 29-31
    as pressure-transmitting medium, 31, 85,
        86, 112, 132, 152, 311
    for processing leather, 20
Glycerin—water transfer compressor, 86
GOI lapping paste, 266
Gold
    in diaphragms, 331, 333
    in packing, 321
Gold—chromium alloy
    composition, 170
    electrical resistance (T,P), 170, 171
Goldman compressibility method, 408, 409
Golubev compressibility installation, 408-410
Gonikberg-et al. Raman-spectra installation,
    431, 432
Gonikberg-Fastovskii-Gurvich pump, 284, 285
Gonikberg-Gurvich-Fastovskii phase-equilib-
    ria apparatus, 308, 310
Gonikberg-Tsiklis-Opekunov wedge intensi-
    fier, 117

Goranson heated press, 128, 130

Gostaflon
    in bearings, 288
    in joints, 255, 256
    mechanical properties (T), 21, 22
    in packings, 21, 22, 211, 241, 242, 314
    pressure limits (ET), 22
    as seals, 243, 314

Graham-Maas adiabatic installation, 391-393

Grain growth prevented by thermobaric treat-
    ment, 5

Graphite
    in asbestos, 23, 211
    from diamond, 5
    as electrical-resistance heater, 118, 121,
    130, 132
    friction (P), 27
    as lubricant, 105, 107
    in lubricating paste, 63
    mechanical properties (P), 14
    as packing, 71

Graphitization of diamond, 5

Gravimetric compressibility methods, 361

Gravimetric measurements, phase equilibria,
    348-351

Griggs gland, 237, 240, 241

Grill-Marshall collar, 255, 256

Grishchenko joint, 255, 256

Hafnium, mechanical and thermal properties,
    444

Hall belt, 72, 73, 120

Hall tetrahedral press, 69, 70

Hanging-drop surface-tension methods, 421-3

Hardenability of steels, 6, 7

Hardness, see also: Mechanical properties of
    solids
    conversion table, Rockwell and Brinell,
    7, 445-447
    conversion table to tensile strength of
    steels, 445-447
    errors, 7-9
    testing, 7

Hardness testers, 7

Healing, see: Self-healing

Heated-gas press, 87

Heated presses, 128-132

Heating at high pressures, 126-133
    by explosives, 130

Heating at high pressures (continued)
    external, 126-128
    internal, 128-133
    by thermite, 130

Heat of melting (P), 175, 176

Heat-resistant metals
    classification of, 444
    mechanical properties (RT), 444
    thermal properties, 444
    steels, 9, 12

Heat treatment
    of components with angles, 6
    of components with holes, 6
    of components with sharp edges, 6
    certified, 7
    distortion-free, 4
    ductility, 5
    low-temperature, 5
    of steels, 2, 4, 5-7, 265, 266, 450, 451
    tempering, 5
    under pressure, 5

Helium
    action on steel, 15
    solubility in gases, 340

Helium embrittlement of steel, 15

High-pressure
    capillary, 197, 198
    circulating pump, 287, 288

High-pressure-piston seals, 243-247

Holes
    in centers of forged steel blocks, 6
    concentricity of, 210
    drilling of, 210, 211
    softness after heat treatment, 6

Hydraulic compressors, 88-92, 94

Hydraulic presses, 122, 123
    solid pressure-transmitting media, 34

Hydraulic seals, 234, 235

Hydraulic-support pressure chamber
    conical die, 70
    design, 64-66

Hydraulic valves, 217, 218
    fillers for, 33

Hydrocarbons
    compression, 82
    removal from gases, 79

Hydrodynamic seal, 237-239

Hydrogen
    chemical properties, 15
    compressor for, 82

Hydrogen (continued)
    effect on steel, 15-17
    solubility in and of liquids, 303, 304, 310
    solution in solids, 130, 204
Hydrogen embrittlement, 15-17
Hydrogen manometer, 169
Hydrogen-resistant steels
    classification of, 16
    compositions, 15, 16
    mechanical properties (T), 16, 17
    resistance to hydrogen pressure, 15, 17
Hydrostatic tension at neck, 43
Hydrostatic weighing
    in installations, 356
    to measure compressibility, 355, 403-410

Impact strength, see: Mechanical properties,
    solids
Impurities
    clean compression, 85, 87
    liquid on manganin, 177
    liquid in mercury, 86
    lubricant in stirred liquids, 272
    oil in gas, 79
    oil in hydrocarbons, 82
    water in gas, 79
Impurity removal, see: Removal of impuri-
    ties
Indium, friction (P), 27
Indrik manometer, 158, 161
Inert gases, compressing devices for, 85
Infrared radiation, pyrometry, 206
Infrared spectra, 433
Installation plan for pressure measurement,
    159, 161-164
Institute of High Pressure Physics press with
    conical anvils, 73
Insulation, see: Electrical (or Thermal)
    properties, solids (or liquids)
Insulators, see: Electrical insulators
Intensifier seals, 243-249
Intensifier manometers, 155, 157, 159-161
Intensifiers, see also: Compressors, Presses,
    Pumps
    100,000-atmosphere intensifier, 110, 112
    autofrettage of, 246
    Bridgman double-support intensifier, 103,
        104, 109
    Bridgman submerged intensifier, 110, 112

Intensifiers (continued)
    Bridgman variable single-support intensi-
        fier, 63
    Butuzov-Shakhovskii-Gonikberg intensi-
        fier, 106, 110, 111
    calibration of, 99, 101
    clamp (restrained) intensifiers, 114-116
    for closures, 235
    deformation of, 246
    with external support, 102, 104-112
    in glands, 239
    in installations, 338
    Itskevich cryogenic intensifier, 127
    Krichevskii-Tsiklis booster, 100, 101, 104
    machining, 247
    in manometers, 155, 157, 159-161
    mechanical finishing, 246
    for oxygen, 87
    piston seals for, 243-249
    Sidorov intensifier, 98, 101
    six-anvil-clamp (restrained) intensifier,
        115, 116
    in valves, 218
    with valves, 100, 101, 104
    Vereshchagin-et al. intensifier, 102,
        104-108
    wedge intensifier, 117-119
    Zhokhovskii packing-support intensifier,
        107, 111, 112
Interference manometer, 183
Interference stress (and strain)
    in autofrettaged pressure chambers, 56
    in external-support pressure chambers, 62
    in interference-fitted pressure chambers,
        59, 69, 75
Internal coils including design, 277-280
Internal-motor stirrer, 271
Invar
    as joint solder, 256
    measuring tape, 139
Inverted-cone electrical leads, 257-263
Iomtov phase-equilibria apparatus, 341, 343
Ipat'ev-et al. installation for determination
    of gas composition, 304, 305
Iron
    Armco, mechanical properties of (P), 14
    chemical properties, 15
    electrical properties (P), 174, 175
    nitrogen embrittlement (P,ET), 15, 204
Iron carbonyl, 15

Iron in
    steels (see: Steel, compositions)
    titanium alloy, 3, 8, 12
Iron oxide, friction (P), 27
Iron nitride, 15
Isobutylene, see: Polyisobutylene
Isolated-stator stirrer, 282-284
Isopentane
    mechanical properties (T,P), 29, 32
    as pressure-transmitting medium, 112,
        132, 179
Itsbevick cryogenic intensifier, 127
Ivanov hydraulic compressor, 88-90
Ivory, 23
    electrical insulator, 23, 258
    electrical properties (T), 23

Jacobs X-ray installation, 437
Johnson-Newhall manometer, 152, 153, 155
Joints, 250-257
    asbestos gland, 253, 254
    capillaries for, 250-255
    conical joints, 251, 252
    Gill-Marshall collar, 255, 256
    Grishchenko joint, 255, 256
    Khalilov joint, 253, 255
    Krichevskii-Efremova joint, 255
    lenses, 252, 253
    lipped joints, 252, 253
    for manometers, 250-252
    metal—glass joints, 253-257
    metal—glass solder joints, 256, 257
    packing for, 250, 251, 253
    packing joints, 251
    pressure limits of, 251-253, 255-256
    shellac—mercury joint, 253, 254
    Sidorov joint, 253, 254
    soft seal, 253, 254
    Teflon joint, 253
    temperature limits of, 253, 255
    tubing for, 250, 251
    for valves, 250-252

Kabalkina-Vereshchagin installation, 438,
    439
Kagan formulas, hydraulic-supported pressure
    chambers, 65, 66
Kalashnikoff electrical lead, 259, 260

Kamzolkin-Bashkirov gas flow meter, 199, 200
Kan freezing-thawing liquid press, 124-126
Kasper-et al. absorption-spectra installation,
    433, 434
Katlenite (variety of pyrophyllite)
    electrical properties, 34
    mechanical properties (P), 14
    pressure limits, 34
    temperature limits, 34
    thermal properties (P), 34
Katsnel'son-Bruns-Gamburg pump, 289
Katsnel'son formula for solenoid force, 279
Katsnel'son-Ikhlov circuit for measuring
    electrical resistance, 172, 173
Kazarnovskii-Sidorov gas-compressibility
    installation, 369-372
Kazarnovskii-Simonov-Aristov gas-com-
    pressibility installation, 365, 367-370
Kerosene
    mechanical properties (T,P), 29, 32
    oil-kerosene pressure-transmitting medi-
        um, 31, 337, 339
    as pressure-transmitting medium, 178,
        243, 339
1Kh13 (steel), mechanical properties (ET), 9
2Kh13 (steel)
    composition, 448, 449
    electrical and magnetic properties, 451
    heat treatment, 451
    mechanical properties (ET), 9
    mechanical properties (RT), 450, 451
    pressure limits (T), 4
    thermal properties, 451
40Kh (steel)
    in electrical lead, 259
    pressure limits, 2
    in support, 104
Khalilov joint, 253, 255
30KhGS (steel)
    composition, 448, 449
    heat treatment, 450, 451
    magnetic properties, 451
    mechanical properties, 450, 451
    pressure limits, 2
    in support sleeve, 105
    thermal properties, 451
30KhGSA (steel), in pressure chamber, 51
30KhMA (steel)
    composition, 448, 449
    heat treatment, 450, 451

30KhMA (steel) (continued)
  magnetic properties, 451
  mechanical properties, 450, 451
  pressure limits, 2
  thermal properties, 451
1Kh18N9 (stainless steel)
  composition, 448, 449
  electrical and magnetic properties (RT),
    4, 451
  heat treatment, 451
  mechanical properties (ET), 9
  mechanical properties (RT), 450, 451
  pressure limits (T), 4
  in stirrers, 4
  thermal properties, 451
18KhNMA (steel)
  composition, 448, 449
  pressure limits, 2
45KhNMF (steel)
  in pressure chambers, 104
  in sealing rings, 241
  tensile strength, 104
45KhNMFA (steel)
  composition, 448, 449
  heat treatment, 450, 451
  magnetic properties, 451
  mechanical properties (RT), 450, 451
  pressure limits, 2
  thermal properties, 451
1Kh18N9T (austenitic stainless steel)
  electrical and magnetic properties, 4, 451
  heat treatment, 451
  mechanical properties (ET,LT), 9
  mechanical properties (RT), 11, 450, 451
  nuts and bolts, 209
  in pressure chamber, 51, 202, 203, 280, 320
  pressure limits (T), 4
  seizing of, 209
  in stirrers, 4
Kh20N77T2YuR (steel), see: É1437B (steel)
18KhNVA (chrome—vanadium steel)
  in chamfer rings, 106
  chemical properties, 15
  composition, 448, 449
  heat treatment, 451
  magnetic properties, 451
  mechanical properties (T), 11, 450, 451
  pressure limits, 2
  in sealing rings, 244
  thermal properties, 451

OKhNZM (steel)
  composition, 448, 449
  heat treatment, 451
  magnetic properties, 451
  mechanical properties, 450
  pressure limits, 2
KhVG (tungsten steel)
  composition, 448, 449
  heat treatment, 450, 451
  magnetic properties, 451
  mechanical properties, 14, 450
  in pistons, 105, 108
  in plugs, 246
  in plungers, 4, 106
  in sealing ring, 246
Kieselguhr friction (P), 27
Kirillin-Ulybin gas-compressibility installa-
    tion, 357, 359
Kiselev manometer, 189
Klyuev absorption-spectra installation, 435,
    436
Kohen-et al. phase-equilibria apparatus, 329
Korndorf single-stage compressor, 80, 83, 84
Kovar, 256, 268
Krichevskii-Efremova
  circulation installation, 310, 311
  joint, 255
  liquid-phase-analysis apparatus, 349-
    351
  pump, 289
  visual-observation phase-transformation
    installation, 316, 317, 319-321
Krichevskii-Gamburg
  condenser, 350, 351
  solubility installation, 303, 304
Krichevskii-Khazanova phase-equilibria ap-
    paratus, 303
Krichevskii-Lebedova solubility-of-solids-
    in-liquids installation, 330-332
Krichevskii-Markov gas compressibility in-
    stallation, 363
Krichevskii-Shurmoskoya-Kal'varskoya
    method, 311-315, 318, 319
Krichevskii-Sorina method, 325-328
Krichevskii-Tsiklis
  booster, 100, 101, 104
  gas compressibility installation, 249,
    382-386, 387
Krichevskii-Zhavoronkov-Tsiklis gas-solu-
    bility apparatus, 301-303, 345-347

Kusakov-Lubman-Koshernik surface-tension
    installation, 421-423

L62 (brass) mechanical properties (T), 11
L80 (brass), mechanical and thermal proper-
    ties, 18
Laboratory compressors, 79, 82, 83
    limitation, 82
Laboratory equipment, 291-298
    compressors, 293
    gas holders, 292
    gas purification, 293
    installation of apparatus, 293, 294
Labyrinth gland, 236, 238
Lacquer, 277
    as electrical installation, 277
Lamé's formula, 44, 45
Lapping compound, 105, 266
Large-diameter seals, 233, 234
Large-piston seals for intensifiers, 247-249
Lavrov method of autofrettage, 55
Lazarev-Kan freezing-water press, 124
Lead
    in babbitt, 20
    electrical properties (P), 174, 175
    in joints, 253, 255
    as lubricant, 63, 107, 108
    mechanical properties (RT), 18
    in packings, 20, 222, 253, 255
    in solders, 256
    temperature limit, 18, 20
    thermal properties, 18
Lead oxide, friction (P), 27
Lead—potash—borosilicate—fluoride glass,
    see: Borosilicate glass
Lead—potash glass, mechanical properties
    (P,T), 24
Leakage, calculation of, through clearance,
    238
Leather
    chemical properties, 21
    classification of, 21 (P)
    as glands, 187, 188
    in joints, 253, 254
    mechanical properties, 21
    in packings, 21, 89, 95, 100, 211
    in seals, 243, 244
Leatheroid, see: Fibra
Lebedeva-Kodeeva ampules, 315

Lens seal, 228, 230, 231, 233, 234
Lenses, 252, 253
Leonidov-Polandrov manometer, 182
Light conductor, of quartz, 206
Likhter galvanomagnetic installation, 441-
    443
Limitations, assemblies
    of compressors, intensifiers, presses, and
        pumps, 79, 80, 86, 87, 101, 127
    of manometers, 138, 148, 153, 159, 166-
        170, 178, 179, 186, 188, 190-192
Links, in compressors and pumps, 88, 89, 94,
    95, 99
Lipped joints, 252, 253
Liquid-compressibility manometers, 169
Liquid-level gauging
    electrical contacts, 83, 85, 86
    resistance, 86
Liquid manometers, 138-140, 149, 150, 152,
    167-169
Liquid phase, analysis, 345-351
Liquids
    action on construction materials,
        16, 17, 20
    circulation of, 284-289, 308, 310, 311
    compressibility, 30, 32, 33, 355, 356,
        397-410
    compression of, 88-112, 124-127
    compressibility, differential coefficient
        of, 397
    compressibility, freezing method, 124-
        127
    density of, 401-410
    metering of, 202-204
    relative volumes of (P,T), 30, 33
    solubility in compressed gases, 303, 304
    surface tension of, 415-429
Lithium
    mechanical properties (P), 13
    as pressure-transmitting medium, 439
Lithium fluoride, 436
Lithium hydride, 436
Lloyd-Hutton-Johnson tetrahedral press, 113,
    116
Load-deformation diagram, see: Stress-strain
    diagram
Low-pressure electrical leads, 257
Low-pressure-piston seals, 247-249
Lubricant
    glycerin as, 83, 105, 107

Lubricant (continued)
  graphite as, 105, 107
  lead as, 107, 108
Lubricating oil, mechanical properties (T,P), 28
Lubricating paste, 63
Lubrication of assemblies
  compressors, 79, 80, 81, 83, 84
  compressors, intensifiers, presses, and pumps, 79-81, 83, 84, 92, 94
  pressure-chamber supports, 63
Lying-drop method, 422, 423

M1 (copper), mechanical properties (T), 11
M3 (copper), thermal properties, 18
M4 (copper), properties, conditions, and temperature limit, 18
Machine oil, mechanical properties (T,P), 28
Machining, see also: Mechanical finishing
  allowance in autofrettaged chamber, 56
  of check valves, 250
  for hardenability test, 7
  of intensifier bores, 247
  of supports, 105
  of threads, 210, 211
  of valve stems, 211
Magnesium
  alloyed with titanium, 5
  mechanical properties (P), 13, 14
Magnesium alloys
  mechanical properties, 14
  melted under pressure, 5
Magnesium perchlorate, see: Anhydrone
Magnetic properties, see: Electrical and magnetic properties, solids
Magnetic
  starters, 275, 276
  stirrers, steels for, 4
Mairanovskii-Gonikberg-Opekunov polarographic installation, 439-441
Manganese dioxide, friction (P), 27
Manganese in steels, 10, 12, 448
Manganese—silver alloy, 169
Manganin
  aging, 177, 178
  annealing, 178
  chemical properties (P), 87
  composition, 170
  electrical properties (P), 169, 170, 178

Manganin (continued)
  in manometers, 169-180, 184, 186, 192, 193
Manganin coils
  calibration of, 173-176
  limitations of, 170, 178, 179
  manufacture of, 176-180
  measurement of electrical resistance of, 170-173
Manganin manometers, 169-180, 184, 186, 192, 193
  in compressors, intensifiers, presses, and pumps, 87, 104, 105, 107, 111, 112, 129, 131, 132
  oxygen pressure, 87
  pressure limits, 87
Manganin transducer, 184
Manifords for fluids, 296
Manometer(s)
  absolute manometers, 137-164
  accuracies of, 148, 150, 156, 159, 162, 164, 166, 168, 170-172
  Amagat manometer, 149, 150
  Aristov manometer, 188, 189
  bellows in, 166, 168, 179
  Bol'shakov manometer, 187, 188
  Bourdon gauges, 164-167, 186
  Bridgman compressimeter, 180, 181
  bulk-modulus manometer, 182
  Bush-Barcas manometer, 183
  calibration of, 139, 140, 156, 166, 172-176, 181, 186
  capacitive manometers, 182, 183
  capillary in, 150, 167-169
  change in volume of gas or liquid in, 168
  clearances in, 140, 141, 145, 147-149, 152, 153, 155, 159, 181, 182
  coil manometers, 170-172, 176-180
  in components, 201, 203
  compressimeter, 180, 181
  constant-clearance manometer, 153, 155, 156
  contactless manometer, 188-191
  controlled-clearance manometer, 152, 153, 155
  corrections to, 141, 149, 150, 168, 169
  crusher manometer, 184-186
  decimal manometric balance, 154, 156
  deformation coefficients of, 141, 143, 144

Manometer(s) (continued)

deformations in, 140, 141, 145, 146, 153, 155, 161, 179

design of, 141, 176-180

diaphragm in, 162-164, 179-181, 192

diaphragm manometers, 179, 181, 183, 184

differential manganin manometers, 192, 193

differential manometers, 186-193

with differential piston, 140, 142, 143, 146, 152, 156-161

Diment-Carson-Charters manometer, 182, 183

with displaced center of gravity, 158, 161

dual differential-piston manometer, 140, 143, 144, 146

dynamic-pressure manometers, 182-186

electrical circuits for, 171-173

electrical leads into, 162, 164, 181, 184, 188, 191, 192

electrical-contact manometer, 187, 188.

electrical manometers, 169-184

electrical-resistance liquid-level gauge in, 169, 186

electrical-resistance manometers, 169-182

with electromagnetic-transducer, 163, 164, 180-182, 189-192

electropneumatic manometer, 189

elastic-electrical manometers, 169-180, 184, 186, 192, 193

elastic-property manometers, 164-169

friction in, 140, 144, 145, 152, 156

Gamburg-Katsnel'son manometer, 189-191

Gielessen manometer, 165, 167

hydrogen manometer, 169

Indrik manometer, 158, 161

installation of, 296, 297

in installations, 309, 312, 323, 324, 334, 338-340, 380-381

installation plan for, 159, 161-164, 296, 297

intensifiers in, 155, 157, 159, 161

interference manometer, 183

Johnson-Newhall manometer, 152, 153, 155

joints in, 250, 252

Kiselev manometer, 189

with lateral-pressure cylinder, 142, 143, 152, 153, 155, 156

Manometers (continued)

Leonidov-Polandov manometer, 182

limitations of, 138, 148, 153, 159, 166-170, 178, 179, 186, 188, 190-192

liquid-compressibility manometers, 169

liquid manometers, 138-140, 149, 150, 152, 167-169

manganin manometers, 169-180, 184, 186, 192, 193

manometric balances, 156

Mendeleev manometer, 138

mercury manometer, 138-140, 149, 150, 163, 164, 167-169, 187, 188

mercury precision multistage manometer, 138

metal U-shaped, 162

Michels manometer, 169

MP-600 manometer, 158, 161

Multistage manometer, 138, 139

Newhall-Abbot manometer, 182

Nogatkin manometer, 165, 167

operation of, 161-164

optical-interference manometer, 183

packings in, 159, 162

piezoelectric manometer, 183, 184

pistons in, 139-162, 164, 185, 191, 192

piston manometer, 140-164, 186

pressure chambers in, 139-147, 149-163, 187-191

pressure limits of, 152, 159, 160, 164, 166-168, 170, 181, 182, 184, 190-192

for rapidly-changing pressures, 184

rate of descent of piston in, 141-145

relative manometers, 137, 164-186

Roebuck-Ibser manometer, 138, 139

rotation of body in, 156

rotation of piston in, 144, 145, 151, 152, 155, 157-162

Rucholz press, 150

Ryabinin crusher manometers, 184-186

seals in, 140, 152, 155, 157, 159, 183, 186

Shelaputin-Voitovich manometer, 184

Shtyukrat balance, 154, 156

with simple piston with lateral pressure, 140, 142, 143, 146, 152

with simple unpacked piston, 140-152

with single differential-piston, 140, 142, 143, 146, 156-161

with single piston, 149-156

sleeve seals in, 155, 157

sleeves in, 151-153, 155-157

Manometers (continued)
  straight-tube manometer, 165, 167, 168
  suppressed-zero manometer, 168
  Tait manometer, 167, 168
  tensimeters, 181
  tensimetric, 181, 182
  Tsiklis-Borodina manometer, 184
  Tsiklis manganin manometer, 78
  tubing in, 25, 165
  tubular-spring manometers, 164, 165,
    167, 168, 185, 189
  upsetting manometer, 184, 186
  valve in, 157, 191, 192
  Vereshchagin-Aleksandrov manometer,
    151-153
  volumetric compression manometers, 182
  water-compressibility manometer, 169
  weight manometer, 140
  windows in, 155, 157-161
  Zelyaev-Shumov-Alekseev manometer,
    181, 182
  Zhokhovskii constant-clearance manom-
    eter, 153, 155, 156
  Zhokhovskii intensifier manometer, 155,
    159, 160
  Zhokhovskii-Konyaev-Levchenko ma-
    nometer, 157, 160, 161
Manometric balances, 156
Martensite, see also: Austenite—martensite
  conversion
  hardenability of steels, 6
Mastic oil, 23
Materials
  behavior under load, 9-20, 39-44
  of construction, 1-27
  elastic limit, 41, 42
  mechanical properties, see specific ma-
    terial
  testing, 7-9
Matte glass, see: Glass
Maximum shear-stress theory, 46
Mean normal stress, see: Average normal
  stress
Measurement of flow of fluids, 197-204
Measurement of high pressures, 137-193
Measurement of temperature of fluids, 204-207
Mechanical agitators, 271, 272
  capacity of, 272
  electrical leads for, 271, 272
  external-motor type, 271

Mechanical agitators (continued)
  in installations, 313, 314, 328
  internal-motor type, 271, 272
  pressure limits, 271
  limitations of, 271
Mechanical finishing, 9
  of angles, 9
  of glands, 209
  for hardenability test, 7
  of intensifier bores, 210, 246
  of nuts and bolts, 209
  of packing surfaces, 9, 209
  of pistons, 210
  of pressure chambers, 210
  of sealing surfaces, 209
  of steels, 6
  of threads, 9, 209
  of valve stems, 210
Mechanical press, 97, 100, 101
Mechanical properties, liquids (P), 28-33, 143
  aviation oil (T,P), 28
  castor oil (T,P), 28
  design, 30
  ethyl alcohol (T,P), 29, 32
  gasoline (P), 32
  glycerin (P), 32
  glycerin (T,P), 29
  isopentane (T,P), 29, 32
  kerosene (P), 32
  kerosene (T,P), 29
  lubricating oil (T,P), 28
  machine oil (T,P), 28
  mercury (P), 30
  mercury (T,P), 29, 32
  normal pentane (T,P), 29, 32
  oil—kerosene mixtures, 31
  petroleum ether (T,P), 29
  silicone oils (T,P), 30, 33
  transformer oil (T,P), 28
  upper cylinder oil (T,P), 28
  vasoline (T,P), 28
  water (T,P), 32
Mechanical properties, solids
  alloy steels (T), 9, 10, 11, 12, 445-447
  aluminum, 18
  Armco iron (P), 14
  asbestos, 22
  austenitic steels (T), 10, 11, 12
  babbitt, 18
  beryllium bronze, 18

Mechanical properties, solids (continued)
  bismuth (P), 13
  brass (T), 11, 18
  bronze (T), 11, 14
  carbon steels (T), 8, 10, 11, 12, 445-447
  cast iron, 14
  cemented carbides (LT), 13
  cemented carbides (P), 14
  cemented carbides (RT), 3, 105
  chrome—nickel steel (ET), 12
  chrome—nickel steel (RT), 445-447
  chrome steel (T), 445-447, 450-451
  copper (P), 13
  copper (T), 11, 18
  duraluminum (T), 11
  fibra, 21
  glass (P,T), 21, 23, 25
  graphite (P), 14
  hydrogen-resistant steels (T), 16, 17
  insulating materials, 21
  katlenite (P), 14
  lead, 18
  leather, 21
  lithium (P), 13
  magnesium alloys, 14
  magnesium (P), 13, 14
  nickel steels (T), 10, 445-447
  nonferrous metals (T), 11, 13
  packing materials, 21
  paronite, 21
  polymers, 19, 22
  porcelain, 21, 24
  potassium (P), 13
  pyrophyllite (P), 14
  quartz glass (P), 21
  rawhide, 21
  rubber, 21
  silver (P), 13
  silver chloride (P), 14
  stainless steels (T), 5, 12
  steels (ET), 9, 12, 17, 204
  steels (LT), 10, 11
  steels (P), 13
  steels (RT), 1, 2, 9-11, 16, 17, 104, 450, 451
  tellurium (P), 13
  textolite, 21
  titanium (T), 3, 8
  titanium alloys (T), 3, 8
  titanium—tungsten (cemented) carbides, 3
  tin (P), 13

Mechanical properties, solids (continued)
  tungsten (cemented) carbides, 3
  zirconium (P), 13
  zinc (P), 13
  zinc (T), 18
Mechanical working
  of St.5 steel, 2
  of supports, 105
Melamine, solubility in substances, 343
Melting point, see also: Thermal properties
      of solids
  assemblies for determining (P), 131, 132
  (P), 175, 176
  pressure scale, 176
Melting under pressure, 5
Membrane (item), see: Diaphragm (item)
Mendeleev manometer, 138
Mercury, see also: Mercury manometers
  chemical properties, 16, 17, 20
  in electrical contacts, 17, 20
  in joints, 254
  mechanical properties (T,P), 29, 30, 32
  oxygen compression, 86
  in pressure-transmitting medium, 83, 85,
      86, 329, 339
  as pressure unit, 137, 448-449
  in rheometers and flow meters, 198, 200,
      203
  safety, 411
  in seals, 17, 319
  thermal properties (P), 29, 33, 173, 174
  transfer compressing device, 83, 85, 86
Mercury-equalizer instrument, 150
Mercury interrupter and relays, 274, 276
Mercury manometers, 138-140, 147-150,
      163, 164, 167-169, 187, 188
  corrections to, 139, 140
Mercury-transfer
  compressor, 83-86
  press, 83, 85, 86
Metal capillary, 197
Metal—glass
  joints, 253-257
  solder joints, 256, 257
Metals
  behavior under experimental conditions,
      9-18
  electrical resistance of (P), 175
  elongation of, 41, 42
  heat-resistant, properties of, 444

Metals (continued)
  mechanical properties (see also specific
    metals) of, 3, 9-13, 18
  melting-point determination, 131, 132
  monocrystals of, 5
Metal tubing, see: Tubing
Meterer, 202-204
Metering of liquid into compressed gas, 312,
    313
  in installations, 322, 323
Methane, 15
  compressor for, 82
  solubility with gases, 337
Methods of analysis, phase-equilibria sam-
    ples, 345-351
Methods of investigating phase-equilibria,
    299-351
  analytic methods, 299-311, 328-333,
    337-351
  sealed-ampule methods, 311-319
  synthetic methods, 299, 311-328, 333-
    337, 340
Methods of measuring high pressures, 137-193
Methods of measuring gas—liquid surface
    tension, 415-424
  hanging-drop method, 421-423
Methyl alcohol
  as pressure-transmitting medium, 127
  solubility of gases in, 301
Methyl methacrylate, see: Polymethyl
    methacrylate
Methylpolysiloxanes, see: Silicone oils
Mica
  as electrical insulator, 130, 258, 259,
    262
  electrical properties, 27
  friction (P), 27
  mechanical properties, 27
Mica-insulated electrical leads, 258, 259
Michels manometer, 169
Microlite (ceramic material), 442
Minimal-clearance ring, 153, 155
Mitchels compressibility installation, 380,
    381
Mixing, see also: Circulation, 271-284
Mixing, liquid-gas, 5, 9, 300, 301
Molybdenum disulfide
  in asbestos, 23
  friction (P), 27
  in Teflon, 22

Molybdenum glass
  capillary, 25
  pressure limit, 25
  tubing, 25
Molybdenum in
  steel, 10, 12, 15, 448
  titanium alloy, 3
Molybdenum trioxide
  friction (P), 27
Molykote, see: Molybdenum disulfide
Monocrystals
  metals, 5
  X-ray investigation, 436, 437, 439
MP-600 manometer, 158, 161
Multilayered pressure chambers
  design, 57-60
  equivalent stresses, 57
  interference stresses, 59
Multistage manometer, 138, 139
  corrections, 139, 140
  windows, 139
Multi-wire electrical leads, 259, 260

Napthalene, solubility in gases, 343
Natural-sealing glands, 235-239
NBS tetrahedral press, 113, 116
Necking, stress state, 43
Needle valves, 191, 192, 216
  in manometer, 191, 192
Newhall-Abbot manometer, 182
Nickel
  chemical properties, 15
  electrical properties (P,T), 205
  as packing material, 121
  in thermocouples, 205
Nickel carbonyl, 15
Nickel in
  babbitt, 20
  steels, 9, 10, 12, 449
Nickel—molybdenum alloy
  electrical properties (P,T), 205
  in thermocouples, 205
Nickel steel
  compositions, 10
  hardness-tensile strength conversion, 445-7
  mechanical properties, 10
Nigrol (lubricant), in compressors, 84
Niobium,
  mechanical and thermal properties, 444
  in steels, 12

Nipple(s)
    procedure for lapping, 264-266
    of ShKh12 steel, 265
    of St. 5 steel, 2
Nitrogen
    chemical properties (P,ET), 15
    compressor for, 82
    embrittlement of iron (P,ET), 15, 204
    as pressure-transmitting medium, 132
Noble gases, see: Inert gases
Node lines, 327
Nonferrous metals and alloys
    mechanical properties (T), 11, 18
    temperature limit, 18
    thermal properties, 18
Nogatkin manometer, 165, 167
Nominal stresses, 43
    definition of, 42
Noncompensated area seal, see: Unsupport-
        ed area seal
Nonmagnetic metals, 4, 451
    in stirrers, 4
Nontransparent
    glass, see: Glass
    quartz glass, see: Quartz glass
Normal atmosphere, 137, 448, 449
Normal stresses, 39, 40
    during necking, 43
Normalizing of steels, 450
Normal pentane, mechanical properties
        (T,P), 29, 32
Notch tensile tests, 6
n-Pentane, see: Normal pentane
Null instruments in manometer installation,
        159, 161, 163
Nut, seizing of, 209
Nylon
    mechanical properties (T), 19, 21
    thermal properties, 19
Nyquist formula, 206
NZhR compressor, 99

Oil catcher, 79
Oil separator, 83, 84, 293
Oils, see also: Methylpolysiloxanes
    in asbestos, 23
    aviation, castor, cylinder, lubricating, ma-
        chine, transformer, and vasoline oils, 28
    gasoline—oil (P), 31

Oils (continued)
    in lubricating paste, 83, 105, 107
    mechanical properties (T,P), 28
    oil—kerosene (P), 31
    oxygen compression, 86
    as pressure-transmitting medium, 85, 86,
        147, 148, 189, 243, 264
    silicone oils, 31
    solubility of gases in, 323
    solution, 79, 82
Oil stable rubber, see: Rubber
Oleinik lead, 260
Operation of, 161-164
Optical-interference manometer, 183
Optical measurements, 431-436
    of absorption spectra, 431-436
    diamond windows for, 431, 433
    by Fitch-Slykhouse-Drickamer absorp-
        tion-spectra installation, 434,
        435
    glass windows for, 431, 432
    by Gonikberg-et al. Raman-spectra in-
        stallation, 431, 432
    by Kasper-et al. absorption-spectra in-
        stallation, 433, 434
    by Klyuev absorption-spectra installa-
        tion, 435, 436
    of Raman-effect spectra, 431, 432
    sapphire windows for, 431
    sodium chloride windows for, 434
    by Weir-Lippincott-Bunting absorption-
        spectra installation, 431, 433
Optical plates, of quartz, 183
Optical pyrometers, 204, 206
Optical windows, see: Windows
Ordinary-piston seals, 243
Organic insulators (fibra, ivory, polymethyl-
        methacrylate), 23
O-rings, 28, 230
    in manometer, 164
    in windows, 267
    of plastics, 164
Oscillogram of compression of nitrogen, 388,
        389
OT-4 (titanium alloy), mechanical proper-
        ties (T), 8
OTCh titanium alloy, mechanical and ther-
        mal properties, 3
OTCh-1 titanium alloy, mechanical proper-
        ties, 3

Oxygen
  chemical properties gases (P), 87
  compressing devices for, 85-87
  compression, 85-87
  compression lubricant for, 83
  combustion with, 297
  unsuited compressor, 82, 83

Packed glands, 235, 237, 239-243
Packing
  in Bridgman anvils, 71
  for collars, 21
  in compressors, intensifiers, pumps, and
      presses, 86, 87, 89, 102, 104, 111, 116,
      121-123, 127, 129, 131, 132, 202, 286
  conical, 71, 72, 73
  for glands, 20, 21, 22, 241
  in Institute of High Pressure Physics de-
      vice, 73
  for joints, 250, 251, 253
  lateral forces in, 239
  in manometers, 159, 162
  in oxygen compression, 87
  plastic flow of, 251
  in press with conical dies, 71-74
  in rheometers and flow meters, 202, 203
  in seals, 20, 222-227, 233, 234, 245, 247
  surfaces, 209
  for valves, 216
  for windows, 263, 267
Packingless seal, 224, 225
Packing materials, 20-23
  aluminum, 20
  asbestos, 22, 23
  babbitt, 20
  classification of, 21, 27
  copper, 20
  fiber, 187
  fibra, 20, 21
  friction (P), 27
  function, 20
  Gostaflon, 22
  lead, 20
  leather, 21, 89, 95, 100
  leatheroid, 20
  low temperature, 20
  mechanical properties (RT), 21
  nickel, 121
  paronite, 21

Packing materials (continued)
  plastics, 95, 164, 245, 249
  polyethylene, 21, 22
  polymers, 21, 22
  polymethylmethacrylate, 22
  polyvinyl chloride, 21, 22, 106
  porcelain, 21
  rawhide, 21
  rubber, 21
  shellac, 254
  steel, 89, 90, 121
  Teflon, 21, 22, 83, 86, 87, 163, 164
  textolite, 21
Palladium, hydrogen solution in, 204
Paronite, mechanical properties, 21
Paste, for lapping, 266
Payne-Sreed-Kent autoclave, 273
Pentane, see: Isopentane, Normal pentane
Permeable wall, 293
Petroleum ether, mechanical properties (T,P), 29
Phase equilibria, 299-351
  analytic methods, see: Analytic methods,
      phase equilibria
  Arons apparatus, 340
  Bol'shakov-Linshits apparatus, 306, 309
  Boomer-Johnson-Argue apparatus, 305,
      307-309, 348, 349
  Copeland-Silverman-Benson apparatus,
      331, 333
  Efremova-Pryanikova apparatus, 342-344
  Gamburg apparatus, 322-325
  Gonikberg-Gurvich-Fastovskii apparatus,
      308, 310
  Ipat'ev-et al. apparatus, 304, 305
  Iomtov apparatus, 341, 343
  Kohen-et al. apparatus, 329
  Krichevskii-Efremova visual observation
      apparatus, 316, 317, 319-321
  Krischevskii-Efremova circulation, 310,
      311, 349-351
  Krichevskii-Gamburg apparatus, 303,
      304, 350, 351
  Krichevskii-Khazanova apparatus, 303
  Krichevskii-Lebedeva apparatus, 330-332
  Krichevskii-Shurmovskaya-Kal'varskaya
      method, 311-315, 318, 319
  Krichevskii-Sorina method, 325-328
  Krichevskii-Zhavoronkov-Tsiklis appara-
      tus, 301-303, 345-347
  polythermic method, 325-328

Phase equilibria (continued)
    Ravich-Borovaya apparatus, 334-337
    Schen-Dodge apparatus, 305
    sealed-ampule apparatus, 311-315, 318,
        319
    synthetic methods, see: Synthetic meth-
        ods, phase equilibria
    thermodynamic phase equilibria, 299
    Tsiklis high-temperature apparatus, 339,
        341
    Tsiklis liquid—gas apparatus, 307
    Tsiklis piezometer, 339
    Tsiklis-Vasil'ev gas apparatus, 340
    visual observation apparatus, 316-323
    volume-change apparatus, 322-325
    Wells-Roof apparatus, 318, 321
    Zel'venskii apparatus, 347
Phase transitions, 129, 130
Phase volumes, 299
Phosphorus in materials, steel, 449
Photoelectrical pyrometers, 204, 206
Photoresistors, 206
Physical atmosphere, 137
Piezoelectric
    manometer, 183, 184
    plates of quartz and seignette salt, 183
    transducer in manometer, 183
Piezometers, 360, 361, 369, 397
    accuracy of, 410, 411
    autofrettaged, 362
    bellows types, 399, 400
    with bottle, 360
    Bridgman types, 399, 400
    constant-volume types, 356-365
    contact type, 365, 367
    of glass, 25
    in installations, 305, 307-311, 330, 331,
        337-339, 356-385, 387, 393-401
    to measure compressibility, 355
    with stirrer, 339
    pycnometer types, 402-403
    Richards type, 397
    variable-volume types, 365-396, 398-403
Pinching-off
    in seals, 226, 229, 246, 294
    stress state, 1, 229
Pipestone, 263
Piston(s)
    adiabatic-compression, 390
    of ball-bearing steel, 152

Piston(s) (continued)
    of cemented carbides, 4, 63, 72, 109,
        120, 155, 244, 247
    in compressors, intensifiers, presses, and
        pumps, 20, 59, 69, 70, 72, 74, 75, 81-
        84, 87-114, 117-121, 123, 125, 127,
        131, 132, 202, 204, 243-249, 284-286, 288
    conical, 73, 74
    deformation of, 140
    differential, see: Differential pistons
    fissures in, 247
    friction, 89, 101
    glands for, see: Glands
    in installations, 308, 382, 386, 389-391
    machining of, 210
    in manometers, 90, 139-164, 185, 191, 192
    seals, 243-249
    of ShKh15 steel, 88
    in valves, 217
    of VK3 tungsten (cemented) carbide, 4
Piston manometers, 140-164, 186
    classification of, 140
    operation of, 140-150
Piston piezometer
    design, 74
    operation, 75, 76
    pistons, 74, 75
    pressure chamber, 74, 75
    pressure limits, 74, 76
    support material, 74, 75
Piston valve, 88
Plastic deformation
    of copper packing, 20
    definition of, 41
    and shear stress, 4
    under pressure, 13
Plastic flow
    criteria for onset of, 46
    of packing materials, 20, 230, 251
    of solid pressure-transmitting media, 34
    through clearances, 34
Plastic zone, 48-51
Plastics, see also: Polymers
    as electrical insulation, 258
    as packing materials, 95, 245, 249
Platinum
    as electrical contact, 188
    in thermocouples, 205
    electrical properties (P,T), 205
    in electrical-resistance heater, 132

Platinum (continued)
    as electrical-resistance liquid-level
       gauge, 169, 186, 305, 308
    hydrogen solution in, 130, 204
    in screen container, 328
Platinum—iridium, in pin, 333
Platinum—rhodium alloy
    electrical properties (P,T), 205
    in electrical-resistance heater winding,
    in thermocouples, 205
Plexiglas, as electrical insulator, 177
Plugs, 222-225, see also: Seals, Plungers
    expanding-section plug, 234
    liquid, 319
    of rubber, 131, 132
Plungers, in installations, 95, 99
    of KhVG steel, 4, 106
    of ShKh12 steel, 4
    of ShKh15 steel, 4
Poiseuille equation, 238
Poisson's ratio, definition of, 43
Polarization, optical, 431
Polarography, 439-441
Pol'di hardness tester, 7
Polycaproamide, see: Nylon
Polyethylene
    chemical properties, 22
    mechanical properties (RT), 19, 22
    in packings, 22
    thermal properties, 19, 22
Polyfluorethylene, see: Teflon
Polyisobutylene, mechanical properties, 19
Polymers (plastics)
    classification of, 19
    as electrical insulation, 277
    mechanical properties (P), 22
    mechanical properties (T), 19, 21
    thermal properties, 19
    thermal properties (P), 22
    as windows, 139
Polymethylmethacrylate
    chemical properties, 22
    as electrical insulator, 23
    in electrical leads, 258
    electrical properties (T), 23
    mechanical properties, 19, 22
    pressure limits, 23
    temperature limits, 23
    thermal properties, 19
    in tubing, 22

Polymethylmethacrylate (continued)
    as windows, 267
Polymorphic transformations, see also: Phase
    transitions
    assemblies for, 129, 130
    bismuth, 174
    cesium, 174
    pressure scale, 174, 175
    thallium, 174
Polypropylene, mechanical properties, 19
Polysiloxanes, see: Silicone oils
Polystyrene, in electrical leads, 259
Polytetrafluorethylene, see: Teflon
Polythermic method, 325-328
Polytrifluorochlorethylene, see: Gostaflon
Polyvinyl chloride
    chemical properties, 22
    in electrical leads, 260
    in installations, 123
    in joints, 255
    mechanical properties (LT), 21
    mechanical properties (RT), 19, 21, 22
    as packing, 106, 211, 241
    as seals, 243
    thermal properties, 19, 22
Porcelain
    in ampules, 129, 130
    in electrical leads, 260, 277
    electrical properties, 27
    mechanical properties, 21, 24
    thermal properties, 27
Portland cement, 261
Potassium
    chemical properties, 22
    mechanical properties (P), 13
    as pressure-transmitting medium, 22
Potassium chloride, friction (P), 27
Potassium hydroxide
    drying agent, 79, 293
    with gas compressors, 79
Potassium tartrate hydrate, see: Seignette salt
Poulter window, 264-266
Press(es), see also: Compressors, Intensifiers,
    Pumps
    adiabatic (dynamic) press, 87, 88
    bellows (sylphon) press, 85-87
    belt, 120, 122
    Boksha-Shakhovskii heated press, 131, 132
    Boyd-England supported-piston and cylin-
      der, 75

Press(es) (continued)
  Bridgman lipped anvils, 71, 72
  Bridgman anvils, 70
  Bridgman shear anvils, 123
  Butuzov-Boksha-Gonikberg heated press,
    129, 130, 132
  Butuzov-Mirinskii-Kats cubic clamp in-
    tensifier, 114, 116
  Butuzov-Mirinskii-Kats 6-anvil clamp
    intensifier, 115, 116
  Daniels-Jones supported-piston and cyl-
    inder, 75
  double-cone packed press, 72, 73
  evaporated-gas press, 87
  Galaktionov tetrahedral press, 113, 116
  Gonikberg-Tsiklis-Opekunov wedge in-
    tensifier, 117
  Goranson heated press, 128, 130
  Hall belt, 72, 73, 120
  Hall tetrahedral press, 69, 70
  heated-gas press, 87
  in installations, 317, 320, 321
  Institute of High Pressures press, 73
  Kan freezing—thawing press, 124-126
  Lazarev-Kan freezing-water press, 124
  Lloyd-Hutton-Johnson tetrahedral press,
    113, 116
  mercury-transfer press, 83, 85, 86
  NBS tetrahedral press, 113, 116
  Ryabinin-Livshits piston piezometer, 74-
  Ryabinin single-cycle unit, 88
  Sidorov mechanical press, 97, 100, 101
  single-cone packed press, 71, 72
  single-support press, 59, 62, 63
  supported piston and cylinder, 74-76
  values, 215, 216
  Vereshchagin-et al. belt, 120
  Vereshchagin-et al. conical belt, 73
  Vereshchagin-et al. hydraulic press, 122,
    123
  Vereshchagin-Likhter-Ivanov conical
    press, 111, 112
  Vereshchagin-Zubova-Shapochkin shear
    anvils, 123-125
  Von Platen cubic press, 118, 119
  Wilson girdle, 73, 75
Pressure, 137
  autofrettaging, 56
  automatic regulator of, 201, 203
  bursting, 48

Pressure (continued)
  change with temperature, 326
  corrections to manometer readings, 149,
    150
  generation of, 79-133
  graphic calculations of working, 49, 50
  and heating, 127-133
  influence on properties of construction
    materials, 13, 14, 39-44
  measurement of, 137-193
  polymorphic transformations, 174
  rupturing, 48
  and shear force, 123-125
  supporting, 60-64
  and temperatures, 54, 55
  thermal, 52
  units, 137, 448, 449
Pressure chambers
  autofrettaged, design, 55-57
  in compressors, intensifiers, presses, and
    pumps, 72, 74, 80-85, 87-112, 114,
    117-120, 122-132, 202-204
  with conical packing, 71, 72
  deformation of, 140
  design of, 47
  design valve pressure limit, 50
  effect of stress on ductility, 2, 6
  electrical leads into, 205, 262
  with hydraulic support design, 64-66
  in installation, 301, 303, 304, 307, 309-
    311, 318, 322, 330, 331, 339, 341,
    436-438
  machining of, 210
  in manometers, 139-147, 149-163, 187-
    191
  multilayered, design, 57-60
  pressure limits in monoblocs, 48, 51
  in rheometers and flow meters, 198-201
  in stirrers, 283
  stress distribution in, 48
  stress state in, 2, 6
  support pressure, 62
  thermal stresses, design, 51-55
  with two conical packings, 72, 73
  wedge, design, 66-69
  with variable external support design,
    59-65
Pressure chambers, materials
  cemented carbides, 121
  É1183 steel, 286

Pressure chambers, materials (continued)
  É1437B steel, 286
  É1698 steel, 286
  1Kh18N9T steel, 286
  45KhNMF steel, 104
  non-magnetic steel, 277
  stainless steel, 99
  tool steel, 48
Pressure generation
  gases, 79-88
  liquids and solids, 88-127
  with shear force, 123-124
  (T), 127-133
Pressure limits, assemblies
  of compressors, intensifiers, presses, and
    pumps, 69, 74, 79, 82-87, 90, 91, 94,
    95, 99-101, 104, 106, 108, 110-112,
    117, 120, 122-124, 126, 127, 130
  of manometers, 152, 159, 160, 164, 166-
    168, 170, 181, 182, 184, 190-192
Pressure limits, components
  autofrettaged pressure chamber, design
    value, 60
  electrical leads, 257-259, 263
  glass capillary, 24, 25
  glass tubing, 24, 25, 26
  joints, 251-256
  monobloc pressure chamber, 48, 51
  monobloc pressure chamber, design val-
    ue, 50
  multilayered pressure chamber, design
    value, 57
  seals, 224, 227, 229, 233, 244, 246, 247
  tetrahedral press, 69
  valves, 213, 216, 217, 221
  variable external-support pressure cham-
    bers, design value, 63, 64
  wedge pressure chambers, design value, 66
  windows, 264, 267
Pressure limits, solids
  asbestos, 22
  cemented carbides, 4
  Gostaflon, 22
  polymethylmethacrylate 22, 23
  steels, 2, 4
  steels (T), 2, 4
  Teflon, 22
Pressure
  equilibrator for liquids, 317
  measurement, 137-193

Pressure maintenance (continued)
  reducer, 213
  vessel, see: Pressure chamber
Pressure regulator
  automatic pressure regulator, 201, 203
  differential pressure regulator, 191
  electrical contact in, 201
  liquids in, 203
  manometer in, 201, 203
  pressure chamber in, 201
  Sidorov pressure regulator, 201, 203
  valve body in, 201
  valve in, 201, 203
Pressure rings of, see also: Chamfer rings,
    Wavy rings
  KhVG steel, 4
  ShKh12 steel, 4
  ShKh15 steel, 4
Pressure scales
  electrical resistance, 175
  melting curves, 176
  polymorphic transformation, 174, 175
  thermodynamic, 175, 176
Pressure-transmitting media, gases, 27, 28,
    30
  nitrogen, 132
  penetration into defects, 210
Pressure-transmitting media, liquids, 27-33,
    83, 85, 86, 106, 112, 124, 126, 127,
    132, 243, 337
  in compressing devices, 85
  in manometers, 152, 168
  penetration into defects, 210
Pressure-transmitting media, solids, 27, 28,
    30, 34
  classification of, 27
  friction (P), 27
  in hydraulic presses, 34
  lithium, 439
  potassium, 22
  pyrophyllite, 34
  rubber, 107, 111, 112
  silver chloride, 34
  sheet materials, 28
  sodium, 22
  sodium chloride, 434
  stress state in, 28
  Teflon, 34
Pressure units, 137, 138, conversion table,
    448-449

Price-Lalos compressibility installation, 389-390, 392
Principal stresses, 39
    in pressurized cylinder, 43
Proportional limit, definition, 41, 42
Proportioning control, see: Dispenser
Proportioning pumps and compressors
    Bruns-Braude pump, 202, 204
    Sidorov pump, 95, 99, 100
Propylene, see: Polypropylene
Pumice, friction (P), 27
Pumping of corrosive media, 94, 99
Pumps, see also: Intensifiers, Compressors,
        Presses, Circulating pumps
    Bruns-Braude, 202, 204
    circulating pumps, 284-289
    in dispenser, 202, 204
    hand pumps, 96
    packing in, 202
    piston in, 202, 204
    pressure chamber in, 204
    proportioning, 202, 203
    rod in, 202
    screw (viscous liquids) pump, 95, 99
    Sidorov (corrosive media and/or mixing)
        pump, 95, 99, 100
    Straub (viscous liquids) pump, 95, 99
    valve in, 202, 204
Punching under pressure, 14
Purification of gases, see: Removal of impurities
Pycnometer in installation, 331, 332
    compressibility installation, 401-403
Pyrex glass
    capillary, 25
    in joints, 256
    mechanical properties (P,RT,ET), 24, 26
    pressure limit, 25, 26
    tubing, 25, 26
Pyrometers, 206
Pyrophyllite
    as clearance seal, 34
    composition, 34
    electrical properties, 34
    as electrical insulation, 258, 262, 263
    friction (P), 27, 34
    mechanical properties, 34
    in presses, 69-71, 116, 117, 121
    pressure properties, 34
    as pressure-transmitting medium, 34, 174

Pyrophyllite (continued)
    thermal properties (P), 34
Pyrophyllite-cone lead, 262, 263

Quartz
    as burrette, 319, 322
    as capillary, 130, 132
    etching of, 256
    in joints, 256
    as light conductor, 206
    as optical plate, 183
    as piezoelectric transducer, 183
    as windows, 263, 264
Quartz glass
    classification of, 24
    mechanical properties (RT), 24, 26
    tubing, pressure limit, 24

Radial stress, 44
Radiation pyrometers, 206
Radiation screens, 378
Raman-effect-spectra installation, 431, 432
Rare gases, see: Inert gases
Rate of descent of piston in manometer, 141-145
Ravich-Borovaya phase equilibria installation, 334-337
Rawhide, see also: Leather
    mechanical properties, 21
    in packings, 21
    in seals, 247
Razumikhin density apparatus, 403-405, 408
Reciprocating-stem valves, 215-217
Reduction in area, see: Mechanical properties of solids
Refraction coefficients, 431
Refrigerant oil
    as pressure medium, 31
    thermal properties, 31
Regulator, pressure, see: Pressure regulator
Relative manometers, 137, 153, 164-186
Remote control, 296
Removable-stem valves, 214
Removal of impurities
    oil, 79, 83, 84, 293
    organic materials, 297
    particles, 204
    solvent vapors, 301
    water, 79, 293

Removal of impurities (continued)
    from gases, 79, 83, 301
    from liquid, 204
    from oxygen, 297
    in compressors, 79, 83, 84
    in valves, 218
    by adsorption, 79
    by condensation cooling, 79
    by drying column, 79
    by oil catcher, 79
    by oil separator, 83, 84, 293
    using activated carbon, 79, 293
    using anhydrone (magnesium perchlorate), 79
    using calcium chloride, 79
    using cotton wool, 293
    using molten potassium hydroxide, 79, 293
    using silica gel, 79
    using silica gel saturated with calcium chloride, 79, 293
    with limitations for hydrocarbons, 79
    with limitations for oil, 293
    with limitations for water, 293
Resins, see: Polymers
Resonant-circuit pump, 289
Rheometers, 197-203
    for air, 197
    Aristov-Sidorov gas flow meter, 200, 202
    Barnet-Jackson-Whitesides rheometer, 198, 199
    Boulestreau gas flow meters, 200
    capillaries in, 197-199
    gland in, 197, 198
    Kamzolkin-Bashkirov gas flow meters, 199, 200
    liquids in, 197, 198, 200, 203
    pressure chamber in, 198, 199
    rotameter, 203
    Vasil'ev-Ustinovich rotameter, 203
    window in, 198, 199
Rhodium, see: Platinum-rhodium alloy
Richards
    compressibility installation, 387
    piezometer, 397
Ring, wavy, 231, 232
Rocking autoclave, 273
Rockwell hardness, see also: Mechanical properties of solids
    conversion table to Brinell hardness and to tensile strength of steels, 445-447

Rockwell hardness (continued)
    errors, 9
Rods, see: Shafts and rods
Roebuck-Ibser manometer, 138, 139
Rooms for high-pressure experimentation, 294-296
Rose window, 266, 267
Rotameter, 203
Rotating autoclave, 272, 273
Rotating-reciprocating stirrer, 281
Rotation of body in manometer, 156
Rotation of piston in manometers, 144, 145, 151, 152, 155, 157-162
Roters formula for solenoid force, 279
Rubber
    chemical properties, 21
    mechanical properties, 21
    in packings, 21, 216, 222, 321
    as plug, 131, 132
    as pressure-transmitting medium, 107, 111, 112
Rubber-packing seal, 233
Rubidium, 174, 175
Rucholz manometer-press, 150
Ruoff electrical lead, 261
Rupture, see: Fracture
Rusching-Gebhart valve, 218
Ryabinin
    crusher manometer, 184-186
    single-cycle "press", 88
Ryabinin-Livshits piston piezometer, 74-76
Ryabinin-Tsiklis compressibility installation, 386-390, 392

S415-32 (cast iron), mechanical properties, 14
Safety
    safety codes, 298
    manometers, 168
    mercury, 411
    safe seals, 234
    safety valves, 297
    in rooms and buildings, 293-296
Sampling for analysis, 344, 345
    direct extraction, 344
    sample separation, 345
Sapphire, as windows, 263, 265, 267
Saturators, 303, 308-311
Scale effect, see: Size
Schen-Dodge phase-equilibria apparatus, 305

Scott-Babb electrical lead, 261

Scratches, 25, 209

Screens, see: Thermal-radiation screens

Screw (viscous-liquids) pump, 95, 99

Sealed-ampule phase-equilibria apparatus, 311-315, 318, 319

Sealing rings
  of brass, 240
  of bronze, 240, 242
  conicity of holes, 211
  of copper, 246
  in installations, 244-246
  machining of, 209
  of polymer, 242
  of steels, 240, 244, 246
  surfaces of, 209

Sealing surfaces
  machining, 209, 210
  of 45NMF Steel, 241

Sealing wax, in electrical leads, 257

Seals, 222-235, 243-249
  of aluminum, 222
  Anson seal, 233
  Babb seal, 230
  Boon-Lok seals, 233, 234
  bracket cap, 223-225
  of brass, 222
  Bridgman unsupported-area seal, 222
  chamber rings in, 245
  classification of, 28
  clearance in, 246
  closures, 222-235
  compensated-area seals, 222-225
  in compressors, 81, 83, 84, 89-91, 94, 97, 98, 100, 102, 105, 110, 131, 132, 243-249
  of copper, 20
  delta-ring seal, 233, 234
  design of, 231, 232
  dynamic seals, 235-249
  edged-sleeve seal, 232, 234
  flat-packing seals, 222-225
  Ganz seal, 223-225
  high-pressure-piston seals, 243-247
  hydraulic seals, 234, 235
  intensifier seals, 243-249
  intensifiers, presses, and pumps, 210
  in installations, 310
  large-diameter seals, 233, 234
  large-piston seals, 247-249

Seals (continued)
  of lead, 20, 222
  of leather and rawhide, 21, 244, 247
  lens seal, 228, 230, 231, 233, 234
  limitations of, 224, 230, 243, 244, 246, 247
  low-pressure-piston seals, 247-249
  in manometers, 140, 152, 155, 157, 159, 183, 186
  of mercury, 17, 27, 28
  of metals, 243, 244-246, 248, 249
  ordinary-piston seals, 243
  O-ring, 230
  packingless seal, 224, 225
  piston seals, 243-249
  plugs, 222-225
  plug closure, 234
  of polymers, 243
  pressure limits of, 224, 227, 229, 233, 243
  of rubber, 222
  rubber-packing seal, 233
  self-energizing seal, 222
  Sidorov ordinary seal, 243
  Sidorov unsupported-area piston seal, 243, 244
  sleeve seal, 84
  small-piston seals, 243-249
  static seals, 222-235
  stationary seals, 222-235
  of Teflon, 22
  Tsiklis seal, 224, 225
  Tsiklis unsupported-area piston seals, 245, 246
  type-I compensated-area seal, 222-225
  type-II compensated-area seal, 222-225
  type-III compensated-area seal, 223-225
  type-IV compensated-area seal, 223-225
  type-I unsupported-area seal, 225-230, 233, 234
  type-II unsupported-area seal, 227, 230
  unsupported-area seal, 222, 225-233, 243-247
  Vereshchagin seal, 233
  wavy-ring seal, 231-233

Second compression stages, 80-85

Seignette salt, as piezoelectric transducer, 183

Seizing
  of aluminum, 20
  of nuts and bolts, 209
  of pistons, 106

Seizing (continued)
    of 1Kh18N9T steel, 209
    of threads, 9, 127, 209, 234
    of valve stems, 211, 213
Self-energizing seals, 28, 222
    in installations, 84, 105
Self-healing, 14
Self-sealing gland, 237, 240, 241
Self-sealing-stem valve, 218, 219
Self-tightening gland, 239-243
Seren windows, 265, 267
Shafts and rods
    in compressors, pumps, intensifiers, and
       presses, 21, 80, 81, 83-86, 88, 89, 91, 92,
          94, 99, 102, 103, 106, 107, 109, 111,
          202, 204, 285
    glands for, 235
    of Kovar, 268
    packing for, 21
Shakhovskii compressibility installation, 400,
    402
Shear
    anvils, 123-125
    deformation (P), 13
    modulus, 43
    strength, see: Mechanical properties, solids
Shear stresses
    under Bridgman anvils, 70
    definitions, 39
    deformation, 40, 41
Shelaputin-Voitovich manometer, 189
Shellac, as packing material, 254
Shellac-mercury joint, 253, 254
ShKh12 (steel)
    aging of, 266
    composition, 448, 449
    heat treatment of, 265, 266, 451
    magnetic properties, 451
    mechanical properties, 4, 14, 450
    in nipples, 5, 265
    in pistons, 105
    in plugs, 246
    in plungers, 4
    in pressure rings, 4
    in sealing rings, 244
    in seals, 244
    in window assemblies, 265, 266
ShKh15 (ball-bearing steel)
    composition, 448, 449
    heat treatment, 450, 451

ShKh15 (ball-bearing steel) (continued)
    magnetic properties, 451
    mechanical properties, 4, 450
    in pistons, 88
    in plungers, 4
    in pressure rings, 4
Shrink-fitting, 57
Shtyukrat balance, 154, 156
Shut-off valves, 213-222
Sidorov
    electrical lead, 258, 259
    intensifier, 98, 101
    joint, 253, 254
    mechanical press, 97, 100, 101
    ordinary seal, 243
    pressure regulator, 201, 203
    (corrosive media and/or mixing) pump,
       95, 99, 100
    unsupported-area seal, 243, 244
    valve stem, 213
Sidorov-Kazarnovskii compressibility instal-
    lation, 371, 373-375, 377
Sidorov-Kazarnovskii-Golydman valve, 215,
    216
Silica gel
    drying agent, 79, 293
    with gas compressors, 79
    saturated with calcium chloride, 79, 293
Silicon in materials
    steels, 10, 12, 449
    titanium alloys, 8, 12
Silicone oils, callasification and mechanical
    properties (T,P), 30, 31, see also:
    Methylpolysiloxanes
Silk, as electrical insulation, 176, 177, 263
Silver, mechanical properties (P), 13
Silver chloride
    as clearance seal, 34
    friction (P), 27, 34
    mechanical properties (P), 14
    as pressure-transmitting medium, 34, 121
    as supported material, 74, 75
Simon equation, 175, 176
Simple piston with lateral pressure manome-
    ter, 140, 142, 143, 146, 152
Simple unpacked piston manometer, 140-152
Single-cone packed press, 71, 72
Single crystals, see: Monocrystals
Single differential-piston manometer, 140,
    142, 143, 146, 156-161

Single-stage compressors, 80, 81, 83-85

Single-support press, 59, 62, 63

Sintering diamond and metals under pressure, 5

Sirota-Mal'tsev pump, 287, 288

Size, effect on ductility, 2

Sleeve(s)
    in compressors, intensifiers, presses, and
        pumps, 81, 84, 89-91, 95, 100
    of hardened steel, 152
    in manometers, 151-153, 155-157
    of steel, 180

Sleeve seal, 81, 84, 89, 90
    in compressors, 81, 84, 89, 90
    in manometers, 155, 157

Small-piston seals for intensifiers, 243-249

Slipping glands, 242, 243

Soda-lime-boron oxide glass, mechanical
    properties (P,T), 24

Soda-potash glass, mechanical properties
    (P,T), 24

Sodium
    chemical properties, 22
    packing material, 20
    as pressure-transmitting medium, 22

Sodium chloride
    friction (P), 27
    optical properties, 434
    as pressure-transmitting medium, 434
    as windows, 263, 434-436

Soft seal, 253, 254

Solders, 256, 257

Solenoid-operated valves, 200, 201, 203, 288

Solidification, liquids, see: Mechanical prop-
    erties, liquids

Solidol (lubricant), in compressors, 84

Solubility, liquid–gas, 299-305
    compressibility apparatus, 364
    gravimetric apparatus, 348-351
    volumetric apparatus, 345-348

Solubility, solid–gas, 341-344

Solubility, solid–liquid, 328-333

Solubility, solid–vapor, 331, 333

Solution
    nitrogen in iron, 204
    oil in gases, 79
    oil in hydrocarbons, 82
    hydrogen in palladium, 204
    hydrogen in platinum, 130, 204

Solvents, 33

Spherical-stem-end valves, 213

Spiral pressure chambers, 60
    presses, 119

SS-I, -II, -III, -IV (methylpolysiloxanes),
    mechanical properties (T,P), 30

St. 45 steel, mechanical properties (T), 11

St. 5 (steel)
    composition, 448-449
    in cones, 252
    creep-resistant, 9
    crystallization under pressure, 5
    defects, 5, 6
    as diaphragms, 183
    ductility requirement, 2
    electrical properties (P), 180, 181
    hardenability, 6, 7
    hardness, 445-447
    heat-resistant, 10, 12
    heat treatment, 2, 5, 6, 450, 451
    hydrogen embrittlement (P), 15
    hydrogen-resistant, 15, 16, 17
    mechanical properties (ET), 9, 17, 204
    mechanical properties (LT), 10, 11
    mechanical properties (P), 13
    mechanical properties (RT), 1, 2, 5, 9,
        10, 11, 16, 17, 445-447
    mechanical working, 5
    mercury action, 16, 17, 20
    nitrogen embrittlement (P,ET), 15
    as packing material, 89, 90, 121
    as piston materials, 152
    pressure limits, 2, 4
    pressure limits (T), 2
    as sealing rings, 240, 244, 246
    as sleeves, 152, 180
    thermal pressure, 52
    heat treatment, 2
    in intensifier piston, 249
    mechanical properties (RT), 1, 450
    in nipples, 2
    pressure limit, 2
    in valve bodies, 2, 213

Stainless steels
    corrosion, 4
    in diaphragms, 163, 164
    in pressure chambers, 99
    pressure limits (T), 4
    in stirrers, 4

Standard atmosphere, 137, 448-449

Static methods, phase equilibria, see: Ana-
    lytic methods, phase equilibria

Static seals, 222-235
Stationary seals, 222-235
Steatite, as electrical insulator, 178
Steels, 1-20, see also: Alloy, Austenitic,
    Carbon, Chrome, Chrome–nickel, Cor-
    rosion-resistant, Nickel, Stainless, and
    Tungsten steels
    action of gases and liquids, 15
    annealing under pressure, 5
    austenite–martensite conversion, 5
    classification of, 9, 11, 448
    compositions, 2, 10, 12, 15, 16, 448, 449
    conversion table, hardness and tensile
        strength, 445-447
    thermobaric treatment, 5
    thermomechanical treatment, 5
Stems, see: Valve stems
Stepanov electrical lead, 262, 263
Stirrers
    Bishnevskii stirrer, 282-284
    drop stirrers, 280, 281
    of ÉI437B stainless steel, 4
    electromagnetic, 274-284
    rotating-reciprocating stirrer, 281
    external-rotor stirrer, 281, 282
    friction of, 271
    glands for, 271
    in installations, 305-307, 309-311, 322,
        324, 330, 331, 339, 340
    isolated-stator stirrer, 282-284
    of 1Kh18N9 stainless steel, 4
    of 1Kh18N9T stainless steel, 4
    limitations, 307
    mechanical stirrers, 271-273
    with motor inside apparatus, 271, 272
    with motor outside apparatus, 271
    packing for, 271
    reversing-rotation stirrer, 281
    Sage-Lacey stirrer, 281, 282
    shafts for, 271
    Tsiklis stirrer, 283, 284
    vibration stirrer, 283, 284
St. 40Kh steel, mechanical properties (T), 11
Straight-tube manometer, 165, 167, 168
Stratification of gases, 337
Straub (viscous liquids) pump, 95, 99
Strength, see: Mechanical properties of
        solids, whether bending, compressive,
        impact, shear, tensile, or yield
        strength

Strength reserve, 46
Stress(es), 39, 57-69
    axial, 46
    classification of, 39
    coefficient of thermal stress, 53
    components, 39, 40, 44
    components in pressurized cylinder, 45
    definition, 39
    distribution in walls of pressure chambers,
        48, 49
    distribution of equivalent stresses in auto-
        frettaged pressure chamber, 56
    distribution of equivalent stresses in
        pressurized chamber, 57
    equivalent, 46
    gradient, 174
    maximum shear-stress theory, 46
    multiplication, 66, 67
    normal, 39
    reduction of radial stresses, 67
    thermal, 51-55
    in thick-walled pressure chambers, 57-69
    true, 43
Stress analysis
    autofrettaged pressure chambers, 55-57
    multilayered pressure chambers, 57-60
    of pressurized thick-walled cylinder, 44-
        55
    in tensile necking, 43
Stress state
    biaxial stress, 2, 6
    effect on tensile strength, 2
    in general, 39
    during pinching-off, 229
    in pressure chamber, 2
    in pressure-transmitting media, 28
    during tensile necking, 43
    in unsupported-area seals, 1, 2
Stress–strain diagram
    definition of, 41, 42
    effect of pressure, 14
    polygonal approximation to, 51
Sulfur in steel, 449
Superficial hardness testing, 7
Supported-piston and cylinder, 74-76
Supported pressure chamber, see: Variable
    external-support pressure chamber
Support(s), fixed, 55-60
    in compressors, intensifiers, presses, and
        pumps, 81, 84, 110, 112, 115, 130-132

Supports (continued)
    in conical anvil device, 73
    in piston piezometer, 74
Support(s), variable, 64-76
    in compressors, intensifiers, presses, and
        pumps, 59, 81, 84, 92, 94, 102-116
    design, 60
    dual, 103, 109
    failure, 247
    hydraulic, 64-66
    supported-piston and cylinder, calcula-
        tions, 74, 76
Suppressed-zero manometers, 168
Surface tension, 415-429
    by Bol'shakov, 418, 420, 421
    by capillary-rise methods, 415-421
    by drop-weight methods, 423, 424
    at gas-gas boundary, 427-429
    by hanging-drop methods, 421-423
    by Kusakov-Lubman-Koshevnik installa-
        tions, 421-423
    at liquid—gas boundary, 415-424
    oil—gas, 82
    oil—hydrocarbons, 82
    at solid-liquid boundary, 424-427
    by Tsiklis-Vasil'ev installation, 427-429
    by Volyak, 417, 419
    by Weinaug-Katz installation, 424
Sylphon, see: Bellows
Synthetic methods, phase equilibria
    definition of, 299
    in gas—gas systems, 340
    limitations, 311
    liquid—gas systems, 311-328
    node lines, 327
    pressure limits, 312, 322, 323, 325
    in solid—liquid systems, 333-337
    temperature limits, 312

Tail ring, 250, 251
Tait manometers, 167, 168
Tangential stress, definition of, 44
Tanned leather, see: Leather
Tantalum, properties of, 444
Technical atmosphere, 137
Teflon
    chemical properties, 22, 271
    as clearance seal, 34
    friction, 34, 271

Teflon (continued)
    in glands, 86, 271
    impregnated, 22
    in joints, 253
    mechanical properties (LT), 21, 22
    mechanical properties (P), 22
    mechanical properties (RT), 19, 21, 22, 271
    in packings, 22, 83, 86, 87, 163, 164,
        319, 321
    pressure limits (ET), 22
    as pressure-transmitting medium, 34
    in sealing rings, 83
    thermal properties (P), 22
    thermal properties, 19, 22
Tellurium
    in babbitt, 20
    mechanical properties (P), 13
Temperature limits, of compressors, intensi-
        fiers, presses, and pumps, 87, 117, 120,
        122, 127, 130, 132
Temperature limits, components
    joints, 253, 255
    windows, 267
Temperature limit, solids
    aluminum, 18
    babbitt, 18
    B90 babbitt, 18
    cemented carbides, 3
    copper, 18
    lead, 18
    titanium—tungsten (cemented) carbide, 3
    tungsten (cemented) carbide, 3
    zinc, 18
Temperature measurement (P), in installa-
        tions, 116, 117, 121, 128-132, see
        also: Measurement of temperature in
        fluids
Temperature measuring means
    electrical-power loss, 206, 207
    electrical-resistance thermometers, 204
    optical pyrometers, 204, 206
    photoelectrical pyrometers, 204, 206
    thermal noise emf, 206
    thermistors, 206
    thermocouples, 204-206
    in thermostats, 204
Tempering of steels, 5, 451
Tensile strength, see also: Mechanical
        properties of solids
    conversion table from hardness, 445-447

Tensile strength (continued)
    definition, 41, 42
    effect of annealing under pressure, 5
    effect of average normal stress, 1
    effect of crystallization under pressure, 5
    effect of stress, state, 2
    effect of thermobaric treatment, 5
    effect of thermomechanical treatment, 5
    theoretical, 4, 5
Tensile test
    description, 41, 42
    under pressure, 6
Tensimeters, 181
Tensimetric-balance compressibility instal-
    lation, 406, 409
Tensimetric manometers, 181, 182
Testing of materials, 7-9
    hardness, 7-9
    notch tensile, 6
    tensile under pressure, 6
Tetrahedral press, 69, 70, 113, 116
    pressure limit, 69
    pyrophyllite packing, 69
Textolite
    in installation, 123
    mechanical properties, 21
TG-0 (titanium alloy), mechanical proper-
    ties (T), 8
Thallium, polymorphic transformation, 174
Theoretical tensile strength, metals, 4, 5
Thermal conductivity of solids, see: Ther-
    mal properties, solids
Thermal expansion of solids, see: Thermal
    properties of solids
Thermal insulating materials
    aluminum oxide, 71
    in Bridgman anvils, 71
    classification of, 27
    in conical dies, 71
    friction (P), 27
Thermal-noise emf (P), 206
Thermal pressure, 52
Thermal properties, liquids
    mercury (P), 29, 33, 173
    refrigerant oil, 31
    transformer oil, 31
Thermal properties, solids
    aluminum, 18
    babbitt, 18
    beryllium bronze, 18

Thermal properties, solids (continued)
    copper, 18
    germanium, 174
    hydrogen-resistant steels, 17
    ivory, 23
    lead, 18
    polymers, 19, 22
    steels, 17, 451
    titanium, 3
    titanium alloys, 3
    zinc, 18
Thermal-radiation screens, in installations,
    128, 130
Thermal stresses
    design of pressure chambers, 51-55
    equivalent, 53, 54
    glass tubing, 26
Thermistors, 206
Thermite, as heat source, 130
Thermobaric treatment of metals, retards
    grain growth, 5
Thermo-compressor, 87
Thermocouples, 204-206
    of alumel, 205
    of chromel, 205
    of constantan, 205
    of copper, 205
    corrections (P), 205
    corrosion, 204
    differential thermocouples, 130-132
    in installations, 130, 132
    of nickel, 205
    of nickel-molybdenum alloy, 205
    of platinum, 204, 205
    of platinum-rhodium alloy, 205
    presses and intensifiers, 121, 126, 128-133
    in pressure chambers, 262
Thermocouple well, 131, 132, 204
Thermodynamic phase equilibria, 299
Thermodynamic pressure scale, 175, 176
Thermomechanical treatment of metals,
    austenite-martensite conversion, 5
Thermostat
    in installations, 301, 304, 309, 329
    for work with sealed ampules, 315, 318, 319
Threads
    centering, 9, 209, 210
    clearances, 209
    design, 9, 209
    machining of, 210, 211

Threads (continued)
    mechanical finishing, 9
    stripping of, 243
    in valves, 213
Throttling valves, 301, 332, 342
Tin
    in babbit, 20
    mechanical properties (P), 13
    in solders, 256
Tin oxide, friction (P), 27
Titanium, see also: Titanium alloys, com-
        position
    in magnesium alloy, 5
    mechanical properties (RT), 3, 8, 444
    mechanical properties (T), 8
    in steel, 15, 449
    thermal properties, 3, 444
Titanium carbide, in cemented carbides, 3
Titanium dioxide, friction (P), 27
Titanium alloys
    chemical properties, 4
    classification of, 3, 8
    composition, 3, 8, 12, 256
    in joints, 256
    mechanical properties (RT), 3, 4, 8
    mechanical properties (T), 8
    melted under pressure, 5
    thermal properties, 3
Titanium-tungsten (cemented) carbides
    composition, 3
    mechanical properties, 3
    temperature limit, 3
    type numbers, 3
T-2 kerosene, mechanical properties (P), 32
T5K10, T15K6, T30K4 titanium-tungsten
        (cemented) carbides
    composition, 3
    mechanical properties, 3
    temperature limit, 3
Toluene, in manometers, 139
Tool steel
    in pressure chamber, 48
    in valve stems, 211
Transducer
    contactless manometers, 189-191
    manganin, 184
Transitions, see: Phase transitions
Transformer oil
    mechanical properties (P, T), 28, 31
    as pressure-transmitting medium, 168

Transparent
    glass, see: Glass
    quartz glass, see: Quartz glass
Triboelectricity, 297
Troostite, 6
"True stress," definition of, 43
Tsiklis
    ampules with valves, 315
    diaphragm valve, 220, 222
    glandless valve, 220, 221
    high-temperature phase-equilibria ap-
            paratus, 339, 341
    hydraulic pump, 93, 95
    manganin manometer, 178
    phase-equilibria apparatus, 307
    piezometer, 339
    seal, 224, 225
    stirrer, 283, 284
    unsupported-area seal, 245, 246
    valve, 218, 219
Tsiklis-Borodina manometer, 184
Tsiklis-Kulikova
    bellows compressor, 87
    compressibility method, 360
Tsiklis-Linshits compressibility installation,
        394-396
Tsiklis-Radle hydraulic pump, 94, 99
Tsiklis-Vasil'ev
    phase-equilibria apparatus, 340
    surface-tension installation, 427-429
Tubing, see also: Capillary
    in ampules, 25
    design, 26
    of glass, 24-26
    joints for, 250, 251
    in manometers, 25, 138, 165
    of metal, 138
    of mica, 27
    of polymethyl methacrylate, 22
    pressure limit, 23, 24, 25
    of quartz glass, 24, 26
Tubular spring(s), in manometers, 164, 165,
        167, 168, 185, 189
Tubular-spring manometers, 164-167, 188, 189
Tungsten
    in electrical resistance heater, 132
    mechanical properties (RT), 444
    thermal properties, 444
    in steel, 9, 10, 12, 15, 449
Tungsten carbide, in cemented carbides, 3, 4

Tungsten (cemented) carbide
    compositions, 3
    mechanical properties, 3, 105
    pressure limits, 4
    temperature limits, 3
    type numbers, 3
Tungsten steel, heat-treatment, 4, 450, 451
Type-I, -II, -III, -IV compensated-area seals,
    222-225
Type-I, -II unsupported-area seals, 225-230,
    233, 234

Ultimate tensile strength, see: Mechanical
    properties of solids
Ultrahigh-pressure valve, 217
Ultraviolet spectra, 433
Unbalanced-bridge circuit, 171
Unsupported-area
    electrical leads, 257-263
    glands, 238, 240, 241
    windows, 264-266
Unsupported-area seal, 28, 222, 225-233,
    243-247
    in installations, 102, 105, 122, 123, 131, 132
    pinching-off, 1, 2
    pressure limits, 2
    stress state, 1
Upper-cylinder oil, mechanical properties
    (T,P), 28
Upsetting manometer, 184-186

Vacuum valve, 221
Valve bodies
    in installations, 84, 100, 104, 201, 286,
    319, 340
    of St.5 steel, 2, 213
    in valves, 212, 214-221
Valve plates
    of glass, 202, 204
    in installations, 90, 91, 95, 202, 204, 250
Valves, 210-222
    for addition of load under pressure, 405
    ampules with valves, 315
    Astrov-Boronely valve, 222
    ball valves, 31
    bellows valves, 222
    Borzunov-Semin valve, 217, 218
    Check valves, 249, 250

Valves (continued)
    in compressors, intensifiers, presses, and
    pumps, 79-85, 88-96, 99-101, 104,
    202, 204, 285, 286, 288
    with conical stem, 211-213
    with cooled-gland, 215, 216
    control valves, 211-213, 218
    with constant volume, 219, 220
    diaphragm valves, 220-222
    drain valve, 218, 219
    fillers for, 33
    with free end, 213, 214
    glandless valves, 220-222
    glands in, 21, 211
    hydraulic valves, 33, 217, 218
    in installations, 301-323, 329, 332-334,
    338-344
    joints in, 250-252
    limitations on, 213, 214, 217, 220
    machining precautions, 210, 211
    in manometers, 157, 191, 192
    membrane valves, 220-222
    needle valves, 191, 192
    non-packed valves, 220-222
    packing for, 20
    piston-valve, 88
    pressure limits of, 213, 216, 217, 221
    press valves, 215, 216
    reciprocating-stem valves, 215-217
    remote control of, 296
    with removable gland, 214
    in rheometers and flow meters, 200, 201, 203
    Rusching-Gebhart valve, 218
    safety valves, 297
    with self-sealing stem, 218, 219
    shut-off valves, 213-222
    Sidorov-Kazarnovskii-Golydman valve,
    215, 216
    solenoid-operated valves, 200, 201, 203
    with spherical tip, 213
    stems in, see: Valve stems
    threads in, 213
    throttling valves, 301, 332, 342
    Tsiklis diaphragm valve, 220-222
    Tsiklis glandless valves, 220-222
    Tsiklis valve, 218, 219
    Vereshchagin press valve, 216
    Vereshchagin-Ivanov valve, 217
    ultrahigh-pressure valve, 217
    with washable stem, 218, 219

Valve stems
    conical stems, 211-213
    cylindricity of, 211
    Daniels stem, 218, 219
    design of, 211
    glands for, 235
    machining of, 210
    removable stems, 214
        with rotating tip, 214
    Rozen stem, 214
    seizing of, 211, 213
    self-sealing stem, 218, 219
    Sidorov stem, 213
    spherical stems, 213
    threads on, 213
    of tool steel, 211
    Tsiklis stem, 218, 219
    in valves, 212, 214-221, 314
    washable stems, 218, 219
Vanadium in
    steel, 449, 12
    titanium alloys, 256
Vane pumps, 284, 285
Variable external-support pressure chamber
    automatic generation of support, 59
    design, 59-65
    support-friction determination, 62
    lubrication, 63
    pressure limits, 63, 64
Variable-stroke compressor, 94, 99
Varnish, as electrical insulation, 176, 177
Vasil'ev-Ustinovich rotameter, 203
Vasoline (oil)
    mechanical properties (P,T), 28, 31
    as pressure-transmitting medium, 168
Vegetable tanned leather, see: Leather
Vereshchagin
    hydraulic compressor, 88-90
    press valve, 216
    seal, 233
Vereshchagin-Aleksandrov manometer, 151-
    153
Vereshchagin-Brandt installation, 437-439
Vereshchagin-et al.
    belt, 120
    conical belt, 73
    hydraulic press, 122, 123
    intensifier, 102, 104-108
Vereshchagin-Galaktionov piezometer and
    compressibility installation, 398, 399

Vereshchagin-Ivanov
    gas compressor, 81, 84, 85
    valve, 217
Vereshchagin-Kabalkina-Ebdokimova instal-
    lation, 436, 437, 439
Vereshchagin-Kuznetsov-Alaeva electrical
    lead, 260
Vereshchagin-Likhter-Ivanov conical press,
    111, 112
Vereshchagin-Preobrazhenskii oxygen com-
    pressor, 86
Vereshchagin-Zubova-Shapochkin shear an-
    vils, 123, 125
Vessel, see: Pressure chamber
Vibration stirrer, 283, 284
Vinyl chloride, see: Polyvinyl chloride
Viscosity, see also: Mechanical properties,
    liquids
    calculation of, 31, 143
    (P,T), 28-31, 33
Visible light, pyrometry, 206
Visual observation phase-equilibrium appara-
    tus, 316-323
VK2, 3, 6, 8, 15 tungsten (cemented) carbide
    composition, 3
    mechanical properties, 3
    temperature limit, 3
Vollbrecht gland, 242, 243
Volume change on melting (P), 175, 176
Volume-change phase-equilibria apparatus,
    322-325
Volumetric capacities, of compressors, in-
    tensifiers, presses, and pumps, 79, 87,
    94, 95, 99
Volumetric compression manometer, 182
Volumetric measurements, phase equilibria,
    345-348
Volumometer, in installations, 322, 324
Volyak surface-tension installation, 417, 419
Von Platen cubic press, 118, 119
VT-1 titanium, see: Titanium
VT-2, 4, 5, 6, 8 (titanium alloy), mechani-
    cal properties (RT), 3, 8
VTU MKhP 2127-4a (silicone oil), viscosity, 33
Vukalovich-Altunin compressibility installa-
    tion, 358

Waggener window, 265, 267
Walls of laboratories

Walls of laboratories (continued)
    break-away, 295
    permeable, 293
Washable-stem valve, 218, 219
Water
    mechanical properties (T,P), 32
    oxygen compression, 86
    as pressure transmitting medium, 85, 86, 124, 127, 311
    in rheometer, 197, 198
Water-compressibility manometer, 169
Water gas holder, 79
Wavy-ring seal, 231-233
Waxpaper, as electrical insulator, 176, 177
Wedge press, 117
Wedge pressure vessels
    design, 66-69
    stress multiplication, 66
Weighing, hydrostatic, 403-410
Weight manometers, see: Piston manometers
Weinaug-Katz surface tension installation, 424
Weir-Lippincott-Bunting absorption-spectra installation, 431, 433
Well-Roof phase equilibria apparatus, 318, 321
Wettability, Bolshakov installation, 424-427
Whalley-Lavergene compressor, 99
Whiskers, 5
Wiebe-Gaddy compressibility installations, 357, 359
Wilson girdle, 73, 75
Windows
    Amagat window, 263, 264
    in autoclaves, 266, 267, 340
    Baldin-Gavrilovskii window, 267, 268
    capillaries as, 267
    compatible liquids, 264
    end windows, 265-268
    in installations, 316, 318
    in manometers, 139, 155, 157-161
    materials in, 431-434
    O-rings in, 267
    packings for, 263-267
    Poulter window, 264-266
    pressure limits, 264, 267
    in rheometers and flow meters, 198, 199
    Rose window, 266, 267
    Seren window, 265, 267
    temperature limits, 267
    Waggener window, 265, 267

Windows (continued)
    water attack on, 263, 264
    unsupported-area window, 264-266
Windows, materials
    glass, 25, 198, 199
    plastic, 139
Wolf-Bowen diaphragm compressor, 82, 85
Work hardening
    definition, 41, 42
    in pressurized cylinder, 48
Working, see: Mechanical working
Wound pressure chambers, 60

Xenon, solubility with gases, 340
X-ray measurements, 436-439
    through B_4C, 436
    through BeO, 436-439
    through BeO, 436
    through BN, 436
    through C, 436
    through diamond, 436
    by Jacobs unit, 437
    by Vereshchagin-Brandt installation, 437-439
    by Vereshchagin-Kabalkina-Ebdokimova installation, 436, 437, 439
    by Kabalkina-Vereshchagin installation, 438, 439
    through LiF, 436
    through LiH, 436
    pressure limits, 437

Yield point, definition of, 42
Yield strength, see also: Mechanical properties of solids
    definition, 41, 42
Young's modulus, see also: Mechanical properties of solids
    definition of, 42, 43
    temperature dependance, of, 52

Zel'venskii phase-equilibria apparatus, 347
Zelyaev-Shumov-Alekseev manometer, 181, 182
Zh69 (steel), mechanical properties (ET), 9
Zhokhovskii
    constant-clearance manometer, 153, 155, 156

Zhokhovskii (continued)
    manometer, 155, 156, 159
    packing-support intensifier, 107, 111, 112
Zhokhovskii-Konyaev-Levchenko manometer,
      157, 160, 161
Zinc
    cast, 18
    mechanical properties (P), 13
    mechanical properties (RT), 18
    rolled, 18

Zinc (continued)
    temperature limit, 18
    thermal properties, 18
Zinc oxide, friction (P), 27
Zinc-soda borosilicate glass, see: Borosilicate
    glass
Zirconium
    mechanical properties (P), 13
    mechanical properties (RT), 444
    thermal properties, 444